SOMATIC EMBRYOGENESIS IN WOODY PLANTS

FORESTRY SCIENCES

Volume 45

The titles published in this series are listed at the end of this volume.

Somatic Embryogenesis in Woody Plants

Volume 1

History, Molecular and Biochemical Aspects, and Applications

Edited by:

S. MOHAN JAIN
Department of Plant Production, University of Helsinki, Helsinki, Finland

PRAMOD K. GUPTA
Weyerhaeuser Inc., Tacoma, Washington, U.S.A.

and

RONALD J. NEWTON
Department of Forest Science, Texas A & M University, College Station, Texas, U.S.A.

KLUWER ACADEMIC PUBLISHERS
DORDRECHT / BOSTON / LONDON

Library of Congress Cataloging-in-Publication Data

Somatic embryogenesis in woody plants / edited by S. Mohan Jain, Pramod K. Gupta, Ronald J. Newton.
p. cm. -- (Forestry sciences ; v. 45)
Includes index.

1. Trees--Micropropagation. 2. Woody plants--Micropropagation. 3. Plant tissue culture. I. Jain, S. Mohan. II. Gupta, Pramod K. III. Newton, Ronald J. IV. Series: Forestry sciences (Dordrecht, Netherlands) ; v. 45.
SD403.5.S66 1995
635.9'77--dc20 94-18512

ISBN 978-90-481-4448-8

Published by Kluwer Academic Publishers,
P.O. Box 17, 3300 AA Dordrecht, The Netherlands

Kluwer Academic Publishers incorporates
the publishing programs of
D. Reidel, Martinus Nijhoff, Dr W. Junk and MTP Press

Sold and distributed in the U.S.A. and Canada
by Kluwer Academic Publishers,
101 Philip Drive, Norwell, MA 02061, U.S.A.

In all other countries, sold and distributed
by Kluwer Academic Publishers Group,
P.O. Box 322, 3300 AH Dordrecht, The Netherlands

Printed on acid-free paper

Printed in the Netherlands

Table of Contents

General Preface

The quality of human life has been maintained and enhanced for generations by the use of trees and their products. In recent years, ever rising human population growth has put tremendous pressure on trees and tree products; growing awareness of the potential of previously unexploited tree resources, and environmental pollution have both accelerated the development of new technologies for tree propagation, breeding and improvement. Biotechnology of trees may be the answer to solve the problems which cannot be solved by conventional breeding methods. The combination of biotechnology and conventional methods such as plant propagation and breeding may be a novel approach to improving and multiplying a large number of the trees and woody plants.

So far, plant tissue culture technology has largely been exploited by commercial companies in the propagation of ornamentals, especially foliage house plants. Generally, tissue culture of woody plants has been recalcitrant. However, limited success has been achieved in tissue culture of angiosperm and gymnosperm woody plants. A number of recent reports on somatic embryogenesis in woody plants such as Norway spruce (*Picea abies*), Loblolly pine (*Pinus taeda*), Sandalwood (*Santalum album*), *Citrus*, Mango (*Mangifera indica*), etc. offer a ray of hope of: (a) inexpensive clonal propagation for large-scale production of plants or "emblings" or "somatic embryo plants"; (b) protoplast work; (c) cryopreservation; (d) genetic transformation; and (e) artificial or manufactured seed production. In the future, with the basic biology for better understanding the genetic control of somatic embryogenesis and in-embryo development and maturation with the help of molecular biology, it may be possible for us to have better control over the induction of somatic embryogenesis. For cost effective large-scale production of elite trees, robotic and automation technology will interface with somatic embryogenesis in the 21st century.

The rapid progress of somatic embryogenesis and its prospects for potential application to improving woody plants prompted us to edit this book in three volumes. Moreover, most of the research information in this field on woody plants is scattered in national and international meeting proceedings, refereed journals, biotechnology books, etc. There is a lack of availability of a comprehensive work on somatic embryogenesis in woody plants including both angiosperms and gymnosperms. We were all convinced that such a treatise was needed and would be extremely useful to researchers and students. Dr Jain contacted Kluwer Academic Publishers, The Netherlands, to work on this book project. The positive response of the publishers encouraged us to proceed further.

In our present endeavour, we have attempted to bring all the research information on somatic embryogenesis in woody plants into three volumes.

In Volume 1, we have included review articles on different aspects and applications of somatic embryogensis such as historical, biochemical, molecular, anatomical and physiological, protoplasts, cryopreservation, manufactured seed (artificial seed), genetic transformation, somaclonal variation, bioreactors, mutation, and future uses in woody plantations. In the introductory chapter, we have taken the views of a forest geneticist on somatic embryogenesis in relation to woody plant improvement. What lies ahead in the application of this technology for commercial purposes and the establishment of germplasm banks, has been discussed. Both Volumes 2 and 3 cover selected important angiosperm and gymnosperm woody plants, respectively. Each selected woody plant has been described in detail covering botany and genetics, importance and geographical distribution, breeding problems, *in vitro* propagation and problems, initiation of embryogenic cultures, culture maintenance, embryo development, embryo germination, and field trials (if any). We have also encouraged the authors to incorporate their recent data including genetic transformation, cryopreservation, etc., in their manuscripts. These volumes are designed as the key reference works, providing detailed information on all aspects of somatic embryogenesis for beginners as well as experienced research workers.

We have invited an international and diverse group of contributors from several countries such as Australia, Austria, Canada, Czech Republic, Denmark, England, Finland, France, India, Ireland, Italy, Japan, Slovakia, Spain, Sweden, Taiwan, U.S.A., and Yugoslavia. Our invited authors belong to academic institutes, universities and industries, and they are leading research scientists in the somatic embryogenesis research work arena.

The selection of chapters and their authors was initially done by Dr S.M. Jain and the final list was prepared by including the suggestions of Dr Gupta and Prof. Newton. Our final list of chapters, with their authors, was reviewed by anonymous reviewers of Kluwer Academic Publishers and the project was finally approved. All the correspondence with the invited authors and the publishers, any further additions or deletions to the chapters, initial and final review of the manuscripts including editing, the selection of reviewers, and finally the mailing of the reviewed manuscripts were done by Dr Jain. All manuscripts have been critically reviewed by one–two persons and revised according to the referees' comments.

S. Mohan Jain
P.K. Gupta
R.J. Newton

Preface to Volume 1

This volume contains 20 review articles on somatic embryogenesis in woody plants covering anatomical, biochemical, historical, molecular, and physiological aspects, and applications including protoplasts, cryopreservation, manufactured or artificial seeds, genetic transformation, somaclonal variation bioreactors, and mutation. In the introductory chapter, the forest geneticist's viewpoint on the problems faced while improving woody plants and how the use of new technologies such as molecular biology and somatic embryogenesis could overcome those problems is presented. Although we have presented the major historical aspects of research conducted on somatic embryogenesis in the woody species of angiosperms and gymnosperms, there are still many questions remaining that need answers. "What makes a somatic cell become embryogenic"? The next chapter describes the anatomical differences and similarities between somatic and zygotic embryos in the developmental embryo stage, lipid and storage protein accumulation. We have included two chapters reviewing at length the progress on somatic embryogenesis in angiosperm woody plants and conifers. Another chapter indicates that somatic embryogenesis is a promising means to clonal propagation and can lead to rejuvenation in mature trees. Rejuvenation of mature trees has not been economical with current horticultural practices and *in vitro* culture methods. One chapter deals with the comparative analysis of conifer somatic and zygotic embryos at the biochemical and molecular levels that has led to the assessment of the quality of somatic embryos. Two chapters on protoplasts in gymnosperm and angiosperm woody plants, respectively, cover at length isolation, regeneration, genetic transformation, and the problems facing protoplast research work. Genetic transformation in angiosperms and gymnosperms have been reviewed exclusively in two chapters addressing the important questions of gene regulation in transgenic plants, gene transfer technologies, and the deployment of transgenic plants. One chapter updates the technology of manufactured seed (artificial seed) in conifer trees and the problems arising in the commercialization of this technology. The potential aspects of bioreactor technology for producing large numbers of somatic embryos cheaply and efficiently, large-scale culture of plant cells, and the problems in commercialization for large-scale embling (plantlets derived from somatic embryos) production have been covered in another chapter. In the following chapter, a review deals in detail with the cryopreservation of plant cells and meristems covering cryogenic protocols and expanding their applicability to a wide range of woody plants.

The role of polyamines in the process of somatic embryogenesis in woody plants and manipulation of polyamine biosynthetic pathways by the use of inhibitors and transgenic techniques can affect somatic embryogenesis through polyamine metabolism in a number of unrelated species; this has been discussed at length in a review chapter. Furthermore, one more chapter deals with biochemistry of conifer embryo development, describing the functions of amino acids, polyamines and storage proteins. In the subsequent two chapters: a) one chapter provides a description of mutation research with somatic embryogenesis in woody plants including ionizing radiations, chemical mutagenesis, *Agrobacterium*-mediated T-DNA mutagenesis, and molecular evaluation of mutations and prevention of producing chimeras; and b) the next chapter covers evaluation of somaclonal variation during somatic embryogenesis. The last two chapters describe the future prospects of using somatic embryogenesis in *Picea abies* and other woody plants for commercial purposes, genetic engineering, and cryopreservation. In the future, the combination of cryopreservation and somatic embryogenesis, and an interface of robotic and automation technology will be ideal for cost-effective clonal propagation.

We are thankful to all the contributing authors for providing their manuscripts within the scheduled time and to the reviewers for their critical comments on the manuscripts.

S. Mohan Jain
P.K. Gupta
R.J. Newton

Acknowledgements

I am grateful to the Finnish Academy of Science for providing me with a grant for editing this book.

My thanks are due to my friends Prof. R.J. Newton and Dr. Pramod K. Gupta, co-editors of this book, for their promptness in responding to me whenever I needed their help. They were extremely helpful in deciding the final list of chapters and their authors, and in reviewing the manuscripts. It has been my greatest pleasure to work with Ron and Pramod on this project, and certainly we have formed an excellent and highly efficient team.

I would like to acknowledge with great appreciation Prof. M.R. Ahuja, Dr. J. Blake, Dr. J.M. Bonga, Dr. J. Finer, Prof. Larry Fowke, Dr. P.K. Gupta, Prof. J. Janick, Prof. S.C. Minocha, Prof. R.J. Newton, Prof. J. Preece, Dr. R.C. Saxena, Prof. R.H. Smith, Dr. David Thompson, and Dr. R.N. Trigiano for critically reviewing the manuscripts promptly, and to all the contributory authors for sending their manuscripts well in time. However, we had some anxious moments due to delay in some manuscripts.

I wish to express my thanks to my colleagues Prof. Eija Pehu, Mr Tapio Poutala, Mr Matti Teittinen, and Mr Tapani Pulkki in our Plant Production Department for their timely help and assistance.

Also, with great love and affection, I want to thank my daughters Sarita and Sonia, and my wife Marja-Liisa for their unceasing patience and understanding while I was working on these volumes.

Finally, I express my deepest sense of appreciation to Adrian C. Plaizier of Kluwer Academic Publishers, The Netherlands, for giving us the opportunity to work on this exciting book project. Adrian has always been cooperative and helpful, and gave me useful advice.

S. Mohan Jain
Book Project Leader

ABBREVIATIONS

2,4-D	2,4-dichlorophenoxyacetic acid
2,4,5-T	2,4,5-trichlorophenoxyacetic acid
2iP	N^6 (2-isopentyl)adenine
ABA	Abscisic acid
BA	N6-benzyladenine
BAP	benzylaminopurine
BLG	Verhagen & Wann (1989)
BM_3, BM_4 and BM_5	Gupta & Pullman (1991)
CBB	Coomassie Brillant Blue
CD	Campbell & Durzan (1975)
CP	cotyledons protoplasts
DCR	Gupta & Durzan (1986)
DMSO	dimethyl sulfoxide
DW	dry weight
EP	embryogenic cell suspension
ESM	Embryonal suspensor masses
FDA	Fluorescein diacetate
FW	fresh weight
G-medium, DCR	Gupta & Durzan medium (1986)
GD	Gresshoff & Doy (1972)
H medium	Halperin medium (1965)
HBLG	half strength Verhagen & Wann (1989)
HLP	half strength von Arnold & Eriksson (1981)
HLM	half strength Litvay *et al.* (1981)
HRH	High relative humidity
IAA	indoleacetic acid
IBA	indolebutyric acid
KIN, K	kinetin
LM	Litvay *et al.* (1981)
LN_2	liquid nitrogen
LP,AE,VE	von Arnold & Eriksson (1981)
MCM	Borman (1981)
MES	2-(N-morpholino)ethanesulfonic acid
MI	mitotic index
MS	Murashige & Skoog
NAA	naphthaleneacetic acid
NEC	non-embryogenic callus
NEPHGE-PAGE	non-equilibrium pH gel electrophoresis
OP	osmotic pressure
P_6	Teasdale *et al.* (1986)

PAR	photosynthetically active radiation
PCV	packed cell volume
PEG	polyethylene glycol
PG	phloroglucinol
6 PGD 1	6-phosphogluconate dehydrogenase
PVP	polyvinylpyrrolidone
RAPD	Random amplified polymorphic DNA
RFLP	Restriction fragment length polymorphism
RPM	revolution per minute
SDS-PAGE	sodium dodecyl sulphate polyacrylamide gel electrophoresis
SE	somatic embryos
SH	Schenck & Hildebrandt (1972)
TAG	triacylglycerols
TCA	trichloroacetic acid
TDZ	thidiazuron
WPMG	Jain *et al.* (1989)
X-GLUC	5-bromo-4-chloro-3-indolyl glucuronide
Zea	zeatin

1. Introduction

Howard B. Kriebel

Contents

1. A Forest Geneticist's Viewpoint

Somatic embryogenesis encompasses a wide array of topics and there are many types of woody plants. In a short introduction, it is impossible to be all-inclusive. This chapter does not cover every aspect of somatic embryogenesis included in this publication, nor does it discuss woody plants in general. It is, as requested, brief and written from the viewpoint of a forest geneticist, touching lightly on various topics that are presented in later chapters in detail. I have pointed out some practical needs, problems faced in meeting these needs, and potential gains from the application of somatic embryogenesis, with specific reference to forest tree improvement. Some of the problems discussed are of particular relevance to the task of breeding a long-lived plant, while others are just as important to anyone concerned with the propagation of woody plants. It is hoped that the discussion may clarify for the plant biotechnologist some of the real problems facing forest geneticists.

Among the woody plants, forest trees present a challenge to the geneticist because of the long reproductive cycle that requires many years of research to achieve significant improvement. The capture of genetic gain and the multiplication of variants via asexual reproduction offer a desirable option in lieu of repeated generations of conventional breeding. However, in forest tree improvement the clonal option is used in connection with a program of selection, progeny testing and controlled breeding. Sexual reproduction usually has a key role in natural forest stand regeneration, where adaptability is critical and the adaptability of individual trees is the product of evolutionary "fine-tuning" over many generations. We know very little about gene frequency distributions in natural forest stands. Because selective cloning may eliminate some of the protective buffering effect of natural variation, cloning entails risks unique to these long life cycle plants. Nevertheless the demonstrated potential for large gains in useful traits from the intensive

S. Jain, P. Gupta & R. Newton (eds.), Somatic Embryogenesis in Woody Plants, Vol. 1, 1–8.

culture of elite material is a strong incentive for mass clonal propagation of many species. The challenge to forestry will be to maintain natural biological diversity in forest ecosystems in combination with intensive clonal culture of carefully diversified genotypes selected for the production of wood and other forest products.

In some species, such as aspen, clonal forests are a natural ecosystem, but in most tree species, pure clonal stands are not the norm. In fact, many tree species are not found in nature in pure stands. The forest is almost always a diverse and complex ecosystem made up of many species of plants and animals. Interactions are not well understood. Thus clonal forestry has many ecological implications when practiced on a large scale.

2. Identification and Preservation of Genetic Material

In forest trees, information on genotypic variation within the species is required prior to the application of somatic embryogenesis to large-scale intensive culture. There is no forest tree for which a genome map provides even rudimentary knowledge of genes coding for adaptive and nonadaptive traits. Thus it is important (1) to preserve samples of the gene pool *ex situ* and also *in situ* if possible, and (2) to undertake research on the structure and composition of the gene pool. The basic assumption is that the tree genome is highly variable. Research has shown this to be the case in nearly every wide-ranging tree species. Variability is assessed in the field by provenance and progeny testing of trees raised from seeds collected by sampling the population. More precise information on individual differences can be obtained by using biochemical and molecular techniques. These techniques include isozyme analysis, RFLP (restriction fragment length polymorphism) analysis or DNA fingerprinting (e.g., Vahala *et al.*, 1991), and RAPD (random amplified polymorphic DNA) analysis. DNA markers can supplement isozyme markers for estimating genetic diversity in breeding populations and for germplasm identification (Neale *et al.*, 1992).

Germplasm of forest trees for gene banks can be collected from either vegetative or reproductive tissues. Vegetative tissues may be buds, stems, leaves, roots or cambium, depending on the regeneration potential of the tissue. Mature seed is usually easily handled and stored for long periods at low temperature and humidity. The tissue most commonly cultured for somatic embryogenesis is the zygotic embryo, although, for forest tree improvement, tissue from mature trees is needed if the mature phenotype is to be reproduced. Purified DNA can be stored easily for genetic transformation of callus cells. The storage method of choice for tissues is cryopreservation, i.e., slow freezing and storage in liquid nitrogen at $-196°C$. followed by carefully-controlled thawing. Although specifics of the technique vary with the species, procedures are well-enough developed to be useful for both gymnosperms and angiosperms. Cryopreservation is relatively simple and inexpensive and may provide safe storage for many years (Millar, 1993). The subject is discussed in detail in this book by Prof. Sakai, Japan.

Since many genotypes can be stored in a small space, *ex situ* preservation of woody plant genomes as stored tissues is possible on a much larger scale than would be possible in seed stands or clone banks. This is particularly important for preservation of the large numbers of population samples that are needed for forest tree gene conservation. The controlled environment provides protection from environmental disasters to which planted trees are exposed. Although seed collection and storage is usually a necessary part of an improvement program, cultures of somatic cells have, at least theoretically, the advantage of maintaining the genotype without change. If we accept the premise that cells in cultures are totipotent, plant regeneration will be possible and valuable genotypes can be preserved.

3. Developmental Genetics Research

When zygotic embryos are used as the source of tissue for the induction of somatic embryogenesis, it is necessary to determine the minimum stage of zygotic embryo development at which somatic embryogenesis can be obtained. Among the conifers, somatic embryogenesis is inducible from mature zygotic embryos of *Picea* as well as from immature embryos (Von Arnold, 1987; Harry and Thorpe, 1991). In contrast, the induction of somatic embryogenesis in *Pinus* requires the culture of early-stage zygotic embryos (Becwar *et al.*, 1988). It is clear that there are fundamental differences in the developmental genetics of these two genera, but we have virtually no understanding of these differences. If we did, it would be possible to bypass extensive trial and error experimentation. Considering the effort that has been applied to the induction of somatic embryogenesis in pines, the cost of this lack of basic scientific information has been very high. Yet there is no near-term prospect that we will have an understanding of the relevant biological differences between *Picea* and *Pinus*. This observation is, moreover, not exclusively a concern in the conifers. *Liriodendron*, for example, is more difficult to regenerate from callus than are some of the other angiosperms. As in pine, the induction of somatic embryogenesis in *Liriodendron* appears to require the use of immature tissue when zygotic embryos are used as source material (Sotak *et al.*, 1991).

We also need research on the relation between cellular mechanisms involved in somatic embryogenesis and those active during zygotic embryogenesis, to determine whether or not the processes are the same and function in parallel. Very few studies of this type have been made on woody plants. Research to date, supported by the results from plantlet establishment and growth, does indicate that these processes are the same, even though the starting points are quite different (De Jong *et al.*, 1993). It will be interesting to learn from molecular analysis whether or not genes expressed during early zygotic embryogenesis, such as those found in *Pinus strobus* (Whitmore and Kriebel, 1987), are also expressed in somatic embryos at the comparable developmental stage.

4. Technology Improvements Needed for Forestry

In contrast with the advanced methodology for mass-production of trees from somatic embryogenesis in some gymnosperm genera, especially *Picea*, much work needs to be done on the large and important genus *Pinus*. Plantlets have been regenerated via somatic embryogenesis in some species, particularly in the Diploxylon pines from southeastern North America. On the other hand, progress has been slower in the Haploxylon pines. Although *Pinus strobus* somatic embryos were cultured to the early cotyledonary stage several years ago (Finer *et al.*, 1989; Kriebel and Finer, 1989), no one has yet succeeded in inducing plantlet formation. The pines are of such economic importance worldwide that they are the focus of extensive genetic research, including breeding, cloning and gene transfer. For this reason, somatic embryogenesis is likely to have an important future role in their intensive culture and an aggressive tissue culture research effort on these conifers appears to be fully justified.

Results of somatic embryogenesis research on a wide range of temperate and tropical angiospermous trees are encouraging. Plant recovery has been obtained in a diversity of genera. In some deciduous trees that are hard to graft and root, e.g., *Juglans* and *Quercus*, somatic embryogenesis and plantlet development have been nevertheless easier to induce than they are in *Pinus*. In *Eucalyptus*, plant regeneration has been achieved from callus in at least 12 species, and plants have also been regenerated from protoplasts (Leroux and Van Staden, 1991). Some progress is also being made with the angiosperms in large-scale plantlet production. Mass-produced *Liriodendron* plantlets derived from somatic embryos have been field-planted in a test of 5500 trees of 9 clones (Merkle *et al.*, 1991).

The genetic stability of callus cultures of trees continues to be a matter of concern. There are encouraging reports of stability for at least one to two years in several species in terms of regeneration potential, ploidy levels or DNA content. Among these species are Norway spruce (Mo *et al.*, 1989), Himalayan poplar (Cheema, 1989), loblolly pine (Franklin *et al.*, 1989) and hybrid larch (Wyman *et al.*, 1992). Conversely, there is evidence of free radical activity in aging plant tissue cultures, including those of *Pinus* and *Eucalyptus* in addition to tobacco, carrot and bean, with implications of oxidative damage (Benson, 1990). More research is needed in this area at the molecular level.

Improvement in the efficiency of large-scale plantlet production is still needed. Cost is an important factor, since clonal propagation by tissue culture is often more expensive than propagation from cuttings or seed. The *in vitro* stages of tissue culture propagation are less easily automated and more labor-intensive than such *ex vitro* operations as media preparation, watering of plantlets, and data management operations. Additional research is needed on automated systems (Aitken-Christie, 1991).

In forestry, we see a special need for more research on the problems involved in the mechanical handling and automated planting of somatic embryos. This includes continued work on the encapsulation of embryos to produce artificial seeds, especially of conifers, which have particularly fragile embryos. It may be possible to create an artificial "endosperm" by capsule modification. Addition of Litvay nutrients with or without sucrose to an alginate-charcoal capsule significantly enhanced root development from white spruce somatic embryos (Lulsdorf *et al.*, 1993). A full discussion of artificial seeds is given in this volume.

The use of zygotic embryo tissue for somatic embryogenesis of forest trees is usually an intermediate step. Its uses are limited because (1) the genetic constitution of the source tissue can be controlled to only a limited degree, and (2) the nature of the phenotype at economic or biological maturity cannot be estimated from observations on juvenile plants. In trees, we often need to be able to induce embryogenesis in undifferentiated somatic cells derived from plants that are several decades old. Recent work is beginning to yield positive results, e.g., from cambial tissue in mature *Robinia* (Han *et al.*, 1993) and *Dalbergia* trees (Kumar *et al.*, 1991), and leaf tissue in *Populus* (Cheema, 1989). The first evidence of somatic embryogenesis from *Picea* leaves was reported by Ruaud *et al.* (1992). The needles were from one-year-old plants.

5. Gene Transfer for Accelerated Tree Improvement

A system for plantlet recovery from mature tree tissues is, however, only part of the technology needed. Although the forest geneticist sees somatic embryogenesis as a very useful tool for the direct propagation of selected individuals and genotypes, other methods of vegetative propagation are also available for the multiplication of valuable selections, including grafting, rooting of cuttings and micropropagation. From a forestry viewpoint, the most pressing reason for the development of somatic embryogenesis technology is its potential for genetic transformation and subsequent multiplication of transgenic plants. Some of the technology useful in agriculture is less easily applied to trees.

Agrobacterium is a useful tool that may be practical for some forest tree species, especially angiosperms. Conifers are, however, less amenable to this transfer system. Recent successful transformations of several conifers with *Agrobacterium rhizogenes* suggest that it merits additional study as an alternative to *A. tumefaciens* (Huang and Tauer, 1993; Diner and Soliman, 1993). However, biolistic techniques are proving to be successful for DNA transfer in species not easily infected with *Agrobacterium* and may be the method of choice for trees in general. Cells of *Picea* can be transformed using a particle gun and regenerated into plantlets. Success has also been obtained with angiospermous trees; foreign genes were introduced into *Liriodendron* and expressed in plants regenerated by somatic embryogenesis (Wilde *et al.*, 1992). Research on other DNA transfer techniques, including targeted delivery via liposomes, DNA soaking and electroporation, does not, at this point, seem to be as promising as the biolistic approach. In comparisons with the biolistic technique, an alternate system has been tested using silicon carbide fibers in a DNA solution to perforate somatic embryo tissues of *Picea mariana*. Transient expression was obtained, although the frequency was low. The system is currently being optimized (Charest *et al.*, 1993).

6. Assessment: Status and Prospects for Tree Improvement

In spite of the paucity of information on the molecular, physiological and biochemical genetics of early zygotic embryogenesis, important advances have been made in recent years that facilitate the practical application of somatic embryogenesis in tree improvement. The hormonal and nutritional requirements of woody plant callus cultures are now much better defined than they were a few years ago. Research on gene transfer, while still at an early stage, is advancing rapidly in both gymnosperms and angiosperms.

There are two major advantages of somatic embryogenesis that make it attractive to the forest geneticist. One is that it facilitates the rapid, large-scale reproduction of pure lines of selected individuals and clones as an integral part of a tree improvement program. The second is that it provides a means of gene transfer and the multiplication of genetically-transformed cells for the mass production of transgenic trees. The end products in both cases are trees that are faster-growing, more resistant to environmental hazards or otherwise restructured to meet specific human needs. The second advantage may be more important to the forest geneticist than the first. As stated, there are optional methods of clonal propagation, but except for pollen transformation (D. Ellis, unpublished) there is no satisfactory alternative technology on the horizon for the mass-production of transgenic trees. Somatic embryogenesis from transformed cells opens the way for a more rapid introduction of intensive culture of forest trees than was previously possible. The propagation of genetically-engineered trees will require a highly-efficient technology for mass-production of somatic embryos and high yields of plantlets that will develop into normal trees. An essential prerequisite is the isolation and molecular cloning of important genes that are either absent in the tree or that exist but need to be blocked from expression, enhanced or replaced. Transferred genes must also be capable of being expressed in the regenerated tree. Some gene expression may be identifiable *in vitro* or in juvenile plants, but genes for other traits may not be expressed for years.

Our lack of knowledge of tree genes continues to be an important problem. The most advanced tree breeding programs only provide information derived from three or four generations of controlled crossing. In addition, gymnosperm nuclear genomes are very large, complicating the job of gene mapping. However, sequence homologies between genes in tree genomes and those of other plants have made possible nuclear gene isolation in the absence of genome maps derived from crossing experiments. Examples are the highly-conserved "housekeeping genes" such as the actins and lectins, and the genes associated with resistance to stress and disease. The encouraging progress being made in the application of marker genes will provide much valuable information for future gene manipulation.

The development of genetically improved trees and other woody plants from transformed cells assumes that techniques will be available when needed for efficient mass propagation by somatic embryogenesis and the large-scale culture of plantlets. In fact, they are now available for some species. The current rate of progress of research on somatic embryogenesis in a large number of other species suggests that for most species we will have the technology on hand to apply when needed for the production of plants from transformed cells.

References

Aitken-Christie, J., 1991. Trends in automation for clonal propagation by tissue culture. In: I. Karube (Ed.), Proc Fourth Toyota Conf, Aichi, Japan, 21–24 October 1990, pp. 235–260. Elsevier, Amsterdam.

Becwar, M.R., S.R. Wann, M.A. Johnson, S.A. Verhagen, R.P. Feirer and R. Nagmani, 1988. Development and characterization of *in vitro* embryogenic systems in conifers. In: M.R. Ahuja (Ed.), Somatic Cell Genetics of Woody Plants, pp. 1–18. Kluwer Academic Publishers, Dordrecht.

Benson, E.E., 1990. Free radicals in stressed and aging plant tissue cultures. In: R. Rodríguez, R. Sánchez Tamés and D.J. Durzan (Eds.), Plant Aging. Basic and Applied Approaches, pp. 269–275. Plenum Publishing Corporation, New York.

Charest, P.J., D. Lachance, C. Jones and Y. Devantier, 1993. Microprojectile and silicon carbide mediated DNA delivery in conifers and recovery of transgenic black spruce (*Picea mariana*). In Vitro Cell. Develop. Biol. 29A(3), Part II: 87A (Congress Abstracts).

Cheema, G.S., 1989. Somatic embryogenesis and plant regeneration from cell suspension and tissue cultures of mature himalayan poplar (*Populus ciliata*). Plant Cell Rep. 8: 124–127.

De Jong, A.J., E.D.L. Schmidt and S.C. de Vries, 1993. Early events in higher-plant embryogenesis. Plant Mol. Biol. 22: 367–377.

Diner, A.M. and K. Soliman, 1993. *Pinus palustris* transformation by *Agrobacterium rhizogenes*. In Vitro Cell. Develop. Biol. 29A(3), Part II: 86A (Congress Abstracts).

Finer, J.J., H.B. Kriebel and M.R. Becwar, 1989. Initiation of embryogenic callus and suspension cultures of white pine (Pinus *strobus* L.). Plant Cell Rep. 8: 203–206.

Franklin, C.I., R.L. Mott and T.M. Vuke, 1989. Stable ploidy levels in long-term callus cultures of loblolly pine. Plant Cell Rep. 8: 101–104.

Han, K.-H., D.E. Keathley and M.P. Gordon, 1993. Cambial tissue culture and subsequent shoot regeneration from mature black locust (*Robinia pseudoacacia* L.). Plant Cell Rep. 12: 185–188.

Harry, I.S. and T.A. Thorpe, 1991. Somatic embryogenesis and plant regeneration from mature zygotic embryos of red spruce. Bot. Gaz. 152: 446–452.

Huang, Y. and C.G. Tauer, 1993. Another tool for gene transfer in pine species: *Agrobacterium rhizogenes*. In Vitro Cell. Develop. Biol. 29A(3), Part II: 65A (Congress Abstracts).

Kriebel, H.B. and J.J. Finer, 1989. Somatic embryogenesis in *Pinus strobus*. Proc. USSR-IUFRO Internat Symp on For Gen, Breeding and Physiol of Woody Plants, Voronezh, USSR, non-pag.

Kumar, A., P. Tandon and A. Sharma, 1991. Morphogenetic responses of cultured cells of cambial origin of a mature tree – *Dalbergia sissoo* Roxb. Plant Cell Rep. 9: 703–706.

Leroux, J.J. and J. van Staden, 1991. Micropropagation and tissue culture of Eucalyptus – a review. Tree Physiol. 9: 435–477.

Lulsdorf, M.M., T.E. Tautorus, S.I. Kikcio, T.D. Bethune and D.I. Dunstan, 1993. Germination of encapsulated embryos of interior spruce (*Picea glauca engelmannii* complex) and black spruce (*Picea mariana* Mill.). Plant Cell Rep. 12: 385–389.

Merkle, S.A., S.E. Schlarbaum, R.A. Cox and O.J. Schwarz, 1991. Mass propagation of somatic embryo-derived plantlets of yellow-poplar for field testing. Proc. 21st Sou For Tree Improvement Conf., pp. 56–68.

Millar, C.I., 1993. Conservation of germplasm in forest trees. In: M.R. Ahuja and W.J. Libby (Eds.), Clonal Forestry II, Conservation and Application, pp. 42–65. Springer Verlag, Berlin.

Mo, L.H., S. von Arnold and U. Lagerkrantz, 1989. Morphogenic and genetic stability in longterm embryogenic cultures and somatic embryos of Norway spruce (*Picea abies* (L.) Karst). Plant Cell Rep. 8: 375–378.

Neale, D.B., M.E. Devey, K.D. Jermstad, M.R. Ahuja, M.C. Alosi and K.A. Marshall, 1992. Use of DNA markers in forest tree improvement research. New For. 6: 391–407.

Ruaud, J.-N., J. Bercetche and M. Pâques, 1992. First evidence of somatic embryogenesis from needles of 1-year-old *Picea abies* plants. Plant Cell Rep. 11: 563–566.

Sotak, R.J., H.E. Sommer and S.A. Merkle, 1991. Relation of the developmental stage of zygotic embryos of yellow-poplar to their somatic embryogenic potential. Plant Cell Rep. 10: 175–178.

Vahala, T., T. Ericsson and P. Engström, 1991. Genetic variability in basket willow (*Salix viminalis*) detected by hybridization to a bacteriophage-M13 DNA. Hereditas 115: 153–161.

Von Arnold, S., 1987. Improved efficiency of somatic embryogenesis in mature embryos of *Picea abies* (L.) Karst. J. Plant Physiol. 128: 233–244.

Whitmore, F.W. and H.B. Kriebel, 1987. Expression of a gene in *Pinus strobus* ovules associated with fertilization and early embryo development. Can. J. For. Res. 17: 408–412.

Wilde, H.D., R.B. Meagher and S.A. Merkle, 1992. Expression of foreign genes in transgenic yellow-poplar plants. Plant Physiol. 98: 114–120.

Wyman, J., N. Brassard, D. Flipo and S. Lalibert, 1992. Ploidy level stability of callus tissue, axillary and adventitious shoots of *Larix* × *eurolepis* Henry regenerated *in vitro*. Plant Sci. (Limerick) 85: 189–196.

2. Historical Aspects of Somatic Embryogenesis in Woody Plants[1]

Subhash C. Minocha and Rakesh Minocha

Contents

1. Introduction

During the next few decades, the world demand for wood products is expected to rise sharply. To meet this growing demand, there will be an increasing need for mass production of improved-quality planting stock of many tree species. The conventional methods of tree improvement and selection offer only limited possibility of meeting the growing demands. Therefore, new and innovative techniques for the creation of new hybrids, early selection and testing of desirable genotypes, rapid vegetative propagation of selected genotypes, and improvement through genetic engineering, etc., must be developed to achieve these goals.

In any tree improvement program, the quality of the product and the economy of propagation are two fundamental measures of success. In addition to the need for genetically uniform stocks of selected genotypes for direct use in planting, these plants can be of enormous value to tree physiologists, pathologists, ecologists, and geneticists for testing their resistance to fungal, bacterial and environmental stress factors. Similarly, an early selection of desirable genotypes could alleviate the need for waiting periods that take several years before field testing can begin. Vegetative propagation has a clear edge over sexual means of reproduction for achieving both rapidity of propagation and assured maintenance of genetic composition of the progeny.

Commercial applications of cell and tissue culture to mass propagation are presently confined mainly to herbaceous plants and to species which can easily be propagated by traditional methods. The potential usefulness of cell and tissue culture techniques for the propagation of forest trees has long been recognized and discussed. However, it is only during the past two

[1]Scientific Contribution Number 1842 from the New Hampshire Agricultural Experiment Station.

S. Jain, P. Gupta & R. Newton (eds.), Somatic Embryogenesis in Woody Plants, Vol. *1*, 9–22.

decades or so that concerted efforts have been made to adapt these methods for the propagation of commercially important tree species. Recent successes with cell and tissue culture of plants like *Pinus lambertiana*, *Pseudotsuga menziesii* and *Picea abies*, although somewhat less than perfect, have generated high hopes for the application of these techniques to mass production of several tree species. The list of plants from which it is now possible to obtain tissue cultures and to induce the formation of plantlets from callus is long and expanding. Nevertheless, the information at hand does not allow for the routine propagation of all species. It is obvious from published literature that the results can seldom be generalized and extrapolated from one species to another with regard to the nutritional requirements for growth and differentiation *in vitro*. Therefore, detailed steps for the successful propagation of a particular species must be worked out individually.

Cell and tissue culture techniques entail the growth of tissue or organ segments on suitable media that stimulate their development along one of several pathways. The most common and so far the most effective pathway is to directly produce whole plants or multiple shoots from stem segments or from excised shoot apices. Depending on hormonal composition of the growth media, these plants may be rootless or rooted. The rootless plantlets can be rooted successfully in most cases. This technique, commonly known as "micropropagation", produces sufficiently large clonal populations of plants and generates the least amount of genetic instability commonly observed in many cell and tissue culture systems. Micropropagation has been successfully used to clone a large number of plant species including fruit trees, conifers, forest trees and other commercial tree crops.

The second pathway, by far the most desirable, involves the regeneration of whole plants from callus and suspension cultures, and is known as somatic embryogenesis. This process is analogous to the development of zygotic embryos, and results in the production of a complete germling with the potential to grow into a whole plant, much as a seedling would. This technique has been successful with some non-woody horticultural plants and its feasibility has been demonstrated in several woody plants (Tautorus *et al.*, 1991). However, large scale routine success has not been achieved with many of the commercially important tree species. Properly controlled somatic embryogenesis carries a low risk of genetic instability, guarantees juvenile plants of normal growth habit, and is amenable to mass production and planting processes. Suspension cultures have been established from various tissues of several tree species, though spontaneous and induced embryogenesis in these suspensions are rare events. However, somatic embryos have been regenerated in more than 50 woody species upon transfer of tissues grown on solid media to suitable growth conditions.

A future advantage of the cell and tissue culture techniques in tree breeding lies in the production of haploid plants from microspores and female gametophytes. Such plantlets can then be induced to produce homozygous diploids by colchicine treatment. This offers the possibility of breeding pure lines. An open-pollinated orchard containing two selected homozygous lines developed from superior individuals would allow all the genetic gain to be captured through the seed.

Research on the formation of protoplasts in tree species has remained far behind that of herbaceous plants. A prerequisite for success with protoplast regeneration and production of somatic hybrids, of course, is the availability of techniques for regeneration of plants from callus. This information is currently limited.

2. Somatic Embryogenesis in Tree Species

The first report of regeneration of adventitious buds from cambial tissue of a woody plant was with *Ulmus campestris* (Gautheret, 1940). This was followed by similar observations from the laboratory of Jacquiot (1949, 1951, 1955) with the same species as well as with *Betula*. In this case some root formation was also noted but complete plantlets were not produced. The first complete plants from tissue culture of a tree species were regenerated by Winton (1970) from leaf explants of aspen. It should be noted that the first tissue cultures of tree species were initiated from cambial explants (Gautheret, 1934, 1948, 1959; Morel, 1948), probably because these explants contained a high content of endogenous auxin and cytokinin. After the discovery of cytokinins, most other explant sources were successfully cultured in vitro for callus and plantlet production. Sommer and Wetzstein (1984) listed more than 100 species of angiospermic woody plants for which shoots or plantlets have been produced from organ/tissue cultures. It is notable from the list that a majority of cases of regeneration involve the use of juvenile tissues such as zygotic embryos and/or young seedlings. Genera with the largest number of species in which regeneration has been obtained by this method include *Eucalyptus*, *Populus*, *Citrus* and *Salix*. Since 1984, the number of hardwood species that have shown regeneration in vitro has nearly doubled.

There has been some confusion regarding the regeneration of somatic embryos from cell/tissue cultures of woody plants. Part of the ambiguity lies in the lack of a clear definition of a somatic embryo. Alternative terms such as "embryoids", "embryo-like structures", "adventitious embryos", etc., have often been used to describe the regenerants. Halperin and Wetherell (1964) emphasized the importance of bipolar organization with distinct root and shoot primordia to distinguish a somatic embryo from a shoot or a plantlet. An additional problem hindering the acceptance of a true somatic embryo relates to the origin of these structures. In some cases the growth of pre-existing zygotic or nucellar embryos, which are a direct product of polyembryonic development, has been confused with the regeneration of somatic embryos. Terms such as apomixis (Nygren, 1954), polyembryony (Webber, 1940), adventive embryony (Schroeder, 1968), nucellar embryony

(Ernst, 1918), and sporophytic embryony or embryogeny (Battaglia, 1963) have all been used to describe such naturally arising structures. Others (Vasil and Hildebrandt, 1966; Haccius and Lakshmanan, 1969) have attempted to distinguish somatic embryos produced in vitro from those of asexual origin in nature by referring to the former as embryoids. The natural occurrence of asexual embryogenesis (cf. somatic embryogenesis) in plants was summarized by Tisserat *et al.* (1979).

Rao (1965) reported the regeneration of embryo-like structures from tissue cultures of *Santalum album* that did not grow into complete plants. The direct production of somatic embryos from the cotyledons of *Ilex aquifolium* was reported by Sussex (1972); again, no mature plants were produced. Several studies from the laboratory of Radojevic described the ultrastructural aspects of somatic embryogenesis in *Corylus avellana* and *Paulownia tomentosa* (Radojevic *et al.*, 1975; Vujicic *et al.*, 1976). A pattern of membrane-bound ribosomes was found in the embryogenic cells that was similar to that seen in fertilized fern eggs and active secretory glands. Histochemical studies revealed that the embryogenic cells were rich in sugars, lipids and protein, and were also involved in starch accumulation.

Sharp *et al.* (1980, 1982) distinguished two different patterns of the origin of somatic embryos from in vitro grown explants: (1) direct production of somatic embryos from the explant cells called the pro-embryonic-determined cells (PEDC), and (2) indirect production of somatic embryos from an unorganized callus/tissue mass called the induced embryogenic-determined cells (IEDC). In the former, somatic embryos are presumed to originate from explant cells that require only an in vitro environment to be released from some suppressive condition imposed by the organization of the explant. By contrast, the IEDC pattern not only requires the release of previously differentiated state through mitotic cell divisions but also an induction of the new pattern of cell divisions to form organized embryos. Often, while the former situation requires no growth regulators, the latter may depend on a sequence of growth regulator treatments first to form callus from which the somatic embryos are regenerated.

Tissues such as nucellus, suspensors, integuments, proembryos, megagametophyte, etc., that are associated with the zygotic embryos generally produce asexual embryos through the PEDC pattern. Conversely, leaf, cotyledon and stem explants often produce somatic embryos via the formation of callus. Some examples of direct somatic embryogenesis in woody angiosperms include various species of *Citrus*, *Ilex aquifolium*, *Malus domestica*, *Mangifera indica*, *Theobroma cacao*, and *Pyrus* spp. Indirect somatic embryogenesis from juvenile and mature tissues of a number of woody plants has been seen both in monocots and dicots; some examples are *Cocos nucifera*, *Chamedorea costaricana*, *Phoenix dactylifera*, *Elaeis guineensis*, *Coffea arabica*, *Corylus avellana*, *Malus pumila*, *Paulownia tomentosa*, *Pyrus communis*, *Santalum album*, *Sapindus trifoliatus* and *Vitis vinifera*.

While the two patterns of somatic embryo development may be distinct and many plants seem to follow either pattern, these are by no means mutually exclusive. Numerous examples exist where the PEDC continue to proliferate in a pro-embryogenic state under appropriate culture conditions. This pro-embryogenic tissue mass can be subcultured through many generations and complete somatic embryos can be generated on transfer to a different medium. Such proliferative tissues have been obtained in *Citrus sinensis*, *C. aurantifolia*, *Mangifera indica*, *Ilex aquifolium*, and *Theobroma cacao*.

Among the gymnosperms, the first reports of somatic embryogenesis in Norway spruce (*Picea abies*) appeared in 1985 (Hakman *et al.*, 1985; Chalupa, 1985). Within a year, other reports were added to the literature, expanding the list to include the genus *Pinus* (Gupta and Durzan, 1986a,b; Von Arnold and Hakman, 1986; Krogstrup, 1986). Since then, *Picea abies* has been used widely as a model experimental system for somatic embryogenesis in conifers (see Roberts *et al.*, 1993, and references therein). Using quite similar treatments, somatic embryogenesis has now been achieved in several additional conifer species.

Somatic embryogenesis in most conifers follows a similar pattern of development. With few exceptions, the embryogenic mass of cells is initiated from immature or mature zygotic embryos. The developmental stage of the zygotic embryo plays a crucial role in the production of embryogenic callus. Whereas in *Larix* the ideal material is young developing embryos (2 to 4 weeks postfertilization), fully mature embryos are suitable for various species of *Picea*. In *Pinus*, pre-cotyledonary stage embryos have generally yielded the best embryogenic tissue. Krogstrup (1986) and Attree *et al.* (1990a) demonstrated the feasibility of obtaining somatic embryos from cotyledons of 7-day-old and 12- to 30-day-old seedlings of *Picea abies* and *Picea mariana*, respectively. These observations were further extended to the use of 1-year-old seedlings of *Picea abies* (Ruaud *et al.*, 1992).

A detailed analysis of *Picea glauca* and *P. abies* revealed that embryogenic tissue originates from epidermal or subepidermal layers of the embryonic axis of zygotic embryos. The embryogenic tissue consists of elongated suspensor-like cells with small densely cytoplasmic clusters of meristematic cells at one end. Based on the similarity of this tissue with the in vivo cleavage polyembryony in conifers, Gupta and Durzan (1986a,b, 1987) suggested the term "somatic polyembryogenesis" for this tissue, which can be proliferated by routine subculture and it still retains its characteristic morphology. It is only upon transfer to a maturation medium, usually supplemented with abscisic acid (ABA), do the pro-embryogenic masses develop into complete embryos (see Dunstan *et al.*, 1991; Tautorus *et al.*, 1991; Roberts *et al.*, 1990a, 1993, and references therein). As opposed to *Picea*, where the embryogenic tissue arises from the hypocotyl zone, somatic embryogenesis in *Pinus* is initiated from the suspensor cells (Tautorus *et al.*, 1991).

In contrast to the dissected embryos from seeds, the use of seedlings offers ease in handling of explant materials, the former being highly labor intensive.

Eventually we must develop procedures to obtain embryogenic tissues from selected mature genotypes in order to take full advantage of proven genetic superiority of mature trees.

Since the first commercial application of somatic embryogenesis was realized for oil palms (Blake, 1983), few commercial plantations with other species have been established. The current status of commercial propagation of *Picea* through artificial seed production has been recently reviewed (Roberts *et al.*, 1993; see also, Lulsdorf *et al.*, 1993, and references therein). A major problem in most cases is the low frequency of maturation and germination of somatic embryos into whole plants. Exceptions include *Coffea arabica*, *Santalum album*, *Liriodendron tulipifera* and *Corylus avellana* for which more than 10 percent of the somatic embryos grew into whole plants (Sita *et al.*, 1979; Merkle and Sommer, 1986; Sondahl and Sharp, 1977; Perez *et al.*, 1983).

Numerous attempts have been made with various species of *Picea* to improve the quality and yield of mature embryos capable of germination into whole plants (Roberts *et al.*, 1993). Lack of production of a well-developed root has been a common problem in the failure of somatic embryos to develop into emblings [a term coined by Libby (1986) to distinguish the plantlets that are produced via somatic embryos from the seedlings that are produced from seeds]. Light has been shown to inhibit the germination of somatic embryos in most cases (Von Arnold and Hakman, 1988). It has been observed that a gradual desiccation under controlled relative humidity conditions substantially improves the efficiency of germination of *Picea* somatic embryos (Gray, 1989; Roberts *et al.*, 1990b).

Field trials of plants of several species of *Picea* produced from somatic embryogenesis have been reported (Becwar *et al.*, 1989; Attree *et al.*, 1990b; Webster *et al.*, 1990). Most have involved relatively small populations; the largest study involved 1200 plants showing a survival rate of more than 80 percent (Webster *et al.*, 1990). Height measurements and cold-hardiness studies showed a great similarity between the emblings and the seedlings. The overall survival rate at the end of the second season was almost 100 percent. These studies have since been expanded to include approximately 10,000 plants of *Picea glauca-engelmannii* complex and 8,000 plants of *P. sitchensis*. Due to the long life span of trees, it will take many years before the overall performance of these plants can be evaluated.

An ideal approach to increasing the efficiency of somatic embryogenesis is the use of large-scale bioreactors that can produce somatic embryos in liquid cultures. While there have been several instances of growth of embryogenic tissues in suspension cultures of gymnosperms (*Abies nordmanniana*, *Picea abies*, *Picea mariana*, *Pinus caribaea*, *Pinus strobus*, *Pseudotsuga menziesii*), sustained production of somatic embryos is not commonly observed (Hakman *et al.*, 1985; Durzan and Gupta, 1987; Gupta and Durzan, 1987; Hakman and Fowke, 1987; Finer *et al.*, 1989; Attree *et al.*, 1989a; Lainé and David, 1990; Tautorus *et al.*, 1990). To date only one case of the production of mature somatic embryos of any conifer in large bioreactors has been reported (Tautorus *et al.*, 1992). Encapsulation of somatic embryos in a variety of gel coatings to produce artificial seeds for direct field planting has also been achieved with limited success (see Roberts *et al.*, 1993; Lulsdorf *et al.*, 1993, and references therein).

A prerequisite for sustained production of superior genotypes by somatic embryogenesis is the ability to field test the product before large-scale commercial planting. This means that successful genotypes should be maintained in an embryogenic state over several years. Cryopreservation is an ideal way to maintain and store highly embryogenic tissues for long periods (Chen and Kartha, 1987). A few attempts have already been made to demonstrate the regeneration of cryopreserved tissues of woody plants (Gupta *et al.*, 1987a,b; Kartha *et al.*, 1988; Ward, 1990; Bercetche *et al.*, 1990). In *Picea glauca*, *Picea abies* and *Pinus taeda*, the tissue remained competent of producing somatic embryos following short term storage in liquid nitrogen. The effect of long-term storage on embryo regeneration still needs to be determined.

3. Haploids and Triploids

Tulecke (1953) first demonstrated the ability of mature pollen grains of *Ginkgo biloba* to produce haploid callus. While limited morphogenesis was seen in the callus, no plantlets were produced. Subsequent work by several laboratories demonstrated the production of haploid callus in a number of gymnosperms including *Taxus brevifolia*, *Taxus baccata* (Tulecke, 1959; Rohr, 1973), *Torreya nucifera* (Tulecke and Sehgal, 1963), *Ephedra foliata* (Konar, 1963) and *Pinus resinosa* (Bonga, 1974). The first reports of haploid callus and plantlet formation from pollen cultures of angiospermic trees date to 1974 (Michellon *et al.*, 1974, for *Prunus amygdalus* and *Prunus persica*; Sato, 1974, for *Populus* spp.).

Numerous attempts have since been made to obtain haploid plants and somatic embryos from anther cultures of other woody plants. Anther cultures of *Malus* domestica yielded early-stage somatic embryos (Milewska-Pawliczuk and Kubicki, 1977) but these embryos failed to develop into plantlets. A mixed ploidy callus obtained from anther cultures of *Vitis vinifera* × *V. rupestris* produced several diploid somatic embryos which developed into plantlets with a high frequency (Rajasekaran and Mullins, 1979). Chen *et al.* (1982) reported the production of haploid somatic embryos from anther cultures of *Hevea brasiliensis*, which developed into whole plants at a relatively low frequency of 3 percent. Numerous haploid plants of *Annona squamosa* were obtained by Nair *et al.* (1983) through regeneration of multiple shoots from anther callus.

In contrast to anther/pollen culture, which has yielded positive results mostly with angiosperms, megagametophytic tissue has been used to produce haploid somatic embryos in a few gymnosperms. Haploid callus capable of producing some roots and shoots was obtained from megagametophytic tissue of *Ginkgo biloba* by Tulecke (1965) and of *Zamia integrifolia* by Norstog (1965). Within a few years, root/shoot regeneration was reported in haploid megagametophytic tissues of *Cycas circinalis* (Norstog and Rhamstine, 1967; Huhtinen, 1972). While in most cases the frequency of regeneration of somatic embryos is extremely low, *Larix decidua* megagametophytic cultures have yielded haploid plants at a relatively high frequency. Rohr (1987) has described in details the development of haploid tissues in a number of gymnosperms.

Triploid plantlets have been produced from the endospermic tissues of some angiospermic woody plants (*Putranjiva roxburghii* – Srivastava, 1973; *Jatropha panduraefolia* and *Leptomeria acida* – Johri & Srivastava, 1973). In none of these has the growth of complete plants in the soil been reported.

4. Protoplasts

The first report on the isolation and culture of protoplasts in woody plants appeared in 1972. Rona and Grignon (1972) demonstrated the growth of protoplasts isolated from suspension cultures of *Acer pseudoplatanus*. While protoplast isolation and culture (to form callus) has been successful in a diverse group of woody angiosperms (McCown and Russell, 1987) and also a few gymnosperms (David, 1987), reports of regeneration of whole plants from single isolated protoplasts are rare. Somatic embryos were first regenerated from protoplasts of embryogenic cells of *Pinus taeda*, and *Picea glauca* (Gupta and Durzan, 1987; Attree *et al.*, 1987, 1989b). Similar results were later reported for several other conifers. The limited range of the explant source from which morphogenetically competent tissues can be obtained is a major reason for such limited success with protoplast culture in trees. The production of somatic hybrids through protoplast fusion has not been demonstrated in any tree species.

5. Major Problems and Future Perspectives

In the preceding pages, we have provided a brief review of the different stages of the research on somatic embryogenesis with tree species. Published information shows that: (1) somatic embryogenesis in some woody plants, especially conifers, has become routine; (2) somatic embryos usually can be grown into whole plantlets (albeit at low-to-moderate frequency) that can be tested for performance in the field; (3) the process of somatic embryogenesis is developmentally similar in most conifers whereas specific requirements for somatic embryo differentiation and maturation may vary among species; (4) the source of the explant plays a decisive role in the ability to produce embryogenic tissue; and (5) the developmental pattern of somatic embryos is analogous to the development of zygotic embryos without the complexity of surrounding gametophytic tissues.

The following are some of the major problems commonly associated with tree tissue culture.

1. Most of the published work has utilized juvenile tissues, particularly zygotic embryos. Tissues from mature trees seldom seem amenable to successful propagation. There is a need for systematic physiological studies on the underlying changes in juvenility to maturity. A knowledge of the process of "phase change" could then be utilized to reverse the process in shoot apices or obtain adventitious juvenile material from mature plants. Rejuvenating treatments currently being investigated include serial grafting of buds onto juvenile rootstock, severe pruning of trees to stimulate latent juvenile meristems, spraying plants with cytokinins, serial subculture of shoot apices *in vitro*, and hormone-induced production of adventitious shoots *in vitro*.
2. Growth of regenerated plants in the greenhouse remains a formidable problem, particularly in conifers. In addition to the use of ABA for maturation of somatic embryos and physical treatments for acclimation of emblings, the usefulness of mycorrhizal associations in the formation and growth of roots should be studied.
3. Production of haploid plants from pollen and microspore cultures has been studied only in a few tree species. The potential advantages of haploid plants through the production of homozygous diploids cannot be overemphasized. As mentioned earlier, for gymnosperms, the female gametophyte should provide a good source of material for haploid plants.
4. There has been little effort to produce somatic hybrids in tree species through protoplast fusion. Although of limited value thus far, this technique has unlimited potential for the production of intraspecific and interspecific hybrids. The best approach will be to work with closely related taxa. Since many of disease/chemical resistance mechanisms reside in cytoplasmic factors, production of cybrids (a fusion product of a nucleated and an enucleated protoplast) can be advantageous in the production of resistant varieties.
5. Recent advances in the techniques of gene transfer and our current knowledge about the regulation of gene activity can be used in tree improvement programs. These techniques are primarily dependent on the availability of easily regenerating cell and tissue cultures.

6. Conclusion

A descriptive analysis of the development of both zygotic embryos and somatic embryos is now available for a number of woody plant species. A major focus of research during the past two decades has been the manipulation of growth media and growth conditions. Another is the testing of a variety of explant sources to obtain somatic embryogenesis in dozens of species of woody angiosperms and gymnosperms. However, only a limited effort has been made to enhance our understanding of the biochemical and molecular basis of somatic embryogenesis; there has been even less work on addressing the question: "What makes a somatic cell become embryogenic?" Perhaps the answer must await a set of breakthroughs in routinely used model experimental systems for somatic embryogenesis such as carrot and alfalfa, or in the biochemical and molecular analysis of developmental mutants in plants like *Arabidopsis* and *Zea mays*. Ultimately, however, the molecular aspects of somatic embryogenesis in conifers will be understood only from direct work with model experimental systems of woody plants such as *Picea* and *Pinus*. The area of developmental mutation research in woody plants, particularly the conifers, is untouched. Obviously, the long lifespan and a long, sometime multiyear, routine for zygotic embryo development are major hurdles to a large-scale analysis of putative mutants. Nevertheless, guidelines provided by research on herbaceous plants should pave the way for the use of modern techniques for genetic, biochemical, and molecular analysis of somatic embryogenesis. This is a prerequisite for achieving full utilization of this approach in the production of genetically improved, economically important woody plants.

While the potential usefulness of haploids, somaclonal variation, and somatic hybridization should not be minimized, these techniques have not yet proven as beneficial as generally presumed. Moreover, genetic transformation with foreign genes has already shown its potential as a viable alternative to somaclonal variation and mutagenesis. Applications of genetic transformation technology hold greater potential for woody plants than even the herbaceous plants because the gains in genetic improvement can be realized over relatively short periods compared to routinely used techniques of hybridization, selection, and mutagenesis which require decades. This area of research depends on at least three factors that must come together for each species or genotype to be improved: (1) availability of specific genes and promoters for transfer and expression; (2) development of reliable techniques for transfer of the genes to target cells and the selection of transformed cells; and (3) regeneration of transformed cells into whole plants which can be mass-propagated, preferably by asexual means.

A variety of potentially useful genes is being characterized and cloned for transfer to agriculturally important plants, many of which also will be useful for the improvement of woody plants. Likewise, a number of different methods of genetic transformation of plants have been developed, some of which have shown applicability to woody plants as well. Stable transfer, integration, and expression of a model gene (such as NPT or GUS) have been demonstrated only in a few woody plant species, and genes regulating cellular metabolism or physiological aspects of growth and development have been tested in still fewer cases. It is hoped that this area of research would soon realize a major spurt of activity.

References

Attree, S.M., F. Bekkaoui, D.I. Dunstan and L.C. Fowke, 1987. Regeneration of somatic embryos from protoplasts isolated from an embryogenic suspension culture of white spruce (*Picea glauca*). Plant Cell Rep. 6: 480–483.

Attree, S.M., D.I. Dunstan and L.C. Fowke, 1989a. Initiation of embryogenic callus and suspension cultures, and improved regeneration of protoplasts, of white spruce (*Picea glauca*). Can. J. Bot. 67: 1790–1795.

Attree, S.M., D.I. Dunstan and L.C. Fowke, 1989b. Plantlet regeneration from embryogenic protoplasts of white spruce (*Picea glauca*). Bio/Technology 7: 1060–1062.

Attree, S.M., S. Budimir and L.C. Fowke, 1990a. Somatic embryogenesis and plantlet regeneration from cultured shoots and cotyledons of seedlings from stored seeds of black and white spruce (*Picea mariana* and *Picea glauca*). Can. J. Bot. 68: 30–34.

Attree, S.M., T.E. Tautorus, D.I. Dunstan and L.C. Fowke, 1990b. Somatic embryo maturation, germination, and soil establishment of plants of black and white spruce (*Picea mariana* and *Picea glauca*). Can. J. Bot. 68: 2583–2589.

Battaglia, E., 1963. Apomixis. In: P. Maheshwari (Ed.), Recent Advances in the Embryology of Angiosperms, pp. 221–264. Intern. Soc. of Plant Morphologists, Univ. of Delhi, Delhi.

Becwar, M.R., T.L. Noland and J.L. Wyckoff, 1989. Maturation, germination, and conversion of Norway spruce (*Picea abies* L.) somatic embryos to plants. In Vitro Cell Develop. Biol. 25: 575–580.

Bercetche, J., M. Galerne and J. Dereuddre, 1990. Efficient regeneration of plantlets from embryogenic callus of *Picea abies* (L.) Karst after freezing in liquid nitrogen. C.R. Acad. Sci. Ser. 3, 310: 357–363.

Blake, J., 1983. Tissue culture propagation of coconut, date, and oil palm. In: J.H. Dodds (Ed.), Tissue Culture of Trees, pp. 29–51. Avi Publishing Co., West Port, CT.

Bonga, J.M., 1974. *In vitro* culture of microsporophylls and megagametophyte tissue of *Pinus*. In Vitro 9: 270–277.

Chalupa, V., 1985. Somatic embryogenesis and plantlet regeneration from cultured immature and mature embryos of *Picea abies* (L.) Karst. Communi. Inst. For. Cech. 14: 57–63.

Chen, F., C. Qian, M. Qin, X. Xu and Y. Xiao, 1982. Recent advances in anther culture of *Hevea brasiliensis* [Muell.-Arg.]. Theor. Appl. Genet. 62: 103–108.

Chen, T.H.H. and K.K. Kartha, 1987. Cryopreservation of woody species. In: J.M. Bonga and D.J. Durzan (Eds.), Cell and Tissue Culture in Forestry, Vol. 2, pp. 305–319. Martinus Nijhoff Publishers, Dordrecht.

David, A., 1987. Conifer protoplasts. In: J.M. Bonga and D.J. Durzan (Eds.), Cell and Tissue Culture in Forestry, Vol. 2, pp. 2–15. Martinus Nijhoff Publishers, Dordrecht.

Dunstan, D.I., T.D. Bethune and S.R. Abrams, 1991. Racemic abscisic acid and abscisyl alcohol promote maturation of white spruce (*Picea glauca*) somatic embryos. Plant Sci. 76: 219–228.

Durzan, D.J. and P.K. Gupta, 1987. Somatic embryogenesis and polyembryogenesis in Douglas-fir cell suspension cultures. Plant Sci. 52: 229–235.

Ernst, A., 1918. Bastardierung als Urache der Apogamie im Pflazenreich. Gustav Fisher, Jena.

Finer, J.J., H.B. Kriebel and M.R. Becwar, 1989. Initiation of embryogenic callus and suspension cultures of eastern white pine (*Pinus strobus* L.). Plant Cell Rep. 8: 203–206.

Gautheret, R.J., 1934. Culture du tissu cambial. Acad. Sci. Paris 198: 2195–2196.

Gautheret, R.J., 1940. Nouvelles recherches sur le bourgeonnement du tissu cambial d'*Ulmus campestris* cultivé *in vitro*. C.R. Acad. Sci. Paris 210: 744–746.

Gautheret, R.J., 1948. Sur la culture indéfinie des tissus de *Salix caprea*. Compt. Rend. Soc. Biol. 142: 807–808.
Gautheret, R.J., 1959. La Culture des Tissus Végétaux. Masson and Cie, Paris.
Gray, D.J., 1989. Effects of dehydration and exogenous growth regulators on dormancy, quiescence and germination of grape somatic embryos. In Vitro Cell Develop. Biol. 25: 1173–1179.
Gupta, P.K. and D.J. Durzan, 1986a. Plantlet regenertion via somatic embryogenesis from subcultured callus of mature embryos of *Picea abies* (Norway spruce). In Vitro Cell Develop. Biol. 22: 685–688.
Gupta, P.K. and D.J. Durzan, 1986b. Somatic polyembryogenesis from callus of mature sugar pine embryos. Bio/Technology 4: 643–645.
Gupta, P.K. and D.J. Durzan, 1987. Somatic embryos from protoplasts of loblolly pine proembryonal cells. Bio/Technology 5: 710–712.
Gupta, P.K., D.J. Durzan and B.J. Finkle, 1987a. Somatic polyembryogenesis in cell masses of *Picea abies* (Norway spruce) and *Pinus taeda* (loblolly pine) after thawing from liquid nitrogen. Can. J. For. Res. 17: 1130–1134.
Gupta, P.K., D. Shaw and D.J. Durzan, 1987b. Loblolly pine: Micropropagation, somatic embryogenesis and encapsulation. In: J.M. Bonga and D.J. Durzan (Eds.), Cell and Tissue Culture in Forestry, Vol. 3, pp. 101–108. Martinus Nijhoff Publishers, Dordrecht.
Haccius, B. and K.K. Lakshmanan, 1969. Adventiv-embryonen – embryoide – adventiv-knospen. Ein beitrag zur klarung der begriffe. Oesterr. Bot. 116: 145–158.
Hakman, I. and L.C. Fowke, 1987. Somatic embryogenesis in *Picea glauca* (white spruce) and *Picea mariana* (black spruce). Can. J. Bot. 65: 656–659.
Hakman, I., L.C. Fowke, S. von Arnold and T. Eriksson, 1985. The development of somatic embryos in tissue cultures initiated from immature embryos of *Picea abies* (Norway spruce). Plant Sci. 38: 53–59.
Halperin, W. and D.F. Wetherell, 1964. Adventive embryony in tissue cultures of the wild carrot, *Daucus carota*. Amer. J. Bot. 51: 274–283.
Huhtinen, O., 1972. Production and use of haploids in breeding conifers. IUFRO Genetics-SABRAO Joint Symposia, Forest Experiment Station of Japan, Tokyo, 1972, D-3(I): 1–8.
Jacquiot, C., 1949. Observations sur la néoformation de bourgeons chez le tissu cambial d'*Ulmus campestris* cultivé *in vitro*. C.R. Acad. Sci. Paris 229: 529–530.
Jacquiot, C., 1951. Action du mesoinositol et de l' adenine sur la formation de bourgeons par le tissu cambial d'*Ulmus campestris* cultivé *in vitro*. C.R. Acad. Sci. Paris 233: 815–817.
Jacquiot, C., 1955. Formation d'organes par le tissu cambial d'*Ulmus campestris* L. et de *Betula verrucosa* Gaertn. cultivés *in vitro*. C.R. Acad. Sci. Paris 240: 557–558.
Johri, B.M. and P.S. Srivastava, 1973. Morphogenesis in endosperm cultures. Z. Pflanzenphysiol. 70: 285–304.
Kartha, K.K., L.C. Fowke, N.L. Leung, K.L. Caswell and I. Hakman, 1988. Induction of somatic embryos and plantlets from cryopreserved cell cultures of white spruce (*Picea glauca*). J. Plant Physiol. 132: 529–539.
Konar, R.N., 1963. A haploid tissue from the pollen of *Ephedra foliata* Boiss. Phytomorphology 13: 170–174.
Krogstrup, P., 1986. Embryolike structures from cotyledons and ripe embryos of Norway spruce (*Picea abies*). Can. J. For. Res. 16: 664–668.
Lainé, E. and A. David, 1990. Somatic embryogenesis in immature embryos and protoplasts of *Pinus caribaea*. Plant Sci. 69: 215–224.
Libby, W.J., 1986. Clonal propagation. J. For. 84: 37–38.
Lulsdorf, M.M., T.E. Tautorus, S.I. Kikcio, T.B. Bethune and D.I. Dunstan, 1993. Germination of encapsulated embryos of interior spruce (*Picea glauca-engelmannii* complex) and black spruce (*Picea marina* Mill.). Plant Cell Rep. 12: 385–389.
McCown, B.H. and J.A. Russell, 1987. Protoplast culture of hardwoods. In: J.M. Bonga and D.J. Durzan (Eds.), Cell and Tissue Culture in Forestry, Vol. 2, pp. 16–30. Martinus Nijhoff Publishers, Dordrecht.

Merkle, S.A. and H.E. Sommer, 1986. Somatic embryogenesis in tissue cultures of *Liriodendron tulipifera*. Can. J. For. Res. 16: 420–422.

Michellon, R., J. Hugard and R. Jonard, 1974. Sur l'isolement de colonies tissulaires de Pêcher (*Prunus persica* Batsch, cultivars *Dixired et Nectared* IV) et d'Amandier (*Prunus amygdalus* Stokes, cultivar Ai) à partir d'anthères cultivées *in vitro*. Compt. Rend. 278: 1719–1722.

Milewska-Pawliczuk, E. and B. Kubicki, 1977. Induction of andorgenesis *in vitro* in *Malus domestica*. Acta Hort. 78: 271–276.

Morel, G., 1948. Recherches sur la culture associée de parasites obligatoires et de tissus végétaux. Ann. Epiphyties 14: 123–234.

Nair, S., P.K. Gupta and A.F. Mascarenhas, 1983. Haploid plants from *in vitro* anther culture of *Annona squamosa* Linn. Plant Cell Rep. 2: 198–200.

Norstog, K., 1965. Induction of apogamy in megagametophytes of *Zamia integrifolia*. Am. J. Bot. 52: 993–999.

Norstog, K. and E. Rhamstine, 1967. Isolation and culture of haploid and diploid cycad tissues. Phytomorphology 17: 374-381.

Nygren, A., 1954. Apomixis in the angiosperms. II. Bot. Rev. 20: 577–649.

Perez, C., B. Fernandez and R. Rodriguez, 1983. *In vitro* plantlet regeneration through asexual embryogenesis in cotyledonary segments of *Corylus avellana* L. Plant Cell Rep. 2: 226–228.

Radojevic, L., R. Vujicic, and M. Neskovic, 1975. Embryogenesis in tissue culture of *Corylus avellana* L. Z. Pflanzenphysiol. 77: 33–41.

Rajasekaran, K. and M.G. Mullins, 1979. Embryos and plantlets from cultured anthers of hybrid grapevines. J. Exp. Bot. 30: 399–407.

Rao, P.S., 1965. *In vitro* induction of embryonal proliferation in *Santalum album* L. Phytomorphology 15: 175–179.

Roberts, D.R., B.S. Flinn, D.T. Webb, F.B. Webster and B.C.S. Sutton, 1990a. Abscisic acid and indole-3-butyric acid regulation of maturation and accumulation of storage proteins in somatic embryos of interior spruce. Physiol. Plant. 78: 355–360.

Roberts, D.R., B.C.S. Sutton and B.S. Flinn, 1990b. Synchronous and high frequencey germination of interior spruce somatic embryos following partial drying at high relative humidity. Can. J. Bot. 68: 1086–1090.

Roberts, D.R., F.B. Webster, B.S. Flinn, W.R. Lazaroff and D.R. Cyr, 1993. Somatic embryogenesis of spruce. In: K. Redenbaugh (Ed.), Synseeds: Applications of Synthetic Seeds to Crop Improvement, pp. 427–449, CRC Press, Inc., Boca Raton, FL.

Rohr, R., 1973. Ultrastructure des spermatozoides de *Taxus baccata* L. obtenus à partir de cultures aseptiques de microspores sur un milieu artificiel. Compt. Rend. 277: 1869–1871.

Rohr, R, 1987. Haploids (Gymnosperms). In: J.M. Bonga and D.J. Durzan (Eds.), Cell and Tissue Culture in Forestry, Vol. 2, pp. 230–246. Martinus Nijhoff Publishers, Dordrecht.

Rona, J.P. and C. Grignon, 1972. Obtention de protoplastes à partir de suspensions de cellules d'*Acer pseudoplatanus* L. C.R. Acad. Sci., Ser. D, Sci. Natur. 274: 2976–2979.

Ruaud, J., J. Bercetche and M. Pâques, 1992. First evidence of somatic embryogenesis from needles of 1-year-old *Picea abies* plants. Plant Cell Rep. 11: 563–566.

Sato, T., 1974. Callus induction and organ differentiation in anther culture of poplars. J. Japan For. Soc. 56: 55–62.

Schroeder, C.A., 1968. Adventive embryogenesis in fruit pericarp tissue *in vitro*. Bot. Gaz. 129: 374–376.

Sharp, W.R., D.A. Evans and M.R. Sondahl, 1982. Application of somatic embryogenesis to crop improvement. In: A. Fujiwara (Ed.), Plant Tissue Culture, pp. 759–62. Proc. 5th Intern. Cong. Plant Tissue and Cell Culture, Japanese Assoc. for Plant Tissue Culture.

Sharp, W.R., M.R. Sondahl, L.S. Caldas and S.B. Maraffa, 1980. The physiology of *in vitro* asexual embryogenesis. Hort. Rev. 2: 268–310.

Sita, G.L., N.V. Raghava Ram and C.S. Vaidyanathan, 1979. Differentiation of embryoids and plantlets from shoot cultures of sandalwood. Plant Sci. Lett. 15: 265–270.

Sommer, H.E. and H.Y. Wetzstein, 1984. Hardwoods. In: P.V. Ammirato, D.A. Evans, W.R.

Sharp and Y. Yamada (Eds.), Handbook of Plant Cell Culture, Vol. 3, pp. 511–540. MacMillan Publishing Co., New York.
Sondahl, M.R. and W.R. Sharp, 1977. High frequency induction of somatic embryos in cultured leaf explants of *Coffea arabica* L. Z. Pflanzenphysiol. 81: 395–408.
Srivastava, P.S., 1973. Formation of triploid plantlets in endosperm cultures of *Putranjiva roxburghii*. Z. Pflanzenphysiol. 69: 270–273.
Sussex, I.M., 1972. Somatic embryos in long-term carrot tissue cultures: histology, cytology, and development. Phytomorphology 22: 50–59.
Tautorus, T.E., S.M. Attree, L.C. Fowke and D.I. Dunstan, 1990. Somatic embryogenesis from immature and mature zygotic embryos, and embryo regeneration from protoplasts in black spruce (*Picea mariana* Mill.). Plant Sci. 67: 115–124.
Tautorus, T.E., L.C. Fowke and D.I. Dunstan, 1991. Somatic embryogenesis in conifers. Can. J. Bot. 69: 1873–1899.
Tautorus, T.E., M.M. Lulsdorf, S.I. Kikcio and D.I. Dunstan, 1992. Bioreactor culture of *Picea mariana* Mill. (black spruce) and the species complex *Picea glauca-engelmannii* (interior spruce) somatic embryos. Growth parameter. Appl. Microbiol. Biotechnol. 38: 46–51.
Tisserat, B., E.B. Esan and T. Murashige, 1979. Somatic embryogenesis in angiosperms. Hort. Rev. 1: 1–78.
Tulecke, W., 1953. A tissue derived from the pollen of *Ginkgo biloba*. Science 117: 599–600.
Tulecke, W., 1959. The pollen cultures of C.D. LaRue: A tissue from the pollen of *Taxus*. Bull Torrey Bot. Club. 86: 283–289.
Tulecke, W., 1965. Haploidy versus diploidy in the reproduction of cell type. In: M. Locke (Ed.), Reproduction: Molecular, Subcellular, and Cellular, pp. 217–241. Academic Press, New York.
Tulecke, W. and N. Sehgal, 1963. Cell proliferation from the pollen of *Torreya nucifera*. Contr. Boyce Thompson Inst. 22: 153–163.
Vasil, I.K. and A.C. Hildebrandt, 1966. Variations of morphogenetic behavior in plant tissue cultures. I. *Cichorium endivia* L. Am. J. Bot. 53: 860–869.
von Arnold, S. and I. Hakman, 1986. Effect of sucrose on initiation of embryogenic callus cultures from mature zygotic embryos of *Picea abies* (L.) Karst (Norway spruce). J. Plant Physiol. 122: 261–265.
von Arnold, S. and I. Hakman, 1988. Regulation of somatic embryo development in *Picea abies* by abscisic acid (ABA). J. Plant Physiol. 132: 164–169.
Vujicic, R., L. Radojevic and M. Neskovic, 1976. Orderly arrangement of ribosomes in the embryogenic callus tissue of *Corylus avellana* L. J. Cell Biol. 69: 686–692.
Ward, C., 1990. Preservation of black spruce clones through tissue culture. In: Proc. Annu. Meet: Black Spruce Clonal For. Prog., pp. 46. OMNR, Timmins, Ontario, Canada.
Webber, J.M., 1940. Polyembryony. Bot. Rev. 6: 575–598.
Webster, F.B., D.R. Roberts, S.M. McInnis and B.C.D. Sutton, 1990. Propagation of interior spruce by somatic embryogenesis. Can. J. For. Res. 20: 1759–1765.
Winton, L.L., 1970. Shoot and tree production from aspen tissue cultures. Am. J. Bot. 57: 904–909.

3. Anatomical Comparision of Somatic and Zygotic Embryogeny in Conifers

R. Nagmani, A.M. Diner, S. Garton and A.E. Zipf

Contents

1. Introduction

Wardlaw (1955, 1968) defined an embryo as "a plant in its initial stage of development". Structurally, an embryo is considered bipolar with root and shoot on opposite ends (Halperin and Wetherell, 1964). These definitions refer to seed or zygotic embryos that result from gametic fusion in higher plants, and can serve as guidelines to identify embryo-like structures often observed in tissue cultures of various plant species. These "embryo-like" structures are referred to as "embryoids" (Haccius and Lakshmanan, 1969; Vasil and Hildebrandt, 1966) or "somatic embryos" and the process of formation and development is termed "somatic or asexual embryogenesis"

S. Jain, P. Gupta & R. Newton (eds.), Somatic Embryogenesis in Woody Plants, Vol. *1*, 23–48.

to both denote their origin from somatic cells and to differentiate them from zygotic embryos.

In conifers, somatic embryogenesis was first documented in diploid cultures of Norway spruce (*Picea abies* (L) Karst) (Hakman *et al.*, 1985) and haploid cultures of European larch (*Larix decidua* L.) (Nagmani and Bonga, 1985). Since then, numerous reports of somatic embryogenesis in tissue cultures of conifer species have appeared (Attree and Fowke, 1991; Tautorus *et al.*, 1991). However, most of these reports do not clearly identify the developmental stage of the explant in relation to its anatomy or morphology at the time of culture, partly due to the lack of a clear understanding of the events in conifer embryology. As a result, when initiation of somatic embryos does occur in cultures, it becomes difficult to trace the origin to any tissue type in the explant.

Also, confusion exists regarding the usage of the terminology for identification and classification of the various developmental stages of conifer somatic embryos. In many reports, developing conifer somatic embryos were described as "globular", "torpedo" or "heart-shaped" embryos, a terminology which is used to describe developing angiosperm embryos. It is inappropriate and confusing to use such terms to describe developing conifer embryos, because the cell division patterns in a conifer zygote during early and late embryogenesis are unique resulting in embryos that are different morphologically and anatomically from angiosperm embryos.

The purpose of this review is (a) to provide an outline of the basal plan of conifer embryo development in Pinaceae along with its accepted terminology, (b) to identify developmental stages of explants at the time of culture, and (c) to classify developing somatic embryos in relation to their zygotic embryo counterparts.

2. Zygotic Embryogeny and Terminology in Pinaceae

A system of classification of developing zygotic embryos in Pinaceae is proposed here, by merging the information from the review of conifer embryology by Dogra (1967) with the classification of embryos of *Pinus ponderosa* in relation to seed size by Buchholz and Steimert (1945). Embryo development in *Pinus* and *Picea* (Dogra, 1967) is described as free nuclear, because the zygote nucleus divides after fertilization, to form 2, 4 or 8 nuclei that migrate to the base of the fertilized archegonium (stages 1–3; Fig. 1A-C). Formation of a free nuclear proembryo is soon followed by wall formation and the proembryo becomes cellular consisting of 8 cells arranged in two tiers of 4 cells each. These tiers are designated as the primary upper tier (pU) and the primary embryonal tier (pE) (stage 4; Fig. 1D). By a process of internal division in both these tiers, a 4-tiered proembryo is formed, with the upper tier of 4 cells (U),the suspensor tier of 4 cells (S),the first embryonal segment (substitute suspensor segments) of 4 cells (E_1) and the embry-

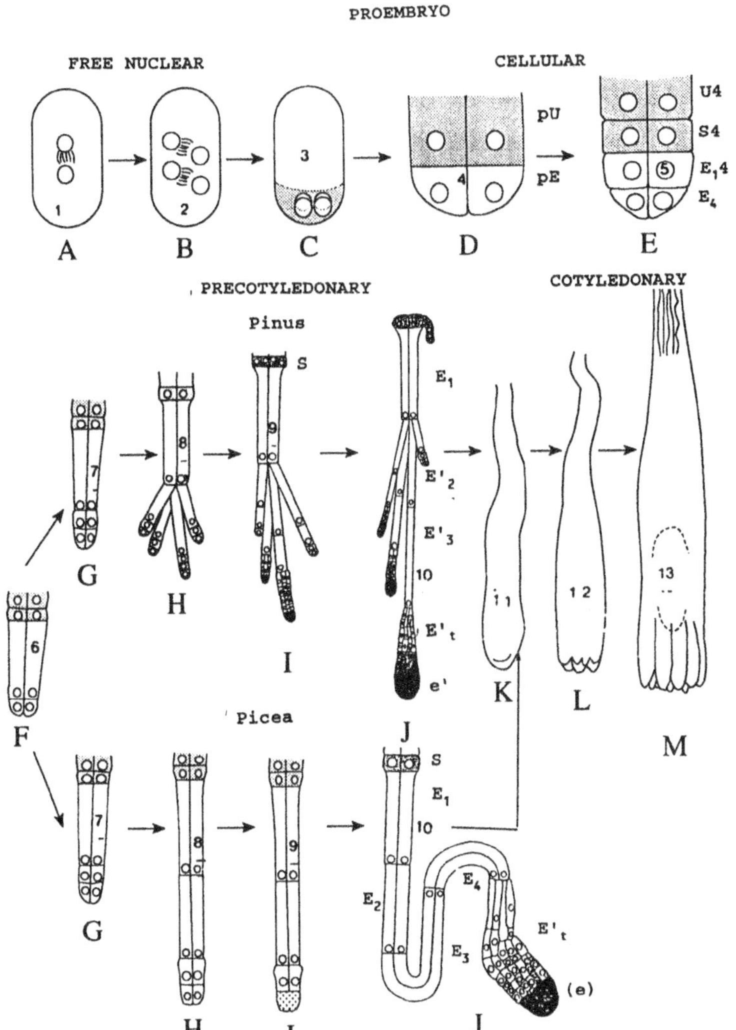

Figure 1. Zygotic embryogeny from free nuclear to mature seed embryo in Pinaceae (after Dogra, 1967; Buchholz and Steimert, 1945). (A-C) Free nuclear proembryos at stages 1–3. (D,E) Cellular proembryo at stage 4, with primary upper tier (pU) and primary embryonal tier (pE) and stage 5, with upper tier of 4 cells (U), suspensor tier (S) of 4 cells, substitue suspensor segements (E_1) of 4 cells, and embryonal cells (E). (F-J) Precotyledonary phase of embryo development at stages 6–10, in cleavage and non-cleavage types, represented by *Pinus* and *Picea*, respectively. Note the proximal cells (E_t), in between embryonal cells and suspensor. (For further explanation see text.) (K) Precotyledonary embryo with an apical dome, at stage 11. (L,M) Cotyledonary embryos at stages 12 and 13.

onal cells (E), (stage 5; Fig. 1E). All these divisions occur within the confines of the fertilized archegonium and proembryos at these early stages of development, cannot be dissected out.

Elongation of the E_1 cells and cell divisions within the "E" tier marks the beginning of yet another phase of zygotic embryogeny, the precotyledonary phase (stages 6,7; Fig. 1F,G).

Embryo development in Pinaceae, follows two different pathways after stage 6; the cleavage type represented by *Pinus*, and the non-cleavage type represented by *Picea*.

In *Pinus*, elongation of the segments of the E_1 tier is followed by divisions in the cells of the "E" tier resulting in the formation of the "E_2" tier (stage 7; Fig. 1G). This is followed by cleavage of the cells of the "E_2" tier into one or several units followed by elongation of these units to various degrees (stage 8; Figs. 1H and 2B,C). Repeated cell divisions followed by subsequent elongation results in the formation of an additional tier of "E_3" cells, both in *Pinus* and *Picea* (stage 9; Fig 1H,I).

In the non-cleavage type, *Picea*, cell divisions and elongation of the segments are similar to those described for *Pinus* (stages 1–7) except that the cleavage of the cells of E_2 tier does not occur.

In both cleavage and non-cleavage types, the segments of the tiers E_1, E_2 and E_3 elongate and constitute the suspensor system (stage 10; Fig. 1J). This is followed by the formation of the proximal cells at the base of the embryonal cells (e), which elongate irregularly and do not form tiers and are designated as "E_t" (stage 10; Fig. 1J).

Further development of cleaved embryos depend on their respective position in the ovule. It is usually the terminal dominant embryo (stage 10; Fig. 1J) that develops and becomes differentiated to the cotyledonary stage and the other accessory embryos are arrested at various stages of development (Figs. 1J and 2D). Development of a dominant embryo in *Pinus* (Fig. 1J; stage 10) and a single embryo in *Picea* (Fig. 1J ; stage 10) beyond stage 10, follows a similar pattern. The cylindrical mass of the embryo becomes dome shaped with the stem primordium bulging at the tip (Buchholz and Steimert, 1945). The stem primordium appears translucent under a dissecting microscope (stage 11; Fig. 1K).

Embryos enter the "cotyledonary" phase with the appearance of a ring of cotyledonary primordia that surrounds the stem tip of the shoulder of the embryo. Since the cotyledons are small, the stem tip at this stage protrudes beyond the length of cotyledons (stage 12; Fig. 1L). Further stages in embryo development are characterized by the increase in the length of hypocotyl and the cotyledons. In mature seeds the cotyledons are long and broad enough to completely obscure the stem tip from view (stage 13; Fig. 1M).

This system of classification coherently identifies and labels the events of zygotic embryogeny as the free nuclear proembryo (stages 1–3), cellular proembryo (stages 4 and 5), precotyledonary (stages 6–11) and cotyledonary embryo (stages 12 and 13) phases.

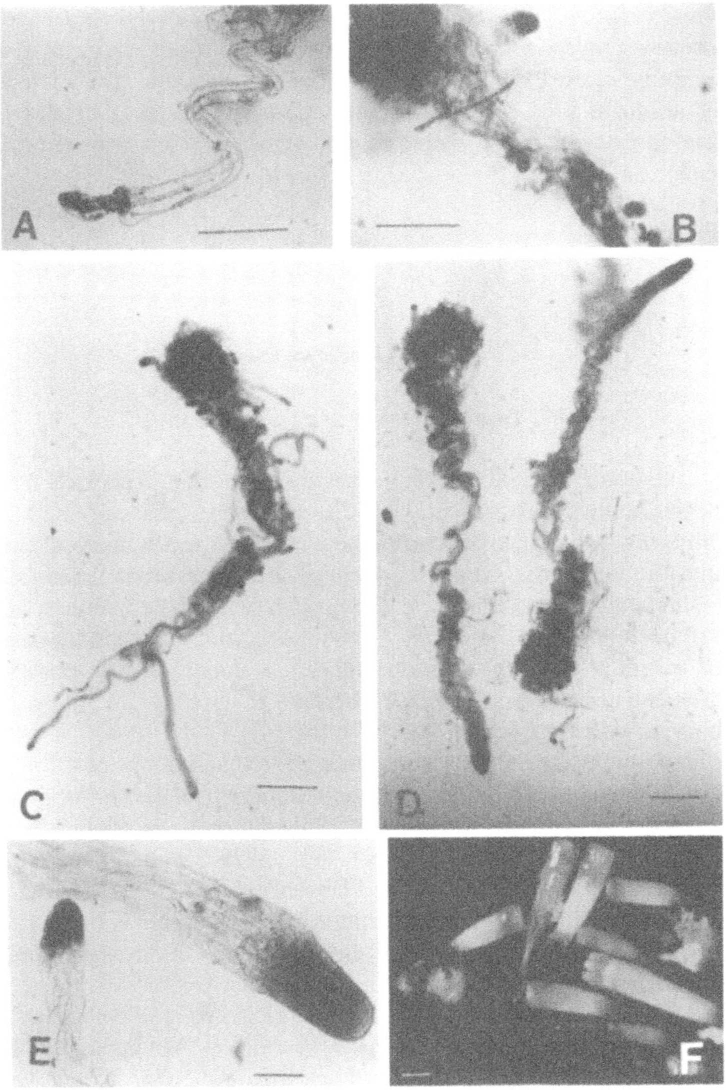

Figure 2. Zygotic embryo explants of *Pinus taeda* and *P. palustris*. (A,B) Precotyledonary embryos of *P. taeda* at stages 9 and 10; with embryonal cells and suspensor. Note cleavage polyembryony in (B). Scale bar = 0.25 mm. (C,D) Precotyledonary embryos of *P. palustris* at stages 9 and 10; showing cleavage polyembryony. Scale bar = 0.25 mm. (E) Embryos of *P. taeda* at stages 10 and 11. Note the cylindrical embryonal mass. Scale bar = 0.25 mm. (F) Cotyledonary embryos of *P. taeda* at stages 12 and 13; with cotyledonary primordia, and shoot apex. Scale bar = 0.25 mm.

The word "proembryo" suggests events, during early embryogenesis, which occur within the fertilized archegonium. As a result, embryos at stages 1–5, cannot be easily isolated or dissected out of the ovule. Precotyledonary embryos at stages 9–11 and cotyledonary embryos at stages 12 and 13 can be separated from the surrounding female gametophytic tissue. Embryos at stages 6–8, can only be separated with difficulty.

3. Somatic Embryogenesis

3.1. *Somatic Embryogenesis In Vivo*

The natural occurrence of asexual embryos has been described in many angiosperm families. Tisserat *et al.* (1979) reviewed reports of somatic embryogenesis *in vivo* in angiosperms. In most of these species, asexual embryos originated predominantly from the nucellus, although other tissues such as integuments, embryo-suspensors, and synergids were also capable of producing asexual embryos. As a result, seeds were found to contain multiple embryos in addition to the zygotic embryo, a condition first observed in orange seeds and was termed "polyembryony".

In conifers, polyembryony is due to either (1) fertilization of more than one archegonium or egg cell by sperm nuclei contained in one or more pollen tubes, or (2) the cleavage of the embryonal segments of one or more zygotic embryos as described above (Section 2). Polyembryony in conifers is of the "simple or archegonial" type (Buchholz, 1918, 1920) when multiple embryos are polyzygotic in origin and of the "cleavage type" when cleavage of a zygotic embryo occurs. It is not uncommon for simple polyembryony to also occur in ovules showing cleavage polyembryonic condition, as in *Pinus* and other genera showing cleavage polyembryony.

As the above definition suggests, it is "cleavage polyembryony" that exemplifies somatic embryogenesis *in vivo*, in conifers. Among the ten genera of the family Pinaceae, only *Cedrus*, *Keteleeria*, *Pinus* and *Tsuga* were reported to show cleavage polyembryony.

3.2. *Somatic Embryogenesis In Vitro*

Since there are several steps involved in the process of somatic embryogenesis in culture, the sequence of events is more easily described under (a) initiation or induction of somatic embryos from primary explants; (b) proliferation or multiplication of embryogenic cultures; and (c) development and maturation of somatic embryos.

3.2.1. *Primary Explant Types Used for Initiation of Somatic Embryos*

Explant types used to initiate embryogenic cultures in conifers vary from haploid tissue such as female gametophytes to diploid tissue such as zygotic embryos and seedling explants.

Primary explants cultured for the initiation of somatic embryos have been (a) female or megagametophytes with no embryos, (b) female gametophytes with zygotic embryos intact, (c) precotyledonary zygotic embryos, (d) cotyledonary embryos from fresh seed cones and stored dry seeds, and (e) seedling explants.

3.2.1.1. *Female Gametophyte as Haploid Explant.* In conifer seeds, the nutritive tissue surrounding the embryo is haploid, whereas in angiosperms the endosperm is triploid, a product of double fertilization. Since the female gametophyte can be easily processed as an explant compared to pollen, it is a preferred source of haploid explants for plant regeneration by tissue culture methods.

One of the first reports of somatic embryogenesis in conifers was from cultured female gametophytic tissue from *Larix decidua* L. (Nagmani and Bonga, 1985). More recently, haploid embryogenic callus was initiated from female gametophytes of *Picea sitchensis* (Bong.) Carr. (Baldursson *et al.*, 1993).

Female gametophytes before fertilization, can be classified as "free nuclear", "alveolar" and "cellular" (Konar and Nagmani, 1980). After fertilization, the anatomy of the female gametophyte changes to suit the requirements of the developing embryo. When gametophytes of *Larix decidua*, were cultured, they appeared opaque and white, and contained a moist cavity termed the "corrosion cavity" by Buchholz (1918).

Anatomically, the majority of the cells of the female gametophyte was larger in size, than the epidermal cells, with large central vacuoles and nuclei often appressed to the cell walls. As they matured, the gametophytes became white and turgid from the accumulation of lipid bodies and storage proteins. The outer layer of the gametophyte adjacent to the megaspore membrane developed as an epidermis (Schopf, 1943) and consisted of cells which were smaller, densely cytoplasmic and contained very little stored food (Hakman, 1993).

Subculturable embryogenic tissue was produced from some of the gametophytic explants in *Larix decidua* (Fig. 3A) (Nagmani and Bonga, 1985).

Von Aderkas and Bonga (1988), observed two pathways of somatic embryogenesis in gametophytic cultures, under the light microscope. In the first type, a mound of tissue developed on the cut surface along the edges of the corrosion cavity. Single, individual, long vacuolate cells often projected from this tissue above the surface of the megagametophyte. Unequal divisions in each of these long cells, eventually resulted in polarized structures which formed somatic embryos.

In the second type, single long cells projected directly from the cut surface

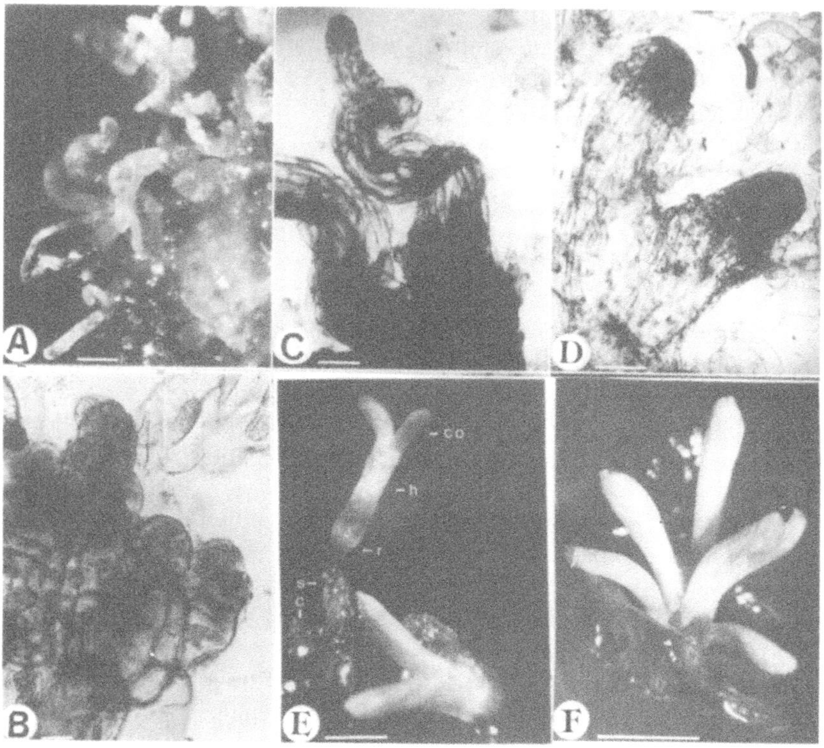

Figure 3. Somatic embryogenesis in subcultured callus of *Larix decidua* (after Nagmani and Bonga, 1985). (A) Embryogenic callus with somatic embryos (se). Scale bar = 1 mm. (B,C) Somatic embryos at stages 9 and 10. Note partial cleavage in (C). Scale bar = 0.1 mm. (D) Somatic embryo at stage 8, showing partial cleavage. Scale bar = 0.2 mm. (E,F) Cotyledonary somatic embryo with root (r), hypocotyl (h) and cotyledons (co); and plantlet. Scale bar = 1 mm.

of the gametophyte. Divisions in these long cells were similar to those described for the first type.

In *Picea sitchensis*, one megagametophytic culture showed the origin of an early stage somatic embryo directly from the surface of the female gametophyte. But this embryogenic cell line could not be maintained (Baldursson *et al.*, 1993).

3.2.1.2. *Female Gametophytes with Intact Zygotic Embryos as Explants.* Female gametophytes with intact zygotic embryos, were cultured in *Pinus radiata* (Smith, 1986, personal communication), *P. lambertiana* (Gupta and Durzan, 1986b), *P. taeda* (Gupta and Durzan, 1987; Becwar *et al.*, 1990), *P. ponderosa* (Becwar *et al.*, 1988) *P. strobus* (Finer *et al.*, 1989), *P. elliottii* (Jain *et al.*, 1989), *P. nigra* (Salajova and Salaj, 1992), *P. caribaea* (Laine and David, 1990), and *P. palustris* (Nagmani *et al.*, 1993). At the time of culture, the intact embryos contained within the female gametophyte, were predominantely precotyledonary, varying from stages 8–11. In all these species, extrusion of embryogenic tissue from the micropylar end of the female gametophyte was observed (Fig. 9A). This embryogenic tissue or callus originated from the suspensor region of the embryo in *P. lambertiana* (Gupta and Durzan, 1986b) and *P. taeda* (Becwar *et al.*, 1990). In *P. elliottii* (Jain *et al.*, 1989), embryogenic callus originated from suspensor cells at the base of the developing embryonal head that measured about one mm in length.

Microscopic examination of the embryogenic tissue revealed the presence of small dense clusters of cytoplasmic cells interspersed with unaggregated suspensor like cells, in *P. taeda* (Becwar *et al.*, 1990).

3.2.1.3. *Precotyledonary Zygotic Embryo as Explant.* Precotyledonary zygotic embryos without the surrounding gametophyte have been cultured at stages 9–11 (Fig. 2A-E).

In *P. taeda* (Becwar *et al.*, 1990), the origin of the embryogenic tissue was by the cell divisions at the "interface" of the suspensor and the embryonal mass. This observation probably refers to the zone of cells (E_t) formed between suspensor and embryonal mass (Fig. 1J).

3.2.1.4. *Cotyledonary Embryos as Explants.* Cotyledonary zygotic embryos at developmental stages 12 and 13 (Fig. 2F) were used as explants to initiate somatic embryos in most of the species. In both *Picea glauca* and *P. abies*, respectively, cotyledonary zygotic embryos at the time of culture consisted of radicle, long hypocotyl and a ring of cotyledons enclosing the shoot tip (Fig. 6A; stage 12) (Nagmani *et al.*, 1987). Sections of the responsive explants showed proliferation of the outer 2–4 layers of the hypocotyl region of the embryo followed by repeated divisions in the cells of the hypocotyl resulting in a mass of translucent mucilaginous embryogenic tissue (Fig. 6B,C). Quantal or unequal cell divison of some of the cells of the hypocotyl callus resulted in a distal, small, semicircular cytoplasmically dense, embryonal initial (ei) and proximal elongated vacuolate suspensor initial (si) (Fig. 6D). This 2-celled structure marked the beginning of early somatic embryogenesis.

In western larch (*Larix occidentalis*), white translucent embryogenic tissue originated from both the cotyledons and upper portion of the hypocotyl (Thompson and Von Aderkas, 1992). Hypocotyl and epicotyl regions of the embryos excised from 5–20 year-old stored seeds of red spruce (*Picea rubens*), produced embryogenic tissue when whole embryos were cultured (Harry and Thorpe, 1991). In *P. abies*, radicle portion of mature embryos excised from 2-year-old mature seeds produced embryogenic callus (Gupta and Durzan, 1986a). In sugar pine (*Pinus lambertiana*) (Gupta and Durzan, 1986b) embryogenic callus was produced from suspensor cells of cotyledonary embryos excised from 5-year-old seeds.

3.2.1.5. *Seedling Explants*. Cotyledons from 7–9 day old seedlings of *P. abies* were used as 12 explants (Krogstrup, 1986). After 3 weeks in culture, nodular, peripheral, meristematic cells were observed on the surface of the cotyledons in contact with the nutrient medium. This was followed by disruption of the epidermal and subepidermal layers "releasing" a mass of translucent cells identified as putative embryogenic callus (PEC). The PEC consisted of filamentous vacuolated cells together with isodiametric densely cytoplasmic cells. Isolation of PEC from the explant and subsequent culture resulted in the differentiation of precotyledonary somatic embryos. In *Picea mariana* and *P. glauca*, respectively, shoot apices with cotyledons or cotyledons alone excised from 12 to 30 day old seedlings produced embryogenic tissue (Attree *et al.*, 1990). The cotyledons first elongated and responsive explants changed color from dark green to light green and finally became white. The cells of the epidermis and subepidermis elongated and became convoluted. Disruption of the epidermal layers occurred within 20 days of culture and the cells produced mucilage and became translucent. Somatic embryos were observed after 3–4 weeks in culture. Separation of the embryogenic callus from the explant and its subculture was essential for its subsequent growth and proliferation.

In californian red wood (*Sequoia sempervirens*) of the family Taxodiaceae, the highest frequency of initiation of embryogenic tissue was observed from seedling cotyledons after 12 weeks in culture, while the frequency of embryogenic induction from hypocotyls was much reduced (Bourgkard and Favre, 1988).

3.2.2. *Multiplication or Proliferation of Somatic Embryos in Culture*

The origin of somatic embryogenic tissue from the primary explants may involve a single cell or group of cells from the primary explant. Once initiation of embryogenic tissue occurs, the callus or tissue has to be subcultured on to fresh media, to sustain proliferation or multiplication.

Observations of embryogenic cultures, maintained either on media gelled with agar or in liquid suspension cultures, indicate that somatic embryos can arise from single cells of the callus as described for *Picea abies* and *P. glauca* (Nagmani *et al.*, 1987; Hakman *et al.*, 1987). Single cell origin of somatic embryos from embryogenic suspension cultures can be demonstrated unequivocally by isolation and culture of protoplasts, which after division produced somatic embryos.

However, in *Pinus taeda*,protoplasts isolated from embryogenic suspension cultures, divided to form microcolonies which differentiated, forming suspensor and embryonal cells. Attree *et al.* (1989) cultured protoplasts isolated from embryogenic cultures of *Picea glauca* and observed cell divisions in the uninucleate protoplasts which produced cell clusters (Fig. 5A,B) Vacuolate suspensor cells projected from these clusters after 8 days in culture, and directly formed somatic embryos (stage 9 and 10; Fig. 5C,D). It is not clear, whether these cells in clusters acted as a mass of embryonal cells and divided

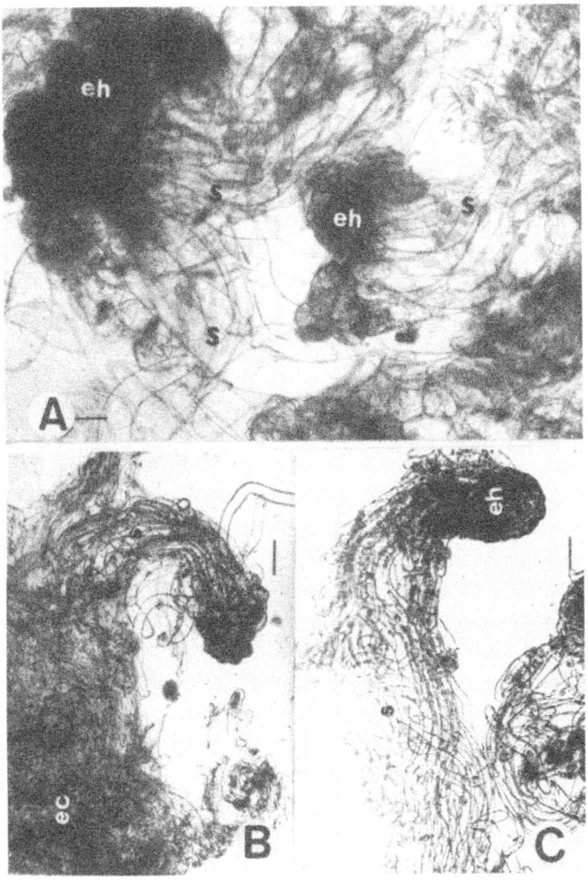

Figure 4. Somatic embryogenesis in subcultured callus of *Pseudotsuga menziessi* (after Nagmani *et al.*, 1991). (A) Somatic embryos at stage 9, showing cleavage during subculture. Note the fusion of embryonal masses or embryonal heads (eh) and suspensors (s). Scale bar = 100 μm. (B,C) Somatic embryos at stages 10 and 11, with bullet shaped embryonal head (eh) and suspensor (s). Scale bar = 100 μm.

to give rise to suspensor, or cells of the cluster already contained suspensor initials that elongated after 8 days in culture. Protoplast derived suspensions, consisting primarily of stage 10 somatic embryos, on transfer to maturation medium, developed into cotyledonary embryos (Fig. 5E-G).

During subculture of embryogenic tissue in *Larix decidua* (Fig. 3C,D); Nagmani and Bonga, 1985) and in *Pseudotsuga menziesii* (Fig. 4A; Nagmani *et al.*, 1991), fusion of several somatic embryos was observed. Durzan and Gupta (1987) also observed a similar phenomenon in embryogenic cultures of *Pseudotsuga menziesii* where several somatic embryos were fused together.

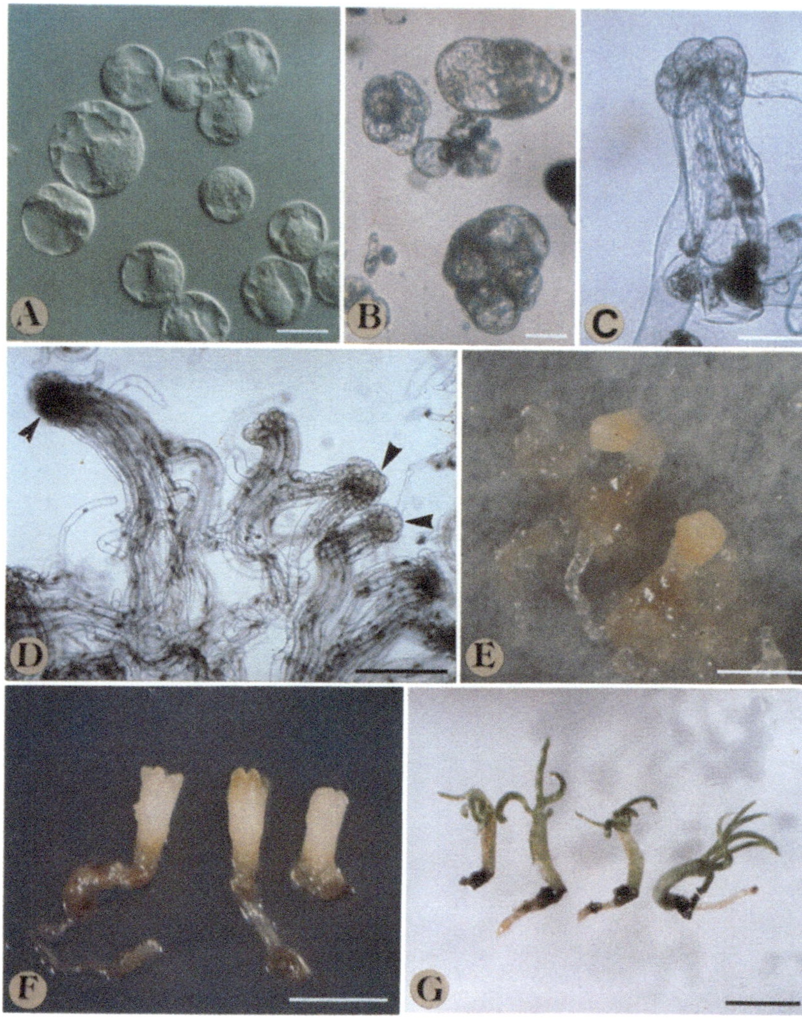

Figure 5. Origin of somatic embryos from protoplasts of *Picea glauca* (after Attree *et al.*, 1989). (A-C) Somatic embryo regeneration from cultured protoplasts. Scale bar = 25 μm, 50 μm, and 100 μm respectively. (D-G) Plantlet development from a protoplast derived embryogenic suspension culture. Scale bar = 300 μm, 1, 2, and 5 mm respectively.

However, these embryos separated from one another on transfer to abscisic acid (ABA) containing medium. These observations suggest that "cleavage polyembryony" occurred in embryogenic callus or tissue cultures, and constituted a method of multiplication of somatic embryos. But unlike the zygotic cleavage polyembryonic condition, cleavage of somatic embryos was not complete.

3.2.3. *Somatic Embryo Development and Maturation*

3.2.3.1. *Terminology.* In embryogenic callus, suspension and protoplast cultures, it has been documented that somatic embryo can be traced to single cells (Nagmani *et al.*, 1987; Hakman *et al.*, 1987; Attree *et al.*, 1989). If these single cells in callus are considered as equivalent or similar to zygotes, then the early stages of somatic embryo development can be described as similar to the developmental stages of zygotic proembryos. Accordingly, the classification and the terminology used to describe zygotic embryo development, will be used here to describe somatic embryo development.

3.2.3.2. *Free Nuclear Somatic Proembryo Phase.* In haploid gametophytic cultures of *Larix decidua* L. Von Aderkas and Bonga (1988) observed that nuclear divisions occurred in some of the long cells which projected above the surface of the gametophytic explant, resulting in four free nuclei. These nuclei migrated towards the base of the cell and wall formation resulted in a 4-celled proembryo (stages 1–4).

Similar observations were made by Gupta & Durzan (1987) in embryogenic suspension cultures of *Pinus taeda* L. where four free nuclei were observed in some cells. The authors did not observe the actual migration of the free nuclei to the base of the cell. Wall formation resulted in cellular proembryos.

Fowke *et al.* (1990) observed that some of the protoplasts which were isolated from embryogenic suspension cultures of *Picea glauca* were multinucleate with 2–10 nuclei. These protoplasts divided synchronously to produce cell clusters which eventually produced polarized embryo-like structures consisting of small cytoplasmic cells and a few elongate highly vacuolated cells. The authors suggested that the formation of somatic embryos from multinucleate protoplasts was similar to the free nuclear phase of the zygotic proembryo.

3.2.3.3. *Cellular Somatic Proembryo Phase.* In most of the embryogenic callus cultures of conifer species, first divisions are quantal (unequal or asymmetrical), resulting in small, densely cytoplasmic cells at one end and large vacuolated cells at the other (Fig. 6D) as reported in *Picea abies* and *P. glauca* (Nagmani *et al.*, 1987); (Hakman and Fowke, 1987); *Pinus elliottii* (Jain *et al.*, 1989),and *P. taeda* (Becwar *et al.*, 1988). Due to the early asymmetric division during somatic embryo development, a demarcation between suspensor and embryonal mass is laid out early on, as opposed to the zygotic embryo pattern, where 4 tiers of 4 cells each are formed first, followed by elongation of the suspensor segments. Further cell divisions in these 2 cells resulted in a 4-celled elongated proembryo as observed in *Picea abies* and *P. glauca* (Fig. 4E ;Nagmani *et al.*, 1987), *Pinus elliottii* (Jain *et al.*, 1989); *P. caribaea* (Laine and David, 1990), and *P. palustris* (Nagmani *et al.*, 1993). An 8-celled somatic embryo was observed in some callus cultures of *P. taeda* (Becwar *et al.*, 1988), and suspension cultures of *P. glauca*

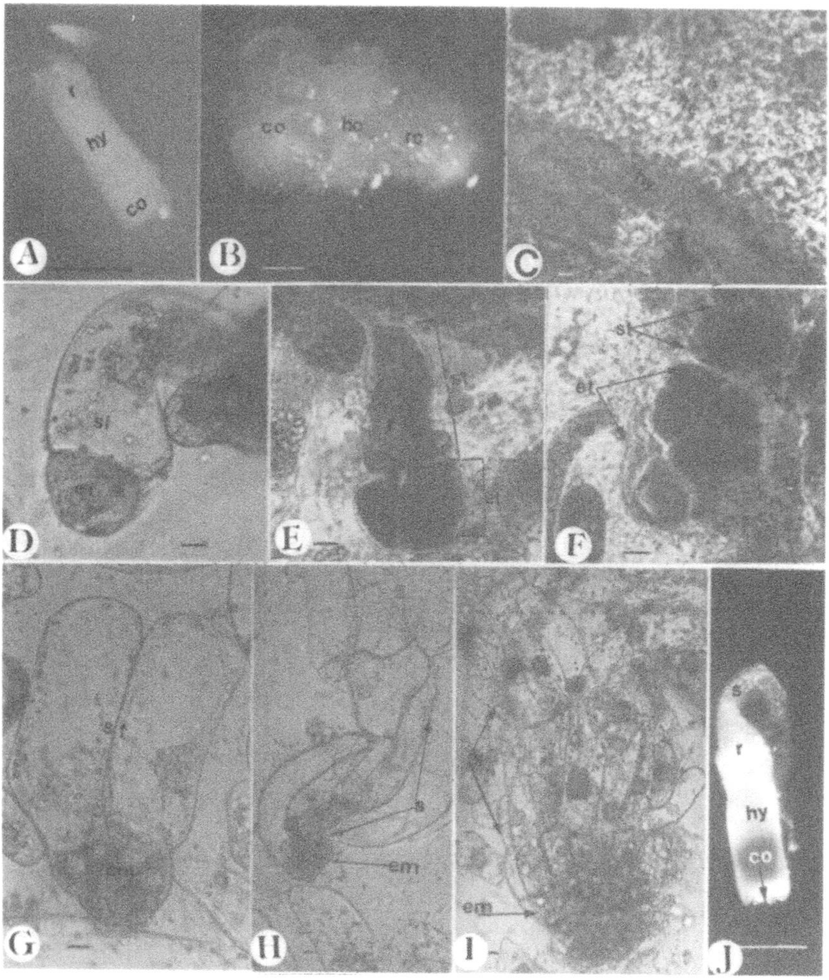

Figure 6. Origin and development of somatic embryos in callus cultures of *Picea glauca* and *P. abies* (after Nagmani *et al.*, 1987). (A) Cotledonary zygotic embryo explant with radicle (r), hypocotyl (hy), and cotyledons (co, at stage 12. scale bar = 1 mm. (B,C) Initiation of embryogenic callus (hc) from hypocotyl. Scale bar = 1 mm and 10 um respectively. (D-F) 2–16 celled cellular proembryos with embryonal initial (ei), suspensor initial, embryonal tier (et), and suspensor tier (st). Scale bar = 10 μm. (G-I) Precotyledonary somatic embryos at stages 9 and 10. Scale bar = 10 μm. (J) Cotyledonary somatic embryo at stage 12. Scale bar = 1 mm.

(Hakman and Fowke, 1987; Fig. 7A), which resulted in the formation of a 16-celled somatic embryo (Fig. 6F). The embryonal mass of some somatic embryos is wedge-shaped with 12–16 cells that are isodiametric with dense cytoplasm and prominent nuclei, subtended by long vacuolate suspensor of 4–8 cells; as observed in *P. taeda* (Fig. 9B; Becwar *et al.*, 1990), *P. palustris*

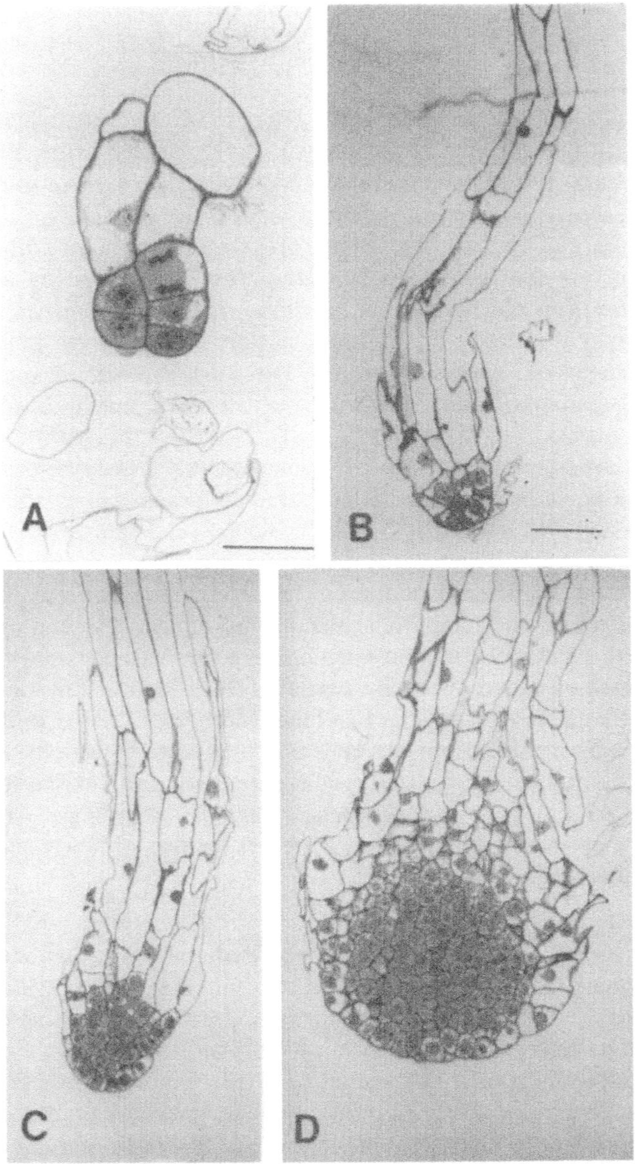

Figure 7. Anatomy of precotyledonary somatic embryos of *Picea glauca* (after Hakman *et al.*, 1987). (A) Somatic embryo at stage 7. Note the mitotic figure in one of the cells. Scale bar = 40 μm. (B-D) Somatic embryos at stages 9–11. Note the densely cytoplasmic, isodiametric cells in the embryonic region and long, vacuolate suspensor. Scale bar = 100 μm.

(Nagmani *et al.*, 1993) and *P. glauca* (Fig. 6G) (Nagmani *et al.*, 1987). These are comparable to stages 5 and 6 of zygotic embryos.

3.2.3.4. *Precotyledonary Somatic Embryo Phase.* Continued transverse divsions in the suspensor cells followed by elongation of the suspensor segments, marks the beginning of the precotyledonary phase of somatic embryo development. As embryos increased in size, the diameter of the suspensor also increased concommitantly (Fig. 6H; stage 9). Somatic embryos at stage 10, observed in *Picea abies*, *P. glauca* (Fig. 6I) (Nagmani *et al.*, 1987), *Pinus taeda* (Fig. 9C; Becwar *et al.*, 1990), *Pseudotsuga menziessii* (Fig. 4C; Nagmani *et al.*, 1991) are morphologically equivalent to precotyledonary zygotic embryos at stages 9 and 10 (Fig. 2B-D). A somatic embryo with a bullet-shaped embryonal head with a broad and elongated suspensor (stage 12) was observed in *P. elliottii* (Fig. 9D; Newton *et al.*, 1993).

Anatomy of developing precotyledonary somatic embryos of *Picea glauca*, at developmental stages 7–11, was studied by examining thin sections of somatic embryos under the light microscope (Hakman *et al.*, 1987). In somatic embryos shown in Fig. 7A, mitotic figure was observed in one of the embryonal cells. Precotyledonary somatic embryos depicted in Fig. 7B-D; consisted of distinct embryonic region and elongate suspensor. The small cells of the embryonic region were densely cytoplasmic, with thin cell walls and contained numerous small vacuoles (Fig. 7A-D). Mitotic activity was observed in embryonic regions of all these embryos. The suspensors consisted of very long highly vacuolated cells and thin layer of peripheral cytoplasm. Mitotic figures were never observed in suspensor cells, except for occasional presence of tiny cells in the suspensor region. According to Hakman *et al.* (1987), these small cells were suggested as sources of new somatic embryos by continued cell division.

3.2.3.5. *Cotyledonary Somatic Embryo Phase.* Differentiation of a ring of cotyledonary primordia, also marks the beginning of the cotyledonary phase of somatic embryo development (Fig. 8A,B; Nagmani *et al.*, 1993).

Cotyledonary somatic embryos of *Picea abies* shown in Fig. 8A,B are morphologically similar to stage 13 of zygotic embryos present in fully ripe seeds as shown in Fig. 8C. At this stage of embryo development, the hypocotyl and cotyledons elongate, and appear opaque and white.

Median longitudinal sections of cotyledonary somatic embryos show well defined shoot apical meristems with central vascular connections (Fig. 8D; Nagmani *et al.*, 1993). The hypocotyl of zygotic embryos are more slender and cotyledons are more linear as compared to somatic embryo counterparts (Fig. 8E).

Upon germination, the somatic plantlets of *Picea abies* (Fig. 8; Nagmani *et al.*, 1993) and *Pinus elliottii* (Newton *et al.*, 1993) consisted of a well formed root, an elongated hypocotyl and cotyledons.

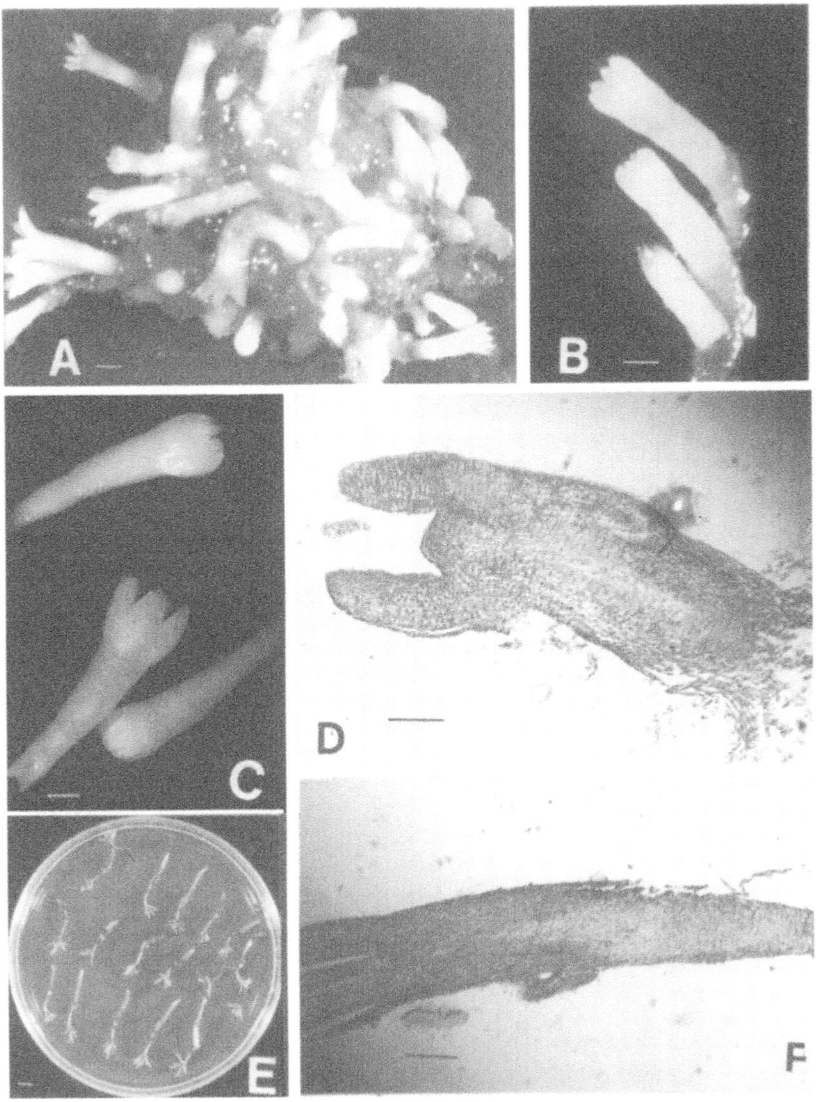

Figure 8. Comparison of cotyledonary somatic and zygotic embryos of *Picea abies* (Nagmani *et al.*, 1993). (A,B) Cotyledonary somatic embryos at stage 13. Scale bar = 1 mm. (C) Mature zygotic embryos at stage 13, from dry seeds. Scale bar = 1 mm. (D) Median longitudinal section of cotyledonary somatic embryo showing shoot apical meristem. Scale bar = 0.5 mm. (E) Germinated somatic plantlets. (F) Peripharal longitudinal section of zygotic embryo. Scale bar = 0.5 mm.

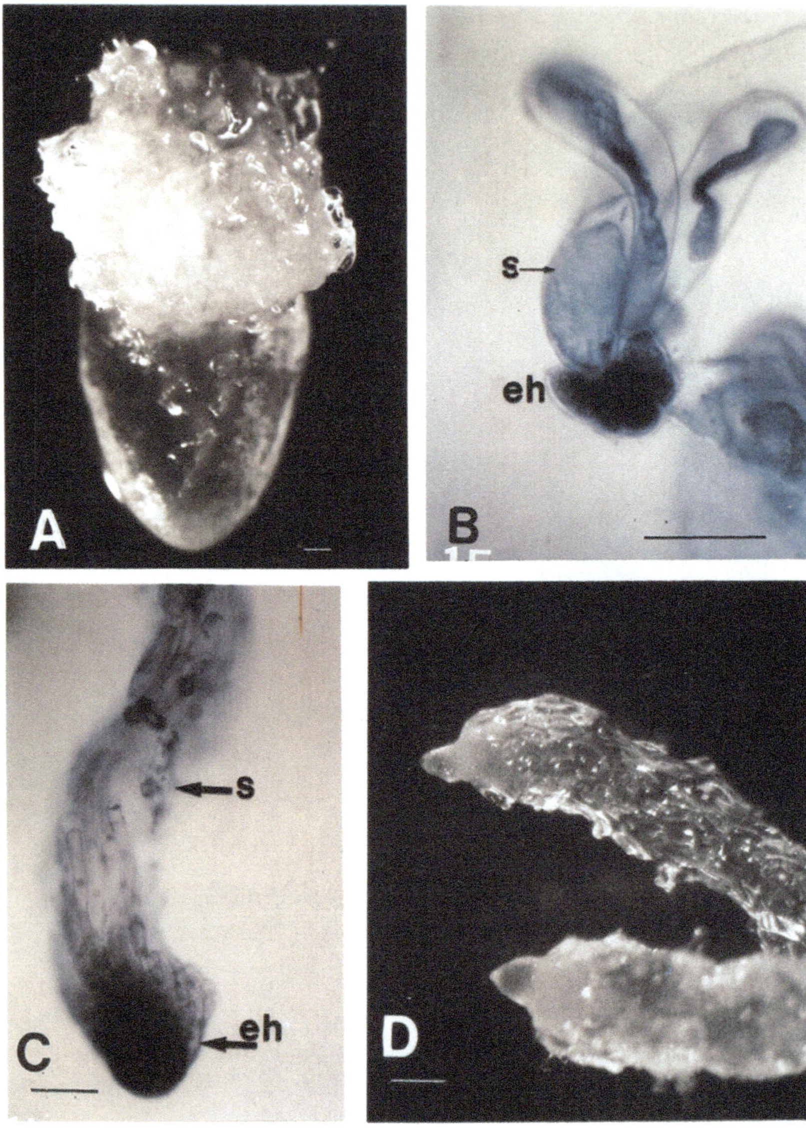

Figure 9. Somatic embryogenesis in *Pinus taeda* and *P. elliottii* (after Becwar *et al.*, 1990; Newton *et al.*, 1995). (A) Extrusion of embryogenic tissue from the micropylar end of the female gametophyte. Scale bar = 0.5 mm. (B) Somatic embryo with densely stained embryonal cells of embryonal head (eh) and elongate suspensor (s). Scale bar = 0.1 mm. (C,D) Precotyledonary somatic embryos at stage 10 (loblollyine) and at stage 11 (slash pine). Scale bar = 0.1 mm.

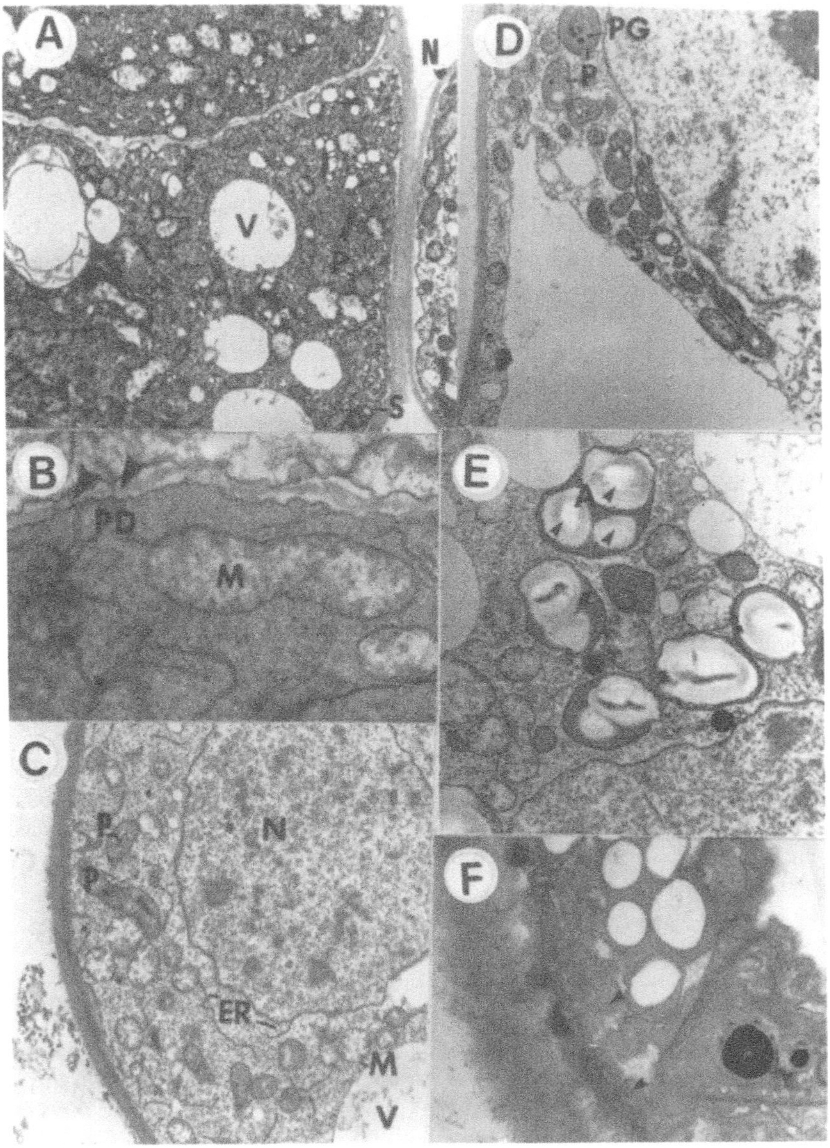

Figure 10. Ultrastructure of haploid somatic embryos of *Larix decidua* (after Rohr *et al.*, 1989). (A-C) General view of short or small cells (of 2 or 4 celled haploid somatic embryo) with dense cytoplasm, plastids (P), starch (S) and vacuoles (V). Note the plasmodesmata on the new cell wall(PD). The mitochondrion (M) has few cristae. A × 4,600, B × 22,000, and C × 6000. (D-F) General view of long cells (of 2 or 4 celled haploid somatic embryo). The amyloplasts (A) show a few dilated membranes. D × 7,260, E × 28,500, and F × 22,000.

4. Ultrastructural Changes in Developing Somatic and Zygotic Embryos

Haploid somatic embryos (*Larix decidua*) consisting of a mound of loose aggregate of spherical short cells, alternating with randomly oriented long cells, were observed under the electron microscope (Rohr *et al.*, 1989). During the early stages of haploid somatic embryo development in *L. decidua* the mitochondria and plastids of both short and long cells showed relatively little differentiation (Fig. 10A-F). In the short cells, plastids were of the leucoplast type and in long cells, that form suspensor tissue, plastids were of the amyloplast type. Dictyosomes and associated secretory activity were observed in all cell types and at all stages of embryo development.

In *Picea glauca*, the cell walls of the embryonal cells contained uniformly distributed plasmodesmata, also observed in *Larix decidua* (Fig. 10B); whereas, in the suspensor cells, plasmodesmata were generally restricted to the transverse cell walls. Numerous mitochondria and plastids were observed in the embryonal mass and suspensor parts of somatic embryos (Hakman *et al.*, 1987).

In *Picea abies* the ultrastructure of the cells in the embryonal mass and the upper suspensor cells of young zygotic and somatic embryos was similar (Hakman, 1993). In both zygotic and somatic embryos, the cells of the embryonal mass were densely cytoplasmic, with a great number of ribosomes, mitochondria, and poorly differentiated plastids. The densely cytoplasmic structure of the cells suggested a high metabolic rate.

As embryos entered the cotyledonary phase, the cells of both somatic and zygotic embryos accumulated lipid bodies and storage proteins. Plastids with starch grains became abundant both in zygotic and somatic embryos, but were reduced in number in the fully ripe embryos of dry seeds as well as the most advanced stages of the somatic embryos (corresponding to stage 13 of the zygotic embryo).

5. Discussion

Primary explant types, which produce somatic embryos in conifers, are restricted to juvenile tissues such as precotyledonary and cotyledonary zygotic embryos, and rarely, to seedling explants (from 7–30 day old seedlings). Explants from mature trees, when cultured, have not demonstrated embryogenic potential. Whether embryogenic potential exists in meristematic tissue types in the mature tree, such as cambium, vegetative and reproductive buds, young female cones (at the time of meiosis of the megaspore mother cell), and prophylls besides apical meristem, remains to be determined.

The zygotic embryos of the family Pinaceae, have been classified into 13 major developmental stages starting from the first division of the zygote nucleus to the cotyledonary embryo in the fully ripe seed (Fig. 1). This classification is a synthesis of information from the review on conifer embryology by Dogra (1967) with the embryo classification system in *Pinus ponderosa* proposed by Buchholz and Steimert (1945). Both *Pinus* and *Picea* are discusssed to illustrate cleavage and non-cleavage types of embryogeny, respectively. By application of this system explants can now be readily assigned to a particular developmental stage based on morphology and anatomy.

This system also replaces the broad and nonspecific terminology that is currently in use, such as, "immature" and "mature" embryos. Earlier, Becwar *et al.* (1988) and Finer *et al.* (1989), characterized zygotic embryo explants based on size, specifically length and the time when the seed cones were removed from the tree including the month and the date. Explant identification using the time of collection,is useful,but arbitrary, because the developmental stage of the explant varies depending on the climate and location of donor tree and position of the seed cone. Embryo development is not always synchronous in a sample of cones collected at a particular time and the length of the explant is not always indicative of the morphological stage of development. A classification based on the morphology and anatomy of an explant at the time of culture, is suggested as it is a more reliable and uniform method of explant identification.

The new scheme of embryo classification also offers a standard by which somatic embryos at various stages of development can be identified and compared to zygotic embryos. Von Arnold and Hakman (1988) classified somatic embryos into four developmental stages: (a) stage 1 embryos consisted of small,densely cytoplasmic cells subtended by long coiled suspensors; (b) stage 2 somatic embryos were still attached to the callus by long suspensor cells and meristematic region was more dense; (c) cotyledonary somatic embryos were assigned to stage 3; and (d) during stage 4, plantlets were formed. Obviously, this classification does not recognize somatic embryos at early stages, and does not account for somatic embryos in suspenson cultures, where the need for attachment of the suspensor of the somatic embryo to the callus is removed. Further, this classification scheme does not enable comparison to zygotic embryo counterparts.

For the first time, a classification of somatic embryos at various stages of development, based on the proposed classification of zygotic embryos along with its terminology, is suggested. However, it is important to note that somatic embryogenic patterns do not follow the exact same developmental patterns as zygotic embryogeny, yet the resulting structures are morphologically similar. For example, the free nuclear phase observed during zygotic embryogenesis is not a regular feauture during early somatic embryogenesis. Only a few authors report a free nuclear phase, as in the haploid and diploid cultures of *Larix decidua* (Von Aderkas and Bonga, 1988), in the embryogenic suspension cultures of *Pinus taeda* (Gupta and Durzan, 1987) and multinucleate protoplasts of *Picea glauca* (Fowke *et al.*, 1990).

Free nuclear and cellular proembryo phases were often found at the same time in somatic cell populations, suggesting that not all embryos go through a free nuclear phase before they become cellular. As a result, the first nuclear division in most callus and suspension cultures was followed by wall formation resulting in a 2-celled proembryo. Even at this early stage of somatic embryo development, polarity was established to demarcate a long vacuolate suspensor from a small, densely cytoplasmic embryonal cell. Further divisions in the suspensor cell resulted in a suspensor that was broader in diameter, in comparision to the suspensor of the zygotic embryo. A similar difference in suspensor morphology was observed in *Picea glauca* between zygotic and somatic embryos (Hakman *et al.*, (1987).

Early development of somatic embryos followed a slightly different route when either uni or multinucleate protoplasts were cultured. Microcolonies or cell clusters were formed from individual protoplasts. Vacuolate suspensor cells emerged from these cell clusters resulting in polarized somatic embryo (Attree *et al.*, 1989; Fowke *et al.*, 1990). However, it was not clear from these reports, whether cells in the microcolonies or cell clusters (1) divided to form embryonic and suspensor cells or (2) divided to form vacuolate suspensor cells or (3) some of the cells in the clusters differentiated into suspensor cells.

In both embryogenic callus and suspension cultures of *Larix decidua* and *Pseudotsuga menziessii* (Nagmani and Bonga, 1985; Nagmani *et al.*, 1991), cleavage of preexisting somatic embryos was observed during subcultures. These embryos could only be separated by transfer to maturation medium. Unlike, the total cleavage of zygotic embryos as observed in *Pinus*, the cleavage of these somatic embryos was not complete.

Regardless of the origin and early development of somatic embryo, at stage 10 of precotyledonary phase, somatic embryos were morphologically similar to stage 10 zygotic embryos (Fig. 9C). Also, cotyledonary somatic embryos were very similar to their zygotic embryo counter parts at stage 13 (Fig. 8A-C).

Anatomical similarities between zygotic and somatic embryos extend to ultrastructural level as well. Electron microscopic observations of somatic and zygotic embryos (Hakman, 1993; Rohr *et al.*, 1989) of spruce and larch have indicated that both have lipid bodies and storage proteins as well as similarities in mitochondrial and plastid structure.

Fig. 11, summarizes the possible routes of in vitro origins of somatic embryos in Pinaceae, both from the primary explants and from embryogenic callus or protoplast cultures. Once the embryogenic cultures are established from the primary explants, the multiplication of somatic embryos might follow any one of the pathways described (Fig. 11).

The embryogenic potential of an explant was observed to be genus specific. For example, in *Picea* and *Larix*, all types of primary explants produced embryogenic tissue, whereas, in *Pinus*, the embryogenic tissue or somatic embryos originated mostly from precotyledonary zygotic embryos between stages 9–11.

At present, expression of the embryogenic potential is limited to juvenile tissues. We have documented the morphological and anatomical similarity

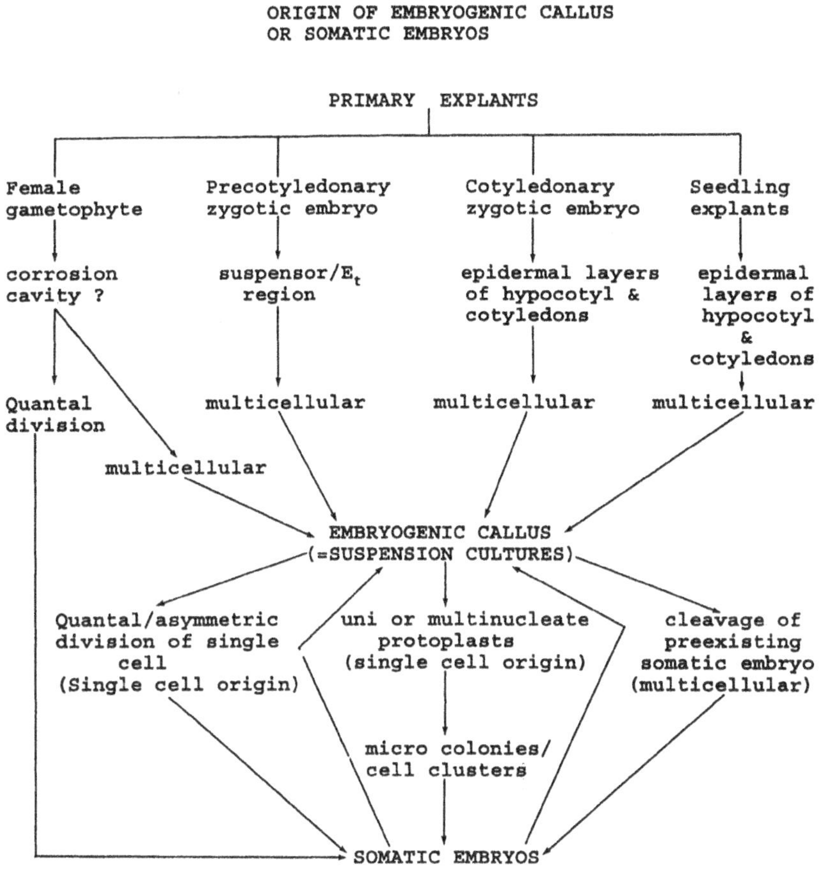

Figure 11. Possible routes of somatic embryogeny on initiation and maintenance media in Pinaceae.

between zygotic and somatic embryos. To extend the expression of embryogenic potential to the explants from mature trees, it is necessary to identify cell or tissue types which are physiologically identical to these juvenile tissues. Use of such mature explants for micropropagation of genetically elite conifers by somatic embryogenesis will be an important addition to clonal forestry programs.

6. Conclusions

The proposed system of zygotic embryo classification in Pinaceae identifies 13 major developmental stages, and zygotic embryo explants can now be readily identified to its developmental stage at the time of its inoculation to the culture medium.

The embryogenic potential of the explant in the production of embryogenic callus in *Pinus* and *Pseudotsuga* is limited mostly to precotyledonary zygotic embryos from stage 7 or 8 to 11, whereas in *Picea*, this potential can be extended to 7–30 day old seedlings as well. *Abies* and *Larix* form an intermediate group between these two extremities, where embryogenic callus can be initiated from cotyledonary zygotic embryos from mature seeds.

A comparision of zygotic and somatic embryos in the family Pinaceae shows that greater similarity is observed among embryos at stages 10 and 11 of precotyledonary phase, and at 12 and 13 of cotyledonary phase. Anatomical similarity is also observed between cotyledonary zygotic and somatic embryos, except that the hypocotyl of zygotic embryo is longer and slender than the cotyledonary somatic embryo. Accumulation of lipids and storage proteins in zygotic and somatic embryos is also similar.

Cleavage polyembryony is observed during multiplication of somatic embryos during subculture in all the genera, and cleavage is not total or complete as observed during zygotic cleavage polyembryony in *Pinus*. Multiplication of somatic embryos by asymmetric divisions in single cells, is also common, in embryogenic cultures.

References

Attree, S.M., D.I. Dunstan and L.C. Fowke, 1989. Plantlet regeneration from embryogenic proptoplasts of white spruce *Picea glauca*. Bio/Technology 7: 1060–1062.

Attree, S.M., S. Budmir and L.C. Fowke, 1990. Somatic embryogenesis and plantlet regeneration from cultured shoots and cotyledons of seedlings from stored seeds of black and white spruce *Picea maraiana* and *Picea glauca*. Can. J. Bot. 68: 30–34.

Attree, S.M. and L.C. Fowke, 1991. Micropropagation through somatic embryogenesis in conifers. In: Y.P.S. Bajaj (Ed.), Biotechnology in Agriculture and Forestry, Vol. 17, pp. 53–70. Springer-Verlag, Berlin.

Baldursson, S., J.V. Norgaard and P. Krogstrup, 1993. Factors influencing haploid callus initiation and proliferation in megagametophyte cultures of Sitka spruce *Picea sitchensis*. Silvae Genetica 42: 79–86.

Becwar, M.R., S.R. Wann, M.A. Johnson, S.A. Verhagen, R.P. Feirer and R. Nagmani, 1988. Development and characterization of in vitro embryogenic systems in conifers. In: M.R. Ahuja (Ed.), Somatic Cell Genetics of Woody Plants, pp. 1–18. Kluwer Academic Publishers, Dordrecht.

Becwar, M.R., R. Nagmani and S.R. Wann, 1990. Initiation of embryogenic cultures and somatic embryo development in loblolly pine *Pinus taeda*. Can. J. For. Res. 20: 810–817.

Bourgkard,F. and J.M. Favre, 1988. Somatic embryos from callus of *Sequoia sempervirens*. Plant Cell Rep. 7: 445–448.

Buchholz, J.T., 1918. Suspensor and early embryo of *Pinus*. Bot. Gaz. 66: 185–228.

Buchholz, J.T., 1920. Polyembryony among Abietineae. Bot. Gaz. 69: 153–167.

Buchholz, J.T. and M.L. Steimert, 1945. Development of seeds and embryos in *Pinus ponderosa* with special reference to seed size. Trans. Illinois Acad Sci. 38: 27–50.

Dogra, P.D., 1967. Seed sterility and disturbances in embryogeny in conifers with particular reference to seed testing and tree breeding in Pinaceae. Studia Forestalia Suecica. 45: 5–97.

Durzan, D.J. and P.K. Gupta, 1987. Somatic embryogenesis and polyembryogenesis in Douglas-fir cell suspension cultures. Plant Sci. 52: 229–235.

Finer, J.J., H.B. Kriebel and M.R. Becwar, 1989. Initiation of embryogenic callus and suspension cultures of eastern white pine (*Pinus strobus* L.). Plant Cell Rep. 8: 203–206.

Fowke, L.C., S.M. Attree, H. Wang and D.I. Dunstan, 1990. Microtubule organization and cell division in embryogenic protoplast cultures of white spruce (*Picea glauca*). Protoplasma 158: 86–94.

Gupta, P.K. and D.J. Durzan, 1986a. Plantlet regeneration via somatic embryogenesis from subcultured callus of mature embryos of *Picea abies* (Norway spruce). In Vitro Cell and Dev. Biol. 22: 685–688.

Gupta, P.K. and D.J. Durzan, 1986b. Somatic polyembryogenesis from callus of mature sugar pine embryos. Bio/technology 4: 643–645.

Gupta, P.K. and D.J. Durzan, 1987. Biotechnology of somatic polyembryogenesis and plantlet regeneration in loblolly pine. Bio/Technology 5: 147–151.

Haccius, B. and K.K. Lakshmanan, 1969. Adventive-embryonen-embryoide-adventive-knospen. Ein beitrag zur klarung der begriffe. Oesterr. Bot. 116: 145–158.

Hakman, I., L.C. Fowke, S. von Arnold and T. Eriksson, 1985. The development of somatic embryos in tissue cultures initiated from immature embryos of *Picea abies* (Norway spruce). Plant Sci. 38: 53–59.

Hakman, I. and L.C. Fowke, 1987. An embryogenic cell suspension culture of *Picea glauca* (white spruce). Plant Cell Rep. 6: 20–22.

Hakman, I., P. Rennie and L.C. Fowke, 1987. A light and electron microscopic study of *Picea glauca* (white spruce) somatic embryos. Protoplasma 140: 100–109.

Hakman, I., 1993. Embryology in Norway spruce (*Picea abies*). An analysis of the composition of seed storage proteins and deposition of storage reserves during seed development and somatic embryogenesis. Physiol. Plant 87: 148–159.

Halperin, W. and D.F. Wetherell, 1964. Adventive embryony in tissue cultures of the wild carrot, *Daucus carota*. Am. J. Bot. 51: 274–283.

Harry, I.S. and T.A. Thorpe, 1991. Somatic embryogenesis and plant regeneration from mature zygotic embryos of red spruce. Bot. Gaz. 152: 446–452.

Jain, S.M., N. Dong and R.J. Newton, 1989. Somatic embryogenesis in slash pine (*Pinus elliottii*) from immature embryos cultured *in vitro*. Plant Sci. 65: 233–241.

Konar, R.N. and R. Nagmani, 1980. Female gametophyte and embryogeny in *Picea smithiana* and *Abies pindrow* (Pinaceae). Bot. Jahrb. Syst. 101: 267–297.

Krogstrup, P., 1986. Embryolike structures from cotyledons and ripe embryos of Norway spruce (*Picea abies*). Can. J. For. Res. 16: 664–668.

Laine, E. and A. David, 1990. Somatic embryogenesis in immature embryos and protoplasts of *Pinus caribaea*. Plant Sci. 69: 215–224.

Nagmani, R. and J.M. Bonga, 1985. Embryogenesis in subcultured callus of *Larix decidua*. Can. J. For. Res. 15: 1088–1091.

Nagmani, R., M.R. Becwar and S.R. Wann, 1987. Single-cell origin and development of somatic embryos in *Picea abies* (L.) Karst. (Norway spruce) and *P. glauca* (Moench) Voss (white spruce). Plant Cell Rep. 6: 157–159.

Nagmani, R., M.A. Johnson and R.J. Dinus, 1991. Effect of explant and media on initiation, maintenance, and maturation of somatic embryos in *Pseudotsuga menziessi* (Mirb.) Franco (Douglas-fir). In: M.R. Ahuja (Ed.), Woody Plant Biotechnology, pp 171–178. Plenum Press, New York.

Nagmani, R., A.M. Diner and G.C. Sharma, 1993. Somatic embryogenesis in longleaf pine (*Pinus palustris*). Can. J. For. Res. 23: 873–876.

Newton, R.J., K.A. Marek-Swize., M.E. Magallanes-Cedeno., N. Dong., S. Sen and S.M. Jain, 1994. Somatic embryogenesis in Slash pine. TAES Tech. Pub. No.

Rohr, R., P. von Aderkas and J.M. Bonga, 1989. Ultrastructural changes in haploid embryoids of *Larix decidua* during early embryogenesis. Am. J. Bot. 76: 1460–1467.

Salajova, T. and J. Salaj, 1992. Somatic embryogenesis in European black pine (*Pinus nigra* Arn.). Biol. Plant. 34: 213–218.

Schopf, J.M., 1943. The embryology of *Larix*. Illinois Biol. Monogr. 19: 1–97.

Tautorus, T.A., L.C. Fowke and D.I. Dunstan, 1991. Somatic embryogenesis in conifers. Can. J. Bot. 69: 1873–1897.

Thompson, R.G. and P. von Aderkas, 1992. Somatic embryogenesis and plant regeneration from immature embryos of western larch. Plant Cell Rep. 11: 379–385.

Tisserat, B., E.B. Esan and T. Murashige, 1979. Somatic embryogenesis in angiosperms. Hort. Rev. 1: 1–78.

Vasil, I.K. and A.C. Hildebrandt, 1966. Variations of morphogenetic behaviour in plant tissue cultures. I. *Chicorium endivia* . Am. J. Bot. 53: 860–869.

Von Aderkas, P. and J.M. Bonga, 1988. Formation of haploid embryoids of *Larix decidua*: early embryogenesis. Am. J. Bot. 75: 690–700.

Von Arnold, S. and I. Hakman, 1988. Regulation of somatic embryo development in *Picea abies* by abscisic acid. J. Plant Physiol. 132: 164–169.

Wardlaw, C.W. (Ed.), 1955. Embryogenesis in Plants. John Wiley and Sons, New York.

Wardlaw, C.W. (Ed.), 1968. Morphogenesis in Plants. Methuen and Co., London.

4. Somatic Embryogenesis in Some Woody Angiosperms

S.V. Kendurkar, R.S. Nadgauda, C.H. Phadke, M.M. Jana, S.V. Shirke and A.F. Mascarenhas*

Contents

1. Introduction

Somatic embryogenesis or asexual embryogenesis in cultured plant cells is the best demonstration of totipotency in plant cells. The first plant somatic embryos were obtained from somatic tissues of carrot (*Daucus carota*) cultured *in vitro* (Reinert, 1958; Steward, 1958) and from a woody angiosperm Sandalwood (*Santalum album*; Rao, 1965). Since then an increasing number

* To whom correspondence should be addressed.

S. Jain, P. Gupta & R. Newton (eds.), Somatic Embryogenesis in Woody Plants, Vol. 1, 49–79.

of plant species have been found amenable to this mode of regeneration. During the last decade somatic embryogenesis is being viewed as an efficient, cost-effective method which facilitates scaling up operations, automation, encapsulation, fluid drilling, and genetic transformations.

The process of embryogenesis follows different patterns such as direct, secondary and through callus stage. Various factors such as genotype, explant, pre-conditioning, media requirements and origin of somatic embryos have already been discussed (Thorpe *et al.*, 1991; Litz and Gray, 1992). Studies at the structural metabolic and molecular levels on the complex sequence of events associated with morphogenesis has been covered (Thorpe and Kumar, 1993).

Despite the difficulties in regeneration of woody species via somatic embryogenesis several forest and fruit tree species have been regenerated and the list is increasing. Reviews on forest trees (Tulecke, 1987; Thorpe *et al.*, 1991); tropical and subtropical fruit crops (Litz & Jaiswal, 1991), perennial fruit and nut crops (Litz and Gray, 1992) temperate zone fruit and nut crops (Zimmerman, 1991), cover the status of somatic embryogenesis and other modes of micropropagation for these crops.

In the following review an attempt has been made to summarise the global status of somatic embryogenesis and protoplast culture in some woody angiosperms (Table 1). Tissue culture studies carried out on two tree species viz. mango (*Mangifera indica* L.) and coconut (*Cocos nucifera* L.) at National Chemical Laboratory have also been described.

The present status indicates that woody plants can no longer be considered as recalcitrant. A number of plant species which could not be propagated via meristem/shoot tip culture have been found amenable to somatic embryogenesis. The important factor contributing to the success is perhaps the right choice of the explant. Immature embryos, and nucellar tissue (fruit spp.) have proven to be the most responsive source of explant because of the presence of predetermined cells for production of adventitious or somatic embryos (Sharp *et al.*, 1980).In most of the successful protoplast regeneration systems source of explant have been organogenetic or embryogenic callus or suspension with high morphogenetic potential. Some of the tree species like citrus, coffee and sandalwood, have exhibited higher morphogenetic potential, citrus and coffee have been used as model system to study the process of somatic embryogenesis. Regeneration of interspecific and intergeneric hybrids have been possible through somatic hybridization in citrus and other species. Recent reports on a new cytokinin like compound thiadizuron show the potential for regeneration of recalcitrant tree species (Huetteman and Preece, 1993).

Table 1. Somatic embryogenesis in some woody angiosperms

Species	Explant	Result, Studies	References
Actinidia chinensis	L	C, PL	Barbieri and Morini, 1987
	L	C, PR, PL	Tsai, 1988
	End	E, PL (triploid)	Huang *et al.*, 1982
Actinidia deliciosa	L	PR, PL	Cai *et al.*, 1991
Aesculus hippocastanum	A	E, PL	Radojevic, 1978, 1980
		SE	Radojevic, 1988
Albizzia richardiana	Hyp	C, E	Tomar and Gupta, 1988
A. lebbeck	Hyp	E, PL	Gharyal and Maheshwari, 1981
	A	E, PL	Gharyal *et al.*, 1983
A. procera	Seedling explants	C, E, PL	Datta, 1987
Annona squamosa	A	SE, PL	Nair *et al.*, 1983
Azadirachta indica	Im Coty	SE, PL	Shrikhande *et al.*, 1993
Bactris gasipaes	ST(J)	C, SE, PL	Valverde *et al.*, 1987
Bambusa arundinacea	Seed	C, SE, PL	Mehta *et al.*, 1982
Bambusa beecheyana	Inf	C, SE, PL	Yeh and Chang, 1986a
Bambusa oldhamii	Inf	C, SE, PL	Yeh and Chang, 1986b
Biota orientalis	Coty	EmB	Konar and Oberoi, 1965
Broussonetia kazinoki	L	PR, PL	Oka and Ohyama, 1985
Cassia siamea	A	C, E, PL	Gharyal *et al.*, 1983
Cassia fistula	A	C, E	Bajaj and Dhanju, 1983
Castanea sativa	Coty	SE	Gonzalez *et al.*, 1985
Citrus spp.	A	A, PL	Chen, 1987
	N	SE, PL	Barlass and Skene, 1982, 1986
	O	E, PL	Gmitter and Moore, 1986
	End	C, PL (triploid)	Gmitter *et al.*, 1990
	EmbC Susp	PR, EmB	Niedz, 1993
	EmbC Susp	PR	Vardi, 1977
Citrus aurantifolia	N	SE, PL	Mitra and Chaturvedi, 1972
Citrus aurantium	N	C, SE	Ben-Hayyim and Neumann, 1983
Citrus aurantium	NC	PR, PL	Vardi *et al.*, 1982
Citrus aurantium	NC	PR	Vardi *et al.*, 1987
× *C. limon*	NC	Somatic cybrid	
Citrus clementina	N	SE, PL	Navarro *et al.*, 1985

Table 1. Continued

Species	Explant	Result, Studies	References
Citrus grandis	N	SE	Rangan *et al.*, 1968
Citrus jambhiri	N	SE	Ben-hayyim and Neumann, 1983
		PR, PL	Li, 1991
Citrus limon	N	SE	Rangan *et al.*, 1968
	NC	PR, PL	Vardi *et al.*, 1982
Citrus madurensis	N	PR, PL	Ling *et al.*, 1989
	A	E, PL	Chen, 1985
Citrus microcarpa	A	E, P	Chen *et al.*, 1980
Citrus mitis	EmbC Susp	PR, DSE, PL	Sim *et al.*, 1988
Citrus nobilis	N	C, SE	Kochba *et al.*, 1982
Citrus paradisi	O, N	E, PL	Kochba *et al.*, 1972
	NC	PR, PL	Vardi *et al.*, 1982
Citrus reticulata	NC	PR, PL	Vardi *et al.*, 1982
	HC	PR, PL	Hidaka and Kajiura, 1988
Citrus sinensis	N	DSE	Rangan *et al.*, 1968
	NC	PR	Galun *et al.*, 1977
	NC	PR	Vardi, 1977
	NC	PR, PL	Vardi *et al.*, 1982
	NC	PR, PL	Kobayashi *et al.*, 1983
	EmbC Susp	PR, PL	Kobayashi *et al.*, 1985
	Hypo. C	PR, PL	Hidaka and Kajiura, 1988
		SE, Cry	Marin and Duran Vila, 1988
	O	SE, Cry, PL	Marin *et al.*, 1993
Citrus sinensis × *Poncirus trifoliata*	Emb Cell susp Seedling L	Somatic hybrid, PL	Grosser *et al.*, 1988a
Citrus sinensis × *Citrus unshiu*	EmbC Susp Leaf	Somatic hybrid, PL	Kobayashi *et al.*, 1988
Citrus sinensis × *Poncirus trifoliata*	Emb. cell susp Seedling L	Somatic hybrid, PL	Ohgawara *et al.*, 1985
Citrus sinensis × *Severina disticha*	Emb. cell susp Epicotyl C	Somatic hybrid, PL	Grosser *et al.*, 1988b
Citrus unshiu	O	Emb C, PR, PL	Ling *et al.*, 1990
	O	PR, SE, PL	Kunitake *et al.*, 1991
Citrus yuko	Hyp. C	PR, PL	Hidaka and Kajiura, 1988
Microcitrus sp.	Emb. C	PR, PL	Vardi *et al.*, 1986
Cocos nucifera	L	E	Pannetier and Buffard-Morel, 1982
	L	E, PL, F	Raju *et al.*, 1984
	L	E, PL, F	Buffard-Morel *et al.*, 1988
	L	E, PL	Verdeil *et al.*, 1989

Table 1. Continued

Species	Explant	Result, Studies	References
	L	E, R	Kendurkar *et al.*, 1989
	L, Inf	E, PL, F	Verdeil *et al.*, 1992
	L	E	Jesty and Francis, 1992
	Inf	E, PL, F	Branton and Blake, 1983a; Smith, 1986
	Inf	E	Gupta *et al.*, 1984; Branton and Blake, 1983b; Sugimura and Salvana, 1989
	A	E	Than-Tuyen and De Guzman, 1983
	A	E	Monfort, 1985
	Im E	E	Karunaratne and Periyapperuma, 1989
	ZE	E	Hornung (pers. comm.)
	MR	C	Fulford *et al.*, 1977
	SR	C	Fulford *et al.*, 1977
Coffee sp.	Emb Cell Susp	PR, SE, PL	Spiral and Petiard, 1991
Coffea arabica	Emb Cell Susp	PR, SE, PL	Yasuda *et al.* 1986
	Emb Cell Susp	PR, PL	Acuna and De Pena, 1991
	L	SE, PL	Sondahl and Sharp, 1977a,b
	Emb Cell Susp	PR, PL	Acuna and De Pena, 1991
	Emb Cell Susp	PR, C, SE, PL, T	Barton *et al.*, 1991
	L	SE	Neuenschwander and Baumann, 1991
C. eugenioides	L	SE	Marques, 1993
C. canephora	M	SE	Sondahl and Sharp, 1979
	SE	PR, PL	Schopke *et al.*, 1987
	St	SE	Staritsky, 1970
C. congensis	ML	SE	Sondahl and Sharp, 1979
C. dewevrei	ML	SE	Sondahl and Sharp, 1979
Dendrocalamus strictus	Seed	E, PL, S	Rao *et al.*, 1985
Duboisia myoporoides	Cell Susp	PR, PL	Kitamura *et al.*, 1989
Elaeis guineensis	L	C, SE, PL	Rabechault and Martin, 1976
	L	C, EM, PL, F	Corley *et al.*, 1976
	L	C, SE, PL	Thomas and Rao, 1985
	L	SE, Cry	Engelmann and Duval, 1986
	L	C, SE, PL	Jones, 1988; Jones and Hughes, 1989
	L	SE, PL, F	Durand-Gasselin *et al.*, 1989

Table 1. Continued

Species	Explant	Result, Studies	References
	L	EmbC Susp, SE, PL	De Touchet *et al.*, 1991
	Im E	SE	Teixeira *et al.*, 1993
	J	SE	Krikorian and Kann, 1986
Eucalyptus citriodora	Seed	E, PL, S, Artificial seeds	Muralidharan and Mascarenhas, 1987; Muralidharan *et al.*, 1989; Mascarenhas *et al.*, 1989
Eucalyptus gunii	Hyp	SE	Boulay, 1987
Euphoria longan	L	SE	Litz, 1988
	A	C, E, P	Yang and Wei, 1984
Euterpe edulis	Im E	SE, PL	Guerro and Handro, 1988
Fagus sylvatica	A	SE	Jorgenson, 1988
	Im E	EmbC Susp, SE, PL	Vieitez *et al.*, 1992
Feijoa sellowiana	ZE	SE, PL	Cruz *et al.*, 1990
Fraxinus americana	ME	SE, PL	Preece *et al.*, 1989
		SE	Bates *et al.*, 1992
Fraxinus spp.	St	SE	Preece *et al.*, 1987
Hevea brasiliensis	Im E	C, SE	Montoro *et al.*, 1993
	Im Seed	C, SE, PL	Hadrami *et al.*, 1991
	A	E, PL, F	Chen *et al.*, 1978
	A	E, PL	Wang *et al.*, 1980
	O	E, PL	Chen *et al.*, 1988
	Im Seed	SE	Carron and Enjalric, 1982
		SE	Paranjothy and Rothman, 1978
Ilex aquifolium	Coty	SE	Hu and Sussex, 1971; Hu *et al.*, 1978
Jacaranda acutifolia	A	C, E	Bajaj and Dhanju, 1983
Juglans regia	Im coty	SE, PL	Tulecke and McGranahan, 1985
	SE	T, PL	McGranahan *et al.*, 1988
	SE	Sec. Emb.	Polito *et al.*, 1989
Juglans nigra	Coty	C, SE	Neuman *et al.*, 1993
Juglans spp.		SE	Cornu, 1988
Juglans spp.	St, L	SE	Park and Son, 1988
Leucosceptrum canum	L	SE, PL	Pal *et al.*, 1985
Liriodendron tulipifera	Im E	SE	Merkle and Sommer, 1986
	EmbC Susp	PR, PL	Merkle and Sommer, 1987

Table 1. Continued

Species	Explant	Result, Studies	References
	Im E	SE, PL	Merkle *et al.*, 1990
	ZE	SE	Sotak *et al.*, 1991
	Im E	SE, PL, S	Merkle *et al.*, 1993
Litchi chinensis	A	SE	Chen, 1987
	A	C, E, PL	Fu and Tang, 1983
Liquidambar styraciflua	ST, Coty	SE, PL	Sommer and Brown, 1980
Malus spp.	A	E	Chen, 1987
	ST, L, F	C, E	Mehra and Sachdeva, 1984
Malus × domestica	S Cult	PR, PL	Patat-Ochatt *et al.*, 1988
Malus domestica	N	E	Eichholtz *et al.*, 1980
Malus prunifohia	A	E, PL	Wu, 1981
Malus plumila	A	E, PL	Lespinasse *et al.*, 1963
Mangifera indica	N	DSE	Litz and Knight, 1983
	N	C, E	Litz *et al.* 1984
	N	C, SE	Litz, 1984b
	N	DSE	Dewald and Litz, 1987
	O	C, SE	Jaiswal, 1990
	N	SE	Sahijram, 1990
	N	DSE, PL	Jana *et al.*, 1993
	N	DSE	Sahijram, 1990
	O	DSE	Litz *et al.*, 1982
	S EmB	T	Mathews *et al.*, 1992
Myrciaria cauliflora	N	DSE	Litz, 1984a
Olea europaea	Im E	SE, PL	Rugini, 1988
Otatea acuminata	ZE	SE, PL	Woods *et al.*, 1992
Paulownia tomentosa	ME	SE, PL	Radojevic, 1979
Persea americana	E	SE, PL	Pliego-Alfaro and Murashige, 1988
	Im E	SE, PL	Mooney and Van Staden, 1987
Phoenix dactylifera		C, E	Reynolds and Murashige, 1979
	ST	SE	Sharma *et al.*, 1984
	L	SE, PL	Sharma *et al.*, 1980
	ST	SE, PL	Tisserat, 1982
	ST, Inf	SE, PL	Bhaskaran and Smith, 1992
	E	SE	Ammar and Benbadis, 1977
	L	SE, PL	Sudhersan *et al.*, 1993
Phyllostachys viridis		Emb, PL	Anas *et al.*, 1987
Poinciana regia	A	C, E	Bajaj and Dhanju, 1983

Table 1. Continued

Species	Explant	Result, Studies	References
Populus spp.	O	SE	Chen *et al.*, 1988
	A	PL	Ho and Ray, 1985
	A	E, PL, F	Lu and Liu, 1990
Populus alba	L	PR, PL	Sasamoto and Hosoi, 1990
Populus alba × *P. grandidentata*	S Cult	PR, PL	Russell and McCown, 1986, 1988
Populus ciliata	L	Cell Susp, SE, PL	Cheema, 1989
Populus deltoides	A	SE, PL	Uddin *et al.*, 1988
Populus nigra	L	C, PR, PL	Lee *et al.*, 1987
Populus nigra × *P. maximowiczii*	L	SE, PL	Park and Son, 1988
Populus nigra × P. trichocarpa (Hybrid poplar)	S Cult	PR, PL	Russell and McCown, 1988
Populus tremula	S Cult	PR, PL	Russell and McCown, 1988
Prunus spp.	Coty	SE	Mante *et al.*, 1989
	A	EmB	Zenkteler *et al.*, 1975
		SE	Raj Bhansali *et al.*, 1990
	R	C, PL	Druart, 1980
	End	S, PL	Zhao, 1983
Prunus avium	Im E	SE	De March *et al.*, 1993
	L	PR, PL	Ochatt, 1991
Prunus avium × P. pseudocerasus	S Cult	PR, PL	Ochatt *et al.*, 1987, 1988a
	L	C, PL	Jones *et al.*, 1984
	S Cult	PR, PL	Power, 1988
Prunus cerasifera	S. Cult	PR, PL	Ochatt, 1992
Prunus cerasus	S Cult	PR, PL	Ochatt and Power, 1988
	R, C	PR, PL	Ochatt, 1990
Prunus incisa × *serrula*	R	SE	Druart, 1990
	RM	SE	Druart, 1981
Prunus persica	E	C, SE, PL	Scorza *et al.*, 1990
Prunus spinosa	S. Cult	PR, PL	Ochatt, 1992
Pyrus spp.	A	E	Jordan, 1975
Pyrus spp.	Im S	E	Janick, 1982
Quercus robur	Im E	SE, PL	Chalupa, 1990
Quercus ruber		SE	Manzanera *et al.*, 1993
Quercus rubra	Im E	SE, PL	Gingas and Lineberger, 1989

Table 1. Continued

Species	Explant	Result, Studies	References
Robinia spp.	O	SE	Wang *et al.*, 1982
Robinia pseudoacacia	Im E	SE, PL	Merkle and Wiecko, 1989
Santalum album	End	E, PL (triploid), F	Laxmisita *et al.*, 1980
	hyp	C, E, F	Bapat and Rao, 1979
	EmbC Susp	PR, SE, PL	Rao and Ozias-Akins, 1985
	St (M)	C, E, PL, F	Laxmisita *et al.*, 1979
		C, susp, PR, E, PL	Bapat *et al.*, 1985
		Synthetic seeds, PL	Bapat and Rao, 1988
Sapindus trifoliatus	ML	C, E, PL	Desai *et al.*, 1986
Sinocalamus latiflora	ZE	C, SE, PL	Yeh and Cheng, 1987
	ZE	SE, PL	Tsai *et al.*, 1990
Syzygium spp.	N	SE	Litz, 1985
Theobroma cacao	Cell Susp	PR, PL	Kanchanapoom and Kanchanapoom, 1991
Tilia cordata	Im E	SE	Chalupa, 1990
Ulmus campestris	L	PR, PL	Dorion *et al.*, 1991
Ulmus × (Hybrid *elm*)	L, C	PR	Sticklen *et al.*, 1986

A – Anthers, C – Callus, Coty – Cotyledon, Cry – Cryopreservation, D – Direct, E – Embryogenesis, EmB – embryoids, Emb – Embryogenic, End – Endosperm, F – Field, Hyp – Hypocotyl, Im E – Immature embryos, Inf – Inflorescence, J – Juvenile, L – Leaves, M – Mature, N – Nucellar, O – Ovules, PL – Plantlet, PR – Protoplast, R – Roots, MR – Mature Roots, SR – Secondary Roots, S – Soil, S Cult – Shoot Cultures, SE – Somatic Embryogenesis, SL – Seedling Leaves, ST – Shoot Tip, St – Stem, susp – suspension, T – Transformation, ZE – Zygotic Embryos.

2. Coconut, *Cocos nucifera* (L.)

2.1. *Economic Importance and World Production*

Coconut (*Cocos nucifera* L.) (family – Palmae) is essentially a tropical plant. It is the only species of the *Cocos* genus. Coconut is now cultivated on nearly 10 million hectares in 80 countries because of its considerable socio-economic importance. This is known as "Tree of life" or "Kalpavriksha" as almost all the products such as water, pulp, oil, husk, leaves, roots, wood, copra are extensively used (Thampan, 1982). The kernel is the main component of the fruit which contains 60–70% of oil. Coconut oil occupies sixth position as a source of oil for human consumption. It is a rich source of short chain fatty acids with 8 to 14 carbon atoms (mainly lauric acid 48 percent) which increases its demand in soap, cosmetics and pharmaceutical industries (Kaufman, 1965).

Asia is the main producing area, accounting for 85 percent of world "copra" production. Philippines, Indonesia and India are the major coconut producing countries. The major coconut producing states in India are Kerala (47 percent) , Tamilnadu (23 percent), Karnataka (14 percent) and Andhra Pradesh (6 percent). Out of a total Indian production, 35 percent is used for religious and edible purposes, 35 percent for coconut oil extraction, 7 percent for coconut water and remaining for dessicated coconut.

2.2. *Pests and Diseases*

Coconut production is affected by various pests and diseases. The major insect pests of the coconut palm are the rhinoceros beetle *Oryctes rhinoceros*, the leaf eating caterpillar *Nephantis serinopera*, the red palm weevil *Rhynchophorus ferrugineus* and the root eating cockchafer *Leucopholis coneophora* (Thangaraj and Muthuswami, 1990).

Coconut trees are affected by various fungal dieases which cause bud rot *Phytophthora palmivora*, leaf rot *Bipolaris halodes*, fruit rot *Phytophthora* spp. and stem bleeding *Theilaviopsis paradora*. The lethal yellowing disease (Jamaica, Tanzania, etc.) and root wilt disease (India) are caused by mycoplasma like organisms (MLOs) (Thangaraj and Muthuswami, 1990).

2.3. *Conventional Methods of Propagation and Need for Tissue Culture*

Coconut propapagation is carried out only by seednuts as coconut has a single apical meristem and axillary meristems form inflorescences, it is unsuitable for propagation by cutting and grafting. The alternative methods of vegetative propagation such as (i) conversion of the floral meristems into vegetative buds (Davis, 1969a,b; Blake and Eeuwens, 1982) and their rooting (Sudarsip *et al.*, 1978; Eeuwens and Blake, 1977), (ii) multiplication by splitting the growing points on seedlings (Balaga, 1975), have been tried on experimental level but cannot be exploited for vegetative propagation of palms. Thus, coconut propagation is solely dependent on seed nuts leading to wide variations in the progeny due to heterozygosity in coconut.

To increase the net productivity it is essential to propagate disease free, high yielding "Super" palms (400 nuts/annum), clonally (Iyer *et al.*, 1979). Somatic embryogenesis is perhaps the only suitable and promising method to achieve clonal propagation of coconuts.

2.4. *Somatic Embryogenesis in Coconut*

2.4.1. *International Status*

The most economically important members of Palmae family are date palm (*Phoenix dactylifera* L.), oil palm (*Elaeis guineensis* Jacq) and coconut palm

(*Cocos nucifera* L.). *In vitro* studies on these palms have been exhaustively reviewed (Blake, 1983, 1990; Branton and Blake, 1989; Brackpool *et al.*, 1986; Pannetier and Buffard-Morel, 1986; Tisserat, 1987; Wooi, 1990; Dublin *et al.*, 1991; Rao and Ganapathy, 1993).

The earliest report on coconut tissue culture dates back to 1954, (Cutter and Wilson, 1954). The work was initiated at various laboratories in the 70s. However information regarding media formulations (Eeuwens, 1976, 1978), growth and regeneration of tissues in vitro started appearing in early eighties. Most of the reports indicate requirements for unusually high amounts of auxins, cytokinins and the presence of activated charcoal in the medium. Although there are several reports on formation of embryo like structures from leaf explants of seedlings and mature trees (Kendurkar *et al.*, 1989), 5-year-old seedlings (Pannetier and Buffard-Morel, 1982), inflorescence explants (Branton and Blake, 1983b; Sugimura and Salvana, 1989), immature embryos (Karunaratne and Periyapperuma, 1989), zygotic embryos (Hornung, personal communication), and anthers (Than-Tuyen and De Guzman, 1983), only a few reports indicated successful regeneration of plantlets from inflorescence (Branton and Blake, 1983a; Smith, 1986) and leaf explants (Raju *et al.*, 1984; Buffard-Morel *et al.*, 1988; Verdeil *et al.*, 1989). Initially only two plants were successfully transferred to soil (Smith, 1986; Buffard-Morel *et al.*, 1988). Recently Verdeil *et al.* (1992) have reported regeneration of plantlets from leaf and inflorescence explants of adult individuals (20–25 yr) of PB-121 hybrid. Regeneration of 20 ramets from 5 clones and their transplantation to soil have been reported.

Presently methods for germination of zygotic embryos of coconut have been successful which are important for exchange and conservation of germplasm (Assy Bah, 1986; Assy Bah and Engelmann, 1993; Rillo and Paloma, 1990, 1991). Germination of embryos from "Makapuno" type of nuts is of great value as they do not germinate under natural conditions (Del Rosario and De Guzman, 1976, 1981; Rillo and Paloma, 1992).

Coconut has generally proved to be recalcitrant *in vitro* possibly due to natural blockages that cannot be removed as yet by any means. The presence of higher levels of zeatin and zeatin riboside in foliar explants (Verdeil, personal communication; Dublin *et al.*, 1991) suggest the peculiar endogenous hormonal constitution of this species.

Studies on cellular responses of coconut leaf explants *in vitro* by Jesty and Francis (1992) reveals specific responses such as (i) complete absence of a gradient of cell division, (ii) low mitotic index i.e., cycling of smaller percentage of cells, (iii) selective accumulation of cells in G1 phase, and (iv) uncontrolled expansion of cell leading to low nuclear to cell area ratio. This uncoupling of cell and nuclear size which disrupts cell co-ordination may be the key contributor to recalcitrant nature of this species *in vitro*. As suggested by the authors, factors stimulating GI/S transition may improve growth rates and perhaps morphogenetic potential. Most of the reports on coconut tissue culture indicate a need for incorporation of activated charcoal as an essential

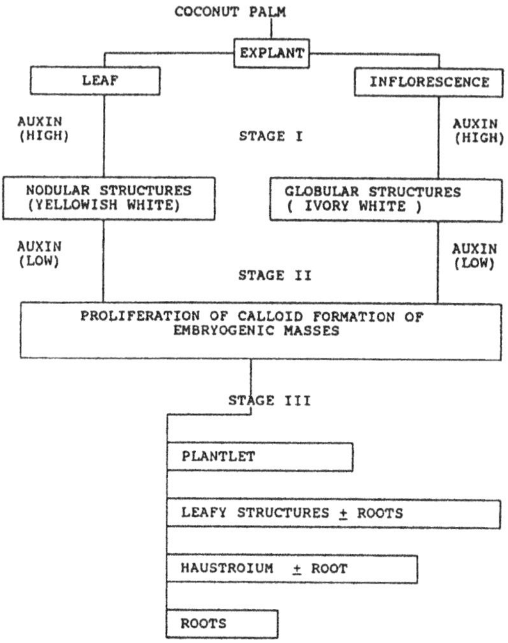

Figure 1. Regeneration in Coconut Palm.

component in the medium to control browning and growth of explants. Analysis of the media containing activated charcoal (Ebert and Taylor, 1990; Ebert *et al.*, 1993) showed that a major proportion of hormones (2,4-dichlorophenoxy acetic acid, and 6-benzylamino purine) were adsorbed by activated charcoal and a very small amount of free hormones was available to the tissue. The possibility of adsorption of endogenous hormones cannot thus be ruled out.

The general pathway for somatic embryogenesis in coconut is described in Fig. 1. During recent years stage I has been repeatable from various explants of different age and origin. New formation and proliferation (stage II) are slow. The development of embryo like structures and maturation in stage III is difficult and often associated with abnormal or incomplete morphogenesis. The embryogenic masses may give rise to roots, leafy structures with or without roots and haustorial tissue with or without roots. Development of normal plantlet is still a rare phenomenon (stage III).

2.4.2. *Status of Work at National Chemical Laboratory, Pune*

At National Chemical Laboratory, Pune work on coconut tissue culture was initiated in early 80s. Initial success obtained with leaf, inflorescence, stem explants and embryos have been reported earlier (Gupta *et al.*, 1984). Recent developments using leaf and inflorescence explants from seedlings and mature trees of three different varieties, viz. D × T, West Coast Tall and Hawaiin supreme are described below.

2.4.3. *Leaf Explant*

2.4.3.1. *Culture Initiation.* Leaf material from seedlings, 2–6 year old and mature palm (10–80 yr) was the source of explant.

Collection of leaf material from mature palms (10–20 yr) was carried out using "Non-destructive method". In this method initially the outermost whorl of leaves was carefully removed to locate the actual growing meristem. The column was then cut at 30–40 cm above the exposed stem. The upper portion was used as the source of material, a mixture of antifungal and insecticidal solutions (Bavistin 2% and Blitox 2%) was applied at the cut end of the tree and covered with a polythene bag to protect the growing tip. These trees regenerated in 6 months time and were ready for a re-collection of material. After sterilization with $HgCl_2$ 0.05% for 10 min leaf folds (0.5–1.0 cm) were inoculated on semisolid Y3 basal medium containing auxin (200–300 mg/L) and 0.25% activated charcoal. The cultures were incubated at 28 ± 2°C under dark conditions.

Initially swelling of the explants was observed followed by initiation of warty outgrowths on the surface within 4–6 weeks. Formation of nodular structures on the surface and at the cut ends was observed after 8–12 weeks (Fig. 2a).

2.4.3.2. *Culture Maintenance.* Proliferation and development of globular embryo like structures was promoted on media containing reduced auxin levels for 2–3 passages (Fig. 2b). After the 4th passage increased browning in the calloid mass reduced the neoformation of globular embryo like structures.

2.4.3.3. *Embryo Development.* The globular structures formed on leaf explants proliferated and developed into shoot bud like protuberances in few cases. In most of the cultures the development of shoot poles was suppressed.

Although the steps of initiation and proliferation of globular structures on leaf explants have been reproducible with all the three genotypes, simultaneous regeneration of the shoot and root is still critical.

2.4.3.4. *Histological Studies.* Leaf explants growing *in vitro* were collected at different stages of development and were fixed in FAA (formalin aceto alchohol solution) for 24 h, passed through grades of alcohol and xylene and finally embedded in paraffin wax. Sections of 10 um thickness were taken and stained with hematoxylin (1%) and eosin (1%). Anatomical sections show initiation of globular structures from the perivascular region consisting of small densely stained meristematic cells (Fig. 2c). Structures with well formed vasculature indicate the possibility of development of root poles.

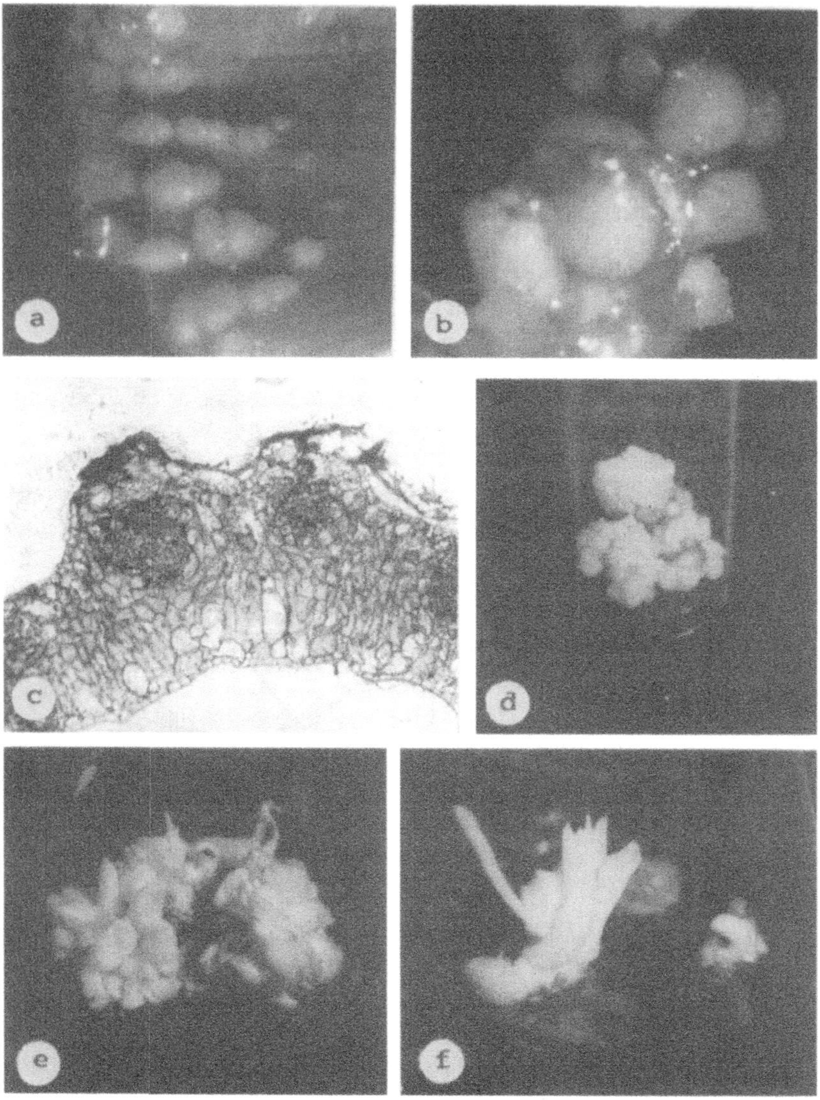

Figure 2. Somatic embryogenesis in Coconut *Cocos nucifera* (L.). (a) Nodular structures on leaf explant. (b) Proliferation and development of globular structures. (c) T.S. of leaf showing globular structures. (d) Formation of "calloid" from inflorescence explants. (e) Development of embryos. (f) Germination of embryos (leafy structures).

2.4.4. *Inflorescence Explants*

2.4.4.1. *Culture Initiation.* Spathes of different sizes (4 to 50 cm) were collected from mature palms (10–80 yr). The spathes were wiped with 70 percent alcohol. The rachillae explants were dissected out and sliced transversely (average 20 to 30 explants/4–10 cm sized spathe) and inoculated on Y3 media (Eeuwens, 1978) formulations with auxins, activated charcoal and sucrose. The cultures were incubated at 28 ± 2°C under dark conditions. Rachillae explants from 4–10 cm size spathes showed minimum browning and best response. These explants swelled considerably during the initial culture period followed by initiation of ivory white, small globular structures of embryogenic nature.

2.4.4.2. *Culture Maintenance.* The gobular structures when shifted to reduced auxin medium proliferated into bigger, globular "calloid" masses (Fig. 2d). The embryogenic cultures could be kept proliferating for more than one year.

2.4.4.3. *Embryo Development and Germination.* The proliferation of calloid was associated with the formation of embryogenic masses which later developed into leafy structures with or without roots on gradual reduction of auxin and incorporation of cytokinins in the medium during subsequent passages. These leafy structures which turned green on shifting to light (117 $\mu Em^{-2}\ S^{-1}$ light intensity of cool white fluorescent tubes) were fused and showed several abnormalities (Fig. 2e,f).

2.4.4.4. *Growth in the Field.* Complete plantlets with fused leafy structures at the shoot pole and well developed normal roots were transferred to soil. Survival and opening of new leaves was not successful.

2.4.5. *Root Explant*

In vitro regenerated roots from leaf explants were subcultured on various media formulations. Auxin (2,4-D) promoted formation of slow growing loose callus whereas long term proliferation of roots to roots have been obtained on media containing NAA. Attempts to grow/regenerate plants using primary, secondary, tertiary roots from the elite palms have not been successful.

3. Mango, *Mangifera Indica* (L.)

3.1. *Economic Importance and World Production*

Mango is a tropical fruit tree belonging to family *Anacardiaceae*. Out of 41 species the edible cultivars belong to *Mangifera indica*. Mango is known as "king of fruit" because of its luscious taste and captivating flavour. The fruit is useful at all stages of ripening and is a rich source of vitamins A and C. Other plant parts such as leaves are used as vegetable in Java and Philippines, leaf ash for burns and scalds, flowers for "otto" preparation, dried flowers for diarrhoea and chronic dysentry. Bark yields magniferin and tannin (16 to 20%), stem and trunk gives gum, wood is used for furniture, flooring, match box, splint, brush backs, boats, oar blades, etc. (Majumdar and Sharma, 1990).

Various cultivars of mango all over the world include both monoembryonic and polyembryonic varieties. Mango is now being cultivated in many other countries besides India. Major producing countries are Mexico, Pakistan, Thailand, China, Brazil, Phillipines and Indonesia. The world production of mango is approximately 15.7 million tonnes/annum and India'a contribution is 9.5 million tonnes. Mexico is the number one exporter contributing 33 percent followed by Phillipines 12 percent, India and Pakistan 10 percent each towards export of mango. Alphonso is one of the finest Indian varieties rated high in India and abroad because of its excellent taste, high pulp:seed ratio and characteristic flavour. It has specific climatic requirements for growth and performs well only on the west coast of Maharashtra, India, which has subtropical, humid, coastal climate.

3.2. *Pests and Diseases*

Major pests of mango are mango hopper, *Amaritodus atlkinsoni* and mealy bug, *Drosicha mangiferae*. Most damaging diseases are powdery mildew *Oidium mangiferae*, anthracnose *Colletotrichum gloesporiodes*, stem end rot *Diplodia* spp., Sooty mold *Meliola mangiferae*, Pink disease *Pellicularia salmonicolor*, etc., and bacterial canker *Pseudomonas mangiferae*. Besides several physiologial disorders such as black tip, leaf scorch, spongy tissue adversely affect the fruits (Majumdar and Sharma, 1990).

3.3. *Conventional Methods of Propagation and Need for Tissue Culture*

Vegetative propagation is the method of choice in mango as it is a highly cross pollinated crop and it has a long juvenile phase. The commonly used methods for vegetative propagation are grafting (Veneer, side, epicotyl or stone grafting), budding (Patch, Shield, Forkert), layering (air layering, pot layering, stooling), and rooting of cuttings. In mango as in some other tree fruit crops work on using selective root stocks for improvement in the progeny have been carried out at Indian Agricultural Research Institute, Delhi, India. Root stocks of various cultivars such as "Totapuri Red small", "Olour", "Kalapadi", "Vellai collumban", etc., appear promising for imparting dwarfness. Olour has been proved to be a dwarfing and productive root stock for "Himsagar", "Langra" and "Alphonso" (Majumdar and Sharma, 1990). In this context tissue culture can be viewed as a complementary method for selective propagation of trees having superior traits, viz. quality of the fruit, yield, pulp to seed ratio, size and disease resistance, etc. Somatic embryogenesis is valuable for understocks of exotic plus trees where propagation by conventional methods have limited success.

3.4. *Somatic Embryogenesis*

3.4.1. *International Status*

There are very few reports on *in vitro* culture of mango. Somatic embryogenesis have been reported from nucellar explants of monoembryonic and polyembryonic cultivars of mango (Litz *et al.*, 1982, 1984; Litz and Knight, 1983; Litz, 1984b; Dewald and Litz, 1987). The nucellus has been the choice of explant for citrus spp. (Barlass and Skene, 1986), jaboticaba (Litz, 1984a), syzigium spp (Litz, 1985) and Loquat (Litz, 1985). In India, there are only 3 reports, one on somatic embryogenesis from ovules of Indian monoembryonic cultivars namely "Langra", "Dasheri" and "Bombay green" (Jaiswal, 1990) where somatic embryos were obtained through callus intervention. Maturation and limited germination of embryos occured in the absence of growth regulators. Sahijram (1990) obtained somatic embryos from nucellus of a monoembryonic cultivar Totapuri. Regeneration of plantlets from economically important monoembryonic cultivars (Alphonso, Mundan, Baneshan) have been successful at National Chemical Laboratory, Pune (Jana *et al.*, 1994). Mathews *et al.* (1992) have reported successful transformation of somatic embryos of mango. The transformation was achieved by co-cultivation of embryogenic masses with *A. tumefaciens* strain pTiT37-SE :: pMON9749 (9749ASE), having genes for Kanamycin resistance and B glucuronidase activity, which was later put on selection medium with 200 mg Kanamycin per L for 5–6 months. Somatic embryos regenerated from these Kanamycin resistant embryogenic masses showed B glucuronidase gene expression which was confirmed by histochemical analyses.

3.4.2. *Work at National Chemical Laboratory, Pune*

Different explants such as leaf, cotyledons, nucellus, were used for developing methods of somatic embryogenesis in mango. Protocol for successful somatic embryogenesis is described below.

3.4.2.1. *Culture Initiation* Immature fruits (20 to 45 days after pollination), of three monoembryonic varieties namely "Alphonso", "Mundan" and "Baneshan" were collected from 20 yr old orchards of National Agricultural Research Projects (NARP). Regional Fruit Research Station Pune, Maharashtra, India. After sterilization of the fruits with 0.1% mercuric chloride for 10 min, nucelli were scooped out aseptically and inoculated on semisolid MS (Murashige and Skoog, 1962) basal medium supplemented with different combinations of 2,4 dichlorophenoxy acetic acid, gibberellic acid, activated charcoal and sucrose. The cultures were incubated in dark at 26°C till the

Figure 3. Somatic embryogenesis in Mango *Mangifera indica* (L.). (a) Structures of globular heart shaped embryos. (b) Maturation of embryos. (c) Germination of embryos. (d) Plantlet in soil.

initiation of embryoids. Induction of somatic embryos occurred within two weeks. Further growth of embryos and development of clusters of globular and heart shaped embryos was observed on enriching the above medium with glutamine within 14–21 days (Fig. 3a). Attempts in inducing somatic embryogenesis on cotyledon and leaf explants were not successful.

3.4.2.2. *Culture Maintenance.* These embryos were multiplied on initiation medium by subculturing at 20–25 days interval. The cultures are now at 25th passage. It was also observed that 80 percent of the embryos on sequential transfer to germination medium gave rise to complete plantlets. Remaining 20 percent of malformed embryos on transfer to induction medium produced secondary embryos. These embryos developed into normal plantlets.

3.4.2.3. *Maturation of Embryos.* Maturation of somatic embryos could be achieved by shifting heart shaped embryos to half strength major, micronutrients and organic compounds of Murashige and Skoog's formulation supplemented with casein hydrolysate, abscisic acid, coconut water and agar. The cultures were incubated in dark (Fig. 3b).

3.4.2.4. *Embryo Germination*. Germination of mature somatic embryos was obtained on MS half strength basal medium supplemented with benzyladenine (BA) after incubation in light for 20 days (Fig. 3c).

3.4.2.5. *Transfer to Soil*. Somatic embryo derived plantlets were transferred to plastic pots containing soil:sand:peat (1:1:1, V:V:V) mixture. Initially high humidity was maintained by covering with polythene sheet (Fig. 3d).

4. Conclusion

Regeneration of tree species through somatic embryogenesis has paved the way for biotechnology in these species. Advances made in protoplast culture (Russell, 1993), encapsulation of somatic embryos (Gupta and Kreitinger, 1993; Mascarenhas *et al.*, 1989; Bapat and Rao, 1988), somaclonal variation (Antonetti and Pinon, 1993; Sondahl and Lauritis, 1992) scaling up (Zamarripa *et al.*, 1991) genetic transformation (Mathews *et al.*, 1992; McGrahanan *et al.*, 1988; Mante *et al.*, 1991) with tree species are promising. Protoplast fusion technology can facilitate intervarietal hybridizations to combine useful characteristics such as fruit size, quality, maturity, dwarfness, regularity of bearing, resistance to pest and diseases in fruit tree species and wood quality, resistance to pest and diseases, etc., in case of forest tree species. Cryopreservation and successful regeneration from frozen tissue (Tisserat *et al.*, 1981; Ulrich *et al.*, 1982; Marin and Duran Vila, 1988) indicates the potential for germplasm preservation.

During recent years, the progress in coconut tissue culture encourages more intensified and multidirectional research aided with modern molecular biology approaches to study the factors responsible for the recalcitrant nature of the tissue and to induce regeneration in this difficult to propagate palm.

In mango, besides developing a suitable method for mass propagation of plus trees, there is a need to develop cultivars suitable for different regions and for differing needs of trade and processing industry. Improved cultivars having resistance against some of the serious diseases and pests are also required. Studies on the underlying mechanisms responsible for biennial bearing in mango which is a major limitation is also a challenging area of research.

Although initiation, maturation and germination of somatic embryos is being controlled, high frequency regeneration, synchronous development of embryos and transplantation to field is not very successful. Refinements in protocols are necessary to get good quality embryos closer to zygotic embryos to facilitate storage, germination and encapsulation of these embryos. Field performance data on these propagules is essential to confirm the feasibility of somatic embryogenesis for large scale propagation.

References

Acuna, J.R. and M. de Pena, 1991. Plant regeneration from protoplasts of embryogenic cell suspensions of *Coffea arabica* L. cv. caturra. Plant Cell Rep. 10: 345–348.

Ammar, S. and A. Benbadis, 1977. Multiplication Vegetative du palmier dattier (*Phoenix dactylifera* L.) par la culture de tissue de heunes plantes tissues de semis. C.R. Acad. Sci. Series D 284: 1789–1794.

Anas, A., E.L. Hasan and P. Debergh, 1987. Embryogenesis and plantlet development in bamboo *Phyllostachys viridis* (young) McClure. Plant Cell Tiss. Org. Cult. 10: 73–77.

Antonetti, P.I.E. and J. Pinon, 1993. Somalclonal variation within Poplar. Plant Cell Tiss. Org. Cult. 35: 99–106.

Assy Bah, B., 1986. *In vitro* culture of coconut zygotic embryos. Oleagineux 41: 321–328.

Assy Bah, B. and F. Engelmann, 1993. Medium term conservation of mature embryos of coconut. Plant Cell Tiss. Org. Cult. 33: 19–24.

Bajaj, Y.P.S. and M.S. Dhanju, 1983. Pollen embryogenesis in three ornamental trees *Cassia fistula*, *Jacaranda acutifolia* and *Poinciana regia* J. Tree Sci. 2: 16–19.

Balaga, Y.H., 1975. Induction of branching in coconut. Phillip. J. Biol. 4: 135–140.

Bapat, V. and P.S. Rao, 1979. Somatic embryogenesis and plantlet formation in tissue cultures of sandalwood. Ann. Bot. 44: 629–630.

Bapat, V.A. and P.S. Rao, 1988. Sandalwood plantlets from "synthetic seeds". Plant Cell Rep. 7: 434–436.

Bapat, V.A., R. Gill and P.S. Rao, 1985. Regeneration of somatic embryos and plantlets from stem callus protoplasts of Sandalwood tree (*Santalum album* L.) Curr. Sci. 54: 978–982.

Barbieri, C. and S. Morini, 1987. Plant regeneration from *Actinidia* callus cultures. J. Hort. Sci. 62: 107–109.

Barlass, M. and K.G.M. Skene, 1982. *In vitro* plantlet formation from *Citrus* species and hybrids. Sci. Hort. 17: 333–341.

Barlass, M. and K.G.M. Skene, 1986. Citrus *Citrus* spp. In: Y.P.S. Bajaj (Ed.), Biotechnology in Agriculture and Forestry Trees, Vol. 1, pp. 207–219. Springer-Verlag, Heidelberg.

Barton, C.R., T.L. Adams and M.A. Zarowitz, 1991. Stable transformation of foreign DNA into *Coffea arabica* plants. XIV International Conference on Coffee Science, San Fransisco, pp. 460–464.

Bates, S., J.E. Preece, N. Navarrette, J.W. Van Sambeek and G.R. Gaffney, 1992. Thidiazuron stimulates shoot organogenesis and somatic embryogenesis in white ash (*Fraxinus americana* L.). Plant Cell Tiss. Org. Cult. 31: 21–30.

Ben-Hayyim, G. and H. Neumann, 1983. Stimulatory effect of glycerol on growth and somatic embryogenesis in citrus callus cultures. Z. Pflanzenphysiol. 110: 331–338.

Bhaskaran, S. and R.H. Smith, 1992. Somatic embryogenesis from shoot tip and immature inflorescence of *Phoenix dactylifera* cv Barhee. Plant Cell Rep. 12: 22–25.

Blake, J., 1983. Tissue culture propagation of coconut, date and oil palm. In: J.H. Dodds (Ed.), Tissue Culture of Trees, pp. 29–50. Croom Helm, London.

Blake, J., 1990. Coconut *Cocos nucifera* (L.) Micropropagation. In: Y.P.S. Bajaj (Ed.), Biotechnology in Agriculture and Forestry, Vol. 10. Legumes and Oil Seed Crops, pp. 538–552. Springer-Verlag, Berlin.

Blake, J. and C.J. Eeuwens, 1982. Culture of coconut palm tissues with a view to vegetative propagation. Symp. Tissue Culture Economically Important Plants, Singapore, April 1981, pp. 145–148.

Boulay, M., 1987. *In vitro* propagation of tree species. In: C.E. Green, D.A. Somers, W.P. Hacket and D.D. Biesboer (Eds.), Plant Tissue and Cell Culture, pp. 367–382. Alan R. Liss Inc., New York.

Brackpool, A.L., R.L. Branton and J. Blake, 1986. Regeneration in palms. In: I.K. Vasil (Ed.), Cell Culture and Somatic Cell Genetics of Plants. Vol. 3, pp. 207–222. Academic Press, London.

Branton, R.L. and J. Blake, 1983a. A lovely clone of coconut. New Scientist 98: 554–557.

Branton, R.L. and J. Blake, 1983b. Development of organized structures in callus derived from explants of *Cocos nucifera* (L.). Ann. Bot. 52: 673–678.

Branton, R.L. and J. Blake, 1989. Date palm, *Phoenix dactylifera* (L.). In: Y.P.S. Bajaj (Ed.), Biotechnology in Agriculture and Forestry, Vol. 5. Trees II, pp. 161–175. Springer-Verlag, Berlin.

Buffard-Morel, J., J.L. Verdeil and C. Pannetier, 1988. Vegetative propagation of coconut palm through somatic embryogenesis, obtention of plantlet from leaf explant. In: Book of Abstracts 8th Intl. Symp. Biotechnology, Paris, p. 177.

Cai, Q.G., Y.Q. Qain, S.Q. Ke and Z.C. He, 1991. Preliminary analysis of regenerated plants derived from protoplasts of *Actinidia deliciosa*. Physiol. Plant 82: A12.

Carron, M.P. and F. Enjalric, 1982. Studies on vegetative micropropagation of *Hevea brasiliensis* by somatic embryogenesis and *in vitro* microcutting. In: A. Fujiwara (Ed.), Plant Tissue Culture 1982. Proc. 5th Int. Cong. Plant Tissue Cell Cult., Tokyo, pp. 751–752.

Chalupa, V., 1990. Plant regeneration by somatic embryogenesis from cultured immature embryos of Oak (*Quercus robur* L.) and Linden (*Tilia cordata* Mill). Plant Cell Rep. 9: 398–401.

Cheema, G.S., 1989. Somatic embryogenesis and plantlet regeneration from cell suspension and tissue culture of mature himalayan poplar (*Populus ciliata*). Plant Cell Rep. 8: 124–127.

Chen, Z., 1985. A study on induction of plants from *Citrus* pollen. Fruit Varieties J. 39: 44–50.

Chen, Z., 1987. Induction of androgenesis in hardwood trees. In: M.R. Ahuja (Ed.), Somatic Cell Genetics of Woody Plants, pp. 39–44. Kluwer Academic Publishers, Dordrecht.

Chen, Z., M.Q. Wang and H.H. Liäo, 1980. The induction of citrus pollen plants in artificial media. Acta Genetica Sinica 7: 189–191.

Chen, Z., Y. Yao and L. Zhang, 1988. Studies on embryogenesis of woody plants in China. In: M.R. Ahuja (Ed.), Somatic Cell Genetics of Woody Plants, pp. 19–25. Kluwer Academic Publishers, Dordrecht.

Chen, Z., W. Li, L. Zhang, X. Xu and S. Zhang, 1988. Production of haploid plantlets in cultures of unpollinated ovules of *Hevea brasiliensis* Muell.-Arg. In: M.R. Ahuja (Ed.), Somatic Cell Genetics of Woody Plants, pp. 39–44. Kluwer Academic Publishers, Dordrecht.

Chen, Z., F. Chen, C. Chein, C. Wang, S. Chang, H. Hsu, Y. Ho and T. Lu, 1978. A process of obtaining pollen plants of *Hevea brasiliensis* Muell-Arg. Sci. Sinica 22: 81–90.

Corley, R.H.V., J.N. Barrett and L.H. Jones, 1976. Vegetative propagation of oil palm via tissue culture. Oil Palm News. 22: 2–7.

Cornu, D., 1988. Somatic embryos in tissue cultures of walnuts (*Juglans nigra*, J. Major and hybrids *J. nigra* × *J. regia*). In: M.R. Ahuja (Ed.), Somatic Cell Genetics of Woody Plants, pp. 45–49. Kluwer Academic Publishers, Dordrecht.

Cruz, G.S, J.M. Canhato and M.A.V. Abreu, 1990. Somatic embryogenesis and plant generation from zygotic embryos of *Feijoa sellowiana* Berg. Plant Sci. 66: 263–270.

Cutter, V.M. and K.S. Wilson, 1954. Effect of coconut endosperm and other growth stimulants upon the development *in vitro* of embryos of *Cocos nucifera*. Bot. Gaz. (Chicago) 115: 234–239.

Datta, S.K., 1987. Tissue culture propagation of forest trees – limitations and perspectives. In: P.K. Khosla and D.K. Khurana (Eds.), Agroforestry for Rural Needs, Vol. 1, Indian Society of Tree Scientists, 1987, pp. 234–241.

Davis, T.A., 1969a. Prospects of clonal propagation of coconut. Ceylon Coconut Plant Rev. 6: 1–5.

Davis, T.A., 1969b. Clonal Propagation of the coconut. World Crops 21: 253–255.

Del Rosario, A.G. and E.V. De Guzman, 1976. The growth of coconut "Makapuno" embryos *in vitro* as affected by mineral composition and sugar level of the medium during the liquid and solid cultures. Phillip. J. Sci. 105: 215–222.

Del Rosario, A.G. and E.V. De Guzman, 1981. The status of plant tissue culture in Philippines. In: A.N. Rao (Ed.), Tissue Culture of Economically Important Plants. Proc. Int. Symp. Singapore, COSTED, ANBS, pp. 292–294.

De March, G., E. Grenier, N. Miannay, G. Sulmont, H. David and A. David, 1993. Potential of somatic embryogenesis in *Prunus avium* immatuare zygotic embryos. Plant Cell Tiss. Org. Cult. 34: 209–215.

Desai, H.V., P.N. Bhatt and A.R. Mehta, 1986. Plant regeneration of *Sapindus trifoliatus* L. (soapnut) through somatic embryogenesis. Plant Cell Rep. 3: 190–192.

De Touchet, B., Y. Duval and C. Pannetier, 1991. Plant regeneration from embryogenic suspension cultures of oil palm (*Elaeis guineensis* Jacq.) Plant Cell Rep. 10: 529–533.

Dewald, S.G. and R.E. Litz, 1987. Somatic embryo in mango *Mangifera indica* (L.). Hort. Sci. 22: 111–117.

Dorion, N., P. Danthu, S. Ohki, C. Preneux, B. Godin and C. Bigot, 1991. Plant regeneration from leaf protoplast of common elm (*Ulmus campestris* Mill.). C.R. Acad. Sci. 313: 467–473.

Druart, P., 1980. Plantlet regeneration from root callus of different *Prunus* sp. Sci. Hort. 12: 339–342.

Druart, P., 1981. Embryogenese somatiques er obtention de plantules chez *Prunus incisa* × *serrula* (GM9) cultive *in vitro*. Bulletin des Recherches Agronomiques de Gembloux 16: 205–220.

Druart, P., 1990. Improvement of somatic embryogenesis of cherry dwarf root stock Inmil/GM9 by the use of different carbon sources. Acta Hort. 280: 125–127.

Dublin, P., F. Enjalric, L. Lardet, M.P. Carren, N. Trolinder and C. Pannetier, 1991. Estate crops. In: P.C. Debergh and R.H. Zimmerman (Eds.), Micropropagation Technology and Application, pp. 337–362. Kluwer Academic Publishers, Netherlands.

Durand-Gasselin, T., V. Le Guen, K. Konan and Y. Duval, 1989. Oil palm *Elaeis guineensis* (Jacq.) plantations in Cote d'Ivoire, obtained through *in vitro* culture, first results. In: Proc. Intl. Conf. on Palms and Palms Products, NIFOR, Benin.

Ebert, A. and H.F. Taylor, 1990. Assessment of the changes of 2,4–dichlorophenoxyacetic acid concentrations in plant tissue culture media in the presence of activated charcoal. Plant Cell Tiss. Org. Cult. 20: 165–172.

Ebert, A., F. Taylor and J. Blake, 1993. Changes of 6–benzylaminopurine and 2,4–dichlorophenoxyacetic acid concentrations in plant tissue culture media in the presence of activated charcoal. Plant Cell Tiss. Org. Cult. 33: 157–162.

Eeuwens, C.J., 1976. Mineral requirements for growth and callus initiation of tissue explants excised from mature coconut palms *Cocos nucifera* and cultures *in vitro*. Physiol. Plant 36: 23–28.

Eeuwens, C.J., 1978. Effects of organic nutrients and hormones on growth and developments of tissue explants from coconut *Cocos nucifera* (L.) and date *Phoenix dactylifera* palms cultured *in vitro*. Physiol. Plant 43: 173–178.

Eeuwens, C.J. and J. Blake, 1977. Culture of coconut and date palm tissues with a view to vegetative propagation. Acta Hort. 78: 277–286.

Eichholtz, D.A., H.A. Robitaille and P.M. Hasegawa, 1980. Adventive embryony in apple. Hort. Sci. 14: 699–700.

Engelmann, F. and Y. Duval, 1986. Cryopreservation d'embryons somatiques de palmier a huile (*Elaeis guineensis* Jacq.) Resultats et perspectives d'application. Oleagineux 41: 169–174.

Fu, L.F. and D. Tang, 1983. Induction of pollen plants of litchi tree (*Litchi chinensis* Sonn.). Acta Genet. Sin. 10(5): 369–374.

Fulford, R.M., S.H.F.W. Justin and A.J. Passey, 1977. Vegetative propagation of coconuts. Rep. E. Malling Res. Stn. for 1976, pp. 81.

Galun, E., D. Aviv , D. Raveh, A. Vardi and A. Zelcher, 1977. Protoplasts in studies of cell genetics and morphogenesis. In: E. Reinhard and A.W. Alfermann (Eds.), Proceddings in Life Science, pp. 301–312. Springer-Verlag, Berlin.

Gharyal, P.K. and S.C. Maheshwari, 1981. *In vitro* differentiation of somatic embryoids in a leguminous tree *Albizzia lebbeck* L. Naturwissenschaften. 67: 379.

Gharyal, P.K., A. Rashid and S.C. Maheshwari, 1983. Production of haploid plantlets in anther culture of *Albizzia lebbeck*. Plant Cell Rep. 2: 308–309.

Gharyal, P.K., A. Rashid and S.C. Maheshwari, 1983. *In vitro* differentiation of *Albizzia lebbeck*. Protoplasma 11: 8–9.

Gingas, V.M. and R.D. Lineberger, 1989. Asexual embryogenesis and plant regeneration in *Quercus*. Plant Cell Tiss. Org. Cult. 17: 191–203.

Gmitter, F.G. Jr., X.B. Ling and X.X. Deng, 1990. Induction of tripliod citrus plants from endosperm calli *in vitro*. Theor. Appl. Genet. 80: 785–790.

Gmitter, F.G. Jr. and G.A. Moore, 1986. Plant regeneration from undeveloped ovules and embryogenic calli of citrus, embryo production, germination and plant survival. Plant Cell Tiss. Org. Cult. 6: 139–147.

Gonzalez, M.L., A.M. Vieitex and E. Vieitex, 1985. Somatic embryogenesis from Chestnut cotyledon tissue cutlured *in vitro*. Sci. Hort. 27: 97–103.

Grosser, J.W. Jr., F.G. Gmitter and J.L. Chandler, 1988a. Intergeneric somatic hybrid plants of *Citrus sinensis* cv. Hamlin and *Poncirus trifoliata* cv. Flying Dragon. Plant Cell Rep. 7: 5–8.

Grosser, J.W. Jr., F.G. Gmitter and J.L. Chandler, 1988b. Intergeneric somatic hybrid plants from sexually incompatible woody species : *Citrus sinesis* and *Severinia disticha*. Theor. Appl. Genet. 75: 397–401.

Guerro, M.P. and W. Handro, 1988. Somatic embryogenesis and plant regeneration in embryo cultures of *Euterpe edutis* mart. (palmae). Plant Cell Rep. 7: 550–552.

Gupta, P.K. and M. Kreitinger, 1993. Synthetic seeds in forest trees. In: M.R. Ahuja (Ed.), Micropropagation of Woody Plants, pp. 107–120. Kluwer Academic Publishers, Dordrecht.

Gupta, P.K., S.V. Kendurkar, V.M..Kulkarni, M.V. Shirgurkar and A.F. Mascarenhas, 1984. Somatic embryogenesis and plants from zygotic embryos of coconut *Cocos nucifera* (L.). Plant Cell Rep. 3: 222–225.

Hadrami, E.L.I., M.P. Carron and J. D'Auzac, 1991. Influence of exogeneous hormones on somatic embryogenesis in *Hevea brasiliensis*. Ann. Bot. 67: 511–515.

Hidaka, T. and I. Kajiura, 1988. Plantlet differentiation from callus protoplasts induced from *Citrus* embryo. Sci. Hort. 34: 85–92.

Ho, R.H. and Y. Ray, 1985. Haploid plant production through anther culture in poplars. For. Ecol. Mgmt. 13: 133–142.

Hu, C.Y. and I.M. Sussex, 1971. *In vitro* development of embryoids on cotyledons of *Ilex aquifolium*. Phytomorphology 21: 103–107.

Hu, C.Y., J.D. Ochs and F.M. Mancim, 1978. Further observations on *Ilex* embryoid production. Z. Pflanzenphysiol. 89: 41–49.

Huang, Z., Y. Huangfu and L. Xu, 1982. Triploid plant from endosperm culture of *Actinidia*. Kexue Tongbao 27: 247–250.

Huetteman, C.A. and J.E. Preece, 1993. Thidiazuron: A potent cytokinin for woody plant tissue culture. Plant Cell Tiss. Org. Cult. 33: 105–119.

Iyer, R.D., E.V.V. Bhaskara Rao and M.P. Govindankutty, 1979. Super yielders in coconut. Indian farming 28: 3–5.

Jaiswal, V.S., 1990. Somatic embryogenesis in nucellus of some Indian cultivars of mango. In: Abstracts of the International Seminar on New Frontiers in Horticulture. Bangalore, p. 33.

Jana, M.M., R.S. Nadgauda and A.F. Mascarenhas, 1993. *In vitro* propagation of monoembryonic varieties of Mango via somatic embryogenesis. In Vitro Cell and Devl. Biol. 30: 155–157.

Janick, J., 1982. Adventive embryony in pear. Acta Hort. 124: 37–39.

Jesty, J.H.F. and D. Francis, 1992. Cellular responses of leaf explants of *Cocos nucifera* (L.) *in vitro*. Plant Cell Tiss. Org. Cult. 28: 235–240.

Jones, O.P., A. Gayner and R. Watkins, 1984. Plant regeneration from callus cultures of the cherry root stock colt (*Prunus avium* × *P. pseudoccerasus*) and the apple rootstock M 25 (*Malus pumila*) J. Hort. Sci. 59: 463–467.

Jones, L.H., 1988. Commercial development of oil palm clones. Fat Sci. Technol. 2: 58–61.

Jones, L.H. and W.A. Hughes, 1989. Oil Palm, *Elaeis guineensis* (Jacq.). In: Y.P.S. Bajaj (Ed.), Biotechnology in Agriculture and Forestry, Vol. 5. Trees II, pp. 176–202. Springer-Verlag, Berlin.

Jordan, M., 1975. *In vitro* Kultur von *Prunus*, *Pyrus* and *Ribes* Antheren. Planta Med. (Suppl.): 59–65.

Jorgenson, J., 1988. Embryogenesis in *Quercus petraea* and *Fagus sylvatica*. J. Plant Physiol. 132: 638–640.

Kanchanapoom, M. and Kanchanapoom, K. 1991. Isolation, culture and regeneration of *Theobroma* protoplasts. Physiol. Plant. 82: A14.

Karunaratne, S. and K. Periyapperuma, 1989. Culture of immature embryos of coconut, *Cocos nucifera* (L.): callus proliferation and somatic embryogenesis. Plant Sci. 62: 247–253.

Kaufman, T.G., 1965. Coconut oil derivatives in cosmetics. Drug Cosmetic Ind. 97: 172–173.

Kendurkar, S.V., R.S. Nadgauda, C.H. Phadke, S.V. Shirke, B.S. Abhyankar and A.F. Mascarenhas, 1989. Tissue Culture of Coconut. Workshop on "Problems and Prospects of Tissue Culture and Related aspects of Oil palm and Coconut" at Bhabha Atomic Research Centre, Bombay, India.

Kitamura, Y., T. Morkawa and H. Miura, 1989. Isolation and culture of protoplasts from cell suspension cultures of *Duboisia myoporoides* with subsequent plant regeneration. Plant Sci. 60: 245–250.

Kobayashi, S., H. Uchimiya and I. Ikeda, 1983. Plant regeneration from "Trovita" orange protoplasts. Japan J. Breed. 33: 119–122.

Kobayashi, S., I. Ikeda and H. Uchimiya, 1985. Conditions for high frequency embryogenesis from orange (*Citrus sinensis* Osb.) Protoplasts. Plant Cell Tiss. Org. Cult. 4: 175–202.

Kobayashi, S., T. Ohgawara, E. Ohgawara, I. Oiyama and S.Ishii, 1988. A somatic hybrid plant obtained proplast fusion between naval orange (*Citrus sinensis*) and Satsuma manderin (*Citrus unshiu*). Plant Cell Tiss. Org. Cult. 14: 63–69.

Kochba, J., P. Spiegel-Roy and H. Safran, 1972. Adventive plants from ovules and nucelli in citrus. Planta 106: 237–245.

Kochba, J., P. Spiegel Roy, H. Neumann and S. Saad, 1982. Effect of carbohydrates on somatic embryogenesis in subcultured nucellar callus of citrus cultivars. Z. Pflanzenphysiol. 105: 359–368.

Konar, R.N. and Y.P. Oberoi, 1965. *In vitro* development of embryoids on the cotyledons of *Biota orientalis*. Phytomorphology 15: 137–140.

Krikorian, A.D. and R.P. Kann, 1986. Oil palm improvement via tissue culture. Plant Breeding Rev. 4: 175–202.

Kunitake, H., H. Kagami and M. Mii, 1991. Somatic embryogenesis and plant regeneration from protoplasts of Satsuma mandarin (*Citrus unshiu* Marc.). Sci. Hort. 47: 27–33.

Laxmisita, G., N.V. Raghava Ram and C.S. Vaidyanathan, 1979. Differentiation of embryoids and plantlets from shoot culture of sandalwood. Plant Sci. Lett. 15: 265–270.

Laxmisita, G., N.V. Raghava Ram and C.S. Vaidyanathan, 1980. Triploid plants from endosperm cultures of sandalwood experimental embryogenesis. Plant Sci. Lett. 20: 63–69.

Lee, J.S., S.K. Lee, S.S. Jang and J.J. Lee, 1987. Plantlet regeneration from callus protoplasts of *Populus nigra*. Inst. For. Gen. Res. Rep. Korea 23: 143–148.

Lespinasse, Y., M. Godecheau and M. Duron, 1963. Potential value and method of producing haploids in the apple tree *Malus pumila* (Mill.). Acta Hort. 131: 223–230.

Li, X.H., 1991. Advance in research of protoplast regeneration of major crops and economic plants. Physiol. Plant. 82: A7.

Ling, J.T., N. Nito and M. Iwamasa, 1989. Plant regeneration from protoplasts of Calamondin (*Citrus madurensis* Lour.). Sci. Hort. 40: 325–333.

Ling, J.T., N. Nito, M. Iwamasa and H. Kunitake, 1990. Plant regeneration from protoplast isolated from embryogenic callus of Satsuma. Hort. Sci. 25: 970–972.

Litz, R.E., 1984a. *In vitro* somatic embryogenesis from jaboticaba *Myrciaria cauliflora* (D.C. Berg.) Callus. Hort. Sci. 1962–1964.

Litz, R.E., 1984b. *In vitro* somatic embryogenesis from nucellar callus of monoembryonic *Mangifera indica* (L.). Hort. Sci. 19: 715–717.

Litz, R.E., 1985. Somatic embryogenesis in tropical fruit trees. In: R.R. Henke, K.W. Hughes, M.P. Constantin and A. Hollaender (Eds.), Tissue Culture in Forestry and Agriculture, pp. 179–193. Plenum, New York.

Litz, R.E., 1988. Somatic embryogenesis from cultured leaf explants of the tropical tree *Euphoria lougan* Stend. J. Plant Physiol. 132: 190–193.

Litz, R.E. and R. Knight, 1983. *In vitro* somatic embryo from nucellar callus of mango. Hort. Sci. 18: 618.

Litz, R.E. and V.S. Jaiswal, 1991. Micropropagation of tropical and subtropical fruits In: P.C. Debergh and R.H. Zimmerman (Eds.), Micropropagation Technology and Application, pp. 247–263. Kluwer Academic Publishers, Dordrecht.

Litz, R.E. and D.J. Gray, 1992. Organogenesis and somatic embryogenesis. In: F.A. Hammerschlag and Litz R.E. (Eds.), Biotechnology of Perrennial Fruit Crops. CAB International, The Netherlands, pp. 3–34.

Litz, R.E., R.J. Knight and S. Gazit, 1982. Somatic embryos from cultured ovules of polyembryonic *Mangifera indica* (L.). Plant Cell Rep. 1: 264–266.

Litz, R.E., R.J. Knight and S. Gazit, 1984. *In vitro* somatic embryogenesis from *Mangifera indica* (L.) callus. Sci. Hort. 22: 233–240.

Lu, Z. and Y. Liu, 1990. Poplar anther culture. In: Z. Chen, D.A. Evans, W.R. Sharp, P.V. Ammirato and M.R. Sondahl (Eds.), Handbook of Plant Cell Culture, Vol. 6. Perennial Crops, pp. 161–190. McGraw-Hill Publishing Company, New York.

Majumdar, P.K. and D.K. Sharma, 1990. Mango. In: T.K. Bose and S.K. Mitra (Eds.), Fruits: Tropical and Subtropical, pp. 1–62. Naya Prakash 206 Bidhan Sarain, Calcutta.

Mante, S., R. Scorza and J.M. Cordts, 1989. Plant regeneration from cotyledons of *Prunus persica*, *Prunus domestica* and *Prunus cerasus*. Plant Cell Tiss. Org. Cult. 19: 1–11.

Mante, S., P.H. Morgens, R. Scorza, J.M. Cordts and A.M. Callahan, 1991. *Agrobacterium* mediated transformation of plum (*Prunus domestica* L.) hypocotyl slices and regeneration of transgenic plants. Bio/Technology. 9: 853–857.

Manzanera, J.A., R. Astorga and M.A. Bueno, 1993. Somatic embryo induction and germination in *Quercus ruber* L. Silvae Genet. 42: 90–92.

Marin, M.L. and N. Duran Vila, 1988. Survial of somatic embryos and recovery of plants of sweet orange (*Citrus sinensis* L.) after immersion in liquid nitrogen. Plant Cell Tiss. Org. Cult. 14: 51–57.

Marin, M.L., Y. Gogorcena, J. Ortiz and N. Duran-Vila, 1993. Recovery of whole plants of sweet orange from somatic embryos subjected to freezing thawing treatments. Plant Cell Tiss. Org. Cult. 34: 27–33.

Marques, D.V., 1993. Induction of somatic embryogenesis in *Coffea eugenioides* Moore by *in vitro* culture of leaf explants. Cafe Cacao 37: 251–255.

Mascarenhas, A.F., R.S. Nadgauda, S.S. Khuspe, P.K. Gupta, B.M. Khan and E.M. Muralidharan, 1989. Biotechnological application of plant tissue culture to forestry species in India. In: V. Dhavan and R. Mott (Eds.), Application of Biotechnology in Forestry and Horticulture. Plenum Press, New York (in press).

Mathews H., R.E. Litz, H.D. Wilde, S.A. Markle and H.Y. Wetzstein, 1992. Stable integration and expression of β-glucuronidase and NPT-II genes in mango somatic embryos. In Vitro Cell. Dev. Biol. 28: 172–178.

McGranahan, G., C. Leslie, S. Uratsu, L. Martin and A. Dandekar, 1988. *Agrobacterium* mediated transformation of walnut somatic embryos and regeneration of transgenic plants. Bio/Technology 6: 800–804.

Mehta, U.I., V.R. Rao and H.Y. Mohan Ram, 1982. Somatic embryogenesis in Bamboo. In: A. Fujiwara (Ed.), Proceedings of the Vth IAPTC Congress on Plant Tissue Culture, Tokyo, pp. 109–110.

Mehra, P.N. and S. Sachdeva, 1984. Embryogenesis in apple *in vitro*. Phytomorphology 34: 26–35.

Merkle, S.A. and H.E. Sommer, 1986. Somatic embryos in tissue cultures of *Litriodendron tulipifera*. Can. J. For. Res. 16: 420–422.

Merkle, S.A. and H.E. Sommer, 1987. Regeneration of *Liriodendron tulipifera* (family Magnoliaceae) from protoplast culture. Amer. J. Bot. 74: 1317–1321.

Merkle, S.A. and A.T. Wiecko, 1989. Regeneration of *Robinia pseudoacacea* via somatic embryogenesis Can. J. For. Res. 19: 285–288.

Merkle, S.A., A.T. Wiecko, R.J. Sotak and H.E. Sommer, 1990. Maturation and conversion of *Liriodendron tulipifera* somatic embryos. In Vitro Cell. Dev. Biol. 26: 1086–1093.

Merkle, S.A., M.T. Hoey, B.A. Watson, S.E. Pauley and S.E. Schlasbaum, 1993. Propagation of *Liriodendron* hybrids via somatic embryogenesis. Plant Cell Tiss. Org. Cult. 34: 191–198.

Mitra, G.C. and H.C. Chaturvedi, 1972. Embryoids and complete plants from unpollinated ovaries and from ovules of *in vitro* grown emasculated flower buds of *Citrus* sp. Bull. Torrey Bot. Club 99: 184–189.

Monfort, S., 1985. Androgenesis of coconut embryos from anther culture. Z. Pflanzenzücht. 94: 251–254.

Montoro, P., H. Etienne, N. Michaux-Ferriere and M.P. Carron, 1993. Callus friability and somatic embryogenesis in *Hevea brasiliensis*. Plant Cell Tiss. Org. Cult. 3: 331–339.

Mooney, P.A. and J. Van Staden, 1987. Induction of embryogenesis in callus from immature embryos of *Persea americana* Mill. Can. J. Bot. 65: 622–626.

Muralidharan, E.M. and A.F. Mascarenhas, 1987. *In vitro* plantlet formation by organogenesis in *Eucalyptus camaldulensis* and by somatic embryogenesis in *Eucalyptus citriodora*. Plant Cell Rep. 6: 256–259.

Muralidharan, E.M., P.K. Gupta and A.F. Mascarenhas, 1989. Plantlet production through high frequency somatic embryogenesis in long term cultures of *Eucalyptus citriodora*. Plant Cell Rep. 8: 41–43.

Murashige, T. and F. Skoog, 1962. A revised medium for rapid growth and bio-assays with tobacco tissue cultures. Physiol. Plant 15: 473–497.

Nair, S., P.K. Gupta and A.F. Mascarenhas, 1983. Haploid plants from *in vitro* anther culture of *Annona squamosa* L. Plant Cell Rep. 2: 198–200.

Navarro, L., J.M. Ortiz and J. Juarez, 1985. Aberrant citrus plants obtained by somatic embryogeneis of nucelli cultured *in vitro*. Hort. Sci. 20: 214–215.

Neuenschwander, B. and T. Baumann, 1991. A novel type of somatic emrbyogenesis in *Coffea arabica*. Plant Cell Rep. 10: 608–612.

Neuman, M.C., J.E. Preece, J.W. Sambeek and G.R. Gaffiney, 1993. Somatic embryogenesis and callus production from cotyledon explants of eastern black walnut (*Juglans nigra* L.). Plant Cell Tiss. Org. Cult. 32: 9–18.

Niedz, R.P., 1993. Culturing embryogenic protoplasts of Hamlin Sweet Orange in calcium alginate beads. Plant Cell Tiss. Org. Cult. 34: 19–25.

Ochatt, S.J., 1990. Plant regeneration from root callus protoplast of sour cherry (*Prunus cerasus* L.). Plant Cell Rep. 9: 268–271.

Ochatt, S.J., 1991. Strategies for plant regeneration from mesophyll protoplasts of the recalcitrant fruit and farmwoodland species *Prunus avium* L. (sweet/wild cherry), Rosaceae. J. Plant Physiol. 139: 155–160.

Ochatt, S.J., 1992. The development of protoplast-to-tree systems for *Prunus cerasifera* and *Prunus spinosa*. Plant Sci. 81: 253–259.

Ochatt, S.J. and J.B. Power, 1988. An alternative approach to plant regeneration from protoplasts of sour cherry (*Prunus cerasus* L.). Plant Sci. 56: 75–79.

Ochatt, S.J., E.C. Cocking and J.B. Power, 1987. Isolation, culture and plant regeneration of Colt cherry (*Prunus avium* × *pseudocerasus*) protoplasts. Plant Sci. 50: 139–143.

Ochatt, S.J., P.K. Chand, E.L. Rech, M.R. Davey and J.B. Power, 1988a. Electroporation mediated improvement of plant regeneration from Colt cherry (*Prunus avium* × *pseudocerasus*) protoplast. Plant Sci. 54: 165–169.

Ohgawara, T., S. Kobayashi, E. Oghawara, H. Uchimiya and S. Ishii, 1985. Somatic hybrid plants obtained by protoplast fusion between *Citrus sinensis* and *Poncirus trifoliata*. Theor. Appl. Genet. 71: 1–4.

Oka, S. and K. Ohyama, 1985. Plant regeneration from leaf mesophyll protoplasts of *Broussonetia kasinoki* Sieb. (Paper Mulberry). J. Plant Physiol. 119: 455–460.

Pal, A., A. Banerjee and K. Dhar, 1985. *In vitro* organogenesis and somatic embryogenesis from leaf explants of *Leucosceptrum canum* Sm. Plant Cell Rep. 4: 281.

Pannetier, C. and J. Buffard-Morel, 1982. First results of somatic embryo production from leaf tissue of coconut, *Cocos nucifera* (L.). Oleagineux 37: 352–353.

Pannetier, C. and J. Buffard-Morel, 1986. Coconut Palm *Cocos nucifera* (L.). In: Y.P.S. Bajaj (Ed.), Biotechnology in Agriculture and Forestry, Vol. 1. Trees-II, pp. 430–450. Springer-Verlag, Berlin.

Paranjothy, K. and R. Rothman, 1978. Embryoid and plantlet development from cell of Hevea. In: T.A. Thorpe (Ed.), Abstr. 4th Int. Cong. Plant Tissue and Cell Cult., Calgony, pp. 42.

Park, Y.G. and S.H. Son, 1988. *In vitro* organogenesis and somatic embryogenesis from punctured leaf of *Populus nigra* × *P. maximowiczii*. Plant Cell Tiss. Org. Cult. 15: 95–105.

Patat-Ochatt, E.M., S.J. Ochatt and J.B. Power, 1988. Plant regeneration from protoplasts of apple rootstocks and scion varieties (*Malus* × *domestica* Borkh.). J. Plant Physiol. 133: 460–465.

Pliego-Alfaro, F. and T. Murashige, 1988. Somatic embryogenesis in avocado (*Persea amerina* Mill.) *in vitro*. Plant Cell Tiss. Org. Cult. 12: 61–66.

Polito, V.S., G. McGranahan, K. Pinney and C. Leslie, 1989. Origin of somatic embryo from repetitively embryogenic cultures of walnut (*Juglans regia* L.) : Implications of *Agrobacterium* mediated transformation. Plant Cell Rep. 8: 219–221.

Power, J.B., 1988. Electroporation-mediated improvement of plant regeneration from Colt cherry (*Prunus avium* × *pseudocerasus*) protoplasts. Plant Sci. 54: 165–169.

Preece, J.E., J.L. Zhao and F.H. Kung, 1987. *In vitro* callus production and somatic embryogenesis of ash, *Fraxinus* sp. Hort. Sci. 22: 1131.

Preece, J.E., J.L. Zhao and F.H. Kung, 1989. Callus production and somatic embryogenesis from white ash. Hort. Sci. 24: 377–380.

Rabechault, H. and J.P. Martin, 1976. Multiplication végétative du palmier a huile (*Elaeis guineensis* Jacq.) a l'aide de cultures de tissue foliares. C.R. Acad. Sci. Paris Sér. 283: 1735–1737.

Radojevic, L., 1978. *In vitro* induction of androgenic plantlets in *Aesculus hippocastanum*. Protoplasma 96: 369–374.

Radojevic, L., 1979. Somatic embryogenesis and plantlets from callus culture of *Paulownia tomentosa*. Z. Pflanzenphysiol. 91: 57–62.

Radojevic, L., 1980. Haploid embryos plantlets and callus formation in woody species. In: D.R. Davis and D.A. Hopwood (Eds.), The Plant Genome, p. 259.

Radojevic, L., 1988. Plant regeneration of *Aesculus hippocastanum* L. (horse chestnut) through somatic embryogenesis. J. Plant Physiol. 132: 322–326.

Raj Bhansali R., A. Driver and D.J. Durzan, 1990. Rapid multiplication of adventitious somatic embryos in Peach and Nectarine by secondary embryogenesis. Plant Cell Rep. 9: 280–284.

Raju, C.R., P. Prakash Kumar, M. Chandramohan and R.D. Iyer, 1984. Coconut plantlets from leaf tissue cultures. J. Plant Crops 12: 75–91.

Rangan, T.S., T. Murashige and W.P. Bitters, 1968. *In vitro* initiation of nucellar embryos in monoembryonic *Citrus*. Hort. Sci. 3: 226–227.

Rao, P.S., 1965. *In vitro* induction of embryonal proliferation in *Santalum album* (L.). Phytomorphology 15: 175–179.

Rao, P.S. and P. Ozias-Akins, 1985. Plant regeneration through somatic embryogenesis in protoplast cultures of sandalwood (*Santalum album* L.). Protoplasma 124: 80–86.

Rao, P.S. and T.R. Ganapathy, 1993. Micropropagation of palms. In: M.R. Ahüja (Ed.), Micropropagation of Woody Plants, pp. 395–421. Kluwer Academic Publishers, Dordrecht.

Rao, I.O., I.V.R. Rao and V. Narang, 1985. Somatic embryogenesis and regeneration of plants in the bamboo *Dendrocamamus strictus*. Plant Cell Rep. 4: 191–194.

Reinert, J., 1958. Morphogenese und ihre kontrolle an Gewebekultures aus carotten. Naturwissenschaften 45: 344–345.

Reynolds, I.F. and T. Murashige, 1979. Asexual embryogenesis in callus cultures of palms. In Vitro Cell. Dev. Biol. 15: 383-387.

Rillo, E.P. and M.B.F. Paloma, 1990. Comparison of 3 media formulations for *in vitro* culture of coconut embryos. Oleagineux 45: 319–323.

Rillo, E.P. and M.B.F. Paloma, 1991. Storage and transport of zygotic embryos of *Cocos nucifera* L. for *in vitro* couture. Plant Genet. Resour. Newsletter 86: 1–4.

Rillo, E.P. and M.B.F. Paloma, 1992. *In vitro* cultures of Makapuno coconut embryos. Coconuts Today 9: 90–108.

Rugini, E., 1988. Somatic embryogenesis and plant regeneration in olive (*Olea europaea* L.). Plant Cell Tiss. Org. Cult. 14: 207–214.

Russell, J.A. and B.H. McCown, 1986. Culture and regeneration of Populus leaf protoplasts isolated from non-seedling tissue. Plant Sci. 46: 133–142.

Russell, J.A., 1993. Advances in the protoplast culture of woody plants. In: M.R. Ahuja (Ed.), Micropropagation of Woody Plants, pp. 67–91. Kluwer Academic Publishers, Dordrecht.

Russell, J.A. and B.H. McCown, 1988. Recovery of plants from leaf protoplasts of hybrid-poplar and aspen clones. Plant Cell Rep. 7: 59–62.

Sahijram, L., 1990. Somatic embryogenesis in mango. In: Abstracts of the International Seminar on New Frontiers in Horticulture, Bangalore, p. 174.

Sasamoto, H. and Y. Hosoi, 1990. Effects of electric fusion on the calli formation and regeneration from the protoplasts of *Quercus* and two *Populus* species. In: Abstracts VIIth IAPTC Congress, Amsterdam, p. 36.

Schopke, C., L.E. Muller and H.W. Kohlenbach, 1987. Somatic embryo- genesis and regeneration of plantlets in protoplast cultures from somatic embryos of coffee (*Coffea canephora* P. ex Fr.). Plant Cell Tiss. Org. Cult. 8: 243–248.

Scorza, R., J.M. Cordts and S. Mante, 1990. Long term somatic embryo production and plant regeneration from embryo-derived peach callus. Acta Hort. 280: 183–190.

Sharma, D.R., R. Kumar and J.B. Chowdhary, 1980. *In vitro* culture of female date palm (*Phoenix dactylifera* L.). Euphytica 29: 169–174.

Sharma, D.R., S. Dawra and J.B. Chowdhary, 1984. Somatic embryogenesis and plant regeneration in date palm (*Phoenix dactylifera* Linn) V.C. Khadrawi through Tissue Culture. Indian J. Exp. Biol. 22: 596–598.

Sharp, W.R., M.R. Sondahl, L.S. Caldas and S.B. Maraffa, 1980. The physiology of *in vitro* asexual embryogenesis, Hort. Rev. 2: 268–310.

Shrikhande, M., S.R. Thengane and A.F. Mascarenhas, 1993. Somatic Embryogenesis and plant regeneration in *Azadirachta indica* L.A. Juss. In Vitro Cell. Dev. Biol. 29: 38–42.

Sim, G.E., C.S. Loh and C.H. Goh, 1988. Direct somatic embryogenesis from protoplasts of *Citrus mitis* Blanco. Plant Cell Rep. 7: 418–420.

Smith, B.G., 1986. Tissue culture of *Cocos nucifera* (L.) -biochemical changes preceding embryogenesis. In: E. Pushparajah and P.S. Chew (Eds.), Cocoa and Coconut; Progress and Outlook, pp. 781–786. Inc. Soc. Plant, Kuala Lumpur.

Sommer, H.E. and C.L. Brown, 1980. Embryogenesis in tissue cultures of Sweetgum. For. Sci. 26: 257–260.

Sondahl, M.R. and W.R. Sharp, 1977a. High frequency induction of somatic embryos in cultured leaf explants of *Coffea arabica* L. In Vitro 13: 146 (Abstr. 14).

Sondahl, M.R. and W.R. Sharp, 1977b. High frequency induction of somatic embryos in cultured leaf explants of *Coffea arabica* L. Z. Pflanzenphysiol. 81: 395–408.

Sondahl, M.R. and W.R. Sharp, 1979. Research in Coffea spp. and applications of tissue culture methods. In: W. Sharp, P.O. Larson, E.F. Paddock and V. Raghavan (Eds.), Plant Cell and Tissue Culture. Principles and Applications, pp. 527–584. Ohio State Univ., Columbus.

Sondahl, M.R. and J.A. Lauritis, 1992. Coffee. In: F.A. Hammerschlag and R.E. Litz (Eds.), Biotechnology of Perennial Fruit Crops. C.A.B. International, UK, pp. 401–420.

Sotak, R.J., H.E. Sommer and S.A. Merkle, 1991. Relation of the developmental stage of zygotic embryos of yellow poplar to their somatic embryogenesis potential. Plant Cell Rep. 10: 175–178.

Spiral, J. and V. Petiard, 1991. Protoplast culture and regeneration in coffee species. In: XIV International Conference of Coffee Science, San Francisco, pp. 383–391.

Staritsky, G., 1970. Embryoid formation in callus cultures of Coffee. Acta Bot. Neerl. 19: 509–514.

Steward, F.C., 1958. Growth and development of cultivated cells. III. Interpretations of the growth from free cell of carrot. Amer. J. Bot. 45: 709–713.

Sticklen, M.B., S.C. Domir and R.D. Lineberger, 1986. Shoot regeneration from protoplasts of Ulmus × Pioneer. Plant Sci. 47: 29–34.

Sudarsip, H., H. Kaat and A. Davis, 1978. Clonal propagation of the coconut via the bulbils. Phillip. J. Coconut Stud. 3: 5–14.

Sudhersan, C., M.M. Abo El-Nil and A. Al-Baiz, 1993. Occurence of direct somatic embryo on the sword leaf of *in vitro* plantlets of *Phoenix dactylifera* L. cultivar barhee. Current Sci. 65: 887–889.

Sugimura, Y. and M.J. Salvana, 1989. Induction and growth of callus derived from rachillae explants of young inflorescences of coconut palm. Can. J. Bot. 67: 272–274.

Teixeira, J.B., M.R. Sondanl and E.G. Kirby, 1993. Somatic embryogenesis from immature zygotic embryos of oil palm. Plant Cell Tiss. Org. Cult. 34: 227–233.

Thampan, P.K., 1982. Hand Book on Coconut Palm. Oxford and IBH Publishing, New Delhi.

Thangaraj, T. and S. Muthuswami, 1990. Coconut. In: T.K. Bose and S.K. Mitra (Eds.), Fruits: Tropical and Subtropical, pp. 336–385. Naya Prakash 206 Bidhan Sarain, Calcutta.

Than-Tuyen, N.T. and E.V. de Guzman, 1983. Formation of pollen embryos in cultured anthers of coconut *Cocos nucifera* (L.). Plant Sci. Lett. 29: 81–88.

Thorpe, T.A. and P.P. Kumar, 1993. Cellular control of morphogenesis. In: M.R. Ahuja (Ed.), Micropropagation of Woody Plants, pp. 11–29. Kluwer Academic Publishers, Dordrecht.

Thorpe, T.A., I.S. Harry and P.P. Kumar, 1991. Application of micropropagation to forestry. In: P.C. Debergh and R.H. Zimmerman (Eds.), Micropropagation Technology and Application, pp. 311–336. Kluwer Academic Publishers, Dordrecht.

Thomas, V. and P.S. Rao, 1985. *In vitro* propagation of oil palm (*Elaeis guineenis* Jacq. var. Tenera) through somatic embryogenesis in leaf derived callus. Curr. Sci. 54: 184–185.

Tisserat, B., 1982. Factors involved in the production of plantlets from date palm callus cultures. Euphytica 31: 201–214.

Tisserat, B., 1987. Palms. In: J.M. Bonga and D.J. Durzan (Eds), Cell and Tissue Culture in Forestry, Vol. 3, pp. 338–356. Martinus Nijhoff Publishers, Dordrecht.

Tisserat, B., J.N. Ulrich and B.J. Finkle, 1981. Cryogenic preservation and regeneration of date palm tissue. HortSci. 16: 47–48.

Tomar, U.K. and S.C. Gupta, 1988. Somatic embryogenesis and organogenesis in callus cultures of tree legume *Albizzia richardiana* King. Plant Cell Rep. 7: 70–73.

Tsai, C.K., 1988. Plant regeneration from leaf callus protoplasts of *Actinidia chinensis* Planch. var. Chinensis. Plant Sci. 54: 231–235.

Tsai, H.S., C.C. Yeh and J.Y. Hsu, 1990. Embryogenesis and plant regeneration from anther culture bamboo *Sinocalamus latiflora* McClure. Plant Cell Rep. 9: 349–351.

Tulecke, W., 1987. Somatic embryogenesis in woody perennials. In: J.M. Bonga and D.J. Durzan (Eds.), Cell and Tissue Culture in Forestry, Vol. 2. Specific Principles and Methods: Growth and Developments, pp. 61–91. Martinus Nijhoff Publishers, Dordrecht.

Tulecke, W. and G. McGranahan, 1985. Somatic embryogenesis and plantlet regeneration from cotyledons of Walnut *Juglans regia*. Plant Sci. 40: 57–63.

Uddin, M.R., M.M. Meyer Jr. and J.J. Jokela, 1988. Plantlet production from anthers of eastern cottonwood (*Populus deltoides*) Can. J. For. Res. 18: 937–941.

Ulrich, J.N., B.J. Finkle and B. Tisserat, 1982. Effects of cryogenic treatment on plantlet production from frozen and unfrozened date palm callus. Plant Physiol. 69: 624–627.

Valverde, R., D. Arias and T.A. Thorpe, 1987. Picloram-induced somatic embryogenesis in pejibaye palm (*Bactris gasipaes* H.B.K.). Plant Cell Tiss. Org. Cult. 10: 149–156.

Vardi, A., 1977. Isolation of protoplasts in *Citrus*. Proc. Int. Soc. Citriculture 2: 167–174.

Vardi, A., P. Speigel-Roy and E. Galun, 1982. Plant regeneration from *Citrus* protoplasts: Variability in methodological requirements among cultivars and species. Theor. Appl. Genet. 62: 171–176.

Vardi, A., D.J. Hutchinson and E. Galun, 1986. A protoplast-to-tree system in Microcitrus based on protoplasts derived from a sustained embryogenic callus. Plant Cell Rep. 5: 412–414.

Vardi, A., A. Breiman and E. Galun, 1987. Citrus cybrids: Production by donor-recipient protoplast-fusion and verification by mitochondrial-DNA restriction profiles. Theor. Appl. Genet. 75: 51–58.

Verdeil, J.L., J. Buffard-Morel and C. Pannetier, 1989. Embryogénèse somatique du Cocotier *Cocos nucifera* (L.) à partir de tissum foliaires et inflorescenciels-Bilan des recherches et perspectives. Oleagineux 44: 403–411.

Verdeil, J.L., C. Huet, F. Grosde manges, A. Rival and E.T.J. Buffard-Morel, 1992. Embryogénèse somatique du cocotier *Cocos nucifera* (L.) Obtention de plusieurs clones de *vitro* plants. Oleagineux 47: 465–470.

Vieitez, F.J., A. Ballester and A.M. Vieitez, 1992. Somatic embryogenesis and plantlet regeneration from cell suspension cultures of *Fagus sylvatica* L. Plant Cell Rep. 11: 609–614.

Wang, Q.Z., Z.X. Wang and X.H. Zhang, 1982. Initial success in the culture of monoploid plants from the unpollinated ovary of *Robinia pseudoacacia*. For. Abstr. 43: 845.

Wang, Z., X. Zeng, C. Chen, H. Wu, Q. Li, G. Fan and W. Lu, 1980. Induction of rubber plantlets from anther of *Hevea brasiliensis* Muell. Ang. *in vitro*. Chinese J. Trop. Crops 1: 16–26.

Woods, S.H., G.C. Phillips, J.E. Woods and G.B. Collins, 1992. Somatic embryogenesis and plant regeneration from zygotic embryo explants in Mexican weeping bamboo *Otatea accuminata aztecorum*. Plant Cell Rep. 11: 257–261.

Wooi, K.C., 1990. Oil palm *Elaeis guineensis* (Jacq) tissue culture and micropropagation. In: Y.P.S. Bajaj (Ed.), Biotechnology in Agriculture and Forestry, Vol. 10. Legumes and Oil Seed Crops I, pp. 569–592. Springer-Verlag, Berlin.

Wu, J., 1981. Obtaining haploid plantlets of crab apple from culture *in vitro*. Acta Hort. Sin. 8: 36.

Yang, Y.Q. and W.X. Wei, 1984. Induction of lougan plantlets from pollens cultured in certain proper media. Acta Genet. Sin. 11: 288–293.

Yasuda, T., M. Tahara, N. Uchida and F. Yamaguchi, 1986. Somatic embryogenesis from coffee callus and protoplast. In: D.A. Somers, B.G. Gengenbach, D.D. Biesboer, W.P. Hackett and C.E. Green (Eds.), Abstracts VIth IAPTC Congress, p. 137. Univ. of Minnesota, Minneapolis.

Yeh, M.L. and W.C. Chang, 1986a. Somatic embryogenesis and subsequent plant regeneration from inflorescence callus of *Bambusa beecheyana* Munro Var. *beecheyana*. Plant Cell Rep. 5: 409–411.

Yeh, M.L. and W.C. Chang, 1986b. Plant regeneration through somatic embryogenesis in callus cultures of green bamboo (*Bambusa oldhamii* Munro). Theor. Appl. Genet. 73: 161–163.

Yeh, M.L. and W.C. Cheng, 1987. Plant regeneration via somatic embryogenesis in mature derived callus cultures in sinocalamus. Plant Sci. 51: 93–96.

Zamarripa, A., J.P. Ducos, H. Tessereau, H. Bollon, A.B. Eskes and V. Petier, 1991. Dévelopment d'un procédé de multiplication en masse du cafeier par embryogénèse somatique en milieu liquide. In: XIV Intl. Conf. of Coffee Sci., San Fransisco, pp. 392–402.

Zenkteler, M., E. Misiura and A. Ponitka, 1975. Induction of androgenic embryoids *in vitro* cultured anthers of various species. Experentia 31: 289–291.

Zhao, H., 1983. Induction of endosperm plant and its ploidy in pear cultivar Jin Feng. Bull. Bot. 1: 38–39.

Zimmerman, R.H., 1991. Micropropagation of temperate zone fruit and nut crops. In: P.C. Debergh and R.H. Zimmerman (Eds.), Micropropagation Technology and Application, pp. 231–246. Kluwer Academic Publishers, Dordrecht.

5. Somatic Embryogenesis in Conifers

Pramod K. Gupta and James A. Grob

Contents

1. Introduction

Somatic embryogenesis is the development of embryos from somatic cells. This is accomplished through a series of developmental stages most of which are similar to zygotic embryogenesis. This process was first reported in *Daucus carota* (Steward *et al.*, 1958). Somatic embryogenesis has now been reported for many plant species and woody perennials which includes both angiosperms and gymnosperms.

Embryogenesis can be induced either: a) directly from an explant without a callus phase, or b) indirectly after a proliferation of callus tissue. Somatic embryogenesis proceeds directly or indirectly after exposure of responsive explants to critical concentrations of exogenously supplied plant growth regulators during the initial culture phase. Somatic embryogenesis offers a tremendous potential for an inexpensive method for large scale propagation of superior genotypes. It potentially provides many production advantages: 1) a large number of plantlets can be produced inexpensively, 2) both root and shoot meristem development occur in the same process step, 3) quick and easy scale-up can be achieved via liquid culture, 4) long term germplasm storage via cryopreservation can be utilized, and 5) manufactured seeds or a direct delivery system can be used for embling (plantlets regenerated from somatic embryos) establishment. Genetic gains of forest trees can be captured through somatic embryogenesis. In forestry, the production of manufactured seeds throughout the year provides a complementary technology which will reduce risks relative to seed orchards where seed production is limited and uncertain.

Recently tremendous success has been achieved in somatic embryogenesis in conifer species (Table 1). Polyembryony, the formation of multiple embryos within the seed *in vivo* is common in conifers (Singh, 1978). There are two types of polyembryony: 1) simple, and 2) cleavage. Simple polyembryony

S. Jain, P. Gupta & R. Newton (eds.), Somatic Embryogenesis in Woody Plants, Vol. 1, 81–98.

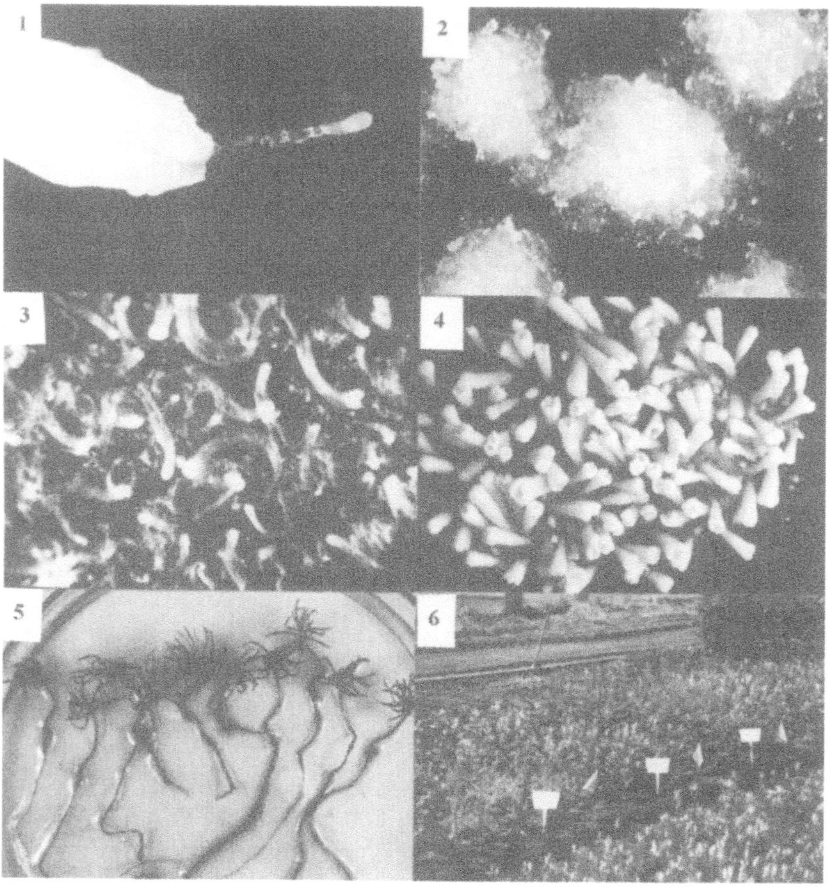

Figure 1. (1) Excised immature embryo (explant) of *Pseudotsuga menziesii*, still attached to the female gametophyte via the suspensor system. (2) Embryonal suspensor masses (ESM) of *Pseudotsuga menziesii* growing on semi-solid maintenance medium. (3) Singulated early stage embryos of *Pseudotsuga menziesii* in liquid medium after ABA treatment. (4) Cotyledonary embryos of *Pseudotsuga menziesii* on a pad soaked in liquid development medium. (5) Somatic embryos of *Pseudotsuga menziesii* germinated on semi-solid medium. (6) Somatic seedlings of several genotypes of *Pseudotsuga menziesii*, in active shoot growth in the spring following fall planting at the Mima Forest Regeneration Center, Washington.

is the process whereby multiple embryos are formed by fertilization of many eggs per female gametophyte. This results in embryos which are genetically different. Cleavage polyembryony is the process whereby one embryo cleaves to form several genetically identical embryos. Cleavage polyembryony can be induced *in vitro*, even in species where it has not been reported in nature, such as *Picea abies* and *Pseudotsuga menziesii* (Durzan and Gupta, 1988; Durzan, 1989).

Table 1. Somatic embryogenesis in conifer species.

Species	Explant	Response	Established in Soil	References
Abies alba	Female gametophyte	SE	–	Schuller *et al.*, 1989
	Mature embryos	SE, PL	–	Hristoforglu *et al.*, 1992
Abies balsamea	Mature embryos	SE	–	Gueven *et al.*, 1992
Abies fraseri	Mature embryos	SE	–	Gueven *et al.*, 1992
Abies nordmanniana	Female gametophyte or immature embryos	SE	–	Norgaard and Krogstrup, 1991
	Mature embryos	SE, PL	–	Norgaard *et al.*, 1992
Agathis australis	Immature embryos	SE, PL	–	Aitken-Christie *et al.*, 1992
Larix decidua	Female gametophyte	SE, PL	–	Nagmani and Bonga, 1985; Von Aderkas and Bonga, 1988a
	Female gametophyte	SE	–	Von Aderkas *et al.*, 1987
	Female gametophyte	SE	–	Von Aderkas *et al.*, 1990
	Immature embryos	SE	–	Von Aderkas *et al.*, 1990
	Female gametophyte	SE	–	Von Aderkas *et al.*, 1990
L. decidua × *L. leptolepis*	Immature embryos	SE	–	Von Aderkas *et al.*, 1990
	Immature embryos	SE, PL	Yes	Klimaszewska, 1989
	Female gametophyte	SE	–	Von Aderkas *et al.*, 1990
Larix leptolepis	Immature embryos	SE	–	Von Aderkas *et al.*, 1990
	Female gametophyte	SE	–	Von Aderkas *et al.*, 1990
L. leptolepis × *L. decidua*	Immature embryos	SE	–	Von Aderkas *et al.*, 1990
	Immature embryos	SE, PL	Yes	Klimaszewska, 1989
	Female gametophyte	SE	–	Simola and Santanen, 1990
Laris occidentalis	Immature embryos	SE, PL	Yes	Thompson and Von Aderkas, 1992
Picea abies	Immature embryos	SE, PL	Yes	Chalupa, 1985
	Immature embryos	SE	–	Hakman *et al.*, 1985
	Immature embryos	SE, PL	–	Hakman and Von Arnold, 1985
	Immature embryos	SE, PL	–	Becwar *et al.*, 1987
	Immature embryos	SE, PL	Yes	Von Arnold and Hakman, 1988
	Immature embryos	SE, PL	–	Hakman *et al.*, 1990
	Mature embryos	SE, PL	Yes	Chalupa, 1985
	Mature embryos	SE, PL	–	Gupta and Durzan, 1986a

Table 1. Continued

Species	Explant	Response	Established in Soil	References
	Mature embryos	SE	–	Von Arnold and Hakman, 1986
	Mature embryos	SE, PL	–	Von Arnold, 1987
	Mature embryos	SE, PL	–	Boulay *et al.*, 1988
	Mature embryos	SE, PL	–	Jain *et al.*, 1988
	Mature embryos	SE, PL	Yes	Von Arnold and Hakman, 1988
	Mature embryos	SE, PL	–	Verhagen and Wann, 1989
	Mature embryos	SE, PL	–	Mo *et al.*, 1989
	Mature embryos	SE, PL	Yes	Gupta *et al.*, 1991
	Stage 3 (cotyledonary) somatic embryos	SE, PL	–	Mo *et al.*, 1989
	Cotyledons	SE	–	Krogstrup, 1986
	Cotyledons	SE, PL	–	Lelu *et al.*, 1987, 1990
	Needles (1 year old emblings)	SE	–	Ruaud *et al.*, 1992
	Buds, Needles (7 year old trees)	Callus	–	Westcott, 1992
Picea glauca	Immature embryos	SE, PL	–	Hakman and Fowke, 1987a,b; Lu and Thorpe, 1987
	Immature embryos	SE, PL	Yes	Attree *et al.*, 1990b; Dunstan *et al.*, 1993
	Mature embryos	SE, PL	Yes	Tremblay, 1990
	Cotyledons	SE	–	Attree *et al.*, 1990a
	Cotyledons	SE, PL	–	Lelu and Bornman, 1990
Picea glauca – engelmannii complex	Immature embryos	SE	–	Roberts *et al.*, 1989
	Immature embryos	SE, PL	–	Webb *et al.*, 1989
	Immature embryos	SE, PL	Yes Yes	Roberts *et al.*, 1990a; Webster *et al.*, 1990
	Stage 3 (cotyledonary) somatic embryos	SE	–	Eastman *et al.*, 1991
Picea glehnii	Mature embryos	SE, PL	Yes	Ishii, 1991
Picea jezoensis	Mature embryos	SE, PL	Yes	Ishii, 1991
Picea mariana	Immature embryos	SE, PL	–	Hakman and Fowke, 1987b
	Immature embryos	SE	–	Tautorus *et al.*, 1990
	Mature embryos	SE	–	Tautorus *et al.*, 1990
	Mature embryos	SE, PL	Yes	Attree *et al.*, 1990b
	Cotyledons	SE, PL	–	Attree *et al.*, 1990a
	Cotyledons	SE, PL	–	Lelu and Bornman, 1990
Picea omorika	Mature embryos	SE, PL	Yes	Budimir and Vujicic, 1992
Picea pungens	Mature embryos	SE, PL	Yes	Afele *et al.*, 1992

Table 1. Continued

Species	Explant	Response	Established in Soil	References
Picea ruben	Mature embryos	SE, PL	Yes	Tremblay and Tremblay, 1991; Harry and Thorpe, 1991
Picea sitchensis	Immature embryos	SE, PL	Yes	Krogstrup *et al.*, 1988
	Immature embryos	SE, PL	Yes	Roberts *et al.*, 1991
	Mature embryos	SE	–	Von Arnold and Woodward, 1988
	Mature embryos	SE, PL	Yes	Krogstrup, 1990
Picea wilsonii	Immature embryos	SE, PL	–	Ying-Hong and Zhong-Shen, 1990
Pinus caribaea	Female gametophyte	SE, PL	–	Lainé and David, 1990
Pinus elliottii	Immature embryos	SE	–	Jain *et al.*, 1989; Newton *et al.*, Chapter from this book; Newton *et al.*, 1993
Pinus lambertiana	Immature embryos	SE, PL	–	Gupta and Durzan, 1986b
	Mature embryos			
Pinus nigra	Immature embryos	SE	–	Salajova and Salaj, 1992
Pinus serotina	Female gametophyte	SE	–	Becwar *et al.*, 1988
Pinus strobus	Female gametophyte	SE	–	Finer *et al.*, 1989
	Immature embryos			
Pinus taeda	Female gametophyte	SE, PL	Yes	Gupta and Durzan, 1987a
	Female gametophyte	SE	–	Becwar *et al.*, 1990
	Immature embryos	SE, PL	Yes	Gupta and Pullman, 1990
Pseudotsuga menziesii	Immature embryos	SE, PL	Yes	Durzan and Gupta, 1987
	Mature embryos			
	Immature embryos	SE, PL	Yes	Gupta and Pullman, 1991
Sequoia sempervirens	Mature embryos	SE, PL	–	Bourgkard and Favre, 1988
	Hypocotyls			
	Cotyledons			

SE – Somatic embryos; PL – Plantlets.

Early on there were a few reports on embryo-like (embryoids) structures and long suspensor-like cells in cultures of *Pinus banksiana* (Durzan and Chalupa, 1976), *Picea glauca* (Durzan, 1980) and *Pseudotsuga menziesii* (Durzan, 1980). However, there were no instances where further development of embryos or plantlets was achieved. The term spheroblasts was also used to distinguish these structures from somatic embryos (Durzan, 1980). Somatic embryogenesis in conifers was first reported in 1985 from immature zygotic embryos of *Picea abies* (Hakman *et al.*, 1985; Chalupa, 1985). In the same year somatic embryogenesis was established from the female gametophyte of *Larix decidua* (Nagmani and Bonga, 1985) and resulted in the production of haploid embryos. Since 1985 several papers have been published on successful regeneration of conifer plantlets produced via somatic embryogenesis (revieved by Attree and Fowke, 1993; Gupta *et al.*, 1991, 1993; Tautorus *et al.*, 1991). Gupta and Durzan (1986b) described somatic polyembryogenesis in *Pinus lambertiana*, followed by *Pinus taeda* (Gupta and Durzan, 1987), *Picea abies* (Gupta and Durzan, 1986a), and *Pseudotsuga menziesii* (Durzan and Gupta, 1987). They termed this embryogenic tissue "embryonal suspensor masses" (ESM) due to its high degree of organization. Embryonal suspensor masses (ESM) of conifers have a distinctly different phenotype compared to callus. The appearance of ESM is usually described as white, translucent, moist or mucilaginous, and consists anatomically of early stage embryos which have an embryonal head and a suspensor system (Gupta and Durzan, 1987). Non-embryogenic callus appears opaque, friable, and may turn green in light. Callus may or may not have anatomical organization.

2. Culture Initiation

There are three different methods for the initiation of embryogenic cultures in conifers: a) through the continuation of natural cleavage polyembryony of embryonal heads of explanted immature embryos (Durzan and Gupta, 1988; Durzan, 1980), b) through cell division in the epidermal and subepidermal layers of the hypocotyl, cotyledons or needles resulting in callus which then rapidly organizes to form embryo suspensor masses (ESM) (Nagmani *et al.*, 1987), c) through cell division of small cells within the suspensor system of the explanted immature embryo (Gupta and Durzan, 1987).

The female gametophyte may play an important role in the induction of ESM. Immature embryos of *Pseudotsuga menziesii* are excised with the female gametophyte still attached via the suspensor system (Fig. 1.1). This culture method produced higher percentages of ESM in *Pseudotsuge menziesii* (Durzan and Gupta, 1987). Initiation of *Pinus* species has been most successful when intact female gametophytes were placed on initiation medium without excision of immature embryos. Embryogenic cultures were later extruded from the micropylar end of the female gametophyte (Becwar *et al.*, 1990).

Media requirements for the initiation of embryogenic cultures do not appear to be very specific. There are several basal media (modified MS, LP, DCR, P6) which have been used for the initiation of embryogenic cultures of *Picea* species, *Pinus* species, *Pseudotsuga menziesii*, and *Larix* species.

ESM has been initiated on several basal media such as DCR (Gupta and Durzan, 1985), modified MS (Gupta and Durzan, 1986a), P6 (Gupta and Pullman, 1990), LP (Hakman *et al.*, 1985) and WTC (Gupta and Pullman, 1991a). However, it has been observed that low percentages of sucrose (1%) produced more ESM (Becwar *et al.*, 1988). Gelrite was found to be a better gelling agent when compared with agar for the initiation of ESM (Finer *et al.*, 1989).

The auxin 2,4-dichlorophenoxyacetic acid (2,4-D) generally has been the preferred auxin used for the initiation of ESM of most conifer species (Gupta *et al.*, 1991; Tautorus *et al.*, 1991). Naphthaleneacetic acid (NAA) has also been used for ESM induction of *Picea abies* (Verhagen and Wann, 1989). However, no difference has yet been reported in ESM proliferation or embryo development in conifers with NAA versus 2,4-D as the sole auxin source. Initiation of ESM cultures from *Abies* sp. were best initiated with cytokinin (BA, N_6-benzyladenine; KIN, kinetin) alone, with auxin found to be inhibitory (Norgaard and Krogstrup, 1991). Initiation of ESM cultures has been achieved with different concentrations of hormones, such as 10 to 110 mg/l 2,4-D (Gupta *et al.*, 1991), up to 5 mg/l NAA, and up to 55 mg/l BA and KIN each (Becwar *et al.*, 1988). Induction of ESM has mostly been done in the dark but it has also been reported to occur in the light (Verhagen and Wann, 1989).

All reports to date of initiation of embryogenic cultures of *Pseudotsuga menziesii* or *Pinus* species resulted from immature embryos at precotyledonary stages (see Table 1). Somatic embryogenesis has been reported from both immature and mature embryos of different *Picea* and *Abies* sp. (see Table 1). Krogstrup (1986) and Lelu *et al.* (1987) have reported ESM induction from cotyledons of germinated seedlings. Attree *et al.* (1990a) initiated ESM from *Picea glauca* and *Picea mariana* seedlings 20–30 days after germination. Recently Ruaud *et al.* (1992) described ESM induction from needles of 14-month-old somatic seedlings (emblings) growing in a greenhouse. Very recently, embryogenic callus was developed from buds and needles of seven year old trees of *Picea abies* (Westcott, 1992).

2.1. *Culture Maintenance*

Maintenance of ESM has been done by transferring cultures on solid medium every 10–12 days onto fresh medium (Fig. 1.2). All cultures were incubated in darkness at 22–25°C. Embryonal suspensor masses were maintained in lower concentrations of hormones (1–2 mg/l 2,4-D and 0.1–0.5 mg/l KIN and BA) when compared with media used for initiation (Durzan and Gupta, 1988).

Cultures of ESM have also been maintained in liquid medium using 250 ml Erlenmeyer flasks, continuously rotated at 100–120 rpm in the dark, and sub-cultured at approximately 7 day intervals (Gupta and Durzan, 1987). Durzan (1989) reported that cultures of ESM grown in this manner multiplied through true-to-conifer type cleavage polyembryony. Free-nuclear and proembryo stages have been observed in suspension cultures of *Pinus taeda*, *Pseudotsuga menziesii* (Gupta and Durzan, 1987) and *Larix* (Von Aderkas and Bonga, 1988a). Fowke *et al.* (1990) observed that early somatic development from protoplasts of *Picea glauca* followed a similar morphological developmental sequence as zygotic embryos. Ultrastructural studies have been made during somatic and zygotic embryogenesis of conifers in *Picea glauca* (Hakman and Fowke, 1987) and *Larix* (Rohr *et al.*, 1989; Von Aderkas and Bonga, 1988b). The embryonal apex region consists of small densely cytoplasmic cells with large nuclei, subtended by elongated and highly vacuolated cells of the suspensor system.

Liquid culture maintenance requires close control of culture density. Krogstrup (1990) reported that culture density is critical and often determined the quality of early stage embryos in suspension culture. Changes in embryogenic potential of ESM cultures and their growth has been reported after several months of weekly subcultures (Dunstan *et al.*, 1993). Differences in yield and morphology of somatic embryos on development and maturation media have been ascribed to variation in embryogenic potential of ESM during liquid culture maintenance (Dunstan *et al.*, 1993). However, improvements have also been made by manipulating the nutritional and hormonal requirements in liquid cultures in bioreactor (Lulsdorf *et al.*, 1992; Smith, 1991). Organic nitrogen sources (Glutamine, other amino acids, casein hydrolysate) have improved the proliferation and development of somatic embryos as compared to inorganic nitrogen sources (Boulay *et al.*, 1988; Finer *et al.*, 1989; Tremblay and Tremblay, 1991).

2.2. *Embryo Development*

Abscisic acid (ABA) has been used for cotyledonary embryo development in many plant species (Skriver and Mundy 1990) including conifers (Durzan and Gupta, 1987). It has been hypothesized that ABA inhibits cleavage polyembryony and allows embryo singulation and further development (Fig. 1.3) (Durzan and Gupta, 1987). The size of the embryonal head may be an important factor which influences ABA treatment. If the early stage embryo head is small, ABA may not promote further development (Gupta and Pullman, 1990; Jalonen and Von Arnold, 1991). Application of ABA to cultures has been made in various combination with or without activated charcoal. Cultures of ESM have been transferred first onto medium without hormones but with activated charcoal for one week (Becwar *et al.*, 1989). It was thought that the charcoal absorbed cytokinins and auxins which caused the cultures to cease cleavage embryony (Gupta and Pullman, 1990). Cultures were then transferred to medium containing ABA (Becwar *et al.*, 1989; Attree *et al.*, 1990b). Other papers described direct transfer of cultures to ABA medium for 2–3 subcultures and then transfer to medium without ABA, with or without activated charcoal (Boulay *et al.*, 1988; Von Arnold and Hakman, 1988).

Abscisic acid has not always been the most critical component for embryo development and maturation. The osmotic potential of the media may also play an important role. Gupta and Pullman (1990) reported the use of increased osmolality of the maintenance medium during early stage embryogenesis by increasing it from 120 mm/kg to 180–250 mm/kg depending upon the species and genotype. This was done by adding myo-inositol, sorbitol, mannitol etc. to the maintenance medium. After several subcultures at this raised osmotic level, early stage embryos were transferred onto a development and maturation medium containing ABA. Without the elevated osmotic levels during maintenance culture, several genotypes of *Pinus taeda* and *Pseudotsuga menziesii* were unable to develop good quality cotyledonary embryos with subsequent ABA treatments alone. Recently, increasing media osmolality in conjunction with ABA addition has been used to advance the size of early stage embryos of *Picea glauca engelmannii* to globular stage before transfer to maturation medium (Roberts, 1991). The limitations for the use of elevated osmotic levels in maintenance medium were reported by Gupta and Pullman (1991b) who found that further increasing the osmolality of maintenance medium to 250–350 mm/kg for long periods of time altered development by stimulating the multiplication of embryonal head cells and inhibiting the growth of the suspensor system.

2.3. *Embryo Maturation*

Abscisic acid has also been used to improve maturation. Different concentrations of ABA (0.1–25 mg/l) have been used with different conifer species for this purpose (Tautorus *et al.*, 1991). Inhibition of precocious germination, accumulation of storage proteins, and maturation of white spruce somatic embryos have been observed with high concentrations (40–60 μM) of ABA-containing medium (Roberts *et al.*, 1990b). It has been shown in both *Picea glauca engelmannii* (Roberts *et al.*, 1990b) and *Picea abies* (Hakman *et al.*, 1990) that somatic embryos matured with ABA in the medium possessed the same storage proteins as zygotic embryos. This conclusion was based on the molecular weight, isoelectric variants, solubility characteristics and disulfide linkages of the isolated proteins. Recently, Mishra *et al.* (1993) reported that some of the major crystalloid and matrix polypeptides were absent from somatic embryos of *Picea glauca* maturing on medium containing ABA and low osmoticum. However, treatment with polyethylene glycol in combination with ABA resulted in the synthesis of a spectrum of storage polypeptides resembling that of mature zygotic embryos (Mishra *et al.*, 1993). However, both *Picea abies* and *Picea glauca engelmannii* contained fewer lipid bodies and lower triglyceride levels compared with zygotic embryos when matured with an optimal ABA concentration (Roberts *et al.*, 1990b; Hakman *et al.*, 1990). Becwar *et al.* (1988) used IBA (indolebutyric acid) with ABA in the maturation medium to improve the quality and germination of somatic embryos of *Picea abies*. Pullman and Gupta (1991) added ABA with activated charcoal in their development and maturation medium. The combination of charcoal (0.1–0.2%) with very high concentrations of ABA (50–100 mg/l) improved the yield and quality of somatic embryos of *Picea abies*. It was thought that the charcoal initially adsorbed the ABA very fast and then more slowly as well as cytokinins and auxins and other phenolic compounds from the maintenance medium. Using this method 50 to 100 high quality somatic embryos of *Picea abies* have been produced from 1 ml settled ESM within six weeks. Select embryos were placed on a filter paper, moistend with few drops of sterilized distilled water, in a petri dish. They were desiccated slowly over a six day period according to the method described by Senaratna *et al.* (1989). Desiccation tolerance to less than 10% water content and 80–90% germination have been achieved from select mature embryos of several genotypes of *Picea abies* (Gupta *et al.*, 1992).

For some conifer species, the presence of ABA in development and maturation medium was not effective in inhibiting precocious germination of somatic embryos (Gupta and Pullman, 1991a). In these cases, increasing the osmolality of the development and maturation medium was necessary to inhibit precocious germination and allow further embryo development and maturation. Polyethylene glycol (PEG 4,000–8,000) was the best osmoticum for conifer embryo maturation (Cornu and Geoffrion, 1990; Gupta and Pullman, 1991a; Attree *et al.*, 1991b). Gupta and Pullman (1991) reported that media which reached an osmolality of 350–600 mm/kg, and contained an ABA concentration of 25–100 mg/l and an activated charcoal concentration of 0.1–0.25% improved the maturation of conifer somatic embryos (Fig. 1.4). Good quality cotyledonary embryos from several genotypes of Douglas-fir and loblolly pine have been produced using this method. Similarly, somatic embryos of white spruce matured on a media containing 16–24 μM ABA and 7.5% PEG 4000 for 8 weeks contained high levels of triglyceride (TAG) and a TAG fatty acid composition similar to that of zygotic embryos (Attree *et al.*, 1992).

2.3.1. *Desiccation Tolerance and Germination*

Germination media is used to promote the vigorous growth and development of fully developed and mature somatic embryos into plants ready to make the transition to autotrophic growth. Germination media consisted of 1–2% sucrose, 0.1–1% activated charcoal and no growth regulators (Fig. 1.5) (Gupta *et al.*, 1991). Somatic embryos germinated most vigorously when first incubated in the dark for seven days and then transferred to light. Becwar *et al.* (1988, 1989) observed that the orientation of the embryos influenced root development on the germination medium. Germination of *Picea abies* was improved by orienting the somatic embryos in a slant or inverted culture tube (Becwar *et al.*, 1989).

High frequency and synchronized germination were achieved after partial drying of both *Picea glauca engelmannii* and *Picea sitchensis* somatic embryos using high (> 85%) relative humidity (HRH) treatments (Roberts *et al.*, 1991). High percentage (81%) germination similar to zygotic embryos has been achieved with *Picea glauca* somatic embryos which were matured for 6–8 weeks on 16 μM ABA, 7.5% PEG 4000 and desiccated in high (43–81%) relative humidity (Attree *et al.*, 1991b, 1992). Desiccation tolerance to less than 10% water content and 80–90% germination have been achieved from select embryos of several genotypes of *Picea abies* (Gupta *et al.*, 1992). These embryos were plated directly from maintenance suspension cultures and matured for 6–7 weeks on medium with 50 mg/l ABA and 0.125% charcoal.

2.3.1.1. *Manufactured Seed.* See Chapter 12 in this book by Carlson and Hartle for a complete description of this technology.

2.3.1.2. *Conversion.* Conversion means the establishment of germinated somatic embryos into soil and the capacity for autotrophic growth. There are several reports on somatic embryogenesis and plantlet regeneration in conifers (Table 1). However, there are very few reports on somatic seedlings (plantlets derived from somatic embryos) established in soil (Webster *et al.*, 1990; Krogstrup, 1990; Gupta *et al.*, 1991; Attree *et al.* 1991a). Becwar *et al.* (1988, 1989) reported very low numbers of *Picea abies* somatic seedlings established in soil. Westvaco Co. has also established over 500 somatic seedlings of *Pinus taeda* in soil (Mike Becwar personal communication). Webster *et al.* (1990) established 1,200 somatic seedlings of *Picea glauca engelmannii* in soil from 71 genotypes. Weyerhaeuser Co., USA, has established in soil over 3,000 somatic seedlings of *Picea abies* from 17 genotypes and over 2,000 somatic seedlings of *Pseudotsuga menziesii* from 10 genotypes (Gupta *et al.*, 1993).

2.3.1.3. *Cryopreservation.* Clonal field testing in forestry species often takes 5–15 years before superior genotypes are identified. Embryogenic cultures multiply very rapidly and have to be sub-cultured biweekly on a fresh solid medium and weekly in liquid medium. Therefore, it is risky to maintain the ESM cultures for long periods of time by continuous subculture due to the following reasons: a) reduction or loss of embryogenic potential, b) loss due to contamination or other external factors, c) loss due to genetic alternation which would cause somaclonal variation, d) loss due to epigenetic alteration which would cause somaclonal variation, e) the high labor cost. In recent years, cryopreservation has been successfully used for long term preservation of plant cells, tissues, and organs (Kartha *et al.*, 1988). Recently it has been demonstrated that ESM cultures are amenable to cryopreservation (Table 2). Plantlets have been regenerated from cryostored ESM cultures of several conifer species and established in soil (Table 2). Plantlets have not displayed any abnormalities when compared to plantlets regenerated from non-cryostored ESM (Klimaszewska *et al.*, 1992). Plantlets of *Picea mariana* and *Larix* from cryostored ESM have been established in a nursery for field evaluation. The plantlets have over wintered well and are showing normal phenology

Table 2. Cryo storage of ESM cultures of conifers.

Conifer Species	Cryoprotectants	Response	Reference
Abies nordmanniana	Sorb, DMSO	SE	Norgaard *et al.*, 1991
Larix × *eurolepi* (Larch)	Sorb., DMSO	SE, PL	Klimaszewska *et al.*, 1992
Picea abies (Norway Spruce)	PEG, Glu, DMSO	SE, PL	Gupta *et al.*, 1987
	Suc., DMSO	SE, PL	Galerne and Dereuddre, 1988
Picea glauca (White spruce)	Sorb., DMSO	SE, PL	Kartha *et al.*, 1988
Picea mariana (Black Spruce)	Sorb., DMSO	SE, PL	Klimaszewska *et al.*, 1992
Picea sitchensis (Sitka spruce)	Sorb., DMSO	SE, PL	Find *et al.*, 1993
Pinus caribaea	Suc., DMSO	SE, PL	Laine *et al.*, 1992
pinus taeda (Loblolly pine)	PEG, Glu, DMSO	SE, PL	Gupta *et al.*, 1987
Pseudotsuga menziesii (Douglas-fir)	Sorb., DMSO	SE, PL	Weyerhaeuser Co., USA

SE – Somatic embryos; Sorb. – Sorbitol; DMSO – Dimethylsulfoxide; PEG – Polyethylene glycol; Suc. – Sucrose; Glu – Glucose; PL – Plantlets.

(Klimaszewska *et al.*, 1992). At Weyerhaeuser Co., USA, cultures of 500 genotypes from superior full-sib families of *Pseudotsuga menziesii* have been cryostored into liquid nitrogen. Plantlets have been regenerated from cryostored ESM of several of these genotypes which have been grown out in an outdoor nursery prior to planting in the field (unpublished).

2.3.1.4. *Field Performance.* The first report on the field performance of the emblings was by Becwar *et al.* (1989) who reported that 29% of *Picea abies* emblings were successfully established in soil. Under greenhouse conditions, these emblings set dormant buds, survived over wintering to − 5°C, renewed vegetative growth synchronously, and appeared phenotypically normal compared to control seedlings grown under the same conditions.

A field trial of 1200 *Picea glauca engelmannii* emblings from 71 genotypes have been established in soil by B.C. Research, Canada (Roberts *et al.*, 1993). At the end of the first season in the nursery, embling survival varied from 80–100% and was genotype dependent. Height measurements showed that emblings and seedlings proceeded through similar growth curves. Emblings and seedlings also showed similar root and shoot morphology and acquired similar levels of cold hardiness. Now emblings, control seedlings and cuttings has been planted at two forest field sites in British Columbia, Canada. All emblings produced new growth in the second growing season, grew normally, and set bud at the end of the season. Overall survival rate was 100% at the end of the second growing season (Roberts *et al.*, 1993).

At Weyerhaeuser Company, survival of emblings in the nursery was similar to seedlings (99%) despite a week of subfreezing temperatures (unpublished). Strikingly uniform growth was observed among emblings within a clone, as compared to the less uniform seedlings. The morphology of all trees appeared normal and 1,100 were selected and planted onto a typical forest regeneration site. Similar growth patterns were observed with emblings of *Picea sitchensis*, *Picea glauca*, and *Picea mariana* (as cited in Tautorus *et al.*, 1991) . Recently, over 600 emblings of *Pseudotsuga menziesii* from eight genotypes have been established in a Weyerhaeuser nursery for later planting on a typical field site (Fig. 1.6) (unpublished).

Conclusions

The ability to initiate, maintain and develop somatic embryos and emblings via somatic embryogenesis has been well established in the literature. The challenge for the future is to deliver that biological potential as a cost effective technology for the large scale and cost effective production of somatic embryos for clonal reforestation purposes. To make the needed technology develop, a more complete understanding of the biological and molecular processes involved in both somatic and zygotic embryogenesis will be required. This view holds that one must intimately understand both the sequence and the order of developmental processes of zygotic embryo in order to develop technology to simulate it in an *in vitro* environment. The blueprints for these processes are present within the genetic material of cells which are cultured. Our job is to provide the needed materials and environment throughout development so that the organism can construct itself.

References

Afele, J.C., T. Senaratna, B.D. Mckersie and P.K. Saxena, 1992. Somatic embryogenesis and plant regeneration from zygotic embryo culture in blue spruce (*Picea pungens*). Plant Cell Rep. 11: 299–303.

Aitken-Christie, J., H. Davies and K. Gough, 1992. Micropropagation of *Agathis australis*. Abstr. in Proc. Intl. Conifer Biotechnology Working Group, Research Triangle Park, N.C. April 23–28, pp. 54–55.

Attree, S.M. and L.C. Fowke, 1993. Embryogeny of gymnosperms: advances in synthetic seed technology of conifers. Plant Cell Tiss. Org. Cult. 35: 1–35.

Attree S.M., S. Budimir and L.C. Fowke, 1990a. Somatic embryogenesis and plantlet regeneration from cultured shoots and cotyledons of seedlings from stored seeds of black and white spruce (*Picea mariana* and *Picea glauca*). Can. J. Bot. 68: 30–34.

Attree, S.M., T.E. Tautorus, D.I. Dunstan and L.C. Fowke, 1990b. Somatic embryo maturation, germination and soil establishment of plants of black and white spruce (*Picea mariana* and *Picea glauca*). Can. J. Bot. 68: 2583–2589.

Attree, S.M., D.I. Dunstan and L.C. Fowke, 1991a. White spruce (*Picea glauca* (Moench) Voss) and black spruce (*Picea mariana* (Mill) B.S.P.). In: Y.P.S. Bajaj (Ed.), Trees III. Biotechnology in Agriculture and Forestry, Vol. 16, pp. 423–455. Springer-Verlag, Berlin.

Attree, S.M., D. Moore, V.K. Sawhney and L.C. Fowke, 1991b. Enhanced maturation and desiccation tolerance of white spruce (*Picea glauca*) somatic embryos: Effect of non-plasmolyzing water stress and abscisic acid. Ann. Bot. 68: 519–525.

Attree S.M., M.K. Pomeroy, L.C. Fowke, 1992. Manipulation of conditions for the culture of somatic embryos of white spruce for improved triacylglycerol biosynthesis and desiccation tolerance. Planta 187: 395–404.

Becwar, M.R., T.L. Noland and S.R. Wann, 1987. Somatic embryo development and plant regeneration from embryogenic Norway spruce callus. Tappi J. 70: 155–160.

Becwar, M.R., S.R. Wann, M.A. Johnson, S.A. Verhagen, R.P. Feirer and R. Nagmani, 1988. Development and characterization of *in vitro* embryogenic systems in conifers. In: M.R. Ahuja (Ed), Somatic Cell Genetics of Woody Plants, pp. 1–18. Kluwer Academic Publishers, Dordrecht.

Becwar, M.R., T.L. Noland and J.L. Wykoff, 1989. Maturation, germination, and conversion of Norway spruce (*Picea abies L.*) somatic embryos to plants. In Vitro Cell Develop. Biol. 25: 575–580.

Becwar, M.R., R. Nagmani and S.R. Wann, 1990. Initiation of embryogenic cultures and somatic embryo development in loblolly pine (*Pinus taeda*). Can. J. For. Res. 20: 810–817.

Boulay, M.P., P.K. Gupta, P. Krogstrup and D.J. Durzan, 1988. Development of somatic embryos from cell suspension cultures of Norway Spruce (*Picea abies* Karst.). Plant Cell Rep. 7: 134–137.

Bourgkard, F. and J.M. Favre, 1988. Somatic embryos from callus of *Sequoia sempervirens*. Plant Cell Rep. 7: 445–448.

Budimir, S. and R. Vujicic, 1992. Benzyladenine induction of buds and somatic embryogenesis in *Picea omorika*. Plant Cell Tiss. Org. Cult. 31: 89–94.

Chalupa, V., 1985. Somatic embryogenesis and plantlet regeneration from cultured immature and mature embryos of *Picea abies* (L.) Karst. Communi. Inst. For. Cech. 14: 57–63.

Cornu, D. and C. Geoffrion, 1990. Aspects de l'embryogenése somatique chez le mélèze. Frill. Soc. Bot. Fr., 137. Actual bot. (3/4): 25–34.

Dunstan, D.I., T.D. Bethune and C.A. Bock, 1993. Somatic embryo maturation from long-term suspension cultures of white spruce (*Picea glauca*). In Vitro Cell. Dev. Biol. 29: 109–112.

Durzan, D.J., 1980. Progress and promise in forest genetics. In: proceedings of the 50th Anniversary Conference, Paper Science and Technology, The Cutting Edge, May 8–10, 1979, pp. 31–60. The Institute of Paper Chemistry, Appleton, WI.

Durzan, D.J., 1989. Physiological aspects of somatic polyembryogenesis in suspension cultures of conifers. Ann. Sci. For. 46 (Suppl.): 101s–107s.

Durzan, D.J. and V. Chalupa, 1976. Growth and metabolism of cells and tissue of jack pine (*Pinus banksiana*). 3. Growth of cells in liquid suspension cultures in light and darkness. Can. J. Bot. 54: 456–467.

Durzan, D.J. and P.K. Gupta, 1987. Somatic embryogenesis and polyembryogenesis in Douglas-fir cell suspension cultures. Plant Sci. 52: 229–235.

Durzan, D.J. and P.K. Gupta, 1988. Somatic embryogenesis and polyembryogenesis in conifers. In: A. Mizrahi (Ed.), Biotechnology in Agriculture, Vol. 9, pp. 53–81. Alan R. Liss Inc., New York.

Eastman, P.A.K., F.B. Webster, J.A. Pitel and D.R. Roberts, 1991. Evaluation of somaclonal variation during somatic embryogenesis of interior spruce (*Picea glauca engelmannii* complex) using culture morphology and isozyme analysis. Plant Cell Rep. 10: 425–430.

Finer, J.J., H.B. Kriebel and M.R. Becwar, 1989. Initiation of embryogenic callus and suspension cultures of eastern white pine (*Pinus strobus* L.). Plant Cell Rep. 8: 203–206.

Find, J.I., F. Franz, P. Krogstrup, J.D. Mollar, J.V. Norgaard and M.M.H. Kristensen, 1993. Cryopreservation of an embryonic suspension culture of *Picea sitchensis* and subsequent plant regeneration. Scand. J. For. Res. 8: 156–162.

Fowke, L.C., S.M. Attree, H. Wang and D.I. Dunstan, 1990. Microtubule organization and cell division in embryogenic protoplast cultures of white spruce (*Picea glauca*). Protoplasma 158: 86–94.

Galerne, M.J. and Dereuddre, 1988. Survie De Cals Embryogenes D'epice' à Abus Congelation à − 196°C. Afocel. 174: 8–31.
Gueven, T.G., V. Micah and E.G. Kirby, 1992. Preliminary observation on cytokinin effects on *in vitro* growth of zygotic Embryos of *Abies fraseri*, and *Abies balsameu*. Abstr. in Proc. Conifer Biotechnology Working Group, Research Triangle Park, N.C., April 23–28, p. 68.
Gupta, P.K. and D.J. Durzan, 1985. Shoot multiplication from mature trees of Douglas-fir (*Pseudotsuga menziesii*) and sugar pine (*Pinus lambertiana*). Plant Cell Rep. 4: 177–179.
Gupta, P.K. and D.J. Durzan, 1986a. Plantlet regeneration via somatic embryogenesis from subcultured callus of mature embryos of *Picea abies* (Norway spruce). In Vitro Cell Devel. Biol. 22: 685–688.
Gupta, P.K. and D.J. Durzan, 1986b. Somatic polyembryogenesis from callus of mature sugar pine embryos. Bio/Technology 4: 643–645.
Gupta, P.K. and D.J. Durzan, 1987. Biotechnology of somatic polyembryogenesis and plantlet regeneration in loblolly pine. Bio/Technology 5: 147–151.
Gupta, P.K. and G.S. Pullman, 1990. Method for reproducing coniferous plants by somatic embryogenesis. U.S. Patent No. 4,957,866.
Gupta, P.K., D.J. Durzan and B.J. Finkle, 1987. Somatic polyembryogenesis in embryogenic cell masses of *Picea abies* (Norway spruce) and *Pinus taeda* (loblolly pine) after thawing from liquid nitrogen. Can. J. For. Res. 17: 1130–1134.
Gupta, P.K., R. Timmis, G. Pullman, M. Yancey, M. Kreitinger, W. Carlson and C. Carpenter, 1991. Development of an embryogenesis system for automated propagation of forest trees. In: I.K. Vasil (Ed.), Scale-Up and Automation on Plant Propagation, pp. 76–90. Academic Press, New York.
Gupta, P.K. and G.S. Pullman, 1991a. Method for reproducing coniferous plants by somatic embryogenesis using abscisic acid and osmotic potential variation. U.S. patent No. 5,036,007.
Gupta, P.K. and G.S. Pullman, 1991b. High concentration enrichment of conifer embryonal cells. U.S. Patent No. 5,041,382.
Gupta, P.K., G. Pullman, R. Timmis, M. Kreitinger, W.C. Carlson, J. Grob and E. Welty, 1992. Scale up somatic embryogenesis of conifers for reforestation. Abstr. in Proc. of 3rd Canadian Workshop on Plant Tissue Culture and Genetic Engineering, University of Guelph, Guelph, Ontario, Canada, 17–20 June, 1992, p. 3.
Gupta, P.K., G. Pullman, R. Timmis, M. Kreitinger, W.C. Carlson, J. Grob and E. Welty, 1993. Forestry in the 21st Century: The biotechnology of somatic embryogenesis. Bio/Technology II: 454–459.
Hakman, I. and L.C. Fowke, 1987a. An embryogenic cell suspension culture of *Picea glauca* (white spruce). Plant Cell Rep. 6: 20–22.
Hakman, I. and L.C. Fowke, 1987b. Somatic embryogenesis in *Picea glauca* (white spruce) and *Picea mariana* (black spruce). Can. J. Bot. 65: 656–659
Hakman, I. and S. Von Arnold, 1985. Plantlet regeneration through somatic embryogenesis in *Picea abies* (Norway spruce) J. Plant Physiol. 121: 149–158.
Hakman, I., L.C. Fowke, S. von Arnold and T. Eriksson, 1985. The development of somatic embryos in tissue cultures initiated from immature embryos of *Picea abies* (Norway spruce). Plant Sci. 38: 53–59.
Hakman, I., P. Stabel, P. Engstrom and T. Eriksson, 1990. Storage protein accumulation during zygotic and somatic embryo development in *Picea abies* (Norway spruce). Physiol. Plant 80: 441–445.
Harry, I.S. and T.A. Thorpe, 1991 Somatic embryogenesis and plant regeneration from mature zygotic embryos of Red Spruce. Bot Gaz. 152(4): 446–452.
Hristoforoglu, K., A. Grahsl and J. Schmidt 1992. Somatic embryogenesis from mature embryo of *Abies alba*. Abstr. in Proc. Conifer Biotechnology Working Group, Research Triangle Park, N.C., April 23–28, p. 77.
Ishii, K., 1991. Somatic embryo formation and plantlet regeneration through embryogenic callus from zygotic embryos of *Picea jezoensis*. J. For. Soc 73(1): 24–28.

Jain, S.M., R.J. Newton. and E.J. Soltes, 1988. Enhancement of somatic embryogenesis in Norway spruce (*Picea abies* L.) Theor. Appl. Genet. 76: 501–506.
Jain, S.M., N. Dong and R.J. Newton, 1989. Somatic embryogenesis in slash pine (*Pinus elliottii*) from immature embryos cultured *in vitro*. Plant Sci. 65: 233–241.
Jalonen, P. and S. von Arnold, 1991. Characterization of embryogenic cell lines of *Picea abies* in relation to thair competence for maturation. Plant Cell Rep. 10: 384–387.
Kartha, K.K., L.C. Fowke, N.L. Leung, K.L. Caswell and I. Hakman, 1988. Induction of somatic embryos and plantlets from cyropreserved cell cultures of white spruce (*Picea glauca*). J. Plant Physiol. 132: 529–539.
Klimaszewska, K., C. Ward and W.M. Cheliak, 1992. Cryopreservation and plant regeneration from embryogenic cultures of larch and black spruce. J. Expt. Bot. 43: 73–79.
Klimaszewska, K., 1989. Plantlet development from immature zygotic embryos of hybrid larch through somatic embryogenesis. Plant Sci. 63: 95–103.
Krogstrup, P., 1986. Embryo like structures from cotyledons and ripe embryos of Norway spruce (*Picea abies*). Can. J. For. Res. 16: 664–668.
Krogstrup, P., 1990. Effect of culture densities on cell proliferation and regeneration from embryogenic cell suspensions of *Picea sitchensis*. Plant Sci. 72: 115–123.
Krogstrup, P., E.N. Eriksen., J.D. Møller and H. Roulund, 1988. Somatic embryogenesis in Sitka spruce (*Picea sitchensis* (Bong)) Carr.) Plant Cell Rep. 7: 594–597.
Laine, E., P. Bade and A. David 1992. Recovery of plants from cryopreserved embryogenic cell suspensions of *Pinus caribaeu*. Plant Cell Rep. 11: 295–298.
Laine, E. and A. David, 1990. Somatic embryogenesis in immature embryos and protoplasts of *Pinus caribaea*. Plant Sci. 69: 215–224.
Lelu, M.A. and C.H. Bornman, 1990. Induction of somatic embryogenesis in excised cotyledons of *Picea glauca* and *Picea mariana*. Plant Physiol. Biochem. 28: 785–791.
Lelu, M.A., M. Boulay and Y. Arnaud, 1987. Formation of embryogenic calli from cotyledons of *Picea abies* (L.) Karst collected from 3 to 7 days old seedlings. C.R. Acad. Sci. Ser. 3. 305: 105–109.
Lelu, M.A., M. Boulay. and C.H. Bornman, 1990. Somatic embryogenesis in cotyledons of *Picea abies* is enhanced by an adventitious bud-inducing treatment. New For. 4: 125–135.
Lu, C.Y. and T.A. Thorpe, 1987. Somatic embryogenesis and plantlet regeneration in cultured immature embryos of *Picea glauca*. J. Plant Physiol. 128: 297–302.
Lulsdorf, M.M., T.E. Tautorus, S.I. Kikcio and D.I. Dunstan 1992. Growth parameter of embryogenic cultures of interior spruce and black spruce. Plant Sci. 82: 227–234.
Mishra, S., S.M. Attree, I. Leal and L.C. Fowke, 1993. Effect of abscisic acid, osmoticum, and desiccation on synthesis of storage proteins during the development of white spruce somatic embryos. Ann. Bot. 71: 11–22.
Mo, L.H., S. von Arnold and U. Laagercrantz, 1989. Morphogenic and genetic stability in long term embryogenic cultures and somatic embryos of Norway spruce (*Picea abies* (L.) Karst). Plant Cell Rep. 8: 375–378.
Nagmani, R. and J.M. Bonga, 1985. Embryogenesis in subcultured callus of *Larix decidua*. Can. J. For. Res. 15: 1088–1091.
Nagmani, R., M.R. Becwar and S.R. Wann, 1987. Single cell origin and development of somatic embryos in *Picea abies* (L.) Karst. (Norway spruce) and *P. glauca* (Moench) Voss (white spruce). Plant Cell Rep. 6: 157–159.
Newton, R.J. *et al.*, Chapter from this book – Embryogenesis of slash pine.
Newton, R.J., Dong, N., Marek-Swize, K. and Cairney, J., 1993. Transformation of slash pine. Proc. 22nd So. For. Trel. Impr. Conf. Holiday-Inn, Buckhead, Atlanta, GA, June 14–17. pp. 390–402.
Norgaard, J.V., S. Baldursson and P. Krogstrup, 1991. Cryopreservation of embryogenic *Abies nordmanniana* cultures. Abstr. in Proc. Trends in the Biotechnology of Woody Plants. 3rd IUFRO Workshop of Somatic Cell genetics, 25–29 Nov., Dehra Dun, p. 13.
Norgaard, V.J., S. Baldursson and P. Krogstrup 1992. Somatic embryogenesis in *Abies nordman-*

niana. Induction and maturation of somatic embryos. Abstr. in Proc. Conifer Biotechnology Working Group, Research Triangle Park, N.C., April 23–28, pp. 90–93.

Norgaard, J.V. and P. Krogstrup, 1991. Cytokinin induced somatic embryogenesis from immature embryos of *Abies nordmanniana* Lk. Plant Cell Rep. 9: 509–513.

Pullman, G.S. and P.K. Gupta, 1991. Method for reproducing coniferous plants by somatic embryogenesis using absorbent materials in the development stage media. U.S. Patent No. 5,034,326.

Roberts, D.R., B.C.S. Sutton and B.S. Flinn, 1990a. Synchronous and high frequency germination of interior spruce somatic embryos following partial drying at high relative humidity. Can. J. Bot. 68: 1086–1090.

Roberts, D.R., W.R. Lazaroff and F.B. Webster, 1991. Interaction between maturation and high relative humidity treatments and their effects on germination of Sitka spruce somatic embryos J. Plant Physiol. 138: 1–6.

Roberts, D.R., B.S. Flinn, D.T. Webb, F.B. Webster and B.C.S. Sutton, 1989. Characterization of immature embryos of interior spruce by SDS-PAGE and microscopy in relation to their competence for somatic embryogenesis. Plant Cell Rep. 8: 285–288.

Roberts, D.R., B.S. Flinn, D.T. Webb, F.B. Webster and B.C.S. Sutton, 1990b. Abscisic acid and indole-3–butyric acid regulation of maturation and accumulation of storage proteins in somatic embryos of interior spruce. Physiol. Plant 78: 355–360.

Roberts, D.R., E.B. Webster, B.S. Flinn, W.R. Lazaroff and D.R. Cyr, 1993. Somatic embryogenesis of spruce. In: K. Redenbaugh (Ed.), SynSeeds: Application of Synthetic Seeds to Crop Improvement, pp. 427–52. CRC Press, Boca Raton, FL.

Roberts, D.R., 1991. Abscisic acid and mannitol promote early development, maturation and storage protein accumulation in somatic embryos of interior spruce. Physiol. Plant 83: 247–254.

Rohr, R., P. von Aderkas and J.M. Bonga, 1989. Ultrastructural changes in haploid embryoids of *Larix decidua* during early embryogenesis. Am. J. Bot. 76: 1460–1467.

Ruaud, J.N., J. Bercetche and M. Paques, 1992. First evidence of somatic embryogenesis from needles of 1 year old *Picea abies*. Plant Cell Rep. 11: 563–566.

Salajova, T. and J. Salaj, 1992. Somatic embryogenesis in European Black Pine (*Pinus nigra*) Biol. Plant 34: 213–218

Schuller, A., G. Reuther and T. Geier 1989. Somatic embryogenesis from seed explants of *Abies alba*. Plant Cell Tiss. Org. Cult. 17: 53–58.

Senaratna T., B.D. McKersie and S.R. Bowley, 1989. Desiccation tolerance of alfalfa (*Medicago sativa* L.) somatic embryos. Influence of abscisic acid, stress pretreatments and drying rates. Plant Sci. 65: 253–259.

Simola, L.K. and A. Santanen, 1990. Improvement of nutrient medium for growth and embryogenesis of megagametophyte and embryo callus lines of *Picea abies*. Physiol. Plant 80: 27–35.

Singh, H. 1978. Embryology of Gymnosperm. Gebriider Bountraeger, Berlin.

Skriver, K. and J. Mundy, 1990. Gene expression in response to abscisic acid and osmotic stress. Plant Cell 2: 503–512.

Smith, D.R., 1991. An automated bioreactor system for mass propagation of *Pinus radiata*. Agricell Rep. 17: 1–2.

Steward, F.C., M.O. Mapes and K. Mears 1958. Growth and organized development of cultured cells. Organization in cultures grown from freely suspended cells. Am. J. Bot. 45: 7095–708.

Tautorus, T.E., S.M. Attree, L.C. Fowke and D.I. Dunstan, 1990. Somatic embryogenesis from immature and mature zygotic embryos, and embryo regeneration from protoplasts in black spruce (*Picea mariana* Mill). Plant Sci. 67: 115–124.

Tautorus, T.E., L.C. Fowke and D.I. Dunstan, 1991. Somatic embryogenesis in conifers. Can J. For Res. 69: 1873–1899.

Thompson, R.J. and P. von Aderkas, 1992. Somatic embryogenesis and plantlet regeneration from immature embryos of Western Larch. Plant Cell. Rep. 11: 379–385.

Tremblay, F.M., 1990. Somatic embryogenesis and plantlet regeneration from embryos isolated from stored seeds of *Picea glauca* Can. J. Bot. 68: 236–242.

Tremblay, L. and F.M. Tremblay, 1991. Carbohydrate requirements for the development of Black Spruce (*Picea mariana*) and Red Spruce (*P. Rubens*) somatic embryos. Plant Cell Tiss. Org. Cult. 27: 95–103.

Verhagen, S.A. and S.R. Wann, 1989. Norway spruce somatic embryogenesis. High frequency initiation from light-cultured mature embryos. Plant Cell Tiss. Org. Cult. 16: 103–111.

Von Aderkas, P. and J.M. Bonga, 1988a. Formation of haploid embryoids of *Larix decidua*: early embryogenesis. Am. J. Bot. 75: 690–700.

Von Aderkas, P. and J.M. Bonga, 1988b. Morphological definition of phenocritical period for initiation of haploid embryogenic tissue from explants of *Larix decidua*. In: M.R. Ahuja (Ed.), Somatic Cell Genetics of Woody Plants, pp. 29–38. Kluwer Academic Publishers, Dordrecht.

Von Aderkas, P., J.M. Bonga and R. Nagmani, 1987. Promotion of embryogenesis in cultured megagametophytes of *Larix decidua*. Can J. For. Res. 17: 1293–1296.

Von Aderkas, P., K. Klimaszewska and J.M. Bonga, 1990. Diploid and haploid embryogenesis in *Larix leptolepis*, *L. decidua*, and their reciprocal hybrids. Can. J. For. Res. 20: 9–14.

Von Arnold, S. 1987. Improved efficiency of somatic embryogenesis in mature embryos of *Picea abies* (L.) Karst. J. Plant Physiol. 128: 233–244.

Von Arnold, S. and I. Hakman, 1986. Effect of sucrose on initiation of embryogenic callus cultures from mature zygotic embryos of *Picea abies* (L.) Karst. (Norway spruce). J. Plant Physiol. 122: 261–265.

Von Arnold, S. and I. Hakman, 1988. Regulation of somatic embryo development in *Picea abies* by abscisic acid (ABA). J. Plant Physiol. 132: 164–169.

Von Arnold, S. and S. Woodward, 1988. Organogenesis and embryogenesis in mature zygotic embryos of *Picea sitchensis*. Tree Physiol. 4: 291–300.

Webb, D.T., F. Webster, B.S. Flinn, D.R. Roberts and D.D. Ellis, 1989. Factors influencing the induction of embryogenic and caulogenic callus from embryos of *Picea glauca* and *P. engelmannii*. Can. J. For. Res. 19: 1303–1308.

Webster, F.B., D.R. Roberts, S.M. McInnis and B.C.S. Sutton, 1990. Propagation of interior spruce by somatic embryogenesis. Can. J. For. Res. 20: 1759–1765.

Westcott, R.J. 1992. Embryogenesis from non-juvenile Norway Spruce (*Picea abies*). Abstr. In Vitro Part II 28(3): 101A.

Ying-Hong, L. and G. Zhong-Shen, 1990. Somatic embryogenesis and plantlet formation of *Picea wilsonii* Mast in different conditions. Acta Bot. Sin. 32: 568–570.

6. Somatic Embryogenesis and Rejuvenation of Trees

J.-N. Ruaud and M. Pâques

Contents

1. Introduction

1.1. *Vegetative Propagation of Trees*

Interest in the clonal propagation of selected trees has been well demonstrated (Franclet, 1963; Bonga, 1982; Libby and Rauter, 1984; Thorpe *et al.*, 1992; Muhs, 1992, Kleinschmit *et al.*, 1993), and its main purpose is to reproduce rapidly the selected individuals while maintaining their morphological and physiological traits including vigour, shape and disease resistance.

Vegetative plant propagation has been achieved on an industrial scale for many woody ornamental and agricultural species through conventional horticultural practices, and through rooted cuttings and grafting. Although this has also been possible for forest trees, most species can easily be propagated vegetatively only during their juvenile phase. Most desirable traits are only expressed in mature trees. As the tree ages, at least until maturity is reached, the rooting ability declines. Furthermore, some specific traits associated with maturity, including growth habit (plagiotropic growth, reduced capacity for height and diameter growth) and reproductive competence are transmitted to propagules. This property can be valuable when the harvest is flowering-dependent such as fruit trees, seed orchard trees, grapevines and ornamentals, but this is a problem for forest trees. For silviculture, vegetative copies must be modified by restoration of the juvenile programme of development.

S. Jain, P. Gupta & R. Newton (eds.), Somatic Embryogenesis in Woody Plants, Vol. 1, 99–118.

1.2. *Maturation and Rejuvenation*

Maturation (ontogenic aging) means the ongoing process of phase change which results in relatively permanent changes. Greenwood (1987) proposed four developmental phases, each characterised by a unique set of morphogenetic competencies: (a) the embryogenetic phase (close to the mature embryo), (b) the seedling phase (close to an ideal juvenile phase), (c) the transition phase (which includes the acquisition of reproductive competence), and (d) the mature phase (reached when reproductive competence is highest, and the capacity for height and diameter growth is lowest. Hackett (1985) defined the juvenile phase as the period where flowering does not occur and cannot be induced by the normal flower-initiating treatment or conditions. Although mechanisms involved in the onset of maturation have been widely studied, they are not yet well-known. Morphological, physiological, biochemical and molecular events which occur during maturation have been reviewed recently (Durzan, 1990; Pierik, 1990; Poethig, 1990; Haffner *et al.*, 1991; Hutchison and Greenwood, 1991; Hutchison *et al.*, 1991; Greenwood, 1992, Hackett *et al.*, 1992; Bonga and Von Aderkas, 1993; Greenwood and Hutchison, 1993).

The reversal of maturation is rejuvenation, the return of tissues from mature or transition phases to the juvenile phase (Greenwood, 1987). The aim of rejuvenation is to reverse the ontogenic programme and restore the morphogenetic competencies of juvenile plants.

Rejuvenation of old trees is of considerable interest to foresters because it would allow the propagation, as young plants, of proven mature genotypes (Franclet, 1982). That is why for many decades breeders and researchers have tried to re-invigorate or to keep stock trees in a "juvenile" state to produce rootable cuttings. As true rejuvenation has rarely been demonstrated, it would be more suitable to use the term reinvigoration (Pierik, 1990).

Some methods that have been used for reinvigoration *in vivo* are pruning, hedging, the use of stool beds, grafting (and serial grafting), the use of root suckers, and spraying a growth regulator such as benzylaminopurine. *In vitro* methods have been developed in addition to *in vivo* methods, and include culture of selected explants such as epicormic buds, repeated subculturing of shoots, micrografting onto juvenile rootstocks, meristem culture, adventitious bud formation and somatic embryogenesis. Successful rejuvenation has been reported using each method (for review see Pierik, 1990), but most of them are labour intensive. Somatic embryogenesis seems to be the most promising tool to achieve rejuvenation with reduced labour (Pierik, 1990; Kleinschmit and Meier-Dinkel, 1990).

1.3. *Rejuvenation and Somatic Embryogenesis*

One of the most convincing reasons to consider somatic embryogenesis as a tool for rejuvenation is based upon the naturally occurring phenomenon of apomixy. Many *Citrus* species are polyembryonic; inside each seed, in addition to the zygotic embryo, there are usually several nucellar embryos. Because the nucellus is a maternal tissue, and despite the fact that some genetic variations have been detected among these embryos (Navarro *et al.*, 1985), nucellar embryos are vegetative copies of the mother plant. Plants regenerated from these embryos display juvenile characteristics such as thorns and delayed flowering. Rejuvenated copies of the mother plant can be obtained through these apomictic embryos. Furthermore, rejuvenated plants were also obtained from nucellus cultures of monoembryonic citrus (see *Citrus* below).

Another decisive factor suggesting that somatic embryogenesis is a rejuvenating process and its closeness to zygotic embryogenesis. Zygotic embryogenesis brings about juvenility whereas somatic embryogenesis mimicks zygotic embryogenesis, which, probably, leads to juvenility. There are numerous reports indicating that several events are closely related between the zygotic and somatic embryogenic pathways. The early events in fertilization are quite similar to rapid responses of "pre-embryogenic" cells to auxins, including signal transduction pathways, initiation of DNA synthesis and cell division (see Dudits *et al.*, 1991). Other events which seem to occur very often in both zygotic and somatic embryogenesis are the asymmetry of the first division and subsequently the polarity of the embryo and the early isolation of the newly formed embryo from the surrounding tissues (nucellus, megagametophyte, other zygotic or somatic embryos, non-embryogenic cells). The daughter cells of the zygote or of the embryonic cell remain enclosed within the mother cell wall (for illustration see David *et al.*, 1992 (*Prunus*); Hakman and Fowke, 1987; Owens and Blake, 1985 (*Picea*); Michaux-Ferrière *et al.*, 1992 (*Hevea*); El Maâtaoui *et al.*, 1990 (*Quercus*)). Early embryos isolated from surrounding cells could in that way follow their own metabolic and morphogenetic pathway (Dubois *et al.*, 1991). In addition, further development of the somatic embryo of numerous species follows closely that of the zygotic embryo, including morphogenesis and protein reserve accumulation (for a review of genes expressed during zygotic and/or somatic embryogenesis see Sterk and De Vries, 1993; Mishra, Volume 1, Chapter 6).

There are many other advantages to somatic embryogenesis (Becwar, 1993) including large-scale mechanised production through bioreactors (Pétiard *et al.*, 1992; Dunstan, Volume 1, Chapter 11) and synthetic seed production (see Redenbaugh *et al.*, 1991; Gray and Purohit, 1991; Carlson, Volume 1, Chapter 10).

During the last decade, the number of woody species inducing somatic embryogenesis, has risen dramatically. However, most of these successes have been reported with juvenile material such as zygotic embryos (Wann, 1988). Thus, the regenerated plants were of unproved genotypes. This does not allow for the use of somatic embryogenesis in a cloning strategy, but it can be a part of genetic improvement programmes for forest trees (Cheliak and Rogers, 1990; Park *et al.*, in press).

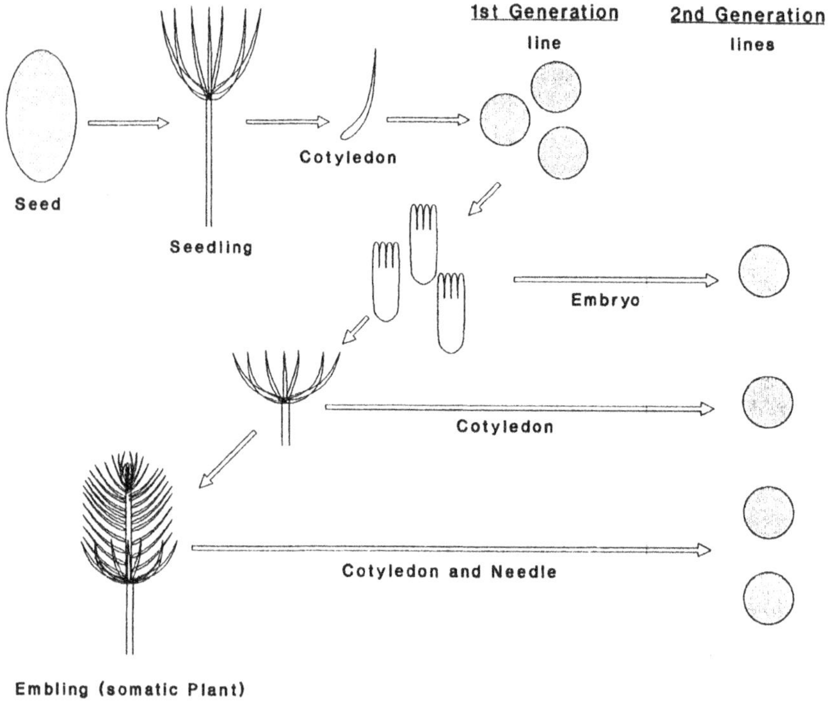

Figure 1. First and second generation somatic embryogenesis.

In this paper, somatic embryogenesis in woody tree species is reviewed in relation to putative rejuvenation; therefore, only studies concerning embryogenesis from ontogenetically older explants than embryos or cotyledons are reported (Table 1). A distinction will be made between "old" vegetative tissues (leaf, bud, root) and "juvenile" somatic reproductive tissues such as petals, nucellus, anther wall and the inner integument of seeds. Reproductive somatic tissues have often been considered to have juvenile physiology, and may be explained by the proximity of cells undergoing meiosis which might have a rejuvenating effect (Nozeran, 1978).

2. Examples of Somatic Embryogenesis from Non-Juvenile Trees

In the present review, some species have been chosen arbitrarily to show the different methods used in research into somatic embryogenesis and the current status of this field. A more exhaustive summary is given in Table 1 (see also chapters of Volumes 2 and 3).

2.1. *Somatic Embryogenesis from "Old" Vegetative Tissues*

2.1.1. *Gymnosperms* (*Pinaceae*)

Some success in rejuvenation using *in vitro* techniques has been reported from aged trees with grafting of shoot tips or meristems onto seedlings (Dumas *et al.*, 1989; Tranvan *et al.*, 1991) or onto rooted juvenile stem cuttings (Huang *et al.*, 1992). Furthermore, axillary or adventitious shoot production have been used for rejuvenate trees, but tissue-cultured plants can quickly lose the reinvigorated morphology during field trials (McKeand, 1985). The evolution of the morphology of tissue-cultured plants toward a mature-appearing morphology was due to the shoot system itself and not due to an indirect effect of plant root system (Anderson *et al.*, 1992). Plants derived from callus cultures, initiated from shoot buds of a 12 year-old *Larix* tree, however, displayed juvenile like growth (Laliberté and Lalonde, 1988).

Somatic embryogenesis seems to induce in a similar way among all of the members of the Pinaceae, with which the most studies have been carried out. Established embryogenic cultures consist of proliferating immature somatic embryos (embryonal suspensor mass or ESM); proliferation occurs through cleavage of embryos or *de novo* embryogenesis from isolated cells of immature embryos (Hakman and Fowke, 1987).

Most of the successes in somatic embryogenesis with conifers were reported from embryos or juvenile plants belonging to the Pinaceae (Tautorus *et al.*, 1991; Roberts *et al.*, 1993). In contrast to angiosperms, the tissue surrounding gymnosperm embryos, the megagametophyte, is the product of the meiosis: it is a maternal haploid tissue, and cannot be used to clone a tree. This kind of gynogenesis was first reported in *Larix* (Nagmani and Bonga, 1985) then in *Picea abies* (Simola and Santanen, 1990), and is potentially highly valuable for tree improvement programmes (Bonga *et al.*, 1988). Large-scale evaluation of conifer somatic embryos derived from juvenile tissue was performed by Webster *et al.* (1990) and Grossnickle *et al.* (1992). They reported that development of emblings (plants from somatic embryos) and seedlings were quite similar after the first growing season, indicating a significant degree of maintenance of juvenility through somatic embryogenesis. Research have been undertaken to improve all the steps involved in somatic embryogenesis and particularly in the initiation of somatic embryogenesis from older plants. Berlyn *et al.* (1991) initiated somatic embryogenesis from needles of 3 month-old *Pinus caribaea* seedlings, but maturation (transition from immature embryos to mature embryo) could not be achieved even though it was possible with immature embryo-derived ESM. Westcott (1992) reported somatic embryogenesis induction and embryo maturation from needles of mature *Picea abies*. Embryo maturation and conversion seemed to occur in a similar way to embryogenic lines derived from juvenile material, suggesting at least a

Table 1. First reports on somatic embryogenesis initiation and embryo conversion from non-juvenile trees.

Species	Explant	Plant Regeneration	Medium	Reference
Aesculus hippocastanum	Filament from 100 y.o.T.	+	WPM 2,4-D BA	Jörgensen, 1989
Bactris gasipaes	Shoot tip callus from 15 month-old plant	+	MS mod PIC BA	Valverde *et al.*, 1987
Betula pendula	Leaf callus from 1 y.o.T.	+	N7 2,4-D Kin	Kurten *et al.*, 1990
Coffea				
canephora	Internode callus	+	MS 2,4-D Kin	Staritsky, 1970
arabica	Leaf	+	MS 2,4-D Kin	Sondahl and Sharp, 1977
	leaf	+	MS 2,4-D Kin	Neuenschwander and Baumann, 1992
Elaeis guineenesis	Leaf callus	+	MS Kin, BA NAA	Rabéchault and Martin, 1976
Eucalyptus grandis	Shoot callus from 4 y.o.T.	–	MS NAA Kin	Lakshmi Sita, 1986
Euphorbia longan	Leaflet from 30 y.o.T.	+	B5 2,4-D BA	Litz, 1988
Hibiscus syriacus	Petiole, filament callus	+	MS 2,4-D BA	Lee *et al.*, 1991
Malus pumila	Petal, flower bud	+	Wh NAA GA3	Mehra and Sachdeva, 1985
Picea abies	Needle from 2 month-old embling	+	MS NAA BA	Ruaud, in press
Pinus caribeae	Needle from 4 month-old seedling	–	MS 2,4-D	Berlyn *et al.*, 1991
Populus				
ciliata	Leaf callus from 40 y.o.T.	+	MS 2,4-D	Cheema, 1989
alba × *grandidentata*	Leaf from stock plant	+	MS 2,4-D BA	Michler and Bauer, 1991
nigra × *maximowiczii*	Leaf from 20 y.o.T.	+	MS 2,4-D BA	Park and Son, 1988
Prunus				
incisa × *serrulata*	Root callus	+		Druart, 1981
Pyrus communis	Petal	–	Wh NAA Kin GA3	Mehra and Jaidka, 1985
Quercus				
bicolor	Catkin	+	MS 2,4-D	Gingas, 1991
ilex	Leaf from 50 y.o.T.	–	MS NAA BA	Féraud-Keller and Espagnac, 1989
petraea	Anther from 10–140 y.o.T.	+	WPM 2,4-D BA	Jörgensen, 1991
rubra	Leaf from 40–100 y.o.T.	–	MS NAA BA	Rancillac and Klinguer, 1991
suber	Internode from 7 month old seedling	–	MS IBA BA	Féraud-Keller *et al.*, 1989
Salix viminalis	Catkin, Pistil	–	MS 2,4-D BA	Grönroos *et al.*, 1989
Santalum album	(inter)node callus 20 y.o.T.	+	MS 2,4-D	Lakshmi Sita *et al.*, 1979
Sapindus trifoliatus	Leaf callus from 60 y.o.T.	+	MS 2,4-D Kin	Desai *et al.*, 1986

y.o.T. – year old Tree; B5 – Gamborg *et al.*, 1968; MS – Murashige and Skoog, 1962; N7 – Simola, 1985; Wh – White, 1943; WPM – Lloyd and McCown, 1981.

2,4-D – 2,4-dichlorophenoxyacetic acid; PIC – picloram; NAA – naphthaleneacetic acid; IBA – indole-butyric acid; Kin – kinetin; BA – benzylaminopurine; GA3 – gibberellic acid.

partial rejuvenation, but information about this work is limited. Using an alternative pathway, we obtained ESM from needles of one year-old plants derived from somatic embryos of *Picea abies* (Ruaud *et al.*, 1992). To study in more detail the initiation of embryogenesis from emblings, we investigated the embryogenic response from emblings aged from 0 (represented by mature somatic embryos) to 56 days and from different tissues (Fig. 1; Ruaud, in press). Embryos from second generation embryogenic lines (Fig. 1) were matured and converted into plants. The rate of embryo conversion and the growth of second generation emblings were similar among the following embryogenic lines: the mother line (derived from a cotyledon of a young seedling), a somatic embryo-derived line, and those from a cotyledon and a needle. The first experiment had shown that emblings from the mother line grew in a similar way to the isolated zygotic embryo-derived plants or seedlings; but development of emblings was delayed, particularly in comparison with seedlings. We concluded that plant age and the nature of the explant producing the ESMs did not influence further development of the embryos. These results suggest the maintenance of juvenility through somatic embryogenesis. Because the emblings from which second generation lines originated, were not older than 56 days, we could not make conclusions about rejuvenation. These results offer hope that somatic embryogenesis can be used in the future to clone elite trees once it is possible to induce somatic embryogenesis from sporophytic tissue of aged trees.

The use of second generation emblings, in combination with cryopreservation, would be very useful for investigating rejuvenation. All second generation emblings are, theoretically, of the same genotype, leading to a reduction of the genotypė variability. Cryopreservation can block age-related changes in morphogenetic potential, metabolic stability, and genetic stability (Benson and Harding, 1990). Thus, freezing mother embryogenic lines and then, thawing, multiplying and maturing ESMs at different times could lead to emblings of different ages without any influence of the age of the mother line. Comparison of many traits including morphology, metabolism and frequency of ESM induction among these different groups of emblings may improve our knowledge about aging. Furthermore, regeneration of second generation emblings will provide information about transmission of mature traits and rejuvenation.

2.1.2. *Angiosperms*

2.1.2.1. *Monocotyledons* (*Palms*)

2.1.2.1.1. *Bactris gasipaes.* Embryogenesis has been initiated from this tropical palm using shoot tips of 15 month-old plants (Valverde *et al.*, 1987) and the emblings were successfully acclimated, but no further details were provided.

2.1.2.1.2. *Elaeis guineensis* (Oil palm). Until the recent advent of tissue culture techniques, oil palm was only propagated from seeds; conventional horticultural methods being inapplicable. Production of oil palm has been developed on a large-scale using somatic embryogenesis (Dublin *et al.*, 1991).

The major steps forward in the development of oil palm propagation have been summarised by Jones and Hugues (1989). Somatic embryogenesis has been achieved from young leaves of mature plants followed by regeneration of emblings (Rabéchault and Martin, 1976). The different steps of somatic embryogenesis were described by Lioret (1981). The process is characterised by two types of callus; a nodular compact callus (NCC) and a fast growing callus (FGC). Only the former is used to establish stable embryogenic lines able expressing high clonal fidelity suitable for use by industry (Durand-Gasselin *et al.*, 1990). No information about rejuvenation is available.

2.1.2.1.3. *Phoenix dactylifera* (Date palm). Using somatic tissues, somatic embryos have been induced in three cultivars (Tisserat, 1979; Sharma *et al.*, 1984; Bhaskaran and Smith, 1992). A more efficient method of plant recovery has been described from shoot tip and immature inflorescence (Bhaskaran and Smith, 1992), but no information was provided on *ex vitro* plant behaviour. (See also chapter in this book by Bhaskaran and Smith).

2.1.2.2. *Dicotyledons*

2.1.2.2.1. *Betula* (Birch). A study of *in vitro* rejuvenation has been performed by Brand and Lineberger (1992a,b) on *Betula papyfera* and *Betula pubescens* × *Betula papyfera*. They noted that *in vitro* micropropagation induced juvenile-like morphological traits, but some of these characteristics are quickly lost *ex vitro*. They concluded that the level of juvenility regained during the *in vitro* phase is not equivalent to that of a seedling. Partial rejuvenation has been confirmed by protein analysis which indicated that micropropagated plants were initially more like seedlings than mature plants.

To our knowledge, Kurten *et al.* (1990) are the first to have induced somatic embryogenesis and to have regenerated plants from non-embryonic explants of *Betula*. Embryogenic calli of *Betula* consisted of small, densely cytoplasmic cells. The embryogenic potential of these calli has to be tested by transferring pieces of callus on another medium. They succeeded from leaves of one year-old *Betula pendula* and 49 plants derived from somatic embryos have been acclimated in the greenhouse, but growth evaluation of emblings was not provided.

2.1.2.2.2. *Coffea arabica*. Selected cultivars are propagated by seeds from a system based on autogamy, or alternatively from cuttings (Dublin *et al.*,

1991). Due to the vegetative dimorphism of coffee (there are orthotropic and plagiotropic shoots), vegetative propagation by cuttings is limited ; only rooted orthotropic shoots can producing normal bearing trees. Under natural conditions, no reversal from plagiotropic to orthotropic habit occurs and it is also exceptional *in vitro* (Dublin *et al.*, 1991).

Alternative solutions include adventitious budding and somatic embryogenesis from internode and leaf explants. Neoformation of buds was first achieved by Dublin (1980), but this method was not as efficient as somatic embryogenesis (Dublin, 1980). Somatic embryogenesis and plant regeneration were first reported by Staritsky (1970) and later by Sondahl and Sharp (1977). These latter authors noted two waves of somatic embryo production which they named LFSE (Low Frequency Somatic Embryo induction) and HFSE (High Frequency Somatic Embryo induction). LFSE involved the formation of a few spontaneously regenerating embryos 11 to 16 weeks after transfer of leaf callus tissue onto an induction medium. HFSE occurred 3 to 6 weeks after LFSE, and was characterised by the production of an almost unlimited number of bipolar embryos, but these embryos did not spontaneously mature. Recently, a third type of somatic embryogenesis was described by Neuenschwander and Baumann (1992), namely SCSE (Self-Controlled Somatic Embryogenesis). SCSE occurs in suspension, goes along with the suppression of HFSE and is characterised by highly synchronised embryo formation. A high percentage (94%) of the resulting somatic embryos were converted into plantlets.

Somatic embryogenesis seems to induce rejuvenation in *Coffea* because all plants regenerated displayed "good growth", even those issued from plagiotropic shoots, indicating that the "ontogenic memory" has been erased. To our knowledge, no comparison has been performed between emblings and seedlings in coffee.

2.1.2.2.3. *Eucalyptus*. Micropropagation through axillary or adventitious shoot proliferation has been successful for numerous species of *Eucalyptus* including some mature trees. Callus cultures have also been established (for a review of *Eucalyptus* tissue culture, see Le Roux and Van Staden, 1991). Several comparisons between tissue cultured plants and seedlings have been achieved. Recently, Bell *et al.* (1993) reported several minor differences between propagules and seedlings, but these differences did not influence major traits such as growth or windthrow resistance. It can be concluded that a partial rejuvenation may occur through micropropagation. Recently, meristematic nodules have been obtained with juvenile explants of *E. grandis* (Warrag *et al.*, 1991), and *E. robusta* (Boxus *et al.*, 1991), and then from a 20 year-old *E. camaldulensis* (Ossor and Boxus, 1992). Although good growth of tissue-cultured plants was described, no information was given about rejuvenation. Nevertheless, this method seems to be highly valuable for large-scale propagation of *Eucalyptus*.

Somatic embryogenesis from non-juvenile material has been reported by two teams (Lakshmi Sita *et al.*, 1986; Qin and Kirby, 1992). Unfortunately, conversion of somatic embryos into plants has only been reported from juvenile tissues (Ouyang *et al.*, 1981; Muralidharan and Mascarenhas, 1987; Watt *et al.*, 1991).

2.1.2.2.4. *Euphorbia longan* (Longan). Somatic embryos and plantlets have been obtained from leaflets of 30 year-old trees of this important tropical fruit tree (Litz, 1988). Emblings displayed normal morphology and a "good growth", but no comparison was made with seedlings.

2.1.2.2.5. *Populus* (Poplars). Initiation of somatic embryogenesis from aged trees was first reported by Park and Son (1988) using punctured leaves of *Populus nigra* × *Populus maximowiczii*, but no data are available about conversion and acclimatation. More recently, embryogenesis was reported from leaves of rooted cuttings taken from an hybrid of *Populus alba* × *Populus grandidentata* (Michler and Bauer, 1991). Experiments are underway to improve the frequency of embryogenesis and the rate of conversion of the somatic embryos. Cheema (1989) succeeded in *Populus ciliata* working with callus and cell suspension cultures of a 40 year-old tree. Although the protocols described to obtain plants through somatic embryogenesis are complex, the results are very interesting since embryos and plants were obtained, and emblings growing in the greenhouse displayed juvenile morphology. However, we cannot conclude from these results that true rejuvenation occurred because only 15 emblings have been planted in the field and comparison with cuttings or seedling was not reported. Nevertheless, these results promise that somatic embryogenesis can be developed for adult poplar.

2.1.2.2.6. *Quercus* (Oaks). This genus has been recalcitrant to tissue culture techniques for many years especially in relation to somatic embryogenesis (first reported by El Maâtaoui and Espagnac, 1987), and only a few species have been investigated. Embryogenic calli of *Quercus* are composed of embryogenic and non-embryogenic cells. The former have thick walls and dense cytoplasms and accumulate starch. Inside the thick walls the protoplasms undergo a series of divisions leading to proembryos and then embryos which are often the site of secondary embryogenesis (El Maâtaoui *et al.*, 1990). Embryogenesis has been achieved indirectly from internodes of 6 to 8 month-old seedlings of *Quercus suber* (Féraud-Keller *et al.*, 1989) and from leaves of old *Quercus ilex* (Féraud-Keller and Espagnac, 1989). The other success from aged tree concerns *Quercus rubra* (Rancillac and Klinguer, 1991). Unfortunately, for the three species, the conversion of somatic embryos into plants remains a limiting step.

2.1.2.2.7. *Santalum album* (Sandalwood). Somatic embryos from non-juvenile sandalwood were first obtained by Lakshmi Sita *et al.* (1979) using 20 to 25 year-old trees. Although many problems still remain with hardening of the root system and variation and change in the ploidy level, emblings have been successfully acclimated. Furthermore, after 3 years, some tissue-cultured plants had grown faster than normal seedlings (Lakshmi Sita, 1986), suggesting a complete rejuvenation. More characteristics than growth habit are needed to conclude that plants were rejuvenated. Unfortunately, no more details such as growth kinetic and reproductive competence were given for better characterisation of emblings.

2.1.2.2.8. *Sapindus trifoliatus* (Soapnut). Somatic embryos have been obtained from undifferentiated, leaf-derived callus of a 60 year-old soapnut tree, after transfer to a medium containing a lower concentration of auxin (Desai *et al.*, 1986). Three hundred and fifty to 400 plants were regenerated from a single 5 mm diameter leaf-disk, but no information is available about acclimatation and further development of the emblings.

2.2. *Somatic Embryogenesis from Somatic Reproductive Tissues*

Somatic embryogenesis has been observed from numerous reproductive tissues including nucellus, ovule, inner tegument of the seed, anther wall and filament in many tree species. Many of these species are tropical and work with them has been reviewed by Litz (1985), Tulecke (1987) and Litz and Jaiswal (1991).

2.2.1.1.1. *Aesculus hippocastanum* (Horsechesnut). Histological and ultrastructural studies of somatic embryogenesis from *Aesculus* have only been performed from juvenile material (Profumo *et al.*, 1986, 1987). Somatic embryogenesis and plant regeneration were obtained by Jörgensen (1989, 1991) from filament callus from 10 to 100 year-old trees. Somatic embryos arose from a whitish, soft callus and could be cloned by secondary embryogenesis. Somatic embryos displayed conformity with the mother plant and were acclimated. Although growth of emblings was recorded, it was not performed in comparison with seedlings.

2.2.1.1.2. *Citrus*. *Citrus* are currently propagated by grafting selected budwood onto cloned rootstocks obtained by cuttings or by seeds. Many *Citrus* species are polyembryonic; inside each seed, in addition of the zygotic embryo, there are usually several nucellar embryos. These embryos are vegetative copies of the mother plant, meaning that for these species, there is unlikely to be any economic impact of *in vitro* vegetative propagation (Duran-Vila and Navarro, 1989). On the other hand, there is a potential application of *in vitro* methods to mono-embryonic species or to seedless poly-embryonic varieties.

Recovery of adventitious embryos from pre-existing nucellar embryos was first achieved by Rangaswany (1961) and later, Rangan *et al.* (1968) were successful in obtaining development of nucellar embryos from nucelli of mono-embryonic seeds. A great advantage of plants derived from nucellar embryos is that they are free of most viruses (see Barlass and Skene, 1986). However, this method is limited in application because nucellar plants display juvenile characteristics, and many years are required before they lose the undesirable juvenile traits and become commercially viable (Duran-Vila and Navarro, 1989). Furthermore, many genetically aberrant plants were obtained from nucellar somatic embryogenesis (Navarro *et al.*, 1985).

2.2.1.1.3. *Hevea brasiliensis* (Rubber). Due to the low rhizogenic potential of selected shoot material and inadequate anchorage of rooted cuttings, *Hevea* is still propagated today by grafting and by "selected" seeds (Carron *et al.*, 1989). To date, vegetative propagation of selected trees by *in vitro* microcuttings has not been reported (Dublin *et al.*, 1991).

Somatic embryogenesis from clones and subsequent regeneration have been reported by Wang *et al.* (1980) from anther wall callus. The inner seed integument from immature seeds has also shown embryogenic potential (Carron and Enjalric, 1982), the origin and ontogeny of somatic embryos, from this tissue, have been described recently by Michaux-Ferrière *et al.* (1992). However, despite of recent progress in embryo conversion (Montoro *et al.*, 1992), this step seems to limit large-scale production as mentioned by Dublin *et al.* (1991). Furthermore, somatic embryogenesis has only been achieved with a few clones.

2.2.1.1.4. *Malus* and *Pyrus* (Apple and Pear). Mehra and Sachdeva (1985) and Mehra and Jaidka (1985) obtained emblings from *Malus pumila* and *Pyrus communis*, respectively, using various organs from juvenile and mature plants. They were successful especially with petal and flower buds. Conformity of emblings with the mother plant was studied by determination of the level of ploidy, but no information was given about growth of emblings.

2.2.1.1.5. *Quercus* (Oak). In addition to leaf disks, somatic embryogenesis has been obtained recently in *Quercus* using catkins of *Quercus bicolor* (Gingas, 1991). In contrast to somatic embryogenesis observed from zygotic embryos, the embryogenic pathway catkins involved a callus phase. The embryos required a desiccation treatment for further development, but no data were available about conversion and subsequent growth.

3. Conclusion

Rejuvenation of mature proven trees permits clonal plantation of elite genotypes, thereby, resulting in a maximum genetic gain in a short period. Until now, for many species, especially forest species, horticultural practices and *in vitro* propagation methods are not sufficiently satisfactory to achieve rejuvenation and low-cost large-scale production. Somatic embryogenesis is promising as a mean of clonal propagation of woody trees because it is amenable to mechanised systems and can lead to rejuvenation.

Somatic embryogenesis has been induced from a wide range of tree species using two kinds of sporophytic explants: maternal tissues associated with zygotic embryogenesis such as nucellus, integuments, and inflorescences and tissues from secondary organs such as leaves, roots and internodes. Can these tissues be considered as juvenile and aged, respectively, meaning that somatic embryogenesis permits maintenance of juvenility in the first case and rejuvenation in the second? As there is no evident link between cell differentiation, and the nature of the tissue and aging, we cannot distinguish between these hypotheses.

Numerous authors have reported the regeneration of emblings displaying juvenile-like traits, but in most cases there is no sufficient proof to confirm the occurence of rejuvenation (or expression of juvenility) through somatic embryogenesis; the comparison between emblings and seedlings is greatly lacking. Until this comparison is made for many traits including growth, branching, flowering and wood quality, and during long periods (several years), the occurence of rejuvenation can not be concluded. Field tests are not only needed to verify that rejuvenation is achieved, but also to check the clonal fidelity of the regenerated plants.

The development of markers of juvenility or maturity at the cell level and the development of immunotechniques will be valuable for the distinction between the rejuvenation hypothesis or the expression of juvenility one. Works of Bon (1988) and Amo-Marco *et al.* (1993) are very promising in this respect; the first author discovered a membranous protein expressed only in meristems of young seedlings or rejuvenated plants of *Sequoiadendron* and the second team found two polypetides ontogenetically only expressed in old material of *Castanea sativa*. The development molecular techniques should also help to discover molecular markers of juvenility and maturity. These techniques will also be very helpful for the early estimation of the conformity of regenerated plants. Until now, techniques involved in this analysis such as morphology, ploidy level and isozyme patterns, are not sufficiently accurate or are too long to implement.

Juvenility can be quickly lost during *ex vitro* culture, as observed with shoots issued from adventitious budding (McKeand, 1985), probably in relation with unsatisfactory *in vitro* and/or *ex vitro* culture conditions. These cultural conditions have to be improved to allow the best growth of embryos *in vitro* and *ex vitro*. The study of biochemical and mechanical relations between the zygotic embryo and its surrounding matrix (megagametophyte and endosperm) will give valuable information which could be adapted to tissue-cultured embryos.

Although somatic embryogenesis seems very promising to achieve large-scale production of "juvenile" plants, there will be no application of this method until a number of problems are solved.

Research is required to:

- improve the basic knowledge of somatic embryogenesis initiation, using model systems such as immature and mature zygotic embryos;
- initiate somatic embryogenesis from vegetative tissues of old trees;
- better understand the preservation of juvenility or rejuvenation by cryopreservation;
- develop biochemical and molecular markers of juvenility and maturation;
- improve tissue culture methods and horticultural practicies allowing harmonious growth of emblings and maintenance of rejuvenation in transplants;
- implement large-scale and long duration field tests of emblings in comparison with seedlings and cuttings.

The progress registered during the last decade continues, we are hopeful that the application of somatic embryogenesis in clonal forestry may be realized in the near future.

Acknowledgements

Deep thanks are due to Dr. N. Hammatt for the critical review of the manuscript. We are also grateful to Mr. A. Franclet for valuable discussions.

References

Amo-Marco, J.B., N. Vidal, A.M. Vieitez and A. Ballester, 1993. Polypeptide markers differentiating juvenile and adult tissues in chestnut. J. Plant Physiol. 142: 117–119.

Anderson, A.B., L.J. Frampton, S.E. McKeand and J.F. Hodges, 1992. Tissue-culture shoot and root system effects on field performance of loblolly pine. Can. J. For. 22: 56–61.

Barlass, M. and K.G.M. Skene, 1986. Citrus (*Citrus* species). In: Y.P.S. Bajaj (Ed.), Biotechnologies in Agriculture and Forestry, Vol. 1, Trees I, pp. 207–219. Springer Verlag, Berlin.

Becwar, M.R., 1993. Conifer somatic embryogenesis and clonal forestry. In: M.R. Ahuja and W.J. Libby (Eds.), Clonal Forestry I, Genetics and Biotechnologies, pp. 201–223. Springer-Verlag, Berlin.

Bell, D.T., P.G. van der Moezel, I.J. Bennet, J.A. McComb, C.F. Wilkins, S.C.B. Marshall and A.L. Morgan, 1993. Comparison of *Eucalyptus camaldulensis* from seeds and tissue culture: root, shoot and leaf morphology of 9-month-old plants grown in deep sand and sand over clay. For. Ecol. Manag. 57: 125–139.

Benson, E.E. and K. Harding, 1990. The control by cryopreservation of age-related changes in plant tissue cultures. In: R. Rodriguez, R. Sanchez Tames and D.J. Durzan (Eds.), Plant Aging, Basic and Applied Approaches, NATO ASI Series, Series A: Life Sciences, Vol. 186, pp. 125–132. Plenum Press, New York.

Berlyn, G.P., S.J. Kohls and A.O. Anoruo, 1991. Caribbean pine (*Pinus caribaea* Morelet). In: Y.P.S. Bajaj (Ed.), Biotechnology in Agriculture and Forestry, Vol. 16, Trees III, pp. 254–268. Springer-Verlag, Berlin.

Bhaskaran, S. and R.H. Smith, 1992. Somatic embryogenesis from shoot tip and immature inflorescence of *Phoenix dactylifera* cv. Barhee. Plant Cell Rep. 12: 22–25.

Bon, M.C., 1988. J 16: An apex protein associated with juvenility of *Sequoiadendron giganteum*. Tree Physiol. 4: 381–387.

Bonga, J.M. and P. von Aderkas, 1993. Rejuvenation of tissues from matures conifers and its implications for propagation *in vitro*. In: M.R. Ahuja and W.J. Libby (Eds.), Clonal Forestry I, Genetics and Biotechnology, pp. 182–199. Springer-Verlag, Berlin.

Bonga, J.M., 1982. Vegetative propagation in relation to juvenility, maturity and rejuvenation. In: J.M. Bonga and D.J. Durzan (Eds.), Cell and Tissue Culture in Forestry, Vol. 2, pp. 387–412. Martinus Nijhoff/W. Junk, The Hague.

Bonga, J.M., P. von Aderkas and D. James, 1988. Potential application of haploid cultures of tree species. In: J.W. Hanover and D.E. Keathly (Eds.), Genetic Manipulation of Woody Plants, pp. 55–77. Plenum Press, New York.

Boxus, P., J.M. Terzi, C. Lievens, M. Pylyser, P. Ngaboyamahina and K. Duhem, 1991. Improvement and perspectives of micropropagation techniques applied to some hot climate plants. Acta Hort. 289: 55–64.

Brand, M.H and R.D. Lineberger, 1992a. *In vitro* rejuvenation of *Betula* (*Betulaceae*): morphological evaluation. Am. J. Bot. 79: 618–625.

Brand, M.H and R.D. Lineberger, 1992b. *In vitro* rejuvenation of *Betula* (*Betulaceae*): biochemical evaluation. Am. J. Bot. 79: 626–635.

Carron, M.P. and F. Enjalric, 1982. Studies on vegetative micropropagation of *Hevea brasiliensis* by somatic embryogenesis and *in vitro* microcutting. In: A. Fujiwara (Ed.), Plant Tissue Culture, pp. 751–752. Mazuren, Tokyo.

Carron, M.P., F. Enjalric, L. Lardet and A. Deschamps, 1989. Rubber (*Hevea brasiliensis* Müll. Arg.). In: Y.P.S. Bajaj (Ed.), Biotechnology in Agriculture and Forestry, Vol. 5, Trees II, pp. 222–245. Springer-Verlag, Berlin.

Cheema, G.S., 1989. Somatic embryogenesis and plant regeneration from cell suspensions and tissue cultures of mature Himalayan poplar (*Populus ciliata*). Plant Cell Rep. 8: 124–127.

Cheliak, W.N. and D.L. Rogers, 1990. Integrating biotechnologies into tree improvement programs. Can. J. For. Res. 20: 452–463.

David, H., J.M. Domon, C. Savy, N. Miannay, G. Sulmont, R. Dargent and A. David, 1992. Evidence for early stages of somatic embryo development in a protoplast-derived cell culture of *Prunus avium*. Physiol. Plant. 85: 301–307.

Desai, H.V., P.N. Bhatt and A.R. Mehta, 1986. Plant regeneration of *Sapindus trifoliatus* L. (soapnut) through somatic embryogenesis. Plant Cell Rep. 3: 190–191.

Druart, P., 1981. Embryogenèse somatique et obtention de plantules chez *Prunus incisata* × *serrulata* (GM9) cultivé *in vitro*. Bull. Rech. Agro. Gembloux 16: 205–220.

Dublin, P., F. Enjalric, F. Lardet, M.-P. Carron, N. Trolinder and C. Pannetier, 1991. Estate crops. In: P.C. Debergh and R.H. Zimmerman (Eds.), Micropropagation, pp. 337–361. Kluwer Academic Publishers, Dordrecht.

Dublin, P., 1980. Induction de bourgeons néoformés et embryogenèse somatique. Café Cacao Thé 24: 121–127.

Dubois, T., M. Guedira, J. Dubois and J. Vasseur, 1991. Direct somatic embryogenesis in leaves of Chicorum. A histological and SEM study of early stages. Protoplasma 162: 120–127.

Dudits, D, L. Bögre and J. Gyorgyey, 1991. Molecular and cellular approaches to the analysis of plant embryo development from somatic cells *in vitro*. J. Cell Sci. 99: 473–484.

Dumas, E., A. Franclet and O. Monteuuis, 1989. Microgreffage de méristèmes primaires caulinaires de Pins maritimes (*Pinus pinaster* Ait.) âgés sur de jeunes semis cultivés *in vitro*. C.R. Acad. Sci. Paris Sér. III 309: 723–728.

Duran-Vila, N. and L. Navarro, 1989. Current status of citrus biotechnologies. In: A. Sasson and V. Costarine (Eds.), Plant Biotechnologies for Developing Countries, pp. 279–287. CTA/FAO, Rome.

Durand-Gasselin, T., V. le Guen, K. Konan and Y. Duval, 1990. Plantations en Côte d'Ivoire de palmiers à huile (*Elaeis guineensis* Jacq.), obtenus par culture *in vitro*. Premiers résultats. Oléagineux 45: 1–9.

Durzan, D.J., 1990. Adult vs. juvenile explants: direct totypotency. In: R. Rodriguez, R.

Sanchez Tames and D.J. Durzan (Eds.), Plant Aging, Basic and Applied Approaches, NATO ASI Series, Series A: Life Sciences, Vol. 186, pp. 19–25. Plenum Press, New York.

El Maâtaoui, M. and H. Espagnac, 1987. Néoformation de structures de type embryons somatiques sur des cultures de tissus de chêne liège (*Quercus suber* L.). C.R. Acad. Sci. Paris Sér. III 304: 83–88.

El Maâtaoui, M., H. Espagnac and N. Michaux-Ferrière, 1990. Histology of callogenesis and somatic embryogenesis induced in stem fragments of cork oak (*Quercus suber*) cultured *in vitro*. Ann. Bot. 66: 183–190.

Féraud-Keller, C. and H. Espagnac, 1989. Conditions d'apparition d'une embryogenèse somatique sur des cals issus de la culture de tissus foliaires du chêne vert (*Quercus ilex*). Can. J. Bot. 67: 1066–1070.

Féraud-Keller, C., M. El Maâtaoui, O. Gouin and H. Espagnac, 1989. Embryogenèse somatique chez trois espèces de chênes méditerranéens. Ann. Sci. For. 46s: 130–132.

Franclet, A., 1963. Amélioration des reboisements d'Eucalyptus par multiplication végétative. In: Proc. Consultation Mondiale sur la Génétique Forestière et l'Amélioration des Arbres, Stockolm, August 23–30, 8 p. FAO/FORGEN, Roma.

Franclet, A., 1982. Rajeunissement des ligneux. In: Proc. IUFRO-AFOCEL Symp. Colloque International sur la Culture *In Vitro* des Essences Forestières, pp. 55–64. AFOCEL (Pub.), Nangis.

Gamborg, O.L., R.A. Miller and K. Ojima, 1968. Nutrients requirements for suspension cultures of soybean root cells. Exp. Cell Res. 50: 151–158.

Gingas, V.M., 1991. Asexual embryogenesis and plant regeneration from male catkins of *Quercus*. HortSci. 26: 1217-1218.

Gray, D.J. and A. Purohit, 1991. Somatic embryogenesis and development of synthetic seed technology. Critic. Rev. Plant Sci. 10: 33–61.

Greenwood, M.S. and K.W. Hutchison, 1993. Maturation as a developmental process. In: M.R. Ahuja and W.J. Libby (Eds.), Clonal Forestry II, Conservation and Application, pp. 14–33. Springer-Verlag, Berlin.

Greenwood, M.S., 1987. Rejuvenation of forest trees. Plant Growth Regul. 6: 1–12.

Greenwood, M.S., 1992. Theoretical aspects of juvenility and maturation. Proc. IUFRO-AFOCEL Meeting: Mass Production Technology for Genetically Improved Fast Growing Trees, pp. 19–25. AFOCEL (Pub.), Nangis.

Grönroos, L., S. von Arnold and T. Eriksson, 1989. Callus production from floral explants of basket willow (*Salix viminalis* L.). J. Plant Physiol. 134: 558–566.

Grossnickle, S.C., D.R. Roberts, J.E. Major, R.S. Folk, F.B. Webster and B.C.S. Sutton, 1992. Integration of somatic embryogenesis into operational forestry: Comparison of interior spruce emblings and seedlings during production of + stock. In: Proc. Intermountain Forest Nursery Association, 1991, pp. 106–113. Fort Collins : U.S.D.A., Forest Service, Rocky Mountain Forest and Range Experiment Station.

Hackett, H.P., 1985. Juvenility, maturation, and rejuvenation in woody plants. Hort. Rev. 7: 109–149.

Hackett, H.P., J.R Murray and A. Smith, 1992. In: Proc. IUFRO-AFOCEL Meeting: Mass Production Technology for Genetically Improved Fast Growing Trees, pp. 83–90. AFOCEL (Pub.), Nangis.

Haffner, V., F. Enjarlic, L. Lardet and M.P. Carron, 1991. Maturation of woody plants: a review of metabolic and genomic aspects. Ann. Sci. For. 48: 615–630.

Hakman, I. and L.C. Fowke, 1987. An embryogenic suspension culture of *Picea glauca* (white spruce). Plant Cell Rep. 6: 20–22.

Huang, L.C., S. Lius, B.L. Huang, T. Murashige, E.F.M. Mahdi and R. Van Gundy, 1992. Rejuvenation of *Sequoia sempervirens* by grafting of shoot tips onto juvenile rootstocks *in vitro*. Plant Physiol. 98 : 166–173.

Hutchison, K.W. and M.S. Greenwood, 1991. Molecular approaches to gene expression during conifer development and maturation. For. Ecol. Manag. 43: 273–286.

Hutchison, K.W., P.B. Singer and M.S. Greenwood, 1991. Gene expression during growth and maturation. In: M.R. Ahuja (Ed.), Woody Plant Biotechnology, pp. 69–75. Plenum Press, New York.

Jones, L.H. and W.A. Hugues, 1989. Oil palm (*Elaeis guineensis* Jacq.). In: Y.P.S. Bajaj (Ed.), Biotechnology in Agriculture and Forestry, Vol. 5, Trees II, pp. 176–202. Springer-Verlag, Berlin.

Jörgensen, J., 1989. Somatic embryogenesis in *Aesculus hippocastanum* L. by culture of filament callus. J. Plant Physiol. 135: 240–241.

Jörgensen, J., 1991. Somatic embryogenesis in *Aesculus hippocastanum* and *Quercus petraea* from old trees (10 to 140 years). In: M.R. Ahuja (Ed.), Woody Plant Biotechnology, pp. 351–352. Plenum Press, New York.

Kleinschmit, J. and A. Meier-Dinkel, 1990. Biotechnology in forest tree improvement: trees of the future. In: R. Rodriguez, R. Sanchez Tames and D.J. Durzan (Eds.), Plant Aging, Basic and Applied Approaches, NATO ASI Series, Series A: Life Sciences, Vol. 186, pp. 319–323. Plenum Press, New York.

Kleinschmit, J., D.K. Khurana, H.D. Gerhold and W.J. Libby, 1993. Past, present, and anticipated applications of clonal forestry. In: M.R. Ahuja and W.J. Libby (Eds.), Clonal Forestry II, Conservation and Application, pp. 9–41. Springer-Verlag, Berlin.

Kurten, U., A.M. Nuutila, V. Kauppinen and M. Rousi, 1990. Somatic embryogenesis in cell cultures of birch (*Betula pendula* Roth.). Plant Cell Tiss. Org. Cult. 23: 101–105.

Lakshmi Sita, G., 1986. Sandalwood (*Santalum album* L.). In: Y.P.S. Bajaj (Ed.), Biotechnology in Agriculture and Forestry, Vol. 1, Trees I, pp. 363–374. Springer-Verlag, Berlin.

Lakshmi Sita, G., N.V. Raghava Ram and C.S. Vaidyanathan, 1979. Differentiation of embryoids and plantlets from shoot callus of sandalwood. Plant Sci. Lett. 15: 265–270.

Lakshmi Sita, G., S. Rani and K.S. Rao, 1986. Propagation of *Eucalyptus grandis* by tissue culture. In: J.K. Sharma, C.T.S. Nair, S. Kedharnath and S. Kondas (Eds.), *Eucalyptus* in India, Past Present and Future. Proc. Natl Sem. Kerala For. Res. Inst., pp. 318–321. Peechi, Kerala.

Laliberté, S. and M. Lalonde, 1988. Sustained caulogenesis in callus cultures of *Larix* × *eurolepis* initiated from short shoot buds of a 12-year-old tree. Amer. J. Bot. 75: 767–777.

Le Roux, J.J. and J. van Staden, 1991. Micropropagation and tissue culture of *Eucalyptus* – a review. Tree Physiol. 9: 435–477.

Lee, B.C., Y.W. Kim, S.K. Lee, W.S. Song and T.S. Kim, 1991. Plant regeneration from adult tree explants via somatic embryogenesis and organogenesis in *Hibiscus syriacus* (Bulsae). Res. Rep. Inst. For. Gen. Korea 27: 85–90.

Libby, W.J. and R.M. Rauter, 1984. Advantages of clonal forestry. For. Chron. 60: 145–149.

Lioret, C., 1981. Vegetative propagation of oil palm by somatic embryogenesis. In: Inc. Soc. Planters (Ed.), Malaysian Intl. Agric. Oil Palm Conf. Rep., p. A13. Kuala Lumpur.

Litz, R.E. and V.S. Jaiswal, 1991. Micropropagation of tropical and subtropical fruits. In: P.C. Debergh and R.H. Zimmerman (Eds.), Micropropagation, pp. 247–263. Kluwer Academic Publishers, Dordrecht.

Litz, R.E., 1985. Somatic embryogenesis in tropical fruit trees. In: R.R. Henke, K.W. Hugues, M.J. Constantin and A. Hollaender (Eds.), Tissue Culture in Forestry and Agriculture, Basic Life Science, Vol. 2, pp. 179–193. Plenum Publishing Co., New York.

Litz, R.E., 1988. Somatic embryogenesis from cultured leaf explants of the tropical tree *Euphobia longan* Stend. J. Plant Physiol. 132: 190–193.

Lloyd, G. and B.H. McCown, 1981. Commercially feasible micropropagation of mountain laurel (*Kalmia latifolia*) by use of shoot tip-culture. Proc. Intl. Plant Prop. Soc. 30: 421–427.

McKeand, S.E., 1985. Expression of mature characteristics by tissue cultured plantlets derived from embryos of loblolly pine. J. Amer. Soc. Hort. Sci. 110: 619–623.

Mehra, P.N. and K. Jaidka, 1985. Experimental induction of embryogenesis in pear. Phytomorphol. 35: 1–10.

Mehra, P.N. and S. Sachdeva, 1985. Embryogenesis in apple *in vitro*. Phytomorphol. 31: 26–36.

Michaux-Ferrière, N., H. Grout and M.P. Carron, 1992. Origin and ontogenesis of somatic embryos in *Hevea brasiliensis* (*Euphorbiaceae*). Amer. J. Bot. 79: 174–180.

Michler, C.H. and E.O. Bauer, 1991. High frequency of somatic embryogenesis from leaf tissue of *Populus* spp. Plant Sci. 77: 111–118.

Montoro, P., H. Etienne, M.P. Carron and A. Nougarède, 1992. Incidence des cytokinines sur l'induction de l'embryogenèse et la qualité des embryons somatiques chez *Hevea brasiliensis* Müll. Arg. C.R. Acad. Sci. Paris Sér. III 315: 567–574.

Muhs, H.J., 1992. Macro- and microvegetative propagation as a tool in tree breeding as demonstrated by case studies of aspen and cashew and regulations for marketing clonal material. In: S.W.G. Baker (Ed.), Rapid Propagation of Fast-Growing Woody Species, CASAVA Report Series III, pp. 71–85. CAB Intl., Walling Ford.

Muralidharan, E.M. and A.F. Mascarenhas, 1987. *In vitro* plantlet formation by organogenesis in *E. camaldulensis* and by somatic embryogenesis in *Eucalyptus citriodora*. Plant Cell Rep. 6: 256–259.

Murashige, T. and F. Skoog, 1962. A revised medium for rapid growth and bioassays with tobacco tissue cultures. Physiol. Plant. 15: 479–497.

Nagmani, R. and J.M. Bonga, 1985. Embryogenesis in subcultured callus of *Larix decidua*. Can. J. For. 15: 1088–1091.

Navarro, L., J.M. Ortiz and J. Juárez, 1985. Aberrant citrus plants obtained by somatic embryogenesis of nucelli cultured *in vitro*. HortSci. 20: 214–215.

Neuenschwander, B. and T.W. Baumann, 1992. A novel type of somatic embryogenesis in *Coffea arabica*. Plant Cell Rep. 10: 608–612.

Nozeran, R., 1978. Polymorphisme des individus issus de la multiplication végétative des végétaux supérieurs, avec conservation du potentiel génétique. Physiol. Vég. 16: 177–194.

Ossor, A. and P. Boxus, 1992. Des agglomérats de domes méristématiques: une nouvelle alternative dans la multiplication clonale d'*Eucalyptus camaldulensis*. In: Proc. IUFRO-AFOCEL Meeting: Mass Production Technology for Genetically Improved Fast Growing Trees, pp. 203–211. AFOCEL (Pub.), Nangis.

Ouyang, Q., H.Z. Peng and Q.Q. Li, 1981. Studies on the development of embryoids from *Eucalyptus* callus. Sci. Silv. Sin. 17: 1–7.

Owens, J.N. and M.D. Blake, 1985. Forest tree seed production. Petawawa Natl. For. Inst. Inf. Rep. PI-X-53.

Park, Y.G. and S.H. Son, 1988. *In vitro* organogenesis and somatic embryogenesis from punctured leaf of *Populus nigra* × *P. maximoxiczii*. Plant Cell Tiss. Org. Cult. 15: 95–105.

Park, Y.S., S.E. Pond and J.M. Bonga, 1994. Initiation of somatic embryogenesis in white spruce (*Picea glauca*): genetic control, culture treatment effects, and implications for tree breeding. Theor. Appl. Genet. (in press).

Pétiard, V., J.P. Ducos, B. Florin, C. Lecouteux, H. Tessereau and A. Zamarripa, 1992. Mass somatic embryogenesis: a possible tool for large-scale propagation of selected plants. In: D. Come and F. Corbineau (Eds.), Proc. Quatrième Rencontre Internationale sur les Semences, Vol. I, pp. 175–192. July 1992, Angers, France.

Pierik, R.L.M., 1990. Rejuvenation and micropropagation. In: H.J.J. Nijkamp, L.H.W. van der Plas and J. van Aartrijk (Eds.), Progress in Plant Cellular and Molecular Biology, pp. 91–101. Kluwer Academic Publishers, Dordrecht.

Poethig, R.S., 1990. Phase change and the regulation of shoot morphogenesis in plants. Science 250: 923–930.

Profumo, P., P. Gastaldo and N. Rascio, 1987. Ultrastructural study of different types of callus from leaf explants of *Aesculus hippocastanum* L.. Protoplasma 138: 89–97.

Profumo, P., P. Gastaldo, R.M. Dameri and L. Caffaro, 1986. Histological study of calli and embryoids from leaf explants of *Aesculus hippocastanum* L.. J. Plant Physiol. 126: 97–103.

Qin, C.L. and E.G Kirby, 1992. *In vitro* responses of cultured explants of *Eucalyptus* spp. In: Proc. IUFRO-AFOCEL Meeting: Mass Production Technology for Genetically Improved Fast Growing Trees, pp. 203–211. AFOCEL (Pub), Nangis.

Rabéchault, H. and J.P. Martin, 1976. Multiplication végétative du palmier à huile (*Elaeis guineensis* Jacq.) à l'aide de tissus foliaires. C.R. Acad. Sci. Paris Sér. D 238: 1735–1737.

Rancillac, M. and A. Klinguer, 1991. Plant biotechnologies applied to a forest tree, the American red oak (*Quercus rubra* L.). Acta Hort. 289: 341–342.

Rangan, T.S., T. Murashige and W.P. Bitters, 1968. *In vitro* initiation of nucellar embryos in monoembryonic citrus. Hortic. Sci. 3: 226–227.

Rangaswany, N.S., 1961. Experimental studies on female reproduction structures of *Citrus microcarpa* Bunge. Phytomorphol. 11: 109–127.

Redenbaugh, K., J. Fujii, D. Slade, P. Viss and M. Kossler, 1991. Artificial seeds – encapsulated somatic embryos. In: Y.P.S. Bajaj (Ed.), Biotechnology in Agriculture and Forestry, Vol. 17, High-Tech and Micropropagation, pp. 395–416. Springer-Verlag, Berlin.

Roberts, D.R., F. Webster, B.S. Flinn, W.R. Lazaroff and D.R. Cyr, 1993. Somatic embryogenesis of spruce. In: K. Redenbaugh (Ed.), Synseeds, Application of Synthetic Seeds to Crop Improvement, pp. 427–449. CRC Press, Boca Raton.

Ruaud, J.N., 1994. Maturation and conversion into plantlets of somatic embryos derived from needles and cotyledons of 7 to 56 day old plants of *Picea abies*. Plant Sci. (in press).

Ruaud, J.N., J. Bercetche and M. Pâques, 1992. First evidence of somatic embryogenesis from needles of 1-year-old *Picea abies* plants. Plant Cell Rep. 11 : 563–566.

Sharma, D.R., S. Dawra and J.B. Chowdhury, 1984. Somatic embryogenesis and plant regeneration in date palm (*Phoenix dactylifera* Linn.) cv. Khadrawi through tissue culture. Ind. J. Exp. Bot. 22: 596–598.

Simola, L.K. and A. Santanen, 1990. Improvement of nutrient medium for growth and embryogenesis of megagametophyte and embryo callus lines of *Picea abies*. Physiol. Plant. 80: 27–35.

Simola, L.K., 1985. Propagation of plantlets from leaf callus of *Betula pendula* F *purpurea*. Scientia Hort. 26: 77–85.

Sondahl, M.R. and W.R. Sharp, 1977. High frequency induction of somatic embryos in cultured leaf explants of *Coffea arabica* L. Z. Planzenphysiol. 81: 395–408.

Staritsky, G., 1970. Embryoid formation in callus tissues of coffee. Acta Bot. Neerl. 19: 509–514.

Sterk, P. and S. de Vries, 1993. Molecular markers for plant embryos. In: K. Redenbaugh (Ed.), Synseeds, Application of Synthetic Seeds to Crop Improvement, pp. 115–132. CRC Press, Boca Raton.

Tautorus, T.E., L.C. Fowke and D.I. Dunstan, 1991. Somatic embryogenesis in conifers. Can. J. Bot. 69: 1873–1899.

Thorpe, T.A., I.S. Harry and P.K. Kumar, 1992. Application of micropropagation to forestry. In: P.C. Debergh and R.H. Zimmerman (Eds.), Micropropagation, pp. 311–336. Kluwer Academic Publishers, Dordrecht.

Tisserat, B., 1979. Propagation of date palm (*Phoenix dactylifera* L.) *in vitro*. J. Exp. Bot. 30: 1275–1283.

Tranvan, H., F. Bardat, M. Jacques and Y. Arnaud, 1991. Rajeunissement chez le *Sequoia sempervirens*: effets du microgreffage *in vitro*. Can. J. Bot. 69: 1772–1779.

Tulecke, W., 1987. Somatic embryogenesis in woody perennials. In: J.M. Bonga and D.J. Durzan (Eds.), Cell and Tissue Culture in Forestry, Vol. 2, pp. 61–91. Martinus Nijhoff Publishers, Dordrecht.

Valverde, R., A. Arias and T.A. Thorpe, 1987. Picloram-induced somatic embryogenesis in pejibaye palm (*Bactris gasipaes* H.B.K.). Plant Cell Tiss. Org. Cult. 10: 149–156.

Wang, Z., X. Zeng, C. Cheng, H. Wu, Q. Li, G. Fan and W. Lu, 1980. Induction of rubber plantlets from anther of *Hevea brasiliensis* Muell-Arg *in vitro*. Chin. J. Trop. Crops 1: 9–13.

Wann, S.R., 1988. Somatic embryogenesis in woody species. Hort. Rev. 10: 153–181.

Warrag, E., M.S. Lesney and D.J. Rockwood, 1991. Nodule culture and regeneration of *Eucalyptus grandis* hybrids. Plant Cell Rep. 9: 586–589.

Watt, M.P, F. Blakeway, C.F. Cresswell and B. Herman, 1991. Somatic embryogenesis in *Eucalyptus grandis*. South Afr. For. J. 157: 59–65.

Webster, F.B., D.R. Roberts, S.M. McInnis and B.C.S. Sutton, 1990. Propagation of interior spruce by somatic embryogenesis. Can. J. For. Res. 20: 1759–1765.
Westcott, R.J., 1992. Embryogenesis from non-juvenile Norway spruce (*Picea abies*). World Congress on Cell and Tissue Culture, Washington, D.C., USA.
White, P.R., 1943. A handbook of Plant Tissue Culture. J. Catell, Lancaster, USA.

7. Molecular Analysis of Zygotic and Somatic Conifer Embryos

Santosh Misra

Contents

1. Introduction

A large number of studies have examined changes in gene expression during embryogenesis and germination in dicot and monocot species (Goldberg *et al.*, 1989). Similar techniques are also being used to examine gene regulation associated with somatic embryogenesis (Choi *et al.*, 1987; Sterk and De Vries, 1993). Indeed, the study of gene expression during the process of somatic embryogenesis is increasingly being used to obtain molecular information on the very early events of embryogenesis and has overcome the difficulties encountered in obtaining sufficient amounts of material from early stages of zygotic embryogenesis. Development of somatic embryos proceeds in the absence of other seed tissues. It is, therefore, possible to: a) manipulate the chemical environment to study the embryogenic process, and b) investigate the way in which the developing embryo may alter its own chemical environment by secreting specific metabolites in to the culture medium (Lindsay and Topping, 1993). Recently, cDNA and antibody probes have been used to identify developmentally regulated genes that exhibit a stage-specific expression pattern in the early stages of carrot (*Daucus carota L.*) somatic embryogenesis. Also genes and proteins have been identified whose expression is strongly correlated with the acquisition and maintenance of embryogenic potential in carrot (*Daucus carota L.*) tissue cultures (Sterk and De Vries, 1993). Based on the expression pattern during zygotic and somatic embryogenesis the potential use of these gene(s) and antibody probes as markers of embryogenic potential and level of somatic embryo maturation has been suggested.

S. Jain, P. Gupta & R. Newton (eds.), Somatic Embryogenesis in Woody Plants, Vol. 1, 119–142.

Somatic embryogenesis in conifers was first described in Norway spruce (*Picea abies*) by Hakman and Von Arnold (1985). Since then, somatic embryogenesis has been reported in more than fifteen species of conifers (Gupta *et al.*, 1991). The main challenge still is to overcome the low frequencies of embryo maturation and their conversion into viable plantlets (Gupta *et al.*, 1993). In order to develop protocols for controlled embryogenesis and production of superior quality somatic embryos, a basic understanding of the biochemical and molecular processes that underlie formation of somatic and zygotic embryos is essential. Although morphology and development of conifer zygotic embryos has been studied for more than 100 years (Singh, 1978), up until recently, studies at the molecular level were non-existent. In the past few years, biochemical and molecular characterization of conifer embryogenesis and germination using recombinant DNA techniques have been initiated (Misra and Green, 1991; Misra *et al.*, 1993; Leal and Misra, 1993a,b; Whitmore and Kreibel, 1987; Newton *et al.*, 1992). It is hoped that similar to angiosperms, molecular characterization of conifer embryogenesis will lead to the development of stage-specific molecular markers which can be used to optimize somatic embryogenesis protocols by fine tuning easily identifiable, developmentally regulated events. This chapter will deal with recent studies on gene expression and ultrastructural development during zygotic and somatic embryogenesis (mainly white spruce (*Picea glauca*) and Douglas fir (*Pseudotsuga menziesii*)). In order to correlate biochemical and molecular changes during embryogenesis with morphological and ultrastructural development of zygotic and somatic embryos, a brief descriptive summary of conifer embryogenesis is also included.

2. Zygotic Embryogenesis

2.1. *Development of Conifer Zygotic Embryos*

The sequence of embryo development in conifers can be divided into two phases:

(1) a short proembryo phase that occurs within the archegonium;
(2) the embryo phase which occurs after the proembryo elongates into the haploid female gametophytic tissue (Singh, 1978; Owens and Moulder, 1984).

Based on the morphology, the embryo phase is commonly separated into several stages; early, mid and late embryogeny.

2.1.1. *Proembryo phase*

The proembryo development begins when the fertilized egg nucleus divides into two, then four nuclei contained within a dense region of cytoplasm. Cell wall formation between the eight nuclei creates two tiers of four cells each.

Each cell divides again forming a sixteen celled proembryo consisting of four tiers of four cells each. The third tier of cells the suspensor tier, elongates and thrusts the distal apical tier out of the archegonial jacket into the female gametophytic tissue which ends the proembryo phase and begins the early embryo phase.

2.1.2. *Embryo Phase*

During early development, cells of the apical tier divide forming the terminal embryonal cells and subterminal suspensor cells. This forces the terminal embryonal cells still deeper into the gametophyte. The embryonal cells divide forming an embryonal mass. The basal cells of the embryonal mass continue to divide and elongate contributing to the thick secondary suspension. Distal cells of the embryonal mass divide to form a club-shaped embryo. The club-shaped early embryo that survives the polyembryonic selection process enlarges rapidly during the mid-embryo stage to fill the corrosion cavity that forms in the female gametophyte. Specific meristematic regions appear which mark the beginning of the late-embryo stage. The late embryo goes through a period of growth and development before cells stop dividing and the embryo matures. During this time the hypocotyl (embryonic stem) and cotyledons elongate and the provascular tissue appear. The stem apical meristems remain small while the root meristem forms a massive cap (Owens and Moulder, 1984). Most conifers undergo two types of polyembryonic process; simple or cleavage polyembryony (Gifford and Foster, 1989). One proembryo is usually more vigorous and continues development while the others degenerate.

2.2. *Ultrastructural Changes during Embryogeny*

Embryogenesis in gymnosperms, proceeding the free nuclear phase involves a series of organized cell-division and differentiation leading to the formation of a differentiated embryo. At the biochemical level, the embryo goes through a series of changes that involve the co-ordinated synthesis and deposition of storage reserves in different parts of the diploid embryo and in the haploid megagametophyte. The accumulation of storage products at the ultrastructural and molecular levels has been extensively studied in angiosperm seeds (Bewley and Black, 1985). Much is known about the physiology and biochemistry of angiosperm seeds as they mature (Bewley and Black, 1985), but there is a less comparable information concerning conifer seeds.

Mature conifer seeds are known to accumulate reserves primarily as lipids and proteins which are mobilized rapidly during germination and subsequent seedling growth (Ching, 1963, 1966; Gifford, 1988; Misra and Green, 1990, 1991; Green *et al.*, 1991; Misra *et al.*, 1993). In *Pseudotsuga menziesii*, lipids made up about 50% and 30% of the megagametophyte and embryo dry weight, respectively, and proteins about 12% and 10% of the megagametophyte and embryo dry weight, respectively (Owens *et al.*, 1993). Soluble sugars made up only 2% to 3% of the dry weight of the megagametophyte and embryo, respectively. Similar distribution of storage reserves was found in mature *Picea glauca* (Misra, unpublished). In order to describe the gene expression and regulation of conifer zygotic and somatic embryos, a careful analysis of the stages of embryogenesis and megagametophytes both histochemically and ultrastructurally is required (Owens *et al.*, 1993). These changes can then be compared to biochemical and molecular changes.

The detailed ultrastructural development of *Picea glauca* (Krasowski and Owens, 1993) and *Pseudotsuga menziesii* (Owens *et al.*, 1993) post-fertilization megagametophytes and embryo were, recently, described. Prior to this, few studies devoted to the conifers examined the ultrastructural features of dormant seeds and changes that occur at germination (Simola, 1974; Durzan *et al.*, 1971; Misra and Green, 1990). The latter study showed ultrastructural features of fully formed isolated protein bodies of mature embryo and megagametophytes which demonstrated the presence of globoids and crystalloids embedded in a buffer soluble amorphous proteinaceous matrix. The globoid regions are known to contain phytin and mineral deposits (Lott, 1980), while the crystalloid inclusions are aggregates of highly insoluble polypeptides. The matrix and crystalloid proteins are often separated in protein extractions on the basis of different solubilities (Gifford, 1988; Misra and Green, 1990). A similar structure of protein bodies was reported in *Pseudotsuga menziesii* (Green *et al.*, 1991) and is similar to that reported in angiosperm seeds. In *Picea glauca* zygotic embryos, storage products accumulated in the following sequence; starch first then followed by lipids and finally proteins (Krasowski and Owens, 1993). The latter study showed that lipids were deposited rapidly from the club-shaped stage until early organogenesis. Major lipid accumulation in the embryos took place during rapid cotyledon development and simultaneously with the build up of protein bodies. Formation of protein bodies in the megagametophyte and embryo was first detected about six and twenty-nine days after fertilization (DAF), respectively. Based on their studies, the sequence of events leading to the formation of protein bodies was as follows: (1) deposition of amorphous protein clumps onto tonoplasts of subdividing vacuoles during early stages of protein body formation; (2) fusion of small cytoplasmic vesicles, possibly derived from the rough endoplasmic reticulum; and (3) the deposition of protein around dense, membrane bound vesicles attached to the tonoplast. The third process was not observed in the embryo. In *Picea abies*, storage protein deposition was not detected until fifteen DAF, whereas in embryo proteins accumulated twenty-one DAF (Hakman, 1993). The most important morphogenetic events in *Pseudotsuga menziesii* occurred during the first forty-three days after fertilization, and the embryo was matured seventy-one days after the fertilization (Owens *et al.*, 1993) (Table 1). During this time, lipid and protein bodies increased rapidly in the megagametophyte. Lipid bodies, starch, then protein bodies became evident in the embryos towards the end of the morphogenetic phase. In the

Table 1. Biochemical and ultrastructural development of conifer embryo[1].

Proembryo	Embryonal Cell Mass	Club Shaped/Early Embryo	Organogenesis	Advanced Organogenesis	Mature Embryo
Collection Date					
June 12	June 26	July 11	July 25	August 8	August 22
DAF					
0–14	14	29	43	57	71
Embryo Morphology					
Proembryo before suspensor elongation.	Embryonal cell mass, suspensor elongated.	Club-shaped embryo, distinct distal and proximal meristems.	Embryo elongation extending to 3/4 of the length of the megagametophyte, short apical meristems between a ring of elongating cotyledons.	Embryo morphologically mature, embryo fully elongated.	No visible morphological change, seeds mature and dry.
Histochemical and Ultrastructural Analysis					
Megagametophyte					
Plastids abundant with one or more large starch grain, no LB or PB, cell division.	LB abundant, protein deposits in vacuoles.	Cells filled with LB, large PB present, little starch.	LB and PB abundant.	Large differentiated PB, some starch, abundant LB.	No change in ultrastructure.
Embryo					
	No LB, electron dense deposits in vacuole periphery, no starch.	Few LB, proteins present in vacuoles.	LB abundant, PB present.	Cells packed w/PB and LB, some starch.	No change.

[1] Based on Owens *et al.*, 1993.

LB – Lipid body; PB – Protein body; DAF – Days after fertilization.

megagametophyte as well as the embryo cells, proteins were first detected fourteen DAF, as electron dense deposits in the vacuoles. Western blot of earlier seeds has detected storage proteins in megagametophytes from earlier collection date (Misra, unpublished). Size and number of megagametophyte lipid and protein bodies, as well as an increase in protein body complexity was observed during the late stages of maturation. Later in the mature seed, starch, lipid bodies and proteins were uniformly distributed in the megagametophyte. At maturity, starch was abundant in some regions of the embryo but not abundant in the megagametophyte. This was also the case in mature seeds of *Picea abies* (Hakman, 1993) and *Picea glauca* (Krasowski and Owens, 1993).

2.3. *Seed Proteins of Conifers*

According to their function seed proteins can be divided into three broad classes (Bewley and Black, 1985): (1) structural proteins, associated with membranes and ribosomes, (2) enzymes e.g., those required for mobilization of storage reserves, (3) storage proteins which are utilized during seed germination and seedling growth, thus supplying the necessary free amino acids and amide nitrogen. Extensive research in angiosperms has led to a detailed understanding of the synthesis, processing, regulation and molecular properties of seed storage proteins (for review see Shotwell and Larkins, 1989). Recent work with gymnosperms and particularly with conifers has demonstrated that their major seed storage proteins have structural homologies with angiosperm seed proteins. Characteristics of the storage proteins such as solubility, molecular weight, sub-unit structure as well as gene sequences have indicated these similarities (Misra and Green, 1990, 1991; Green *et al.*, 1991; Leal and Misra, 1993b; Gifford, 1988; Jensen and Berthold, 1989; Hakman, 1993; Roberts *et al.*, 1990a; Newton *et al.*, 1992; Allona *et al.*, 1992; Jensen and Lixue, 1991). Traditionally, the storage proteins have been classified as albumins, globumins, glutelins or prolamins, based on their solubility in solutions of water, salt, acid or alkali, or aqueous alcohols, respectively (Higgins, 1984). In legumes, the major seed storage proteins are of the globulin type while in cereals prolamins predominates. The globulins of legumes are split into major families based roughly on their molecular mass and sedimentation co-efficients (S valves). These are the 7S (vicilins) proteins and llS (legumin) proteins. Within the llS globulins, two broad classes of proteins can be distinguished, and they are; a) completely salt soluble (e.g., legumins of pea and soybean), and b) require the addition of SDS, urea or other denaturing substances to solubilize them (e.g., the crystalloids) (Bewley and Black, 1985). The storage proteins are initially synthesized as precursor polypeptides from large multigene families, and may undergo post-translational modifications such as signal peptide cleavages, glycosylation, oligomeric association, and combinations of intermolecular and intramolecular disulphur bridging (Shotwell and Larkins, 1989). Once processed, the proteins are either deposited in the cytoplasm or are transported via the Golgi apparatus to membrane-bounded organelles called protein bodies. In the conifers examined thus far (Misra and Green, 1990; Green *et al.*, 1991; Krasowski and Owens, 1993; Owens *et al.*, 1993) the protein bodies consist of globoid and crystalloid inclusions embedded in a buffer-soluble amorphous proteinaceous matrix. The globoid regions are predominantly phytin and mineral depositions (Lott, 1980), while the crystalloids are aggregates of highly insoluble polypeptides. The matrix and crystalloid proteins are often separated in protein extractions on the basis of differential solubilities (Misra and Green 1990; Green *et al.*, 1991).

Studies of seed proteins from a number of conifer species have shown that the crystalloids are the major storage proteins e.g., *Pinus* species (Gifford, 1988; Jensen and Lixue, 1991), *Picea glauca*, *Picea engelamannii*, *Picea mariana* (Misra and Green 1990, 1991), *Pseudotsuga menziesii* (Green *et al.*, 1991), *Picea glauca engelmannii complex* (Roberts *et al.*, 1990a), *Picea abies* (Hakman, 1993). In mature seeds of *Picea glauca* and *Pseudotsuga menziesii* these proteins accounted for 70 to 80% of the storage proteins and were localized in protein bodies. The SDS-soluble proteins in the protein bodies, isolated from megagametophytes and embryos, consisted of a complex of 55–63 kDa which under reducing conditions appeared as two major groups with molecular weights of 20–23 kDa and 32–35 kDa (Misra and Green, 1990; Green *et al.*, 1991). The predominant buffer-soluble proteins had an apparent molecular weight of 45–47 kDa. Similar subunit structures was reported in high salt-soluble globulin fraction of *Picea abies* (Hakman, 1993) and, recently, in urea soluble glutelin fraction isolated from megagametophytes of *Pinus pinaster* (Allona *et al.*, 1992). Based on the antigenic homology, the crystalloid proteins appear to have homology with the 11S legumin proteins of angiosperms (Misra and Green, 1993). Molecular cloning of legumin cDNA of *Pseudotsuga menziesii* (Misra and Leal, 1993; Leal and Misra, 1993b) as well as 7S vicilin of *Picea glauca engelmannii complex* (Newton *et al.*, 1992) were recently described. The characterization of these sequences demonstrated homology to corresponding sequences in angiosperms. A survey of conifer seeds using anticrystalloid antibodies has confirmed that, with the exception of *Abies*, crystalloids are widespread in conifer seeds (Misra and Green, 1993). The conclusion can be drawn from these studies that the conifer seed proteins that require SDS, urea, or high salt-extractions belong, primarily, to the legumin class of storage proteins and are the major component of storage proteins of most conifer seeds.

2.4. *Gene Expression during Zygotic Embryogeny*

The characterization of gene expression during embryo development, maturation and germination has led to the identification of distinct subsets of developmentally regulated genes in angiosperms (Dure, 1985; Galau *et al.*, 1987; reviewed in Goldberg *et al.*, 1989; Skriver and Mundy, 1990; Delseny *et al.*, 1993). These genes can be divided into five major subsets as follows: (1) constitutively expressed subsets of genes whose products are present at all stages, (2) germination specific subsets, (3) genes highly expressed during early embryogenesis, (4) a subset representing storage protein mRNAs, prevalent during expansion phase of cotyledons (mid-embryogenesis) and declining sharply at the end of this developmental phase, (5) a set of genes expressed abundantly in late embryogenesis until seed maturation which are stored in the dry seed and are rapidly degraded following imbibition and germination, generally referred to as the late embryogenesis abundant or Lea proteins.

Storage protein gene expression of angiosperms has been a subject of investigation for several years (reviewed in Higgins, 1984; Shotwell and Larkins, 1989; Goldberg *et al.*, 1989; Bewley and Marcus, 1990; Skriver and Mundy, 1990; Delseny *et al.*, 1993). These studies have shown that storage protein gene expression is under developmental, genetic and environmental control (Higgins, 1984). The expression is regulated mainly at the transcriptional level. However, post-transcriptional and translational regulation also contributes to regulating levels of these proteins (Quatrano, 1986).

Another set of genes characterized in detail from angiosperm embryos are the Lea genes which are preferentially expressed during late embryogenesis (Galau *et al.*, 1987; Skriver and Mundy, 1990; Lane, 1991; Dure *et al.*, 1989; Delseny *et al.*, 1993). Under experimental conditions, the level of these maturation-associated proteins can be manipulated using abscisic acid (ABA) – or – an osmoticum for example; during imbibition, as well as by premature drying of seeds (Skriver and Mundy, 1990). The onset or rise in level of these polypeptides coincides with an increase in ABA levels within the seed. In developing zygotic seed, the mRNAs encoding Lea proteins are detected in embryos around mid-embryogenesis, highest levels expressed during late desiccation phase. These mRNAs are stored in mature dry seed but are rapidly degraded upon imbibition (Skriver and Mundy, 1990). Many of these Lea genes have now been characterized from various plant species (Mundy and Chua, 1988; reviewed in Delseny *et al.*, 1993). Based on the sequence homology among a number of these genes, three major homology groups have been identified; Group I e.g., *Triticum aestivum* Em, *Raphanus sativus* p8B6, and *Gossypium hirsutum* D19; Group II Lea genes, e.g., Rab 16, Rab 21, *Gossypium hirsutum* D11, and Group III (Em-like genes) e.g., *Hordeum vulgare* B19, *Gossypium hirsutum* D132 (Delseny *et al.*, 1993). Generally, these proteins are hydrophilic and contain a large number of uncharged as well as hydroxylated amino acids arranged in conserved protein domains. They are believed to stabilize other proteins and possibly membranes, thus protecting seed tissue during desiccation and dormancy or possible cellular disruption upon subsequent rehydration (Dure *et al.*, 1989).

Most of the Lea genes can also be induced in other plant parts by exogenous ABA application (rab genes – ABA responsive genes) in the absence of water stress. Other stress treatments e.g., wounding, salt, cold (Mundy and Chua, 1988) are also capable of eliciting expression of these genes. Therefore, the function of these proteins may reflect a common protective role in plant cells during stress situations.

The gene expression studies in conifers have focused mainly on storage proteins and, recently on Lea proteins (Leal and Misra, 1993a). Storage protein synthesis in developing seeds of *Picea glauca* (Misra and Green, 1991), *Picea glauca engelmannii complex* (Flinn *et al.*, 1991a,b) and *Picea abies* (Hakman *et al.*, 1990; Hakman, 1993), *Pseudotsuga menziesii* (Misra, unpublished) has been investigated using SDS-PAGE and immunoblot analysis (Misra and Green, 1991). The latter study showed that 57 kDa crystalloid complex could be detected in the megagametophyte a few days after fertilization and its level increased significantly during the following 14 days. In this initial period of accumulation, the 35 kDa protein was the most prominent component of the 57 kDa complex. The most dramatic changes in protein accumulation occurred between the stage of mid-embryogenesis to the beginning of late embryogenesis. In *Picea glauca* this period corresponds to the enlargement of club-shaped embryo to fill the corrosion cavity formed in the megagametophyte. This was also the time corresponding to the early maturation stages of cotyledonary differentiation. The increase in protein content of the embryo coincided with an increase in fresh weight as well as the dry weight of the embryo. Similar trends were observed in developing *Pseudotsuga menziesii* seeds (Misra, unpublished) and in *Picea glauca engelmannii complex* seeds (Flinn *et al.*, 1991a). In *Picea abies*, storage proteins in the megagametophyte began accumulating as the undifferentiated embryo began to grow into the surrounding tissue whereas in the embryo storage proteins accumulated during later stages of rapid growth and differentiation. This was also the time corresponding to the early maturation stages of cotyledonary differentiation.

Unlike *Picea abies*, in megagametophytic tissue of *Picea glauca*, the synthesis of storage proteins was first detected six days after fertilization, coinciding with the initiation of the proembryo phase. These changes were also observed at the mRNA levels (Leal and Misra, 1993a). The transcripts hybridizing to a crystalloid cDNA probe began accumulating in the megagametophytic tissue soon after fertilization and accumulated to high levels as early as the embryonal mass stage (Fig. 1A-D). In the embryonic axis, crystalloid transcripts reached a maximal level in the club-shaped stage when embryonic axis were easily dissected from the megagametophyte tissue. Similar pattern of crystalloid (legumin-like storage protein) gene expression was observed in *Pseudotsuga menziesii* (Leal and Misra, 1993b). In *Pseudotsuga menziesii* megagametophyte Pseudotsugin (crystalloid) mRNAs started accumulating two weeks after fertilization and were abundant in the early to mid-stages of embryogenesis, corresponding to the club-shaped (29 DAF) embryos through to cotyledonary stages (43 DAF). In embryonic axis these

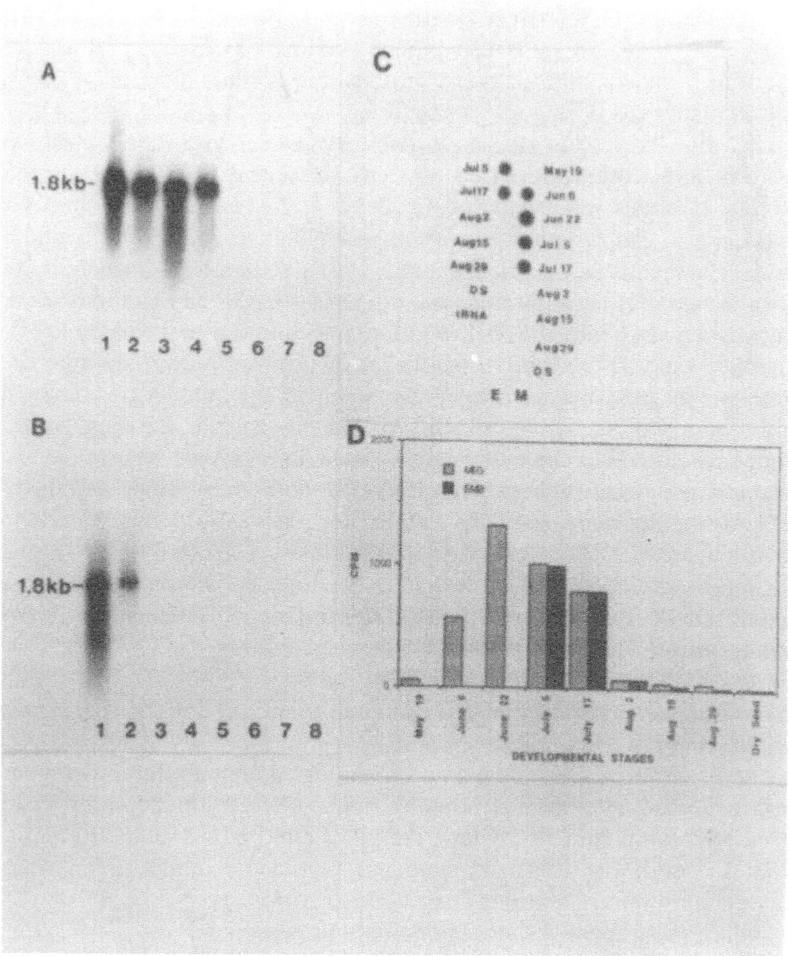

Figure 1. Analysis of crystalloid legumin transcript levels in *Picea glauca* RNA isolated from: (A) megagametophytes at the following developmental stages 6 DAF (1): 15 DAF (2); 29 DAF (3); 43 DAF (4); 58 DAF (5); 70 DAF (6); 84 DAF (7); mature seed (8). (B) embryos at the following developmental stages: 29 DAF (1); 43 DAF (2); 58 DAF (3); 70 DAF (4); 84 DAF (5); mature seed (6) and from germinated seedlings: leaves (7) and roots (8). (C) Dot-blot. Analysis of crystalloid transcripts in embryo (E) and megagametophytes (M). (D) The graph below shows the increase in the accumulation level of transcripts as measured by the amount of radioactivity incorporated (adapted from Leal and Misra, 1993a. Plant Science 88: 25–37).

transcripts also accumulated to high levels during early-to-mid stages of embryonic development corresponding to 43 to 57 DAF. The levels of these transcripts decreased rapidly, during the late stages of maturation.

In each organ, the maximal levels of crystalloid transcript accumulation preceded the period of rapid accumulation of crystalloid storage proteins

(Leal and Misra, 1993a). Albumin storage protein gene expression using a cloned cDNA probe from *Pseudotsuga menziesii* embryonic cDNA library showed a similar pattern of gene expression (Misra and Chatthai, unpublished). Only a few other studies have characterized embryo-specific gene expression of conifers at the transcriptional and post-transcriptional level. Whitmore and Kreibel (1987) studied the expression of a gene transiently expressed in ovules and developing embryos of *Pinus strobus*. The temporal expression of crystalloid storage protein mRNAs in *Picea glauca* differed from the 23 kDa protein mRNA described in *Pinus strobus*. Also, unlike the storage protein gene expression, the expression of the 23 kDa protein gene was limited to the seed coat and nucellus tissue.

Contrary to storage protein gene expression, expression of Lea genes in conifer embryogenes is comparable to the developmental pattern reported in *Raphanus sativus* (Raynal *et al.*, 1989) and in other angiosperms (Galau *et al.*, 1987; Skriver and Mundy, 1990). In Northern blot analysis of *Picea glauca* RNA from developing embryos and megagametophytes, the p8B6 Lea probe from *Raphanus sativus*, hybridized to a 600 nucleotide mRNA class which began accumulating just prior to the onset of desiccation (late cotyledonary phase), reached maximal levels in the fully mature embryo and was stored in the dry seed (Fig. 2A-C). The cross-hybridization of *Raphanus sativus* desiccation cDNA (Group I Lea gene) to corresponding sequences in conifers suggests that there are sequence similarities between the Lea genes of angiosperms and gymnosperms and that cellular mechanisms operating during stress in two groups may be comparable.

3. Somatic Embryogenesis

3.1. *Development of Somatic Embryos*

A problem encountered in the analysis of synthetic events during the development of somatic embryos is the difficulty in making direct comparisons with the developmental stages of zygotic embryogeny (Krochko *et al.*, 1992). In conifers, somatic embryogeny in general entails three stages: induction, proliferation, development and maturation. In the first phase embryogenic cultures can be initiated from several explant types e.g., immature and mature zygotic embryos, young seedlings, and megagametophytes (Gupta *et al.*, 1991). Initiation of embryogenic cultures usually requires the presence of auxin, cytokinin and low osmotic conditions such as those provided by 1–3% sucrose in the medium. Occasionally callus tissue undifferentiated parenchyma cells forming on the explant may give rise to somatic embryos e.g. *Picea glauca engelmannii complex* (Tautorus *et al.*, 1991). Most often, in conifers, however, this tissue is an organized mass of compact embryonal cells associated with long suspensors, comparable to the early embryo of conventional zygotic development. These structures are referred to as embry-

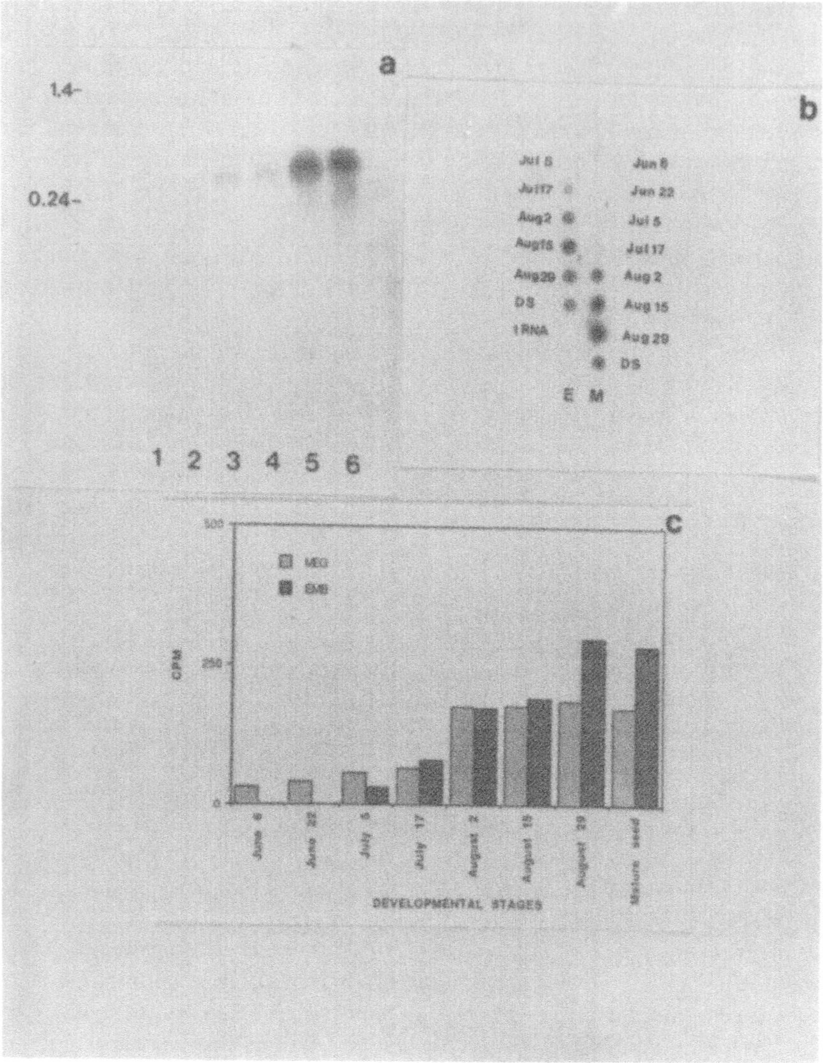

Figure 2. Analysis of desiccation (Lea) transcripts in developing seeds of *Picea glauca*. A) Northern blot of RNA isolated from zygotic embryos of the following developmental stages: 29 DAF (1); 43 DAF (2); 58 DAF (3); 70 DAF (4); 84 DAF (5); mature seed (6). (B) Dot-blot analysis of desiccation transcripts in developing embryo (E) and megagametophytes (M), using desiccation p8B6 cDNA clone from *Raphanus sativus* as a probe. The graph below shows the increase in the accumulation level of transcripts as measured by the amount of radioactivity incorporated (data after Leal and Misra, 1993a. Plant Science 88: 25–37).

onal masses; in the literature, they are designated stage 1 somatic embryos (Von Arnold and Hakman, 1988), embryonal suspensor masses (Gupta *et al.*, 1991), and proembryo and embryogenic calli (Tautorus *et al.*, 1991). During the proliferation phase of embryogenesis the embryonal masses undergo cleavage polyembryony analogous to that observed in zygotic development (Gupta *et al.*, 1991). Thus, the induction and proliferation phase of conifer somatic embryogenesis are analogous to the early phase of zygotic embryogeny, which is normally preceded by a proembryo phase *in situ*. Development of the embryonal masses to form bipolar embryos with cotyledons, hypocotyl, and radicle is then induced by modification of the culture medium. These modifications include exogenous application of ABA (Gupta *et al.*, 1991; Hakman *et al.*, 1990; Roberts, 1991; Roberts *et al.*, 1990a) and possibly by increased osmotic concentration of the medium (Attree *et al.*, 1992; Gupta *et al.*, 1991, 1993).

3.2. *Role of ABA and Osmoticum*

Abscisic acid (ABA) (Skriver and Mundy, 1990) and osmotica (Kermode, 1990) are two factors thought to be involved in regulation of storage polypeptide synthesis in angiosperm species examined, thus far. The two factors are not independent since ABA inhibits water intake in developing seeds (Goffner *et al.*, 1990) and osmotic stress increases endogenous levels of ABA (Skriver and Mundy, 1990). Abscisic acid is postulated to play a role in many processes of embryogenesis (Quatrano, 1986), which include the onset and maintenance of dormancy, inhibition of precocious germination (Kermode, 1990), embryo morphogenesis, and desiccation tolerance (Skriver and Mundy, 1990; Hetherington and Quatrano, 1991).

The effect of ABA and osmotica in development and maturation of zygotic embryo of conifers remains largely unknown. However, their central role in conifer embryogeny is emerging from various studies on maturation of somatic embryos (Gupta *et al.*, 1993; Attree *et al.*, 1992; Misra *et al.*, 1993). Studies of conifer somatic embryogenesis have shown that in conifers ABA is essential for: (a) inhibiting cleavage polyembryony (Gupta *et al.*, 1991, 1993), leading to separation and singulation of cotyledonary embryos, (b) accumulation of storage reserves (Roberts *et al.*, 1990a; Misra *et al.*, 1993; Hakman, 1990), and (c) synchronized maturation of somatic embryos. Abscisic acid and osmotic stress are known to be important factors for the maturation of seeds of many angiosperm species (Hetherington and Quatrano, 1991; Kermode, 1990). In general, ABA accumulates in the seed during mid-to-late-stages of seed development and prevents the developing embryos from germinating precociously and is, probably, involved with the acquisition of desiccation tolerance by the embryo (Kermode, 1990). In most seeds, this decrease in ABA occurs prior to water loss and then continues to decrease to low levels during the desiccation phase of the seed. The decrease of ABA in the mature dry seed allows the mature embryos to germinate upon imbibition. Consistent with the decrease in ABA levels during later seed development, there is an evidence for a decrease in tissue sensitivity to ABA (Hetherington and Quatrano, 1991). In conifer zygotic embryos, changes in the levels of ABA (40 μM) are required to prolong the maturation phase (Roberts, 1991) and prevent precocious germination. Increased osmolarity in development and maturation media (in combination with low concentration of ABA) has been found to be an effective treatment to inhibit greening and precocious germination of somatic embryos of conifers, which allow extended maturation of high quality embryos, and the capability of withstanding desiccation to below 10–30% moisture content (Gupta *et al.*, 1993; Attree *et al.*, 1992). For example, Gupta and Pullman (1990, 1991) used a combination of osmolarity (ranging from 350–600 mm/kg PEG 8000) ABA (2.5–10.0 ppm) and charcoal (0.05–0.25%) for maturation of *Picea abies* and *Pseudotsuga menziesii* embryos. A combination of ABA and charcoal allowed for a gradual decrease in levels of ABA. The high quality of cotyledonary somatic embryos produced using these protocols could be desiccated to 10% moisture content with 80 to 90% germination rate from several genotypes of *Picea abies*. Similarly, Attree *et al.* (1992) used 5–10% PEG 4000 with 16 μM ABA for maturation of *Picea glauca* somatic embryos. These conditions led to three fold increase in maturation and produced somatic embryos with well developed cotyledons, increased storage products (Misra *et al.*, 1993; Attree *et al.*, 1992) and ability to withstand desiccation below 30% moisture. Use of polyethylene glycol (PEG), as non-plasmolyzing osmotic stress in maturation media, is more effective than sucrose and mannitol treatments. For example, Roberts *et al.* (1990b) reported an increase in the frequency of globular stage embryos of *Picea glauca engelmannii complex* when a single week pulse of treatment of 2–6% mannitol was included in the media along with 3–4% sucrose and 40–80 μM ABA. However, prolonged culture with additional mannitol inhibited the subsequent cotyledonary stage embryos development.

3.3. *Biochemical and Molecular Analysis of Somatic Embryos*

Traditionally, morphological descriptions have been used to evaluate the developmental stage and conditions of somatic embryos, however this method may not accurately reflect culture vigor, and may be subjective (Nomura and Komamine, 1986). Somatic embryos which have an outwardly normal appearance and serve the basic function of propagules may display deviant cellular or tissue differentiation, biochemical or genetic abnormalities. Analysis at the cellular level using nucleic acid and protein analysis could potentially reveal molecular markers that are indicative of or are specific to particular stage(s) of embryo development. They may provide consistency and reliability in the assessment and may also lead to a greater understanding of the processes, control and genetics of somatic embryogenesis. Several workers have followed the accumulation of storage reserves in maturing somatic embryos of conifers at the ultrastructural level (Hakman, 1993; Misra *et al.*, 1993; Attree *et al.*, 1992) and at biochemical level (Feirer *et al.*, 1989; Hakman *et al.*, 1990; Misra *et al.*, 1993; Attree *et al.*, 1992; Roberts *et al.*, 1990a; Flinn *et al.*, 1991b) (Table 2).

A comparative study of somatic and zygotic embryos of *Picea abies* (Hakman *et al.*, 1990; Hakman, 1993) and *Picea glauca engelmannii* (Flinn *et al.*, 1991b) showed that the developmental pattern of storage protein accumulations were similar. The latter study showed that in both somatic and zygotic embryos major matrix protein accumulated during the later stages of embryo development whereas, crystalloids began to accumulate during the early stages of embryogenesis. Following germination and early growth of somatic embryos, the storage proteins are mobilized in a similar manner to zygotic embryos (Misra *et al.*, 1993; Cyr *et al.*, 1991).

Studies have shown that the accumulation of storage reserves is influenced by ABA and a high concentration of osmoticum. In *Picea abies* SDS-PAGE analysis showed that the protein composition of somatic embryos maturing on media containing just 7.5 μM ABA and low osmoticum was very similar to that of zygotic embryos (Hakman *et al.*, 1990). In *Picea glauca engelmannii complex*, a high level of ABA (up to 40 μM) was required to promote storage protein synthesis, typical of zygotic embryos, and to prevent precocious germination under low osmotic conditions (Roberts, 1991; Flinn *et al.*, 1991b). In four-week-old *Picea glauca*, somatic embryos maturing on 10–16 μM ABA alone, some of the major storage polypeptides e.g., 42 kDa protein subunit were missing or underexpressed (Joy *et al.*, 1991; Misra *et al.*, 1993). The crystalloid stórage protein profile of these somatic embryos resembled the protein profile of immature zygotic embryos (Misra and Green, 1991). This lack of maturity displayed by the *Picea glauca* somatic embryos under low osmotic conditions and low ABA concentrations was also reflected in the structure of the protein bodies where only diffuse deposits of proteins were observed near the periphery of the vacuole. In contrast to this, the full spectrum of storage polypeptides was synthesized after just four weeks when *Picea glauca* somatic embryos were cultured under high osmoticum (Fig. 3A-B). Under these conditions, storage protein levels continued to increase during an eight week maturation period (Misra *et al.*, 1993). The ultrastructure of osmotically treated somatic embryos closely resembled that of mature zygotic embryos.

A combination of ABA and high osmoticum also stimulated accumulation of triglycerides (TAG) in *Picea glauca* somatic embryos (Attree *et al.*, 1992). The somatic embryos matured with PEG 4000 and 16 μM ABA demonstrated nine-fold increase in the amount of TAG accumulated per embryo. This is about five times more than in the zygotic embryo levels of TAG (Attree *et al.*, 1992). After eight-weeks-of maturation under these conditions, the fatty acid composition of somatic embryos was comparable to that of mature zygotic embryos. Contrary to this, somatic embryos of *Picea glauca* and *Picea abies* matured with ABA, and low osmoticum accumulated at much lower

Table 2. Storage protein analysis in developing somatic embryos of conifers.

Species	Treatment Used for Maturation	Maturation Period (Weeks)	Major Polypeptides (kDa)	References
Picea glauca Moench Voss. (White spruce)	2.5–10.0% PEG 4000, 16 μM ABA, 3% sucrose	4–8	42, 35, 23, 8	Misra *et al.*, 1993
Picea glauca Moench Voss. (White spruce)	10 μM ABA, 3% sucrose	4	35, 22	Joy *et al.*, 1991
Picea abies L. Karst (Norway spruce)	7.6 μM ABA, 90 mM sucrose	4–5	42, 33, 28, 22	Hakman *et al.*, 1990
Picea glauca/Picea engelmannii (Interior spruce)	30–40 μM ABA, 0.1–10 μM IBA, 3–4% sucrose	–	41, 33, 24, 22	Roberts *et al.*, 1990a
Picea glauca Moench Voss/Picea engelmannii parry (Interior spruce)	40 μM ABA, 1 μM IBA, 3–4% sucrose	7–9	41, 35, 33, 24, 22	Flinn *et al.*, 1991b
Picea glauca Moench Voss/Picea englemannii Parry (Interior spruce)	40 μM ABA, 1 μM IBA, 3–4% sucrose, 2–6% mannitol (1 week pulse)	3–4	41, 33, 22	Roberts, 1991
Pseudotsuga menziesii Mirb. Franco (Douglas fir)	19–26% PEG 8000, 2% sucrose, 0.1% activated charcoal (5–10 ppm ABA pulse for 3 weeks)	4–6	47, 35, 22, 14	Misra *et al.*, unpublished; Gupta *et al.*, 1990

IBA – Indole butyric acid; PEG – Polyethylene glycol.

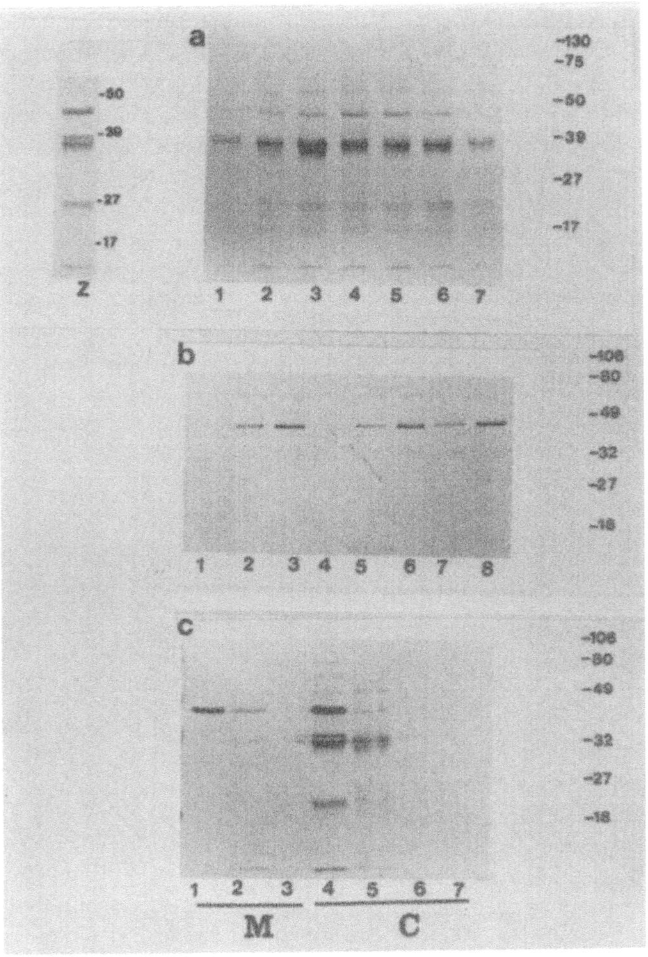

Figure 3. Western blot (immunoblot) analysis of storage protein profiles in *Picea glauca* somatic and zygotic embryos (Z). (A) Effect of osmoticum and osmoticum followed by a subsequent desiccation treatment on crystalloid protein profiles of *Picea glauca* somatic embryos matured for 4 weeks on 16 μM ABA and different concentrations of PEG. Lane 1, 2.5% PEG + 80% RH; 2, 5% PEG + 80% RH; 3, 7.5% PEG + 80% RH; 4, 2.5% PEG; 5, 5% PEG; 6, 7.5% PEG; 7, 10% PEG. (Z) Proteins isolated from mature zygotic embryos. (B) Effect of osmoticum followed by a subsequent desiccation (81% RH) or osmoticum alone on matrix (soluble) protein profiles of white spruce somatic embryos matured as described in (A). Lane 1, 2.5 PEG + desiccation; 2, 5% PEG + dessiccation; 3, 7.5% PEG + desiccation: 4, 2.5 PEG; 5, 5% PEG; 6, 7.5% PEG; 7, 10% PEG; 8, mature zygotic embroys. (C) Western blot of matrix (M) and crystalloid (C) proteins extracted from: lanes 1 and 5, desiccated somatic embryos; 2 and 6, rehydrated somatic embryos; 3 and 7, 2–week-old plantlets; 4) mature dry zygotic embryos. Somatic embryos were matured for 4 weeks on 16 uM ABA with 7.5% PEG then desiccated for 2 weeks at 81% RH prior to transfer to phytohormone-free plantlet regeneration medium for plantlet development (after Misra *et al.*, 1993. Ann. Bot. 71: 11–22).

levels of TAG when compared to their mature zygotic counterparts (Feirer *et al.*, 1989; Joy *et al.*, 1991; Attree *et al.*, 1992). Higher osmoticum concentration and ABA also yielded a three fold higher total protein content compared to somatic embryos matured with ABA and low osmoticum conditions (Misra *et al.*, 1993). Analysis of the *Picea glauca* mRNAs by *in vitro* translation showed that the crystalloid protein mRNA profile of zygotic embryos was similar to those of somatic embryos matured with ABA and low osmoticum (Misra *et al.*, 1993). The legumin transcript levels analyzed by Northern hybridization using a legumin cDNA clone as a probe (Leal and Misra, 1993b) were first detected in two-week-old somatic embryos matured with polyethylene glycol (PEG) and ABA, and continued to increase until six weeks of maturation (Fig. 4). After that, there was a decline in the level of legumin transcripts. Similar trend is seen for albumin transcripts (Misra, unpublished). In ESM cultures, the transcripts were not detected. The level of storage protein transcripts in somatic embryos matured with ABA (16 μM) alone was much lower than that observed with ABA and PEG (Misra, unpublished). These studies indicate that in differentiated conifer somatic embryos storage protein genes are induced in response to ABA treatment, however, high osmoticum is required for continued accumulation of the legumin transcripts as well as storage polypeptides. Barratt and Clark (1989) observed a cumulative effect of ABA and osmoticum in promoting maturation of pea embryos. Recent reports indicate that the effect of osmoticum is not mediated through a rise in endogenous levels of ABA and that each treatment imparts a distinct metabolic response on the developing embryo (Goffner *et al.*, 1990; Xu *et al.*, 1990; Rivin and Grudt, 1991). For example, Xu *et al.* (1990) showed that ABA and high osmotic potential prevented precocious germination of developing embryos in culture; however, only the osmoticum promoted the maintenance of a protein pattern which was typical of developing embryos. The cumulative effect of ABA and osmoticum on the maintenance of storage protein synthesis in conifers is similar to that observed in angiosperms, indicating common regulatory mechanisms in these two distant groups of plants.

3.4. *Desiccation and Plantlet Regeneration*

For most seeds, maturation drying is the terminal event in development, leading to a state of metabolic quiescence (Bewley and Black, 1985). Following imbibition of mature dry seeds, the metabolic activity associated with seed germination is initiated. It has been proposed that maturation drying or desiccation plays some role in turning off of the developmental gene expression and activating a germination/growth oriented gene expression (Misra *et al.*, 1985; Kermode, 1990). The switch can be affected by either partial or complete desiccation treatment of developing seeds during the desiccation-tolerant phase (Kermode, 1990).

Partial drying of mature somatic embryos resulted in a 3–5% moisture

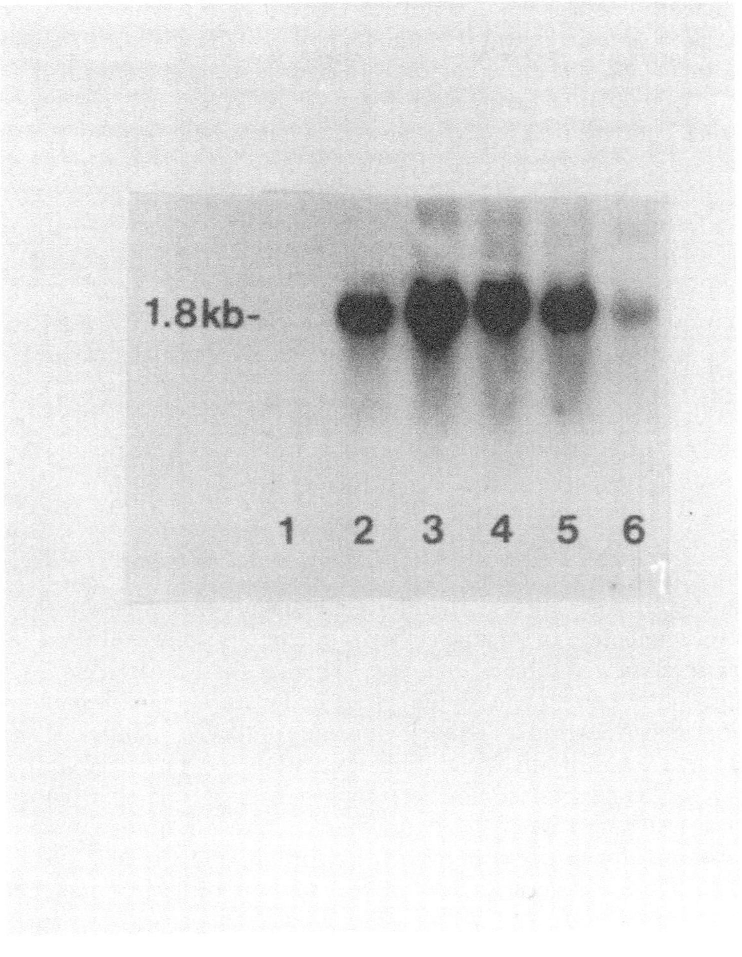

Figure 4. Analysis of crystalloid transcripts in somatic embryos of *Picea glauca*: Embryos were matured with 16 μM ABA and 7.5% PEG. Lane 1 ESM; 2, 1 week; 3, 2 week; 4, 4 week; 5, 6 week; 6, 8 week old somatic embryos. A *Picea glauca* legumin cDNA clone (WS2) isolated from zygotic embryos was used as a probe.

loss in *Picea glauca engelmannii complex* (Roberts *et al.*, 1990b). Conversion of somatic embryos was considerably improved (from 14–65%) by this mild desiccation treatment at 97% relative humidity (Roberts *et al.*, 1991). Recent reports show that the somatic embryos of conifers matured with high osmoticum and ABA can be desiccated to a moisture content below 10–30% (Gupta *et al.*, 1991, 1993). Desiccated somatic embryos of *Picea abies* displayed 80–90% germination rates from several genotypes (Gupta *et al.*, 1993). Desiccation of *Picea glauca* somatic embryos to 32% moisture content was achieved in a two stage process (Attree *et al.*, 1991, 1992). Initially, maturation of somatic embryos on medium containing PEG 4000 and ABA lead to mature somatic embryos with hydrated moisture content reduced from about 60% in controls to about 45–50%. Further, as reduction in moisture content to 32% was achieved by placing the mature somatic embryos under relative humidity 81–43% for two weeks. Following imbibition, the somatic embryos developed to plantlets at frequencies over 80%.

Analysis of protein patterns showed that the desiccated *Picea glauca* somatic embryos displayed metabolic events similar to those reported in desiccated embryos of angiosperms (Misra *et al.*, 1985; Kermode, 1990). The protein profile of *Picea glauca* desiccated somatic embryos was similar to that of nondesiccated somatic embryos and desiccated zygotic embryos (Fig. 3A-C). Thus, PEG treatment prevented the metabolism of storage polypeptides during maturation as well as desiccation. During plantlet regeneration, the storage proteins and lipids of the somatic embryos were rapidly mobilized in a similar manner to that in zygotic embryos and germination related gene activity was observed (Misra *et al.*, 1993; Attree *et al.*, 1992). Based on the findings on agronomic species, accumulation of storage reserves is likely to be associated with an ability to tolerate desiccation to less than 10% water content (Kermode, 1990). Therefore, treatments that promote storage reserve accumulation in somatic embryos of conifers, may contribute to desiccation tolerance of somatic embryos. Whether these treatments lead to accumulation of Lea proteins during prolonged maturation of somatic embryos on PEG and ABA and a declining moisture content, remains to be analyzed.

So far, studies of mRNAs during desiccation and germination of somatic embryos of conifers have not been reported. In zygotic embryos of angiosperms desiccation treatment is required to turn off storage proteins gene expression (Finkelstein *et al.*, 1985; Finkelstein and Crouch, 1986) and to activate genes for germination. In the absence of such a treatment, the developmental and germination gene activity appear to overlap. Such a situation in non-quiescent somatic embryos may be the cause of poor postgerminative vigor, whereas partial or complete drying of somatic embryos, which leads to synchronization of germination, may be effective in turning off of the developmental associated gene activity. Dot-blot analysis of legumin transcripts in *Picea glauca* somatic embryos showed that desiccation treatment resulted in turning off of storage protein gene expression (data not shown). Indeed, a decrease in level of legumin transcripts was observed in eight-week-old somatic embryos matured with PEG and ABA (Fig. 4). This decline could be attributed to a gradual moisture loss of somatic embryos matured with PEG. These results are similar to those of angiosperms which show that storage protein gene expression is turned off by a desiccation treatment.

Conclusion

In conclusion, the comparative analysis of conifer somatic and zygotic embryos at the biochemical and molecular levels has provided valuable means of assessing the quality of somatic embryos produced *in vitro* using a combination of cultural conditions. This information has helped in the production of high quality somatic embryos with an ability to tolerate desiccation to below 10% moisture content, a property of potential value for storage and delivery systems. The studies described above strongly support the use of well characterized seed storage proteins and their cloned cDNAs as potential markers to gauge the maturity of somatic embryos of conifers. The Lea genes, associated with the desiccation stage as well as the genes associated with early stages of embryogenesis of conifers, need to be cloned and characterized further. The latter could provide useful markers of embryogenic potential of conifer cultures (Egertsdotter *et al.*, 1993). The continued progress in molecular genetics of conifer somatic and zygotic embryogenesis will lead to a better understanding of gene regulation in response to cultural conditions *in vitro*. Recent developments in stable transformation of conifer somatic embryos (Ellis *et al.*, 1993) will permit analysis of regulatory sequences in transgenic tissue. Molecular cloning and characterization of embryonic cDNAs and their corresponding genomic sequences will provide a repertoire of gene sequences of potential use in the genetic engineering of conifers.

References

Allona, I., R. Casado and C. Aragoncillo, 1992. Seed storage proteins from *Pinus pinaster* Ait.: homology of major components with 11S proteins from angiosperms. Plant Sci. 87: 9–18.

Attree, S.M., D. Moore, V.K. Sawhney and L.C. Fowke, 1991. Enhanced maturation and desiccation tolerance of white spruce (*Picea glauca* [Moench.] Voss) somatic embryos: Effects of a non-plasmolyzing water stress and abscisic acid. Ann. Bot. 68: 519–525.

Attree, S.M., M.K. Pomeroy and L.C. Fowke, 1992. Manipulation of conditions for the culture of somatic embryos of white spruce for improved triacylglycerol biosynthesis and desiccation tolerance. Planta 187: 395–404.

Barratt, D.H.P. and J.A. Clark, 1989. Proteins arising during the late stages of embryogenesis in *Pisum sativum* L. Planta 184: 14–23.

Bewley, J.D. and A. Marcus, 1990. Gene expression in seed development and germination. Prog. Nucleic Acids, Res. Mol. Biol. 38:165–193.

Bewley, J.D. and M. Black, 1985. Seed development and maturation. In: J.D. Bewley and M. Black (Eds.), Seeds: Physiology of Development and Germination, pp. 29–88. New York, Plenum Press.

Ching, T.M., 1963. Fat utilization in germinating Douglas fir seed. Plant Physiol. 38: 722–728.

Ching, T.M., 1966. Compositional changes of Douglas fir seed during germination. Plant Physiol. 41: 1313–1319.

Choi, J.H., L.S. Liu, C. Borkird and Z.R. Sung, 1987. Cloning of genes developmentally regulated during plant embryogenesis. Proc. Natl. Acad. Sci USA 84: 1906–1910.

Cyr, D.R., F.B. Webster and D.R. Roberts, 1991. Biochemical events during germination and

early growth of somatic embryos and seed of interior spruce (*Picea glauca/engelmanii* complex). Seed Sci. Res. 1: 91–97.
Delseny, M., P. Gaubier, G. Hull, J. Saezunsquez, P. Gallois, M. Raynal, R. Coóke and F. Grellet, 1993. Nuclear genes expressed during seed desiccation: Relationship with responses to stress. In: A.A. Basra (Ed.), Stress. Induced Gene Expression. Harwood Academic Publishers, Reading (in press).
Dure, III L., 1985. Embryogenesis and gene expression during seed formation. Oxford Surv. Plant Mol. Cell Biol. 2: 179–197.
Dure, III L., M. Crouch, J. Harada, J. Ho, T.D. Mundy, R. Quatrano, T. Thomas and Z.R. Sung, 1989. Common amino acid sequence domains among the Lea proteins of higher plants. Plant Mol. Biol. 12: 475–486.
Durzan, D.J., A.J. Mia and P.K. Ramaiah, 1971. The Metabolism and subcellular organization of the jack pine embryo (*Pinus banksiana*) during germination. Can. J. Bot. 49: 927–938.
Egertsdotter, U., L.H. Mo and S. von Arnold, 1993. Extracellular proteins in embryogenic suspension cultures of Norway spruce (*Picea abies*). Physiol. Plant. 88: 315–321.
Ellis, D.D., D.E. McCabe, S. McInnis, R. Ramachandran, D.R. Russell, K.M. Wallace, B.J. Martinell, D.R. Roberts, K.F. Raffa and B.H. McCown, 1993. Stable transformation of *Picea glauca* by particle acceleration. Bio/Technology 11: 84–89.
Feirer, R.P., J.H. Conkey and S.A. Verhagen 1989. Triacylglycerides in embryogenic conifer calli: a comparison with zygotic embryos. Plant Cell Rep. 8: 207–209.
Finkelstein, R.R., K.M. Tenbarge, J.E. Shumway and M.L. Crouch, 1985. Role of ABA in maturation of rapeseed embryos. Plant Physiol. 78: 630–636.
Finkelstein, R.R. and M.L. Crouch, 1986. Rapeseed embryo development in culture on high osmoticum is similar to that in seeds. Plant Physiol. 81: 907–912.
Flinn, B.S., D.R. Roberts, D.T. Webb and B.C.S. Sutton, 1991a. Storage protein changes during zygotic embryogenesis in interior spruce. Tree Physiol. 8: 71–81.
Flinn, B.S., D.R. Roberts and I.E.P. Taylor, 1991b. Evaluation of somatic embryos of interior spruce. Characterization and developmental regulation of storage proteins. Physiol. Plant 82: 624–632.
Galau, G.A., N. Bijaisoradat and D.W. Hughes, 1987. Accumulation kinetics of cotton late embryogenesis abundant (Lea) mRNAs and storage protein mRNAs: Co-ordinate regulation during embryogenesis and the role of abscisic acid. Dev. Biol. 124: 198–242.
Gifford, D.J., 1988. An electrophoretic analysis of the seed proteins from *Pinus monticola* and eight other species of pine. Can. J. Bot. 66: 1808–1812.
Gifford, E.M. and A.S. Foster, 1989. Comparative Morphology of Vascular Plants, 3rd Edition. W.H. Freeman and Co., San Francisco.
Goffner, D., P. This and M. Delseny, 1990. Effects of abscisic acid and osmotica on helianthinin gene expression in sunflower cotyledons *in vitro*. Plant Sci. 66: 211–219.
Goldberg, R.B., S.J. Barker and L. Perez-Grau, 1989. Regulation of gene expression during plant embryogenesis. Cell 56: 149–160.
Green, M.J., J.K. McLeod and S. Misra, 1991. Characterization of Douglas fir protein body composition by SDS-PAGE and electron microscopy. Plant Physiol. Biochem. 29: 49–55.
Gupta, P.K. and G.S. Pullman, 1990. Method for reproducing coniferous plants by somatic embryogenesis using abscisic acid and osmotic potential variation. U.S. Patent No. 5,036,007.
Gupta, P.K. and G.S. Pullman, 1990. Methods for reproducing coniferous plants by somatic embryogenisis. U.S. Patent No. 4,957,866.
Gupta, P.K. and G.S. Pullman, 1991. Douglas fir somatic embryogenisis. U.S. Patent No. 5,036,007.
Gupta, P.K., R. Timmis, G. Pullman, M. Yancey, M. Kreitinger, W. Carlson and C. Carpenter, 1991. Development of an embryogenic system for automated propagation of forest trees. In: I.K. Vasil (Ed.), Authors – Cell Culture and Somatic Cell Genetics of Plants, Vol. 8, pp. 75–90, Academic Press Inc, New York.
Gupta, P.K., G. Pullman, R. Timmis, M. Kreitinger, W.C. Carlson, J. Grob and E. Welty,

1993. Forestry in the 21st Century: The biotechnology of somatic embryogenesis. Bio/Technology 11: 454–459.

Hakman, I. and S. von Arnold, 1985. Plantlet regeneration through somatic embryogenesis in *Picea abies* (Norway spruce) J. Plant Physiol. 87: 148–159.

Hakman, I., 1993. Embryology in Norway spruce (*Picea abies*). An analysis of the composition of seed storage proteins and deposition of storage reserves during seed development and somatic embryogenesis. Physiol. Plant. 87: 148–159.

Hakman, I., P. Stabel, P. Engstrom and T. Eriksson, 1990. Storage protein accumulation during zygotic and somatic embryo development in *Picea abies* (Norway spruce). Physiol. Plant. 80:441–445.

Hetherington, A.M. and R.S. Quatrano, 1991. Mechanisms of action of abscisic acid at the cellular level. Tansley Review No. 31, New Phytol. 119: 9–32.

Higgins, T.J.V., 1984. Synthesis and regulation of major proteins in seeds. Annu. Rev. Plant Physiol. 35: 191–221.

Jensen, U. and C. Lixue, 1991. Abies seed protein profile divergent from other Pinaceae. Taxon 40: 435–440.

Jensen, U. and H. Berthold, 1989. Legumin-like proteins in gymnosperms. Phytochemistry 28: 1389–1394.

Joy, R.W., E.C. Yeung, L. Kong and T. Thorpe, 1991. Development of white spruce somatic embryos: 1. Storage product deposition. In Vitro Cell. Dev. Biol. 27P: 32–41.

Kermode, A.R., 1990. Regulatory mechanisms involved in the transition from seed development to germination. CRC Crit. Rev. Plant Sci. 9: 155–195.

Krasowski, M.J. and J.N. Owens, 1993. The ultrastructural and histochemical post-fertilization megagametophyte and zygotic embryo development of interior spruce. Can. J. Bot. 71: 98–112.

Krochko, J.E., S.K. Pramanik and J.D. Bewley, 1992. Contrasting storage protein synthesis and messenger RNA accumulation during development of zygotic and somatic embryos of alfalfa (*Medicago sativa* L.) Plant Physiol. 99: 446–53.

Lane, B.G., 1991. Cellular desiccation and hydration: developmentally regulated proteins and the maturation and germination of seed embryos. FASEB J. 5: 2893–2901.

Leal, I. and S. Misra, 1993a. Developmental gene expression in conifer embryogenesis and germination. III. Analysis of crystalloid protein mRNAs and desiccation protein mRNAs in the developing embryos and megagametophyte of white spruce (*Picea glauca* [Moench] Voss). Plant Sci. 68: 163–173.

Leal, I. and S. Misra, 1993b. Molecular cloning and characterization of a legumin-like storage protein cDNA of Douglas fir seeds. Plant Mol. Biol. 21: 709–715.

Lindsay, K. and J.F. Topping, 1993. Embryogenesis: A question of Pattern. J. Expt. Bot. 44: 359–374.

Lott, J.N.A., 1980. Protein bodies In: N.E. Tolbert (Ed.), The Biochemistry of Plants, Vol. I, pp. 589–623. Academic Press, New York.

Misra, S. and M.J. Green, 1990. Developmental gene expression in conifer embryogenesis and germination. I. Seed proteins and protein composition of mature embryo and the megagametophyte of white spruce (*Picea glauca* [Moench] Voss). Plant Sci. 68: 163–173.

Misra, S. and M.J Green, 1991. Developmental gene expression in conifer embryogenesis and germination. II. Crystalloid protein synthesis in the developing embryo and megagametophyte of white spruce (*Picea glauca* [Moench] Voss). Plant Sci. 78: 61–71.

Misra, S. and I. Leal, 1993. Molecular cloning of crystalloid storage protein cDNAs of Douglas fir. In: D. Come and I. Corbineau (Eds.), Basic and Applied Aspects of Seed Biology, Vol. I, pp. 21–27. Imprimerie de Brabanter, Saint-Denis.

Misra, S. and M. Green, 1993. Legumin-like storage polypeptide of conifer seeds and their antigenic homology with 11S globulins from angiosperms (submitted for publication).

Misra, S., A. Kermode and J.D. Bewley, 1985. Maturation drying as the "switch" that terminates seed development and promotes germination. In: L. van Vloten-Doting, G.S.P. Groot and T.C. Hall (Eds.), Molecular Form and Function of the Plant Genome, NATO ASI Series, pp. 113–128. Plenum Press, New York.

Misra, S., S.M. Attree, I. Leal and L.C. Fowke, 1993. Effect of abscisic acid, osmoticum, and desiccation on synthesis of storage proteins during the development of white spruce somatic embryos. Ann. Bot. 71: 11–22.

Mundy, J. and N.H. Chua, 1988. Abscisic acid and water stress induce the expression of a novel rice gene. EMBO J. 7: 2279–2286.

Newton, C.H., B.S. Flinn and B.C.S. Sutton, 1992. Vicilin-like seed storage proteins in the gymnosperm interior spruce (*Picea glauca/engelmanii*). Plant Mol. Biol. 20: 315–322.

Nomura, K. and A. Komamine, 1986. Molecular mechanisms of somatic embryogenesis In: B.J. Mifflin (Ed.), Oxford Surv. Plant Mol. Cell Biol., Vol. 3, pp. 456–466. Oxford University Press, Oxford.

Owens, J.N. and M. Moulder, 1984. The reproductive cycle of interior spruce. Province of British Columbia, Ministry of Forests Information Services Branch, Victoria.

Owens, J.N., S.J. Morris and S. Misra, 1993. The ultrastructural, histochemical and biochemical development of the post-fertilization megagametophyte and the zygotic embryo of *Pseudotsuga menziesii* (Mirb.) Franco Can. J. For. Res. 23: 816–827.

Quatrano R.S., 1986. Regulation of gene expression by abscisic acid during angiosperm embryo development. Oxford Serv. Plant. Mol. Cell Biol. 3: 467–477.

Raynal M., D. Depigny, R. Cooke and M. Delseny, 1989. Characterization of a radish nuclear gene expressed during late seed maturation. Plant Physiol. 91: 829–836.

Rivin, C.J. and T. Grudt, 1991. Abscisic acid and the developmental regulation of embryo storage proteins in maize. Plant Physiol. 95: 358–365.

Roberts, D.R., B.S. Flinn, D.T. Webb, F.B. Webster and B.C.S. Sutton, 1990a. Abscisic acid and indole-3-butyric acid regulation of maturation and accumulation of storage proteins in somatic embryos of interior spruce. Physiol. Plant. 78: 355–360.

Roberts, D.R., B.C.S. Sutton and B.S. Flinn, 1990b. Synchronous and high frequency germination of interior spruce somatic embryos following partial drying at high relative humidity. Can. J. Bot. 68: 1086–1090.

Roberts, D.R., W.R. Lazaroff and F.B. Webster, 1991. Interaction between maturation and high relative humidity treatments and their effects on germination of sitka spruce somatic embryos. J. Plant Physiol. 138: 1–6.

Roberts, D.R., 1991. Abscisic acid and mannitol promote early development, maturation and storage protein accumulation in somatic embryos of interior spruce. Physiol. Plant. 83: 247–254.

Shotwell, M.A. and B.A. Larkins, 1989. The biochemistry and molecular biology of seed storage proteins. In: A. Marcus (Ed.), The Biochemistry of Plants, Vol. 15. Molecular Biology, pp. 297–345. Academic Press, New York.

Simola, L.K., 1974. The ultrastructure of dry and germinating seeds of *Pinus silvestris L.* Acta Botanica Fenn 103: 1–31.

Singh, H., 1978. Embryology of Gymnosperms. Begrüder Borntraeger, Berlin Struttgart.

Skriver, K. and J. Mundy, 1990. Gene expression in response to abscisic acid and osmotic stress. Plant Cell 2: 503–512.

Sterk, P. and S.C. de Vries, 1993. Molecular markers for plant embryos. In: K. Redenbaugh (Ed.), Synseeds: Applications of Synthetic Seeds to Crop Improvements, pp. 115–132. CRC Press, London.

Tautorus, T.E., L.C. Fowke and D.I. Dunstan, 1991. Somatic embryogenesis in conifers. Can. J. Bot. 69: 1873–1899.

Von Arnold, S. and I. Hakman, 1988. Regulation of somatic embryo and development in *Picea abies* by abscisic acid (ABA). J. Plant Physiol. 132: 164–169.

Whitmore, F.W. and H.B. Kreibel, 1987. Expression of a gene in *Pinus strobus* ovules associated with fertilization and early embryo development. Can. J. For. Res. 17: 408–412.

Xu, N., K.M. Coulter and J.D. Bewley, 1990. Abscisic acid and osmoticum prevent germination of developing alfalfa embryos, but only osmoticum maintains the synthesis of developmental proteins. Planta 182: 382–390.

8. Progress in Protoplast Technology for Woody Angiosperms

A. Tibok, J. B. Power and M. R. Davey*

Contents

1. Introduction

The improvement of most woody plants through conventional breeding is a slow and lengthy process, due to their long juvenile phase and their generally large size combined with high levels of self-incompatibility and heterozygosity. Many woody species reach maturity only after one or more decades. Consequently, the transfer of specific traits, such as resistance to disease, insects and cold, through introgression of foreign genes, requires several generations of crossing and repeated back-crossing. Conventional breeding also involves the testing of large numbers of progeny in order to recover the desired trait, at the same time ensuring that the existing elite genotype is retained. The application of innovative biotechnological methods has gained attention for accelerating the improvement of woody species. These methods can be employed for mass propagation, the exposure of somaclonal variation, *in vitro* selection for desirable agronomic traits and genetic manipulation. Such approaches are being utilised routinely to improve non-woody agricultural crops and are now progressively being applied to woody plants.

One core technology for woody plant improvement involves the use of protoplasts as unique populations of single cells for the generation of inter-

* To whom correspondence should be addressed.

S. Jain, P. Gupta & R. Newton (eds.), Somatic Embryogenesis in Woody Plants, Vol. 1, 143–166.

specific and inter-generic somatic hybrids, for direct uptake of foreign genes and for the generation of stable, heritable genetic variability.

2. Protoplast Technology for Woody Plant Improvement

Protoplasts are cells from which the surrounding walls have been removed, usually by enzymatic digestion. Isolated protoplasts are amenable to genetic manipulation procedures that are not possible with intact cells, particularly somatic hybridization through chemically or electrically-induced fusion and transformation by direct DNA uptake. As in herbaceous genera, the regeneration of plants from protoplasts is a prerequisite, for the general applicability of these approaches to woody plant improvement, with plant regeneration from at least one parental species being essential in somatic hybridization.

Until comparatively recently, protoplasts from most woody plants were considered to be recalcitrant in the context of sustainable division coupled with shoot regeneration, with the notable exceptions of a few species in the genera *Citrus*, *Picea* and *Prunus*. However, during the last five years, considerable progress has been achieved, with isolated protoplasts, from an increasing number of woody plants, being regenerated to plants via organogenesis or, occassionally, via somatic embryogenesis. Nevertheless, the number of species for which protoplast-to-plant systems have been fully developed is still small as compared with other agronomic groups of plants. Indeed, as a generality, this still remains a limiting factor in the application of protoplast technology to woody plant improvement.

3. Plant Regeneration from Protoplasts of Angiospermous Forest Trees

The isolation and culture of protoplasts of Angiospermous forest trees has been described for several genera, including *Alnus* (Huntinen *et al.*, 1982; Tremblay *et al.*, 1985), *Betula* (Tremblay, 1988), *Eucalyptus* (Teulières and Boudet, 1991), *Fagus* (Lang and Kohlenbach, 1988), *Populus* (Russell and McCown, 1986), *Salix* (Vahala and Eriksson, 1991), *Sorbus* (Jorgensen and Binding, 1984) and *Ulmus* (Sticklen *et al.*, 1985). Sasamoto and Hosoi (1992) have produced callus from protoplasts isolated from embryogenic cell cultures of *Quercus serrata*, a species used to supply timber for culturing Shiitake mushrooms in Japan.

As a generality, the most common source of protoplasts has been leaf mesophyll tissue, especially from young leaves of axenic, cultured shoots that were notionally in their juvenile stage of growth. Many studies have been performed with *Populus*, where the release of mesophyll protoplasts was enhanced by grinding leaves prior to enzymatic digestion, followed by thorough washing of digested material to facilitate protoplast release (Russell

Table 1. Examples of shoot regeneration from protoplasts of forest trees

Species	Protoplast source	Result obtained	Reference
Angiosperms			
Crotalaria juncea	Cots	SE-P	Rao *et al.*, 1982
Eucalyptus saligna	LiS	C-P	Ito *et al.*, 1990
Liriodendron tulipefera	ZE	SE-P	Merkle and Sommer, 1987
Pithecellobium dulce	L	C-P	Saxena and Gill, 1987
Populus alba	L	C-P	Sasamoto and Hosoi, 1990
Populus alba × *glandulosa*	L	C-P	Park and Son, 1988
Populus alba × *grandidentata*	L	C-P	Russell and McCown, 1988
Populus glandulosa	L	C-P	Park *et al.*, 1990
Populus nigra	C	C-P	Lee *et al.*, 1987
Populus nigra × *maximowiczii*	L	C-P	Park and Son, 1992
Populus nigra × *trichocarpa*	L	C-P	Russell and McCown, 1988
Populus sieboldii	L	C-P	Sasamoto *et al.*, 1990
Populus tremula	L	C-P	Russell and McCown, 1988
Santalum album	cS	SE-P	Rao and Ozias-Akins, 1985
Sesbania spp.	C	C-P	Zhao, 1993
Ulmus campestris	L	P	Dorion *et al.*, 1991
Ulmus × *Pioneer*	C	C-P	Sticklen *et al.*, 1985

C, callus; Cots, seedling cotyledons; cS, cell suspensions; eC, embryogenic callus; ecS, embryogenic cell suspensions; L, leaves; LiS, leaves from *in vitro* grown shoots; P, plants; SE, somatic embryos; ZE, zygotic embryos.

and McCown, 1988). Leaf callus has been used as source material in *Ulmus* × "Pioneer" (Sticklen *et al.*, 1985), as have embryogenic cell cultures in, for example, *Santalum album* (sandalwood; Rao and Ozias-Akins, 1985) and *Coffea* species (Schopke *et al.*, 1987; Acuna and Pena, 1991). In the case of *Santalum album*, the embryogenic cell cultures were derived from stem explants of a mature, twenty year-old tree. The culture of protoplasts in a medium with reduced ammonium levels was also essential for success in sandalwood. Coffee protoplasts were derived from both embryogenic callus and embryogenic cell suspensions.

Plant regeneration has followed successful culture of isolated protoplasts in several cases (Table 1). The first leguminous tree to be regenerated from protoplast-derived cells was *Crotalaria juncea* (sunnhemp) by Rao *et al.* (1982), who isolated protoplasts from seedling cotyledons. Such protoplasts produced callus when cultured in the light, again, with a low level of ammonium ions in the medium. Similarly, a low ammonium medium was also employed by Saxena and Gill (1987) to culture seedling leaf protoplasts of the tree legume *Pithecellobium dulce*. Recently, Zhao (1993) regenerated plants from protoplasts of the legume *Sesbania*. For most of the woody angiosperms that have been regenerated to plants, the route has been via organogenesis. In general, somatic embryogenesis has been a rare event, although it has been the pathway for plant recovery in *Crotalaria juncea*

(Rao *et al.*, 1982). *Santalum album* (Rao and Ozias-Akins, 1985) and *Liriodendron tulipefera* Merkle and Sommer, 1987). It is noteworthy that, in this latter case, zygotic embryos, excised from seeds, were enzymatically degraded to release the protoplasts. All these investigations indicated a requirement, as in studies with other woody genera, for a reduced ammoniun concentration in the medium, especially for the successful culture of *Liriodendron* protoplasts. In most examples where plant regeneration from isolated protoplasts has been reported, the procedures which have been developed have been highly genotype-dependent, especially for species in the genus *Populus*.

4. Plant Regeneration from Protoplasts of Fruit Trees

4.1. *Tropical and Sub-Tropical Fruit Trees*

Citrus was the earliest example of a woody plant in this category to be regenerated from protoplasts (Vardi *et al.*, 1982). In fact, this genus of the family Rutaceae provided the only example in which protoplast-to-tree systems in fruit trees and, indeed, in woody plants generally, were available for some period of time. Members of the genus *Citrus* remain unusual amongst woody plants in their general amenability to protoplast isolation, culture and plant regeneration. Following the isolation and culture of protoplasts from nucellar callus of orange (*C. sinensis* cv Shamouti) with subsequent somatic embryo production (Vardi *et al.*, 1975), protocols were eventually established for plant regeneration for nucellar callus-derived protoplasts of a range of *Citrus* cultivars (Vardi *et al.*, 1982). Whilst most reports quote nucellar-derived tissues as source material, *Citrus* protoplasts have also been isolated enzymatically from leaves (Grosser and Chandler, 1987) and from fruit juice vesicles (Echeverria, 1987).

Success with *Citrus* was extended to other genera such as *Coffea canephora* (Schopke *et al.*, 1987) and *C. arabica* (Yasuda *et al.*, 1986; Acuna and Pena, 1991). Protoplasts were isolated from embryogenic callus (Yasuda *et al.*, 1986) or from somatic embryos which had been induced on callus of leaf explants (Schopke *et al.*, 1987). Recently, a number of other tropical fruit trees, that can be regenerated from protoplasts, have been added to the list (Table 2), including *Theobroma cacao* (Kanchanapoon and Kanchanapoon, 1991), *Carica papaya* × *C. cauliflora* (Chen and Chen, 1992) and *Passiflora edulis* (d'Utra Vaz *et al.*, 1993). Since the report of the latter authors, experiments with *Passiflora* have been extended to other species, including *P. incarnata*, *P. seemannii*, *P. suberosa* and the F1 hybrid *P. maliformis* × *P. serrulatus* (Otoni *et al.*, 1994b).

In the tropical and sub-tropical fruit trees currently regenerable from protoplasts, the recovery of plants is principally through somatic embryogenesis and, consistent with non-fruit tree genotypes, often on a medium with a reduced ammonium level, as in *Coffea*. An exception is the genus *Passiflora*, where organogenesis is the main pathway of shoot and plant formation. Embryogenic cell cultures have been the source of protoplasts in *Citrus*, *Coffea* species (Schopke *et al.*, 1987; Acuna and Pena, 1991; Spiral and Petiard, 1991) and in *Carica papaya* × *C. cauliflora* (Chen and Chen, 1992), while in *Passiflora*, protoplasts were isolated from leaves of glasshouse-grown seedlings. It is interesting to note that the behaviour in culture of protoplasts of *Passiflora edulis* was influenced by seedling age, with protoplasts undergoing sustained division only when isolated from seedlings prior to tendril development. The enriched protoplast culture medium of Kao and Michayluk (1975) as modified by Gilmour *et al.* (1989) and designated as KM8P, was employed. Protoplast division was enhanced by plating in medium made semi-solid with agarose, an observation consistent with reports for protoplasts of other woody fruits (Ochatt and Caso, 1986; Ochatt *et al.*, 1987; Ochatt and Power, 1988a, 1992; Vardi *et al.*, 1986). In addition, the inclusion of the cephalosporin-group antibiotic cefotaxime ("Claforan") at 250 mg L^{-1} in the culture medium was essential for sustained protoplast division (d'Utra Vaz *et al.*, 1993).

4.2. *Temperate Fruit Trees*

Protoplasts from leaves of *in vitro* germinated seedlings and a culture medium with reduced ammonium ions, enabled plants to be recovered, by organogenesis, in paper mulberry (*Broussonetia kazinoki*; Moraceae), providing the first example of plant regeneration from mesophyll protoplasts in temperate fruit trees (Oka and Ohyama, 1985; Oka, 1987). Considerable effort has been focussed on members of the Rosaceae (Table 2). Ochatt and Caso (1986) reported a protoplast-to-plant system for wild pear (*Pyrus communis* var *pyraster L.*), protoplasts being isolated from leaves of fifteen year-old field-grown plants and/or from axenic shoot cultures. The latter were initiated from the same field-grown material. Interestingly, leaves of *in vitro* shoots released twice as many viable protoplasts as field-grown plants. The use of an ammonium-free medium and high light intensity were essential for successful protoplast culture. In contrast to the situation in *Citrus*, protoplast-derived shoots of pear exhibited protoclonal variation with respect to callus proliferation and leaf phenotype, such changes being correlated with a lack of rooting ability in abnormal shoots (Ochatt, 1987). During the same year, Ochatt *et al.* (1987) also reported plant regeneration from protoplasts of the cherry rootstock Colt (*Prunus avium* × *pseudocerasus*). Although tissues derived from mesophyll protoplasts underwent organogenesis, tissues from protoplasts isolated from cell suspensions only exhibited shoot formation following electrical stimulation of the freshly isolated protoplasts prior to culture (Ochatt *et al.*, 1988). This work provided the first report of plant regeneration, via organogenesis, from protoplasts isolated from tissues other than leaf mesophyll of a woody species. Electroporation of protoplasts from

Table 2. Examples of plant regeneration from protoplasts of perennial fruit trees and other woody plants

Species	Common name, variety	Protoplast source	Result obtained	Reference
Sub-Tropical and Tropical Fruit Trees				
Actinidia chinensis Kiwifruit		C	C-SH-P	Tsai, 1988
		C	C-SH-P	Mii and Ohashi, 1988
Actinidia deliciosa Kiwifruit		C	C-SH-P	Oliviera and Pais, 1991
Carica papaya × *C. cauliflora*	Papaya hybrid	ecS	C-SE-P	Chen and Chen, 1992
Citrus aurantium Sour orange		NCcS	C-SE-P	Vardi *et al.*, 1982
Citrus jambhiri Rough lemon		NCcS	C-SE-P	Vardi and Galun, 1988
Citrus limon	Villafranca	NCcS	C-SE-P	Vardi *et al.*, 1982
		NCcS	C-SE-P	Vardi and Galun, 1988
Citrus madurensis Calamondin		HC	C-SE-P	Ling *et al.*, 1989
Citrus miti		NCcS	C-SE-P	Sim *et al.*, 1988
Citrus paradisi Grapefruit	Duncan	NCcS	C-SE-P	Vardi *et al.*, 1982
Citrus reticulata Mandarin	Dancy	NCcS	C-SE-P	Vardi *et al.*, 1982
	Murcott	NCcS	C-SE-P	Vardi *et al.*, 1982
	Ponkan	NCcS	C-SE-P	Vardi *et al.*, 1982
		NC	C-SE-P	Hidaka and Kajiura, 1988
Citrus *sinensis* Orange	Shamouti	NC-C	C-SE-P	Vardi *et al.*, 1975, 1982
	Trovita	NCcS	C-SE-P	Kobayashi *et al.*, 1983, 1985
		NCcS	C-SE-P	Kobayashi, 1987
	Navel	NC	C-SE-P	Hidaka and Kajiura, 1988
		cS	C-SE-P	Wang *et al.*, 1986
Citrus unshiu Satsuma		NCcS	C-SE-P	Ling *et al.*, 1990
Citrus yuko		NCcs	C-SE-P	Hidaka and Kajiura, 1988
Coffea arabica Coffee	Arabica	eC	C-SE-P	Yasuda *et al.*, 1986
		ecS	C-SE-P	Acuna and Pena, 1991
		ecS	C-SE-P	Spiral and Petiard, 1991
		ecS	C-SE-P	Barton *et al.*, 1991
Coffea canephora Coffee	Robusta	SE	C-SE-P	Schopke *et al.*, 1987
		ecS	C-SE-P	Spiral and Petiard, 1991
Coffea canephora × *C. arabica*	Arabusta	ecS	C-SE-P	Spiral and Petiard, 1991
Microcitrus spp.		CcS	C-SE-P	Vardi *et al.*, 1986
Passiflora edulis fv. flavicarpa	Passionfruit	L	C-SH-P	d'Utra Vaz *et al.*, 1993
Passiflora incarnata	Passionfruit	L	C-SH-P	Otoni *et al.*, 1984
Passiflora maliformis × *P. serrulatus*		L	C-CH-S	Otoni *et al.*, 1994
Passiflora seemannii	Passionfruit	L	C-SH-P	Otoni *et al.*, 1994
Passiflora suberosa	Passionfruit	L	C-SH-P	Otoni *et al.*, 1994

Theobroma cacao	Cocoa	cS	C-SE	Kanchanapoom & Kanchanapoom, 1991
Temperate Fruit Trees				
Broussenetia kazinoki	Paper mulberry	L	C-SH-P	Oka and Ohyma, 1985
Diospyros kaki	Japanese persimmon	C	C-SH-P	Tao *et al.*, 1991
				Oka, 1987
Malus × *domestica* Apple	Rootstock M.9	L	C-SH-P	Patat-Ochatt *et al.*, 1988
	Rootstock MM.106	L	C-SH	Patat-Ochatt *et al.*, 1988
			C-SH-P	Patat-Ochatt and Power, 1990
	Spartan	L	C-SH-P	Patat-Ochatt *et al.*, 1988
	Bramley's seedling	L	C-SH-P	
	[Cox × Court Pendu Plat × Wijick McIntosh]	L	C-SH-P	Wallin and Johansson, 1990
	Haploid Golden Delicious	L St	C-SH-P	Patat-Ochatt *et al.*, 1993
Prunus avium	Bird cherry	CotCcS	C-SE	David *et al.*, 1992
Prunus avium × *pseudocerasus*	Cold cherry	RCcS	C-SH-P	Ochatt *et al.*, 1987, 1988
Prunus cerasifera		L	C-SH-P	Ochatt, 1992
Prunus cerasus	CAB 4D	L	C-R-SH-P	Ochatt and Power, 1988b
		RC	C-SH-P	Ochatt, 1990b
		RC	C-R-SH-P	Ochatt, 1990b
	CAB 5H	L	C-SH-P	Ochatt and Power, 1988b
		L	C-R-SH-P	Ochatt and Power, 1988b
Prunus spinosa		L	C-SH-P	Ochatt, 1992
Pyrus communis	Pear. Conference	eC	C-R-SH-P	Ochatt, 1990a
	Williams Bon			
	Chretien (Bartlett)	L	C-SH-P	Ochatt and Power, 1988a
Pyrus communis var. pyraster	Wild pear	L	C-SH-P	Ochatt and Caso, 1986
Medicinal plants, ornamentals, fibre and root crops				
Gossypium hirsutum	Cotton	ecS	C-SE-P	She *et al.*, 1989
Lycium barbarium		L	C-SH-P	Ratushnyak *et al.*, 1989
Manihot esculentum	Cassava	L	C-SH-P	Shahin and Shepard, 1980
Oxalis glaucifolia		C	C-SH-P	Ochatt *et al.*, 1989
Panax ginseng	Ginseng	eC	C-SE-P	Arya *et al.*, 1991
Rosa persica × *xanthina*	Rose	ecs	C-SE-P	Matthews *et al.*, 1991
Solanum dulcamara	Woody nightshade	cS	C-SH-P	Chand *et al.*, 1988

C, callus; cS, cell suspensions; Cots, seedling cotyledons; eC, embryogenic callus; ecS, embryogenic cell suspensions; HC, hypocotyl callus; L, leaves; NC, nucellar callus; P, plants; RC, root callus; SE, somatic embryos; SH, shoots; st, stems.

Table 3. Transient gene expression and transgenic plant production following direct gene transfer into protoplasts

Species	Gene transfer method	Gene expressed	Result	Reference
Actinidia deliciosa	PEG	*cat*	TGE	Oliveira *et al.*, 1991
Alnus incana	Electroporation	*gus*	TGE	Seguin and Lalonde, 1988
Citrus jambhiri	PEG	*cat*	TP	Vardi *et al.*, 1990
Citrus sinensis	PEG	*npt*II	TC	Kobayashi and Uchimiya, 1989
Coffea arabica	Electroporation	*npt*II	TP	Barton *et al.*, 1991
Eucalyptus citriodora	Electroporation	*cat*	TGE	Manders *et al.*, 1992
Eucalyptus gunnii	PEG; Electroporation	*cat,gus*	TGE	Teulières *et al.*, 1991
Eucalyptus saligna	Electroporation	*npt*II	TP	Kawazu *et al.*, 1990
Malus × *domestica*	Electroporation	*cat*	TGE	Manders *et al.*, 1990

TC, transformed callus; TGE, transient gene expression; TP, transgenic plants; PEG, polyethylene glycol.

Prunus cell suspensions was shown to stimulate not only shoot formation, but also protoplast division. These observations were consistent with the electro-stimulation of protoplast division and plant regeneration in the woody medicinal plant *Solanum dulcamara* (Chand *et al.*, 1988).

In Sour Cherry, the pathway of plant regeneration involved an intermediate stage of rhizogenesis. Interestingly, protoplast-derived tissues of the plant clones CAB4D and CAB5H produced roots; the latter developed shoot buds at their cut surfaces, following excision from the parent tissue and culture of the excised roots. Such buds developed into shoots already attached to a primary root. More recently, Ochatt (1992) regenerated plants from mesophyll protoplasts of *Prunus cerasifera* and *P. spinosa* using a modified MS-based medium supplemented with 2,4-dichlorophenoxyacetic acid (2,4-D) and 6-benzyladenine (BA) (for *P. cerasifera*, clone P2032), α-naphthaleneacetic acid (NAA) and zeatin (for *P. cerasifera*, clone P2980) and NAA, with BA and zeatin for *P. spinosa* clone P51. Shoots were obtained from clones P2980 and P51 on MS-based medium with double the original concentration of group B vitamins and supplemented with NAA, BA and zeatin. This constitutes the first report of plant regeneration from protoplasts of *Prunus* species belonging to the subgenus *Prunophora*, section *Euprunus* (prunes and plums). As Ochatt (1992) emphasised, the development of protoplast-to-plant systems for these species will permit somatic hybridisation to be used to examine theories on the origin of the European prune (*P. domestica*). The latter is believed to be a natural hybrid between diploid *P. cerasifera* and tetraploid *P. spinosa*, followed by spontaneous doubling of chromosomes of the F1 hybrid.

Success with protoplasts of Colt cherry was followed by the development of leaf mesophyll protoplast-to-plant systems for other temperate fruit trees, including clones of *Prunus cerasus* L. (sour cherry; Ochatt and Power,

1988b), *Pyrus communis* L. (Williams Bon Chretien pear; Ochatt and Power, 1988a) and *Malus* × *domestica* (apple). In apple, viable protoplasts were isolated from leaves of the rootstocks M.9 and MM.106, and the scion cultivar "Spartan" (Patat-Ochatt *et al.*, 1988), with callus developing shoot buds and the subsequent production of rooted plants of "Spartan" and M.9, followed by MM.106 (Patat-Ochatt Power, 1990). Other workers have reported success with apple, Wallin and Johansson (1990) regenerating plants from mesophyll protoplasts of the columnar apple cultivar A1583 [Cox × Court Pendu Plat) × Wijick McIntosh].

Some attention has been given to haploid apple. Recently, Patat-Ochatt *et al.* (1993) isolated protoplasts from *in vitro*-grown leaf and stem tissues of a haploid clone of the apple scion cultivar Golden Delicious. In agreement with the results for other apple genotypes, mesophyll protoplast-derived tissues produced buds on medium with increased B vitamins. However, the shoots which developed from these buds were weak and vitrified and failed to survive. In contrast, shoots from stem protoplasts could be propagated *in vitro*, although they failed to root. Failure to root was also observed in micro-propagated shoots of the same haploid clone, but has been circumvented by grafting shoot tips onto appropriate apple rootstocks.

Protoplast systems of other fruit trees which have been studied include those of kiwifruit (*Actinidia chinensis* var *chinensis* Planch; Tsai, 1988; Mii and Ohashi, 1988) and *A. deliciosa* var *delisiosa* cv Hayward (Oliveira and Pais, 1991), where plants were regenerated from protoplasts isolated from leaf callus, the latter being a long-established culture in the cultivar Hayward. In Japanese persimmon (*Diospyros kaki* L. cv Jiro), one of the most important fruit crops in Japan, protoplasts were isolated from callus initiated from leaf primordia excised from dormant winter buds (Tao *et al.*, 1991). Agarose-solidified KM8P medium and a protoplast plating density greater than 1.0×10^5 mL^{-1} were required for initiation of protoplast division, with shoots being induced following transfer of protoplast-derived tissues to MS-based medium with NAA and zeatin.

To date, there have been no reports of shoot regeneration in temperate fruit trees through somatic embryogenesis. Although Kouider *et al.* (1984) described the induction of embryo-like structures on protoplast-derived callus of *Malus* × *domestica*, plant regeneration was not reported. David *et al.* (1992) provided evidence for the early stages of somatic embryo development from protoplasts of *Prunus avium*, but, again, shoots were not recovered from these structures.

5. Protoplasts of Woody Medicinal Plants, Ornamentals, Fibre and Root Crops

Several examples are represented in this diverse, multi-purpose group. Chand *et al.* (1988) regenerated plants from tissues derived from cell suspension protoplasts of the woody medicinal plant *Solanum dulcamara* (woody nightshade). In agreement with the report for Colt cherry (Ochatt *et al.*, 1988), electrical treatment of isolated protoplasts stimulated growth and morphogenesis of protoplast-derived tisssues. Regenerated shoots also rooted more readily and developed more prolific root systems than shoots from untreated protoplasts. These observations with both *Solanum* and *Prunus* have important implications in other woody genera in which difficulty is experienced in the induction of organogenesis and the rooting of regenerated shoots. *Panax ginseng* (ginseng; Araliaceae) is an important medicinal herb, the root of which is used as a vitalising and stimulatory agent, with its pharmacological properties being attributed to triterpene saponins and ginsenosides. The three year juvenile period of the plant has hampered conventional breeding programmes to improve the plant genetically. Consequently, an efficient protoplast-to-plant system would make this herb amenable to transformation and asymmetric somatic hybridisation with, for example, carrot for altered ginsenoside properties. Such hybrids may also be easier to grow than ginseng itself. Using somatic embryos obtained from a four year-old embryogenic cell line, itself derived from a zygotic embryo, Arya *et al.* (1991) isolated protoplasts, of which ten percent divided in an agarose-gelled medium with, unusually, six per cent (w/v) myo-inositol as osmoticum. Protoplast-derived callus produced somatic embryos which regenerated into plants on a medium supplemented with gibberellic acid.

An unusual remnant of the Jurassic era is *Ginkgo biloba* (order Ginkgoales, which produces ginkgolide B. The latter is used for treating cardiovascular disorders, thrombosis and inflammatory reactions. Although *Ginkgo* is a gymnosperm, reference to this woody plant is included here because of current interest in this plant. In an elegant study, Laurain *et al.* (1993) removed juvenile ovules from a female tree and isolated haploid protoplasts from prothalli (female gametophytes). Such protoplasts divided to form microcolonies which developed directly into embryos when the protoplasts were cultured in medium lacking ammonium ions but supplemented with glutamine, BA and NAA, or in medium without synthetic growth regulators but containing coconut milk. An interesting observation was that nuclei surrounded by a limited amount of cytoplasm were ejected from some of the protoplasts, and these structures also developed into microcolonies. To date, plant regeneration has not been reported for haploid protoplasts of *Ginkgo*, although this may soon be achievable.

Experiments have been initiated aimed at utilising tissue culture as a tool to improve understanding of the cytogenetics and evolution of the genus *Oxalis*. Following their success in regenerating plants from explant-derived callus of the shrubby, medicinal species *Oxalis glaucifolia* and *O. rhombeo-ovata*, Ochatt *et al.* (1989a) employed the same tissue as a source of protoplasts. Plants were obtained from protoplast-derived callus of *O. glaucifolia* on MS-based medium with NAA, kinetin and gibberellic acid, although nodular callus of *O. rhombeo-ovata* failed to undergo organogenesis.

Ratushnyak *et al.* (1989) have recovered plants from isolated protoplasts of *Lycium barbarum* while Zhao *et al.* (1991) described their results for *Hibiscus syriacus* (Malvaceae), a perennial woody plant used in China for medicinal products and for fibre production. Although protoplasts isolated from callus underwent sustained division, shoot regeneration was not achieved. In contrast, plants have been regenerated from protoplasts of cotton (*Gossypium hirsutum*), another economically important member of the Malvaceae (She *et al.*, 1989). As Zhao *et al.* (1991) have indicated, the generation of *Hibiscus-Gossypium* somatic hybrids might combine the useful economic attributes of both parents.

Protoplast-to-plant systems have been reported for *Manihot esculentum* (cassava; Shahin and Shepard, 1980) and an ornamental rose, *Rosa persica* × *Xanthina* (Matthews *et al.*, 1991). As in ginseng, somatic embryogenesis was the pathway of regeneration in cotton and rose, following protoplast isolation from embryogenic cultures. In the case of cassava, well-expanded leaves from plants grown in an environmentally controlled chamber were the source of viable protoplasts. Excised leaves were floated on a solution containing 1 mM $CaCl_2$, 1 mM NH_4NO_3, 1 ppm NAA and 5 ppm BA for 48 h, surface sterilised, and soaked overnight at 4 °C in a mineral salts solution prior to protoplast isolation. Shoot formation from mesophyll protoplasts was occasionally observed in the cultivar Mexico No. 35, but, to date, other workers have been unable to repeat these results. This is unfortunate, in view of the urgent requirement to apply somatic cell techniques to the genetic manipulation of cassava. Promising in this respect, for cassava, is the use of oxygenated perfluorocarbons and non-ionic surfactants as protoplast media supplements (Anthony, Davey, Power and Lowe, unpublished).

6. Somatic Hybridization and Cytoplasmic Hybridization (Cybridization) in Woody Plants

6.1. *Fusion of Isolated Protoplasts*

The isolation of protoplasts and the regeneration of plants from protoplast-derived tissues offer the possibility to generate novel somatic hybrid plants. Until the introduction of this somatic cell technology, hybridisation in plants was possible only through sexual crossing and, rarely, graft hybridisation. Thus, somatic hybridisation is a valuable adjunct to conventional approaches, particularly for woody species. Protoplast fusion can be induced by the use of chemical fusogens, such as polyethylene glycol (PEG), by electrofusion, or by a combination of these procedures (Lynch *et al.*, 1993; Marchant *et al.*, 1993). Protoplast fusion permits the production of both nuclear and cytoplasmic hybrids (cybrids), thereby overcoming the limitations (biparental inheritance excepted) of maternal inheritance of cytoplasmic traits normally associated with sexual hybridisation. Consequently, somatic hybridisation offers the possibility of obtaining unique genetic combinations that are not possible sexually. An additional feature of protoplast fusion is that it enables polygenic and characteristics as yet genetically ill-defined, to be transferred without the need for detailed molecular information of such traits.

Inter-specific and inter-generic somatic hybrid plants resulting from the fusion of diploid protoplasts may be infertile and, as a consequence, of limited use in breeding. However, in some cases, fertility can be reinstated through chromosome doubling in plants that can tolerate high ploidy levels, or by reducing the ploidy by half. The failure of some species to tolerate high ploidy levels can be circumvented through the fusion of haploid protoplasts. The production of fertile somatic hybrid plants is important in enabling them to be incorporated into conventional breeding programmes for further improvement to produce elite individuals, or to be repeatedly backcrossed with one or both of the original parents for the introgression of desired traits.

6.2. *Somatic Hybrids*

Almost all of the somatic hybrids produced, to date, in woody plants have been between *Citrus* species and their close relatives. The *Citrus* hybrid plants generated, to date, have been reviewed in detail and listed by Gmitter *et al.* (1992), including the relevant literature. In general, somatic hybridisation in *Citrus* has been promoted by the extensive citrus industry worldwide and, fortuitously, with the ease with which plants can be regenerated from *Citrus* protoplasts compared to those of other woody genera. As discussed by Gmitter *et al.* (1992), somatic hybridisation in *Citrus* enables heterozygous tetraploid hybrids to be generated, both inter-specifically and inter-generically. Several of these combinations are known to be sexually compatible (Ohgawara *et al.*, 1989), with somatic hybridisation providing the opportunity to generate new and additional, nuclear-cytoplasmic combinations. The fertility of *Citrus* somatic hybrids has been assessed (Kobayashi *et al.*, 1991). Some somatic hybrids combine alleles for fruit colour with acceptable fruit size and maturity, as in sweet orange-satsuma mandarin hybrids. Others combine the performance and quality of "Femminello" lemon with cold hardiness and tolerance to the systemic "mal secco" fungal disease (incited by *Phoma tracheiphila*), exhibited by "Valencia" and "Hamlin" sweet oranges or by the lemon hybrid "Milam". Additionally, agronomically useful traits that may be combined through inter-specific and inter-generic somatic hybridisation of sexually compatible parents include resistance to *Phytopthora* rot, resistance to tristeza virus and *Citrus* nematode, a reduced juvenile growth phase, blight tolerance and increased vigour. In some cases, somatic hybrids may have limited direct use as new scion cultivars, because their fruit generally exhibits thick rinds, irregular shape and coarse texture. However, they are of value as tetraploid breeding accessions in crosses with diploids to generate seedless, triploid progeny (Tusa *et al.*, 1992). Such tetraploids may also be used as rootstocks, where, of course, fruit quality is irrelevant (Louzada *et al.*, 1992).

The considerable genetic diversity exhibited by *Citrus* relatives, which provides resistance to *Citrus* blight, a prevalent disease in humid regions of Florida, Brazil and South Africa, has prompted the generation of somatic hybrid plants between *Citrus* and genera such as *Severinia disticha* and *S. buxifolia*, *Citropsis gilletiana* (Grosser *et al.*, 1990) and *Atalantia ceylanica* (Louzada *et al.*, 1993). The latter genera are sexually incompatible with *Citrus*.

In addition to the examples in the family Rutaceae, somatic hybrid plants have also been reported in the family Rosaceae between, for example, wild pear (*Pyrus communis* var pyraster) and Colt cherry (*Prunus avium* × *pseudocerasus*) by Ochatt *et al.* (1989b). Such a somatic hybrid may be useful as a rootstock for both pear and cherry (Ochatt, 1990a). A novel somatic hybrid has been generated recently in the Passifloraceae between the tropical woody species *P. edulis* and *P. incarnata* (Otoni *et al.*, 1994a). Fusion involved leaf protoplasts isolated from glasshouse-grown seedlings of both parents. As was the case in most of the somatic hybrids described for *Citrus*, a complex hybrid cell selection system was also unnecessary in *Passiflora*. The faster growing protoplast-derived colonies (a manifestation of heterosis), which first appeared on the culture plates, were the ones which ultimately produced somatic hybrid shoots. Presumably, hybrid vigour enabled such hybrid cells to outgrow cells derived from homokaryons and unfused protoplasts. Back crossing of such *Passiflora* somatic hybrids to the parental species resulted in fruit set. Such somatic hybrids may prove useful not only in the genetic manipulation of *Passiflora* for improved fruit and fruiting characteristics, but they may also be of equal value as novel ornamentals because of their larger flowers, as compared with those of the parents.

Protoplast fusion may also be used to generate triploid trees of interest to the forestry industry. In nature, triploid trees have been found to exhibit desirable attributes, such as rapid biomass accumulation in Aspen (Lakshimi Sita, 1987). However, the natural occurrence of triploids is rare. Protoplast technology provides, at least in theory, the opportunity to produce such triploid trees through the fusion of diploid somatic protoplasts with haploid gametic pollen tetrad protoplasts (gametosomatic hybridisation; Pirrie and Power, 1986), or with protoplasts from haploid plants, as is possible for herbaceous species. The recent report of the regeneration of haploid plants from stem protoplasts of haploid apple (Patat-Ochatt *et al.*, 1993), should facilitate triploid plant production in the future.

6.3. *Cybrids*

Protoplast fusion has also been employed for the unidirectional transfer of chloroplasts and mitochondria between plants. This method of cytoplasmic genome transfer involves the elimination of the nuclear genome of the cytoplasmic donor, usually by irradiation, and transient inhibition of the metabolism of unfused recipient protoplasts by treatment with chemicals such as iodoacetate. Both treatments are carried out prior to protoplast fusion. Using this procedure, cybrid (cytoplasmic hybrid) plants with unique nuclear-cytoplasmic combinations have been generated. Again, *Citrus* provides excellent examples in which cybrid trees have been generated (Vardi *et al.*, 1989, 1990a,b). Such cybrids are also listed by Ochatt *et al.* (1992), together with the relevant references. They include those generated between *Citrus aurantium* and *C. limon* with *Poncirus trifoliata* (donor), and *C. aurantium* and *C. jambhiri*, both with *Microcitrus* species (donor). Organelles of the donors were transferred into protoplasts of the recipients. Cybrid plants had the morphological characters of the recipient plants, but the organellar genomes of the donor.

In some "donor-recipient" combinations, both chloroplasts and mitochondria could be transferred independently between species, since an early sorting out of the organelles occurs post-fusion. In other cases, sorting out of chloroplasts was delayed beyond sexual reproduction and there were indications that in certain "donor-recipient" combinations, only one of the two types of organelles (e.g., chloroplasts) could be transferred. Enucleated protoplasts can also be used to foster organelle transfer. The transfer of cytoplasmic characteristics is important in terms of crop improvement, since organellar genomes control traits such as herbicide resistance (encoded by chloroplast DNA) and cytoplasmic male sterility (encoded by mitochondrial DNA). In several *Citrus* somatic hybrid and cybrid combinations, comparatively large populations of plants have been generated, enabling field trials to be initiated, especially in Florida. Continuing genetic, cytogenetic and physiological studies will clarify the long-term value of such somatic hybrids and cybrids to the citrus industry.

6.4. *Characterization of Somatic Hybrids and Cybrids*

Several approaches are available which, collectively, confirm the hybrid/cybrid nature of plants derived from fusion-treated protoplasts. Morphologically, somatic hybrids frequently exhibit characteristics intermediate to those of their parents. For example, Ochatt *et al.* (1989b) compared plant height and the shape and size of leaves, petioles and stipules, together with the presence or absence of petiolar glands and trichomes of plants of *Pyrus communis* var pyraster (+) *Prunus avium* × *pseudocerasus* with the same characteristics of parental material. Cytological studies showed that, despite minor differences in plant morphology, somatic hybrids had, with the exception of one aneuploid plant, 58 chromosomes. This chromosome complement was equivalent to the summation of the chromosome number of *Pyrus communis* var pyraster ($2n = 2x = 34$) and *Prunus avium* × *pseudocerasus* ($2n = 3x = 24$). Such plants were designated $2n = 2x + 3x = 58$.

Flow cytometric characterisation of DNA content is more rapid than conventional cytological analysis and is being employed by several laboratories, which have this facility, to determine the ploidy of protoplast-derived plants.

Chopping of leaves in a suitable buffer containing ethidium bromide releases nuclei whose fluorescence is proportional to their DNA content. The relative linear fluorescence values of G_0/G_1 nuclear peaks, compared with the fluorescence of standards, provides a measure of the ploidy of the plant (Hammatt *et al.*, 1991). Such an approach has been used to determine the ploidy of somatic hybrids of *Passiflora edulis* with *P. incarnata* (Otoni *et al.*, 1994a).

Isozyme banding patterns are useful for indicating hybridity. Ochatt *et al.* (1989b) analysed leaf extracts of *Pyrus communis* var pyraster (+) *Prunus avium* × *pseudocerasus* for esterases, phosphoglucomutase and malate dehydrogenase, while Otoni *et al.* (1994) determined the banding patterns for esterase, leucine amino peptidase and alcohol dehydrogenase in *Passiflora* somatic hybrids. The appearance of new bands, in addition to parental bands, was evidence for hybridity.

Molecular analysis is becoming increasingly important for characterising somatic hybrids and cybrids. Vardi *et al.* (1987) verified the cybrid nature of *Citrus* plants by their mitochondrial DNA restriction profiles, while Otoni *et al.* (1994a) applied random amplified polymorphic DNA (RAPD) analysis to *Passiflora* somatic hybrids. Parental species exhibited different DNA profiles when analysed using specific DNA primers. Putative hybrids exhibited some amplification products typical of both parents, together with additional DNA bands. Similarly, restriction fragment length polymorphism (RFLP) analysis, using specific DNA probes, should be useful in characterising somatic hybrids of woody species.

7. Transformation of Protoplasts of Woody Plants

Genetic transformation is probably the most powerful approach (not withstanding environmental concerns) for the improvement of woody plants, since it permits the transfer, integration and expression of foreign genes for specific traits in recipient plant cells. Theoretically, this method should result in the least possible disruption to target genomes when compared with somatic hybridisation. Aspects of gene transfer to trees have been reviewed (Manders *et al.*, 1992). Whilst *Agrobacterium*-mediated gene delivery has been the method of choice for transgenic plant production in *Malus* × *domestica* (apple; James *et al.*, 1993), *Populus alba* × *grandidentata* (poplar; Filliatti *et al.*, 1987), *Juglans regia* (walnut; McGranahan *et al.*, 1988), *Rubus* spp. (Graham *et al.*, 1990), *Actinidia deliciosa* (kiwifruit; Uematsu *et al.*, 1991), *Prunus domestica* (plum; Mante *et al.*, 1991), *Passiflora edulis* (passionfruit; Manders *et al.*, 1994) and *Eucalyptus urophylla* (Tibok, Davey and Power, unpublished), the direct transfer of foreign genes into isolated protoplasts has also been employed to transform woody species. The uptake of foreign genes, carried on suitable vectors, can be induced by treatment of protoplast-plasmid mixtures with chemicals such as PEG, by electroporation involving high voltage electrical pulses, or by a combination of chemical and physical procedures (Davey *et al.*, 1989).

Several reports have described transient gene expression in protoplasts following direct DNA uptake into isolated protoplasts, with a number of parameters being investigated which influence gene uptake and expression. The chloramphenicol acetyltransferase (*cat*) and β-glucuronidase (*gus*) genes have been used most extensively as the readily assayable reporter systems. For example, Seguin and Lalonde (1988) evaluated *gus* gene expression in protoplasts of *Alnus incana* (alder). In early experiments to genetically manipulate kiwifruit, Oliveira *et al.* (1991) used PEG to induce plasmid uptake into protoplasts of petiole-derived callus. Heat-shocking of the protoplasts, prior to DNA delivery, and the use of PEG 4000 at thirty per cent (w/v), resulted in optimal *cat* gene expression.

Eucalyptus is a target plant for genetic manipulation because of its economic value as a source of timber, oils and cellulose fibre for paper manufacture. Teulières *et al.* (1991) compared PEG-induced plasmid uptake with electroporation into protoplasts from callus and cell suspensions of *E. gunnii*. PEG-mediated DNA uptake was stimulated by heat-shock pretreatment of isolated protoplasts and was more efficient than electroporation; the efficiency of electroporation depended on electrical parameters, pH and the source of protoplasts. Later, Manders *et al.* (1992) reported success with protoplasts from seedling cotyledons of *E. citriodora*. Fifty seven per cent of protoplasts were viable after electroporation, with forty seven per cent conversion of ^{14}C-chloramphenicol to its acetylated forms. Gene expression increased with plasmid concentrations up to sixty μg/ml. The use of sheared salmon sperm DNA as a carrier and treatment of protoplasts with forty per cent (w/v) PEG 6000 after electroporation, also promoted gene expression. Such transient gene expression studies are a prelude to stable transformation, although it does not necessarily follow that the conditions optimised for transient expression will be the best for stable transformation.

In studies of stable transformation, Kobayashi and Uchimiya (1989) first reported the transformation of *Citrus* protoplasts, with the formation of protoplast-derived cells, but these workers were unable to convert these cells to transgenic plants. In subsequent work with *Citrus jambhiri* and PEG-mediated delivery of pCAP212, followed by culture of PEG-treated protoplasts in the presence of feeder cells, Vardi *et al.* (1990a) detected CAT activity in protoplasts three days after DNA uptake. This result gave confidence to continue with longer-term gene expression experiments, protoplast-derived colonies which expressed the *npt*II gene being selected on medium containing paromomycin sulphate. Nine stably transformed embryogenic clones were selected, of which two gave transgenic plants. The production of transgenic plants of *Coffea arabica* (Barton *et al.*, 1991) through direct DNA transfer into protoplasts, has also been reported, with electroporation being the preferred DNA delivery method in these experiments. The wider application of direct DNA uptake for the genetic improvement of woody species awaits improved and more reproducible plant regeneration systems from protoplasts than are currently available for woody species.

8. Exposure of Somaclonal Variation Through Protoplast Culture

Plants regenerated from somatic cells following a culture phase often express genetic variability, constituting the phenomenon of somaclonal variation (Scowcroft *et al.*, 1987; Larkin and Scowcroft, 1991). The occurrence of somaclonal variation is undesirable if the production of true-to-type plants is required. However, it can be exploited as a small proportion of somaclonal variants may exhibit desirable, stable and heritable attributes that may be used in plant improvement. Cloning of plants through the isolation and culture of protoplasts has been found to enhance their genetic variability, which may be a reflection of an extended time-frame in culture required for shoot regeneration from protoplasts.

The selection of somaclonal variants has been used to produce poplar trees that are tolerant to herbicides (Michler and Haissig, 1988) and resistant to the leaf spot fungus *Septoria musiva* (Ostry and Skilling, 1988). Additionally, enhanced resistance to leaf rust in cottonwood (Prakash and Thielges, 1989), improved resistance to the bacterial pathogen *Xanthomonas campestris* in peach (Hammerschlag, 1992) and resistance to fireblight in apple (Donovan, 1991), have all been claimed. These somaclonal variants, together with many other examples quoted by Hammerschlag (1992), have been generated from explants and cultured cells. Reports of protoplast-derived somaclonal variants in woody plants are rare. Ochatt (1987) described plants of *Pyrus communis* which differed in their leaf morphology and rootability. Rosaceous fruit trees of the genus *Prunus* are commonly subjected to drought and salinity stress under orchard conditions. In Colt cherry (*Prunus avium* × *pseudocerasus*), Ochatt and Power (1989) recovered salt and mannitol-tolerant cell lines following direct recurrent selection derived from protoplasts of leaves and root cell suspensions. Protoplast-derived calli were cultured on semi-solid medium with 0, 25, 50, 100 or 200 mN NaCl, KCl or Na_2SO_4 or iso-osmotic (with NaCl) concentrations of mannitol (0, 8.7, 17.1, 33.3, or 66.6 g L^{-1}). Tissues which survived six transfers on the same salt or mannitol-containing medium were subjected to three cycles of direct recurrent selection, each consisting of two, three-week periods on medium with salt or mannitol. Subsequently, salt and drought-resistant plants of Colt cherry were regenerated from these selected tissues.

9. Concluding Remarks

Considerable potential is offered by protoplast technology to the improvement of woody species, especially those with long and complex life cycles.

Techniques based on the isolation and culture of protoplasts have already been exploited with success in non-woody agricultural crops and such approaches are also likely to have impact on woody plants. In general, the application of somatic hybridisation and transformation to woody plants is currently restricted by the recalcitrance of many species to shoot regeneration from protoplast-derived tissues. The fundamental processes of plant regeneration from protoplasts are not understood, reflecting the diverse methods currently in use for culturing protoplasts isolated from a variety of source tissues. The occurrence of somaclonal variation in protoplast-derived plants is a potential limitation if true-to-type plants are required. However, in some cases, such variation can be of benefit to the breeder.

In spite of existing limitations in the culture of protoplasts of woody species, considerable progress has been achieved in protoplast technology, particularly in recent years, since increasing numbers of woody plants can now be readily regenerated from protoplasts. This will allow such species and genera to become amenable to genetic manipulation through complete and partial genome transfer by protoplast fusion or direct DNA transfer technologies. The applicability of this technology to woody plant improvement has already been demonstrated in some species which can be regenerated with comparative ease, as in the Rutaceae and, to some extent, in the Rosaceae and Passifloraceae. The extension of somatic cell approaches to other woody species will, undoubtedly, follow the development of protoplast-to-plant regeneration systems.

References

Acuna, J.R. and M. Pena, 1991. Plant regeneration from protoplasts of embryogenic cell suspensions of *Coffea arabica* L. cv. Caturra. Plant Cell Rep. 10: 345–348.

Arya. S., J.R. Liu and T. Eriksson, 1991. Plant regeneration from protoplasts of *Panax ginseng* (C.A. Meyer) through somatic embryogenesis. Plant Cell Rep. 10: 277–281.

Barton, C.R., T.L. Adams and M.A. Zarowitz, 1991. Stable transformation of foreign DNA into *Coffea arabica* plants. In XIV Internatl. Conf. Coffee Science, San Francisco, pp. 460–464.

Chand, P.K., S.J. Ochatt, E.L. Rech, J.B. Power and M.R. Davey, 1988. Electroporation stimulates plant regeneration from protoplasts of the woody medicinal species *Solanum dulcamara* L. J. Exp. Bot. 39: 1267–1274.

Chen, M.H. and C.C. Chen, 1992. Plant regeneration from *Carica protoplasts*. Plant Cell Rep. 11: 404–407.

Davey, M.R., E.L. Rech and B.J. Mulligan, 1989. Direct DNA transfer to plant cells. Plant Mol. Biol. 13: 273–285.

David, H., J.M. Domon, C. Savy, N. Miannay, G. Sulmont, R. Dargent and A. David, 1992. Evidence of early stages of somatic embryo development in protoplast-derived cell cultures of *Prunus avium*. Physiol. Plant. 85: 301–307.

Donovan, A., 1991. Screening for fire blight resistance in apple (*Malus primula*) using leaf assays from *in vitro* grown material. Ann. Appl. Biol. 119: 59–68.

Dorion, N., P. Danthu, S. Ohki, C. Preneux, B. Godin and C. Bigot, 1991. Plant regeneration from leaf protoplasts of common elm (*Ulmus campestris* Mill.). Compt. Rendus de L'Acad. Sci. Paris, Serie III. 313: 467–473.

d'Utra Vaz, F.B., A.V.P. dos Santos, G. Manders, E.C. Cocking, M.R. Davey and J.B. Power, 1993. Plant regeneration from leaf mesophyll protoplasts of the tropical woody plant, passionfruit (*Passiflora edulis* fv flavicarpa Degener.): the importance of the antibiotic cefotaxime in the culture medium. Plant Cell Rep. 12: 220–225.

Echeverria, E., 1987. Preparation and characterisation of protoplasts from *Citrus* juice vesicles. J. Am. Soc. Hort. Sci. 112: 393–396.

Filliatti, J.J., J. Sellmer, B. McCown, B. Haisig and L. Comai, 1987. *Agrobacterium*-mediated transformation and regeneration of *Populus*. Molec. Gen. Genet. 206: 192–199.

Gilmour, D.M., T.J. Golds and M.R. Davey, 1989. *Medicago* protoplasts: Fusion, culture and plant regeneration. In: Y.P.S. Bajaj (Ed.), Biotechnology in Agriculture and Forestry, Vol. 8. Protoplasts and Genetic Engineering, pp. 370–388. Springer-Verlag, Heidelberg.

Gmitter, F.G. Jr., J.W. Grosser and G.A. Moore, 1992. *Citrus*: In: F.A. Hammerschlag and R.E. Litz (Eds.), Biotechnology of Perennial Fruit Crops, pp. 335–369. C.A.B. International, Wallingford, Oxon.

Graham, J., R.J. McNicol and A. Kumar, 1990. Use of the GUS gene as a selectable marker for *Agrobacterium*-mediated transformation of *Rubus*. Plant Cell, Tiss. Org. Cult. 10: 35–39.

Grosser, J.M. and J.L. Chandler, 1987. Aseptic isolation of leaf protoplasts from *Citrus*, *Poncirus*, *Citrus* × *Poncirus* hybrids and *Severinia* for use in somatic hybridization experiments. Scientia Hortic. 31: 253–257.

Grosser, J.W., F.G. Gmitter, N. Tusa and J.L. Chandler, 1990. Somatic hybrid plants from sexually incompatible woody species: *Citrus reticulata* and *Citropsis gilletiana*. Plant Cell Rep. 8: 656–659.

Hammatt, N., N.W. Blackhall and M.R. Davey, 1991. Variation in the DNA content of *Glycine* species. J. Exp. Bot. 42: 659–665.

Hammerschlag, F.A., 1992. Somaclonal variation. In: F.A. Hammerschlag and R.E. Litz (Eds.), Biotechnology in Agriculture, Vol. 8, Biotechnology of Perennial Fruit Crops, pp. 35–55. C.A.B. International, Wallingford, Oxon.

Hidaka, T. and L. Kajiura, 1988. Plantlet differentiation from callus protoplasts induced from *Citrus* embryo. Scientia Hortic. 34: 85–92.

Huntinen, O., J. Honkanen and L.K. Simola, 1982. Ornithine and putrescine supported divisions and cell colony formation in leaf protoplasts of alders (*Alnus glutinosa* and *A. incana*). Plant Sci. Lett. 28: 3–9.

Ito, K., K. Doi, Y. Tatemichi and M. Shibata, 1990. Plant regeneration from protoplasts of *Eucalyptus*. In: Abstr. VIIth Internatl. Congr. Plant Tissue and Cell Cult., Amsterdam, Abstract A1-65, p. 19.

James, D.J., S. Uratsu, J. Cheng, P. Negri, P. Viss and M. Dandekar, 1993. Acetosyringone and osmoprotectants like betaine and proline synergistically enhance *Agrobacterium*-mediated transformation of apple. Plant Cell Rep. 12: 559–563.

Jorgensen, J. and H. Binding, 1984. Callus regeneration with protoplasts of *Sorbus aucuparia* L. Z. Pflanzenphysiol. 113: 371–372.

Kanchanapoom, M. and K. Kanchanapoom, 1991. Isolation, culture and regeneration of *Theobroma* protoplasts. Physiol. Plant. 82A: 14.

Kao, K.N. and M.R. Michayluk, 1975. Nutritional requirements for growth of *Vicia hajastana* cells and protoplasts at very low population densities in liquid media. Planta 126: 105–110.

Kawazu, T., K. Doi, T. Ohta, Y. Shinohara, K. Ito and M. Shibata, 1990. Transformation of Eucalyptus (*Eucalyptus saligna*) using electroporation. In: Abstr. VIIth Internatl. Congr. Plant Tissue and Cell Cult., Amsterdam, Abstract A2-80, p. 64.

Kobayashi, S., 1987. Uniformity of plants regenerated from orange (*Citrus sinensis* Osb.) protoplasts. Theor. Appl. Genet. 74: 10–14.

Kobayashi, S., I. Ikeda and H. Uchimiya, 1985. Conditions for high frequency embryogenesis from orange (*Citrus sinensis* Osbeck.) protoplasts. Plant Cell Tiss. Org. Cult. 4: 249–259.

Kobayashi, S., S. Oiyama, K. Yoshinaga, T. Ohgawara and S. Ishii, 1991. Fertility in an intergeneric somatic hybrid plant of Rutaceae. HortSci. 26: 207.

Kobayashi, S. and H. Uchimiya, 1989. Expression and integration of a foreign gene in orange (*Citrus sinensis* Osb.) protoplasts by direct DNA transfer. Japan J. Genet. 64: 91–97.

Kobayashi, S., H. Uchimiya and I. Ikeda, 1983. Plant regeneration from 'Trovita' orange protoplasts. Jap. J. Breed. 33: 119–122.

Kouider, M., R. Hauptmann, J.M. Widholm, R.M. Skirvin and S.S. Korban, 1984. Callus formation from *Malus* × *domestica* cv Jonathan protoplasts. Plant Cell Rep. 3: 142–145.

Lang, H. and H.W. Kohlenbach, 1988. Callus formation from mesophyll protoplasts of *Fagus sylvatica* L. Plant Cell Rep. 7: 485–488.

Laine, E. and A. David, 1990. Somatic embryogenesis in immature embryos and protoplasts of *Pinus caribaea*. Plant Sci. 69: 215–224.

Lakshimi Sita, G., 1987. Triploids. In: J.M. Bonga and D.J. Durzan (Eds.), Cell and Tissue Culture in Forestry, Vol. 2, Specific Principles and Methods: Growth and Development, pp. 269–284. Martinus Nijhoff Publishers, Dordrecht.

Larkin, P.J. and W.R. Scowcroft, 1991. Somaclonal variation – a novel source of variability from cell cultures for plant improvement. Theor. Appl. Genet. 60: 197–214.

Laurain, D., J-C. Chénieux and J. Trémouillaux-Guiller, 1993. Direct embryogenesis from female haploid protoplasts of *Ginkgo biloba* L., a medicinal woody species. Plant Cell Rep. 12: 656–660.

Lee, J.S., S.K. Lee, S.S. Jang and J.J. Lee, 1987. Plantlet regeneration from callus protoplasts of *Populus nigra*. Res. Rep. Inst. For. Gen. Korea 23: 143–148.

Ling, J.T., N. Nito and M. Iwamasa, 1989. Plantlet regeneration from protoplasts of calamondin (*Citrus madurensis* Lour). Scientia Hortic. 40: 325–333.

Ling, J.T., N. Nito, M. Iwamasa and H. Kunitake, 1990. Plant regeneration from protoplasts from embryogenic callus of satsuma. HortSci. 25: 970–972.

Louzada, E.S., J.W. Grosser and F.G. Gmitter Jr., 1993. Intergeneric somatic hybridization of sexually incompatible parents: *Citrus sinensis* and *Atlantia ceylanica*. Plant Cell Rep. 12: 687–690.

Louzada, E.S., J.W. Grosser, F.G. Gmitter Jr., X.X. Deng, N. Tusa, B. Nielsen and J.L. Chandler, 1992. Eight new somatic rootstocks with potential for improved disease resistance. HortSci. 27: 1033–1036.

Lynch, P.T., M.R. Davey and J.B. Power, 1993. Protoplast fusion and somatic hybridisation. In: N. Duzgunes (Ed.), Methods in Enzymology: Membrane Fusion Techniques, Vol. 221, pp. 379–393. Academic Press, London.

Manders, G., M.R. Davey and J.B. Power, 1992a. New genes for old trees. J. Exp. Bot. 43: 1181–1190.

Manders, G., T.J. Golds, E.L. Rech, M.R. Davey and J.B. Power, 1990. Transient gene expression in electroporated protoplasts of Bramley's seedling apple. Abstr. VIIth Internatl. Congr. Plant Tissue and Cell Cult., Amsterdam, p. 67.

Manders, G., W. C. Otoni, F.B. d'Utra Vaz, N.W. Blackhall, J.B. Power and M.R. Davey, 1994. Transformation of passionfruit (*Passiflora edulis* var. flavicarpa Degener.) using *Agrobacterium tumefaciens*. Plant Cell Rep. 13: 697–702.

Manders, G., A.V.P. dos Santos, F.B. d'Utra Vaz, M.R. Davey and J.B. Power, 1992b. Transient gene expression in electroporated protoplasts of *Eucalyptus citriodora* Hook. Plant Cell Tiss. Org. Cult. 30: 69–75.

Mante, S., P.H. Morgans, R. Scorza, J.M. Cordts and A.M. Callahan, 1991. *Agrobacterium*-mediated transformation of plum (*Prunus domestica* L.) hypocotyl slices and regeneration of transgenic plants. Bio/Technol. 9: 853–857.

Marchant, R., M.R. Davey and J.B. Power, 1993. Protoplast fusion. In: J. Bryant (Ed.), Methods in Plant Biochemistry, Vol. 10. Molecular Biology, pp. 187–205. Academic Press, London.

Matthews, M., J. Mottley, I. Horan and A.V. Roberts, 1991. A protoplast to plant system in roses. Plant Cell Tiss. Org. Cult. 24: 173–180.

McGranahan, G.H., C.A. Leslie, S.I. Urasu, L.A. Martin and A.M. Dandekar, 1988. *Agrobacterium*-mediated transformation of walnut somatic embryos and regeneration of transgenic plants. Bio/Technol. 6: 800–804.

Merkle, S.A. and H.E. Sommer, 1987. Regeneration of *Liriodendron tulipefera* (family Magnoliaceae) from protoplast culture. Amer. J. Bot. 74: 1317–1321.

Michler, C.H. and B.E. Haissig, 1988. Increased herbicide tolerance of *in vitro* selected hybrid Poplar. In: M.R. Ahuja (Ed.), Somatic Cell Genetics of Woody Plants, pp. 183–189. Kluwer Academic Publishers, Dordrecht.

Mii, M. and H. Ohashi, 1988. Plantlet regeneration from protoplasts of kiwifruit, *Actinidia chinensis* Planch. Acta Hortic. 230: 167–170.

Ochatt, S.J., 1987. Coltura di protoplasti come metodo per il miglioramento genetico nelle piante da frutto. Frutticoltura 49: 58–60.

Ochatt, S.J., 1990a. Protoplast technology and top-fruit tree breeding. Acta Hortic. 280: 215–226.

Ochatt, S.J., 1990b. Plant regeneration from callus protoplasts of sour cherry (*Prunus cerasus* L.). Plant Cell Rep. 9: 268–271.

Ochatt, S.J., 1992. The development of protoplast-to-tree systems for *Prunus cerasifera* and *Prunus spinosa*. Plant Sci. 81: 253–259.

Ochatt, S.J. and O.H. Caso, 1986. Shoot regeneration from leaf mesophyll protoplasts of wild pea (*Pyrus communis* var pyraster L). J. Plant Physiol 122: 243–249.

Ochatt, S.J., P.K. Chand, E.L. Rech, M.R. Davey and J.B. Power, 1988. Electroporation-mediated improvement of plant regeneration from Colt cherry (*Prunus avium* × *pseudocerasus*) protoplasts. Plant Sci. 54: 165–169.

Ochatt, S.J., E.C. Cocking and J.B. Power, 1987. Isolation, culture and plant regeneration of Colt cherry (*Prunus avium* × *pseudocerasus*) protoplasts. Plant Sci. 50: 139–143.

Ochatt, S.J., A.S. Escandon and A.J. Martinez, 1989a. Isolation, culture and plant regeneration from protoplasts of shrubby *Oxalis* species from South America. J. Exp. Bot. 40: 493–496.

Ochatt, S.J., E.M. Patat-Ochatt and J.B. Power, 1992. Protoplasts. In: F.A. Hammerschlag and R.E. Litz (Eds.), Biotechnology in Agriculture, Vol. 8, Biotechnology of Perennial Fruit Crops, pp. 77–103. C.A.B., International, Wallingford, Oxon.

Ochatt, S.J., E.M. Patat-Ochatt, E.L. Rech, M.R. Davey and J.B. Power, 1989b. Somatic hybridization of sexually incompatible top-fruit tree rootstocks, wild pear (*Pyrus communis* var pyraster L.) and Colt cherry (*Prunus avium* × *pseudocerasus*). Theor. Appl. Genet. 78: 35–41.

Ochatt, S.J. and J.B. Power, 1988a. Plant regeneration from leaf mesophyll protoplasts of Williams Bon Chretien (syn. Bartlett) pear (*Pyrus communis* L.). Plant Cell Rep. 7: 587–589.

Ochatt, S.J. and J.B. Power, 1988b. An alternative approach to plant regeneration from protoplasts of sour cherry (*Prunus cerasus* L.). Plant Sci. 56: 75–79.

Ochatt, S.J. and J.B. Power, 1989. Selection for salt-drought tolerance using protoplasts and explant-derived tissue cultures of Colt cherry (*Prunus avium* × *pseudocerasus*). Tree Physiol. 5: 259–266.

Ochatt, S.J. and J.B. Power, 1992. Plant regeneration from cultured protoplasts of higher plants. In: M.W. Fowler, G.S. Warren and M. Moo-Yong (Eds.), Plant Biotechnology. Comprehensive Biotechnology Second Supplement, pp. 99–127. Pergamon Press, Oxford/New York/Seoul/Tokyo.

Ohgawara, T., S. Kobayashi, S. Ishii, K. Yoshinaga and I. Oiyama, 1989. Somatic hybridisation in *Citrus*: navel orange (*C. sinensis* Osb.) and grapefruit (*C. paradisi* Macf.). Theor. Appl. Genet. 78: 609–612.

Oka, S., 1987. Isolation, culture and organogenesis of leaf mesophyll protoplasts of paper mulberry (*Broussonetia kazinoki* Sieb.). Jap. Agric. Res. Quart. 21: 8–14.

Oka, S. and K. Ohyama, 1985. Plant regeneration from mesophyll protoplasts of *Broussonetia kazinoki* Sieb. (paper mulberry). J. Plant Physiol. 119: 455–460.

Oliveira, M.M., J. Barroso and M.S.S. Pais, 1991. Direct gene transfer into *Actinidia deliciosa* protoplasts: analysis of transient expression of the *CAT* gene using TLC autoradiography and a GC-MS-based method. Plant Molec. Biol. 17: 235–242.

Oliveira, M.M. and M.S.S. Pais, 1991. Plant regeneration from protoplasts of long term callus cultures of *Actinidia deliciosa* var. deliciosa cv. Hayward (Kiwifruit). Plant Cell Rep. 9: 643–646.

Ostry, M.E. and D.D. Skilling, 1988. Somatic variation in resistance of *Populus* to *Septoria musiva*. Plant Disease 72: 724–727.

Otoni, W.C., N.W. Blackhall, F.B. d'Utra Vaz, V.W.D. Casali, J.B. Power, and M.R. Davey, 1994a. Somatic hybridisation of the *Passiflora* species *P. edulis* fv. flavicarpa and *P. incarnata*. J. Exp. Bot. (submitted).

Otoni, W.C., F.B. d'Utra Vaz, V.W.D. Casali, M.R. Davey and J.B. Power, 1994b. Plant regeneration from mesophyll protoplasts of wild passionfruit species (*Passiflora suberosa* L., *P. seemannii* Griseb., *P. incarnata*) and the F1 hybrid *P. maliformis* × *P. serrulatus*. Plant Cell Rep. (in preparation).

Park, G.Y., M.S. Choi and J.H. Kim, 1990. Plant regeneration of *Populus glandulosa* from mesophyll protoplasts. Korean J. Plant Tiss. Cult. 17: 189–199.

Park, G.Y. and S.H. Son, 1988. Culture and regeneratlon of *Populus alba* × *glandulosa* leaf protoplasts isolated from *in vitro* cultured explant. J Korean For. Soc. 77: 208–215.

Park, G.Y. and S.H. Son, 1992. *In vitro* shoot regeneration from leaf mesophyll protoplasts of hybrid poplar (*Populus nigra* × *P. maximowiczii*). Plant Cell Rep. 11: 2–6.

Patat-Ochatt, E.M., J. Boccon-Gibod, M. Duron and S.J. Ochatt, 1993. Organogenesis of stem and leaf protoplasts of haploid golden delicious apple clone (*Malus* × *domestica* Borkh.). Plant Cell Rep. 12: 118–120.

Patat-Ochatt, E.M., S.J. Ochatt and J.B. Power, 1988. Plant regeneration from protoplasts of apple rootstocks and scion varieties (*Malus* × *domestica* Borkh.). J. Plant Physiol. 133: 460–465.

Patat-Ochatt, E.M. and J.B. Power, 1990. Advances in plant regeneration from apple protoplasts. Acta Hortic. 280: 285–288.

Pirrie, A. and J.B. Power, 1986. The production of fertile, triploid somatic hybrid plants (*Nicotiana glutinosa* [n] + *N. tabacum* [2n]) via gametic:somatic protoplast fusion. Theor. Appl. Genet. 72: 48–52.

Prakash, C.S. and B.A. Thielges, 1989. Somaclonal variation in Eastern cottonwood for race-specified partial resistance to leaf rust disease. Phytopathol. 79: 805–808.

Rao, P.S. and P. Ozias-Akins, 1985. Plant regeneration through somatic embryogenesis in protoplast cultures of sandalwood (*Santalum album*). Protoplasma 124: 80–86.

Rao, I.V.R., U. Mehta and H.Y.M. Ram, 1982. Whole plant regeneration from cotyledonary protoplasts of *Crotalaria juncea*. In: A. Fujiwara (Ed.), Plant Tissue Culture 1982. Proc. 5th Internatl. Congr. Plant Tissue and Cell Cult. Japanese Assoc. Plant Tissue Culture, Tokyo, pp. 595–596.

Ratushnyak, Y.I., N.M. Piven and V.A. Rudas, 1989. Protoplast culture and plant regeneration in *Lycium barbarum* L. Plant Cell, Tiss. Org. Cult. 17: 183–190.

Russell, J.A. and B.H. McCown, 1986. Culture and regeneration of *Populus* leaf protoplasts isolated from non-seedling tissue. Plant Sci. 46: 133–142.

Russell, J.A. and B.H. McCown, 1988. Recovery of plants from leaf protoplasts of hybrid-poplar and aspen clones. Plant Cell Rep. 7: 59–62.

Sasamoto, H. and Y. Hosoi, 1990. Effects of electric fusion on calli formation and regeneration from protoplasts of *Quercus* and two *Populus* species. Abstr. VIIth Intl. Congr. Plant Tissue and Cell Cult., Amsterdam, Abstract A1–131, p. 36.

Sasamoto, H. and Y. Hosoi, 1992. Callus proliferation from the protoplasts of embryogenic cells of *Quercus serrata*. Plant Cell, Tiss. Org. Cult. 29, 241–245.

Sasamoto, H., Y. Hosoi, K. Ishii, T. Sato and A. Saito, 1990. Callus formation and plantlet regeneration from mesophyll protoplasts of *Populus sieboldii*. Trans. 100th Meeting Japan Forest Soc. 100: 531–532.

Saxena, P.K. and R. Gill, 1987. Plant regeneration from mesophyll protoplasts of the tree legume *Pithecellobium dulce* Benth. Plant Sci. 53: 257–262.

Schopke, C., L. E. Muller and H. W. Kohlenbach, 1987. Somatic embryogenesis and regeneration of plantlets in protoplast cultures from somatic embryos of coffee (*Coffea canefora* P. ex Fr.). Plant Cell, Tiss. Org. Cult. 8: 243–248.

Scowcroft, W.R., R.I.S. Brettell, S.A. Ryan, P.A. Davies and M.A. Pallotta, 1987. Somaclonal variation and genomic flux. In: C.E. Green, D.A. Somers, W.P. Hackett and D.D. Biesboer (Eds.), Plant Biology, Vol. 3. Plant Tissue and Cell Culture, pp. 275–286. Alan R. Liss, New York.

Seguin, A. and M. Lalonde, 1988. Gene transfer by electroporation in Betulaceae protoplasts: *Alnus incana*. Plant Cell Rep. 7: 367–370.

Shahin, E.A. and J.F. Shepard, 1980. Cassava mesophyll protoplasts: isolation, proliferation and shoot formation. Plant Sci. Lett. 17: 459–465.

She, J., J. Wu, H. Wang, H. Zhou, Z. Chen, S. Li and J. Yue, 1989. Somatic embryogenesis and plant regeneration from protoplast culture of cotton. Jiangsu J. Agric. Sci. 5: 63–68.

Sim, G.E., C.S. Loh and C.J. Goh, 1988. Direct somatic embryogenesis from protoplasts of *Citrus mitis* Blanco. Plant Cell Rep. 7: 418–420.

Spiral, J. and V. Petiard, 1991. Protoplast culture and regeneration in coffee species. XIV Intl. Conf. Coffee Science, San Francisco, pp. 383–391.

Sticklen, M.B., S.C. Domir and R.D. Linberger, 1985. Shoot regeneration from protoplasts of *Ulmus* × "pioneer". Plant Sci. 41: 29–34.

Tao, R., M. Tamura, K. Yonemori and A. Sugiura, 1991. Plant regeneration from callus protoplasts of adult Japanese persimmon (*Diospyros kaki* L.). Plant Sci. 79: 119–125.

Teulières, C. and A.M. Boudet, 1991. Isolation of protoplasts from different *Eucalyptus* species and preliminary studies on regeneration. Plant Cell, Tiss. Org. Cult. 25: 133–140.

Teulières, C., J. Grima-Pettenati, C. Curie, J. Teissie and A.M. Boudet, 1991. Transient foreign gene expression in polyethylene/glycol treated or electropulsated *Eucalyptus gunnii* protoplasts. Plant Cell, Tiss. Organ Cult. 25, 125–132.

Tremblay, F.M., 1988. Callus formation from protoplasts of *Betula papyrifera* March. cell suspension culture. J. Plant Physiol. 113: 247–251.

Tremblay, F.M., J.B. Power and M. Lalonde, 1985. Callus regeneration from *Alnus incana* protoplasts isolated from cell suspensions. Plant Sci. 41: 211–216.

Tsai, C.K., 1988. Plant regeneration from leaf protoplast callus of *Actinidia chinensis* var chinensis. Plant Sci. 54: 231–235.

Tusa, N., J.W. Grosser, F.G. Gmitter Jr. and E.S. Louzada, 1992. Production of tetraploid somatic hybrid breeding parents for use in lemon cultivar improvement. HortSci. 27: 445–447.

Uematsu, C., M. Murase, H. Ichikawa and J. Imamura, 1991. *Agrobacterium*-mediated transformation and regeneration of kiwifruit. Plant Cell Rep. 10: 286–290.

Vahala, T. and T. Eriksson, 1991. Callus production from willow (*Salix viminalis* L.) protoplasts. Plant Cell Tiss. Org. Cult. 27: 243–248.

Vardi, A., P. Arzee-Gonen, A. Frydman-Shani, S. Bleichman and E. Galun, 1989. Protoplast-fusion-mediated transfer of organelles from *Microcitrus* into *Citrus* and regeneration of novel alloplasmic trees. Theor. Appl. Genet. 78: 741–747.

Vardi, A., S. Bleichman and D. Aviv, 1990a. Genetic transformation of *Citrus* protoplasts and regeneration of transgenic plants. Plant Sci. 69: 206.

Vardi, A., A. Breiman and E. Galun, 1987. *Citrus* cybrids: production by donor-recipient protoplast-fusion and verification by mitochondrial DNA restriction profiles. Theor. Appl. Genet. 75: 51–58.

Vardi, A., A. Frydman-Shani, E. Galun, P. Gonen and S. Bleichman, 1990b. *Citrus* hybrids: transfer of *Microcitrus* organelles into *Citrus* cultivars. Acta Hortic. 280: 239–245.

Vardi, A. and E. Galun, 1988. Recent advances in protoplast culture of horticultural crops: *Citrus*. Scientia Hortic. 37: 217–230.

Vardi, A., D.J. Hutchinson and E. Galun, 1986. A protoplast-to-tree system in *Microcitrus* based on protoplasts derived from a sustained embryogenic callus. Plant Cell Rep. 5: 412–414.

Vardi, A., P. Spielgel-Roy and E. Galun, 1975. *Citrus* cell culture: isolation of protoplasts, plating densities, effect of mutagens and regeneration of embryos. Plant Sci. Lett. 4: 231–236.

Vardi, A., P. Spiegel-Roy and E. Galun, 1982. Plant regeneration from citrus protoplasts: variability in methodological requirements among cultivars and species. Theor. Appl. Genet. 62: 171–176.

Wallin, A. and L. Johansson, 1990. Plant regeneration from leaf mesophyll protoplasts of *in vitro* cultured shoots of a columnar apple. J. Plant Physiol. 135: 565–570.

Wang, D., M.R. Sondahl and D.A. Evans, 1986. Somatic embryogenesis and plant regeneration from cell suspension-isolated protoplasts of *Citrus sinensis*. Abstr. VIth Intl. Congr. Plant Tissue and Cell Culture, Minneapolis, p. 339.

Yasuda, T., M. Tahara, N. Uchida and T. Yamaguchi, 1986. Somatic embryogenesis from coffee callus and protoplasts. Abstr. VIth Intl. Congr. Plant Tissue and Cell Culture, Minneapolis, p. 137.

Zhao, Y.X., 1993. Protoplast culture and regeneration from *Sesbania* species. Proc 13th NERC Tree Biotechnology Liaison Group Meeting, April 1993, Coventry Univ., UK.

Zhao, Y-X., D-Y. Yao and P.J.C. Harris, 1991. Isolation and culture of protoplasts of callus tissue of *Hibiscus syriacus* L. Plant Cell Tiss. Org. Cult. 25: 17–19.

9. Gymnosperm Protoplasts

Faouzi Bekkaoui, Tom E. Tautorus
and David I. Dunstan

Contents

1. Introduction

Protoplasts are plant cells deprived of their wall by a combination of cell wall-digesting enzymes. The removal of the cell wall facilitates increased uptake of media constituents. Initially, protoplasts were used as a biological tool to study various aspects of cell biology and physiology, i.e., membrane structure and function, cell compartmentation, cytoskeleton (Cocking, 1972; Galun, 1981; Fowke and Constabel, 1985). Also, protoplast isolation permitted cell fusion and somatic hybridization (Gleba and Sytnic, 1984). During the past few years, with the development of molecular biology techniques, the focus has been mostly on germplasm improvement using protoplasts as a tool in plant transformation (Bajaj, 1989). In gymnosperms, considered as recalcitrant to *in vitro* manipulations, efforts have largely been aimed at identifying the conditions for protoplast regeneration. This step is critical because regeneration to an entire plant is a prerequisite to genetic manipulations based on somatic hybridization or direct DNA uptake.

Protoplast isolation from gymnosperms was first obtained with callus of

S. Jain, P. Gupta & R. Newton (eds.), Somatic Embryogenesis in Woody Plants, Vol. 1, 167–191.

Pseudotsuga menziesii derived from vegetative shoots (Winton *et al.*, 1975). Later, protoplasts from cotyledons of *Pseudotsuga menziesii* (Kirby and Cheng, 1979) and *Pinus pinaster* (David and David, 1979) were shown to grow into callus. Regeneration of an organized tissue from protoplasts was possible only after the development of somatic embryogenesis cultures in *Picea abies* (Chalupa, 1985; Hakman *et al.*, 1985), and was observed in *Pinus taeda* (Gupta and Durzan, 1987) and *Picea glauca* (Attree *et al.*, 1987). Subsequently, plantlet regeneration from protoplasts was obtained in *Picea glauca* (Attree *et al.*, 1989b) and *Larix* × *eurolepis* (Klimaszewska, 1989). The capacity of gymnosperm protoplasts to regenerate to plants seems to depend on the type of source tissue used as starting material.

2. Factors Affecting Isolation, Culture and Regeneration of Protoplasts

Successful regeneration of plants from gymnosperm protoplasts involves a series of sequential culture steps, most of which have to be determined empirically. These steps are similar to those developed for other species (Bajaj, 1989) and regarding conifers have been discussed in previous reviews (David, 1987; Kirby *et al.*, 1989). In most cases, it is the combination of several critical factors that will lead to regeneration of plants, as described in the following sections.

2.1. *Protoplast Isolation and Culture*

2.1.1. *Enzymes and Purification Protocols*

The choice of the concentration and the nature of the cell wall-degrading enzymes (cellulase, hemicellulase, pectinase) depend on the species type and source tissue used as starting material. Usually, optimization experiments are necessary for each new investigation. For example, for *Picea glauca* with 0.5% Cellulase R-10 (Onozuka), 0.25% each of pectinase and driselase (Sigma) in 5 mM $CaCl_2$ and 0.5 M mannitol, a yield of 0.8–1.0 × 10^6 protoplasts per g of fresh weight and 75–85% viability were achieved (Bekkaoui *et al.*, 1988). In most cases, commercial enzymes are used directly. However, desalting the enzymes may be beneficial as shown with *Pinus pinaster* protoplasts (David *et al.*, 1986). Adding bovine serum albumin (BSA) during cell wall digestion has led to a higher degree of protoplast viability (Lainé *et al.*, 1988), confirming earlier results observed by Teasdale and Rugini (1983). BSA is thought to be a substrate for proteases and these may be present as impurities in commercial enzyme preparations and otherwise adversely affect protoplast membranes. However, in *Picea glauca*, the absence of BSA during enzymatic treatment, did not affect subsequent development of the protoplasts into callus or embryos (Attree *et al.*, 1987; Bekkaoui *et al.*, 1987).

The purification of protoplasts is an important procedure, the result of which may determine subsequent development of protoplasts. For example,

Picea glauca embryogenic suspension cultures contain several types of cells (Hakman and Fowke, 1987; Dunstan *et al.*, 1988). Protoplasts from these may separate and develop differently according to the purification procedure used. After preliminary filtration through a nylon membrane (e.g., 65 μm pore size), three general methods have been used with protoplasts from conifer cultures:

Method 1 – Repeated centrifugation at 100 g and resuspension in culture media (David and David, 1979; Kirby and Cheng, 1979).
Method 2 – A sucrose gradient method (Teasdale and Rugini, 1983; Patel *et al.*, 1984; Hakman *et al.*, 1986; Bekkaoui *et al.*, 1987).
Method 3 – A Ficoll gradient method (Gupta and Durzan, 1986; Attree *et al.*, 1987; Lainé *et al.*, 1988; Klimaszewska, 1989).

The 3rd method, although more laborious, has given interesting results. For example, Klimaszewska (1989) obtained 4 populations of protoplasts from *Larix* × *eurolepis* somatic embryos cultures, among them a population of large protoplasts which originated from suspensor-like cells and a population of small protoplasts which originated from cells in the head of the embryo. The fraction with the highest specific gravity contained many cytoplasm-rich protoplasts and exhibited the highest plating efficiency. The fraction with the lowest specific gravity contained mainly highly vacuolated protoplasts and had a low plating efficiency.

After protoplasts have been purified, viability can be monitored by a dye exclusion method such as Evan's blue (Gaff and Okang'o-Ogola, 1971) or phenosafranine (Widholm, 1972). Alternatively, fluorescein diacetate can be used to monitor membrane integrity (Widholm, 1972). Phenosafranine was routinely used with protoplasts from *Picea glauca* (Bekkaoui *et al.*, 1988, 1990) and *Picea mariana* (Tautorus *et al.*, 1989).

2.1.2. *Culture Media and Method*

A wide variety of culture media has been used for protoplast culture (Table 1). Addition of auxin is necessary for protoplast division; however, the addition of a cytokinin is required only in certain cases (Table 1). In *Picea glauca*, regeneration from embryogenic cultures was improved by changing specific medium components (Attree *et al.*, 1989a). The composition of macroelements affected the plating efficiency and protoplast development. Protoplasts from certain embryogenic cultures of *Picea glauca* differentiated into proembryos only when the macroelement concentration was reduced (Attree *et al.*, 1989a). In terms of nitrogen requirement, glutamine has been shown to have a positive effect on the plating efficiency and, therefore, has usually been included (David, 1987; Kirby *et al.*, 1989). For example, in *Picea glauca*, microcalli were recovered only in the presence of 5 mM glutamine (Bekkaoui *et al.*, 1987). The polyamines, L-ornithine and putrescine were shown to promote division in *Pinus oocarpa* and *Pinus patula* (Lainé *et al.*, 1988).

Myo-inositol has been used as a plasmolysing agent in lieu of or in combination with other carbohydrates such as sucrose, glucose, mannitol and sorbitol. Myo-inositol has been observed to improve the regeneration of somatic embryos from conifer protoplasts (Gupta and Durzan, 1987; Attree *et al.*, 1989a), though this was inconsistent (Attree *et al.*, 1987; Lang and Kohlenbach, 1989). With protoplasts from callus cultures established from shoot tips of mature trees of *Pinus sylvestris*, Hohtola and Kvist (1991) have used a new floating agent Nycodenz at 9–10% and have shown that it improved the viability and subsequent growth of the cell when used in conjunction with an osmoticum.

Several culture methods have been used, the most common has been culture of protoplasts in droplets of 50–100 μl medium or embedding in agarose. In *Picea glauca*, culture in liquid medium in a multiwell plate was shown to lead to a better protoplast plating efficiency, probably due to reduced aggregation of the protoplasts as compared to culture in liquid droplets in Petri dishes (Bekkaoui *et al.*, 1987). Agarose embedded protoplasts may have certain advantages over culture in liquid media. In the immobilized state of agarose culture, plating efficiency is higher possibly because the protoplasts do not aggregate. Furthermore, the protoplasts can be monitored more easily. This latter system was used successfully with *Picea glauca* (Attree *et al.*, 1987) and *Abies alba* protoplasts (Lang and Kohlenbach, 1989). Kirby (1980) used a polyester fleece as support for protoplasts which may improve oxygenation. Finally, nurse culture systems have been shown to promote protoplast plating efficiency in several conifers (Bekkaoui *et al.*, 1987; Tautorus *et al.*, 1990b; Hohtola and Kvist, 1991).

2.2. *Explant Origin*

The starting material, i.e., the source tissue, from which gymnosperm protoplasts are isolated is of major importance and affects both their isolation, and the ability of protoplasts to regenerate (Kirby *et al.*, 1989).

2.2.1. *Cotyledons*

One of the most widely used source tissues for gymnosperm protoplast isolation has been seedling cotyledons (Table 1). Several factors were identified which critically affected protoplast yield and survival. These included pre-treatment and age of cotyledons (David and David, 1979; Kirby and Cheng, 1979; David *et al.*, 1982; Kirby, 1980, 1988; Tautorus *et al.*, 1990b). Additional factors were, the nitrogen source in the culture medium (David, 1987), medium osmolarity (David *et al.*, 1982), and plating density (Lainé *et al.*, 1988). For example, pre-treatment of *Pinus pinaster* and *Pseudotsuga menziesii* cotyledons (e.g., benzyladenine, BA + α-naphthalene acetic acid, NAA treatment for 8 days) increased mitotic activity (Kirby and Cheng, 1979; Kirby, 1988). Such preculture may stimulate cell division in cotyledons prior to protoplast isolation or may alter the chemical nature of the cell wall so that it becomes more susceptible to enzymatic degradation (Kirby, 1988). Cold pretreatment of *Pinus banksiana* seedlings prior to protoplast isolation contributed to improved protoplast viability (Tautorus *et al.*, 1990b). The effects of cold conditioning may be important in stabilizing the protoplast membrane (Alain David, personal communication).

2.2.2. *Callus and Suspension Cultures*

2.2.2.1. *Non-Embryogenic Cultures.* Callus and suspension cultures provide cell populations which are relatively homogeneous, undifferentiated, and are easily accessible to enzyme action (David, 1987). Protoplasts have been prepared from cell suspensions derived from cotyledons and hypocotyls (2-week-old) of *Pinus taeda* (Teasdale and Rugini, 1983; Verma and Wann, 1983). Cell divisions were observed after 10 days in culture and callus was regenerated from 10% of plated *P. taeda* protoplasts (Teasdale and Rugini 1983).

Using the same culture conditions which permitted regeneration of callus derived from cotyledonary protoplasts of *Pinus banksiana*, protoplasts from cell suspension cultures of *P. banksiana* regenerated to small cell clusters, but these became brown and died within 14 days (Tautorus *et al.*, 1990b). Protoplast yield and viability varied with different cell lines. Additional studies are needed to determine whether a different growth factor is needed and (or) whether release of an inhibitory substance(s) is responsible for the lack of continued development of protoplasts derived from suspension cultures of *Pinus banksiana*. Recalcitrance in growth and development was also observed in protoplasts of cereals and has been recently reviewed (Cutler *et al.*, 1991). The browning which is sometimes observed in protoplast cultures is the result of phenolic oxidation which is an oxidative stress, antioxidants may be required to reduce this effect as described by Cutler *et al.* (1991).

2.2.2.2. *Embryogenic Cultures.* Somatic embryogenesis and plantlet regeneration was first reported for *Picea abies* in 1985. Zygotic embryos had been used as explants to establish the cultures (Chalupa, 1985; Hakman *et al.*, 1985; for review see Tautorus *et al.*, 1991). As demonstrated for monocot plants, somatic embryos represent a source of morphogenetic material suitable for protoplast isolation (Vasil, 1988). Conifer protoplast regeneration to somatic embryos was first reported for *Pinus taeda* (Gupta and Durzan, 1987) and *Picea glauca* (Attree *et al.*, 1987) (Table 1). In both cases, suspension cultures with somatic embryos were used as the starting material. Since then, protoplasts derived from embryogenic suspension or tissue cultures have been isolated and regenerated to somatic embryos or plantlets from 8 different gymnosperm species (Table 1). It has been our experience that the use of suspension culture in an exponential growth phase is the most suitable for protoplast isolation and culture.

Differences in protoplast regeneration capacity amongst various embryogenic culture lines has been shown for several species. For example,

Table 1. Protoplast culture of gymnosperms.

Species	Explant	Protoplast Source	Medium*	Yield (/g.f.w.)	Results	Reference
Abies alba	Immature zygotic embryo	Embryogenic tissue	KM; 4.8 μM NAA	7.5×10^5	Embryo-like structures regenerated	Lang and Kohlenbach, 1989
	Immature zygotic embryo	Embryogenic suspension	KM; 2.2 μM BA, 0.9 μM 2,4-D.	4×10^5	Stage 2 embryos regenerated	Hartmann *et al.*, 1992
Biota orientalis	Seeds	Cotyledons	< half > MS; 0.7 M glucose, 15 μM BA, 0.5 μM NAA		Protoplast isolation	David *et al.*, 1981
Cupressus arizonica	Male cones	Pollen (haploid)	0.5 M sucrose	Not reported	Protoplast isolation	Duhoux, 1980
Larix decidua	Mega-gametophyte	Embryogenic tissue (haploid)	< half > LM	Not Reported	Plantlets produced	Von Aderkas, 1992
Larix × *eurolepis*	Immature zygotic embryo	Embryogenic tissue and suspension	MSG; 9 μM 2,4-D, 2.25 μM BA	6.6×10^5 tissue; 1.3×10^6 suspension	Plantlets produced	Klimaszewska, 1989
Picea abies	Mature zygotic embryos	Buds, shoots	Kao; 0.5 μM 2,4-D, 5 μM NAA, 0.4 μM BA	Not reported	Protoplast isolation	Hakman *et al.*, 1984
	Seeds	Megagametophyte (haploid)	W5	Not reported	Protoplast isolation	Hakman *et al.*, 1986
	Seeds	Hypocotyls	I	Not reported	Protoplast isolation and fusion	Ivanova, 1986
	Seeds	Cotyledons	Not reported	1×10^6	Protoplast isolation and fusion	Kirsten *et al.*, 1987
Picea abies	Mature zygotic embryo	Embryogenic suspension	< half > LP; 9 μM 2,4-D, 4.4 μM BA	$3–5 \times 10^5$	Somatic embryos regenerated	Egertsdotter and Von Arnold, 1993

Picea excelsa	Not reported	Callus	MS	2.4×10^6	Protoplast isolation	Kakoniova *et al.*, 1987
Picea glauca	Immature zygotic embryo	Embryogenic suspension	LP; 9 μM 2,4-D, 4.4 μM BA	1.2×10^5 – 2.3×10^6	Somatic embryos regenerated	Attree *et al.*, 1987, 1989a
	Immature zygotic embryo	Embryogenic suspension	LP; 4.5 μM 2,4-D	4.5×10^5	Callus formation	Bekkaoui *et al.*, 1987
	Immature zygotic embryo	Embryogenic suspension	LP; 9 μM 2,4-D, 4.4 μM BA	$1–5 \times 10^5$	Plantlets produced	Attree *et al.*, 1989b
	Immature zygotic embryo	Embryogenic suspension	LP; 4.5 μM 2,4-D	6.4×10^5	Somatic embryos regenerated (permeabili-zation study)	Bekkaoui and Dunstan, 1989
	Immature zygotic embryo	Embryogenic suspension	LP; 9 μM 2,4-D, 4.4 μM BA	$0.5–1 \times 10^6$	Somatic embryos regenerated (microtubule study)	Fowke *et al.*, 1990
Picea mariana	Mature zygotic embryo	Embryogenic suspension	LP; 9 μM 2,4-D, 4.5 μM BA	$0.5–4 \times 10^5$	Somatic embryos regenerated	Tautorus *et al.*, 1990a
Pinus banksiana	Mature zygotic embryo	Cell suspension	C; 10 μM NAA, 7 μM BA	$0.7–5 \times 10^5$	Colony formation	Tautorus *et al.*, 1990b
	Seeds	Cotyledons	C; 10 μM NAA, 7 μM BA	$2–7 \times 10^5$	Callus formation	Tautorus *et al.*, 1990b
	Bark	Xylem	CPW	1.5×10^5	Protoplast isolation	Leinhos and Savidge, 1993
Pinus caribaea	Immature zygotic embryo	Embryogenic suspension	C; 10 μM NAA, 7 μM BA	$7–10 \times 10^4$	Stage 3 embryos regenerated	Lainé and David, 1990

Table 1. Continued.

Species	Explant	Protoplast Source	Medium*	Yield (/g.f.w.)	Results	Reference
Pinus contorta	Mature zygotic embryo	Cell suspension	KAO; 0.5 μM 2,4-D, 5 μM NAA, 0.4 μM BA	Not reported	Colony formation	Hakman and Von Arnold, 1983
Pinus coulteri	Seeds	Cotyledons	B5; 2.3 μM 2,4-D, 2.7 μM NAA, 1.4 μM GA, 2.2 μM BA, 2.3 μM ZEA	Not reported	Colony formation	Patel *et al.*, 1984
Pinus lambertiana	Lateral branches of mature trees, seeds	Callus, suspension culture, needles	DCR; 4.5 μM 2,4-D, 2.7 μM NAA, 0.9 μM BA	0.5×10^4 needles	Colony formation	Gupta and Durzan, 1986
Pinus nigra	Seeds	Cotyledons	Not reported	Not reported	Protoplast isolation	Martinović *et al.*, 1985
Pinus oocarpa	Seeds	Cotyledons	C; 10 μM NAA, 7 μM BA	1.5×10^6	Callus formation	Lainé *et al.*, 1988
Pinus patula	Seeds	Cotyledons	C; 10 μM NAA, 7 μM BA	1×10^6	Callus formation	Lainé *et al.*, 1988
Pinus pinaster	Seeds	Roots	Knop; 5 μM BA, 15 μM NAA	6×10^3	Colony formation	Faye and David, 1983
	Seeds	Cotyledons	MS; 16.1, 32.3 μM NAA, 4.4, 8.9 μM BA	Not reported	Colony formation	David and David, 1979
	Seeds	Cotyledons	< half > MS; NAA, BA	1.5×10^7	Colony formation	David *et al.*, 1982
	Seeds	Cotyledons	Medium G	9×10^6	Colony formation	David *et al.*, 1984
	Seeds	Apical, axillary buds	Medium G	2.5×10^4	Callus formation	David *et al.*, 1986
	Seeds	Cotyledons	Medium B; 15 μM NAA, 5 μM BA	Not reported	Callus formation	David *et al.*, 1989

Pinus strobus	Bark	Xylem	CPW	1.5×10^5	Protoplast isolation	Leinhos and Savidge, 1993
Pinus sylvestris	Seeds	Cotyledons	C2	Not reported	Protoplast isolation (IAA study)	Sundberg *et al.*, 1985; Sandberg and Hällgren, 1985; Sandberg *et al.*, 1990
Pinus sylvestris	Seeds	Cotyledons	Not reported	1.3×10^6	Protoplast isolation and fusion	Kirsten *et al.*, 1987
	Shoot tips of mature trees	Callus	MS; 1 μM 2,4-D, 0.9 μM BA	2×10^5	Colony formation (Microcallus)	Hohtola and Kvist, 1991
Pinus taeda	Seedling hypocotyl	Cell suspension	V	Not reported	Callus formation	Teasdale and Rugini, 1983
	Immature zygotic embryo	Embryogenic suspension	< half > MS; 5 μM 2,4-D, 2 μM BA	Not Reported	Somatic embryos regenerated	Gupta and Durzan, 1987
Pseudotsuga menziesii	Vegetative shoot	Callus	BL; 24.8 μM NOAA	Not reported	Protoplast isolation	Winton *et al.*, 1975
	Seeds	Cotyledons	Cheng; 5μM BA, 15 μM NAA	5.5×10^4	Colony formation	Kirby and Cheng, 1979
	Seeds	Cotyledons	Cheng; 5 μM BA, 15 μM NAA	Not reported	Callus formation	Kirby, 1980
	Immature zygotic embryo	Embryogenic suspension	DCR; 5 μM 2,4-D, 2 μM KIN, 2 μM BA	Not Reported	Somatic embryos regenerated	Gupta *et al.*, 1988

* Some of these basal media have been modified, refer to particular references for details.

B5 – Gamborg *et al.*, 1968; Medium B, C – Lainé *et al.*, 1988; C2 – Sundberg *et al.*, 1985; BL – Brown and Lawrence, 1968; Cheng – Cheng, 1977; CPW – Frearson *et al.*, 1973; DCR – Gupta and Durzan, 1985; Medium G – David *et al.*, 1984; I – Ivanova, 1986; Kao – Kao *et al.*, 1974; KM – Kao and Michayluk, 1975; Knop – Gautheret, 1959; LM – Litvay *et al.*, 1981; LP – Von Arnold and Eriksson, 1981; MS – Murashige and Skoog, 1962; MSG – Becwar *et al.*, 1987; V – Verma *et al.*, 1982; W5 – Menczel *et al.*, 1981; 2,4-D – 2,4-dichlorophenoxyacetic acid; BA – benzyladenine; NAA – α-naphthalene acetic acid; Kin – Kinetin; ZEA – zeatin; NOAA – naphthoxyacetic acid.

regeneration of somatic embryos from protoplasts of *Picea mariana* (Tautorus *et al.*, 1990b), *Picea glauca* (Attree *et al.*, 1989a), and *Picea abies* (Egertsdotter and Von Arnold, 1993) only occurred if protoplasts were isolated from suspension cultures which contained very few somatic embryos or from older cultures which may have lost some embryogenic capacity. This type of behaviour was also observed in two lines of *Picea glauca* (Attree *et al.*, 1987, Bekkaoui *et al.*, 1987). For comparison, protoplasts isolated from *Larix* × *eurolepis* regenerated only from highly embryogenic lines (Klimaszewska, 1989). The differences in response to culture of protoplasts from different lines for any one species is probably due to an interaction of genotypic differences combined with non-optimal culture conditions (Attree *et al.*, 1989a).

Egertsdotter and Von Arnold (1993) recently characterized embryogenic cultures of *Picea abies* and divided them into two main groups, based on their growth habit and the appearance of the embryos. Group A cultures grew well on solid medium and the embryonal head regions of the somatic embryos were dense and compact. Group B cultures grew well in liquid suspension and the embryonal region of the somatic embryos consisted of loosely packed isodiametric cells and single suspensor cells. These two groups were subsequently used for protoplast isolation and culture. The highest proportion of dividing cells was obtained from cultures belonging to group A cultured on solid medium. However, only group B cultures developed further into somatic embryos and only if cultured in liquid media. These results emphasize the influence of the cell line and the culture method.

2.2.3. *Haploid Material*

Although there is a great interest in the production of haploid cultures in gymnosperms (Bonga *et al.*, 1988), there are only three reports concerning the use of haploid material as a source of protoplasts. Pollen of *Cupressus arizonica* was incubated in a high concentration of cell-wall-digesting enzymes (12.5% w/v) for more than 20 h (Duhoux, 1980). A pre-hydration step was required in order to disrupt the exine layer. Complete removal of the intine layer of the pollen wall was difficult, as it continued to be resynthesized during protoplast isolation. In contrast, haploid protoplasts of *Picea abies* were easily obtained from megagametophytes provided that the tissue was used between the stages of archegonium initiation and the beginning of embryo development (i.e., when cells of the gametophyte were devoid of storage materials) (Hakman *et al.*, 1986). However, in neither of these two cases was the division of cultured protoplasts reported.

Recently, Von Aderkas (1992) isolated haploid protoplasts of *Larix decidua* from embryogenic tissue derived from megagametophytes. The regeneration protocol was similar to that used by Attree *et al.* (1987, 1989b) with *Picea glauca*. Cells regenerated from protoplasts began dividing within 2 days and colonies with embryoids began to form within 1 week. Once embryoids possessing an embryonal mass and a suspensor were present, cleavage polyembryogenesis and development of embryoids from suspensor cells greatly increased the production of embryogenic tissue. Regeneration of somatic embryos from protoplasts was achieved in 3 different lines. All cultures had the expected chromosome number of 12.

2.2.4. *Mature Explants*

Because of the potential of cell technology methods in breeding (Franclet *et al.*, 1987), it would be desirable to regenerate plants from protoplasts of mature trees. David (1987) suggested that regeneration of plants from protoplasts derived from tissues of mature trees may lead to cellular rejuvenation. Protoplasts have been isolated from needles of mature trees of *Pinus lambertiana* (Gupta and Durzan, 1986). Specific conditions were osmotic preconditioning in 10% mannitol and the use of MES buffer. First cell divisions were reported after 12 days of culture but cells failed to develop any further. First divisions were promoted by dimethyl sulfoxide which is known to promote cortical microtubule assembly (see Simmonds, 1991). Hohtola and Kvist (1991) used callus from mature *Pinus sylvestris* trees as source for protoplast isolation. Protoplasts were regenerable to a microcallus stage but developed no further.

2.2.5. *Other Tissues*

Ivanova (1986) described the isolation of protoplasts from hypocotyls of *Picea abies*. These protoplasts were used for fusion studies with polyethylene glycol (PEG), but no division of protoplasts was reported. Elongating primary leaves (needles; 5–7 mm) from apical buds, and from shoots obtained from *in vitro*-induced axillary buds of *Pinus pinaster* seedlings have provided viable protoplasts in great numbers (David *et al.*, 1986). First cell divisions were observed after 8–10 days of culture, and callus formation was observed after 6–7 weeks of culture. Here, glutamine served as the sole source of nitrogen. Protoplasts have also been isolated from adventitious buds and shoots of *Picea abies* (Hakman *et al.*, 1984). Protoplasts were also produced from 20 mm long root segments of *Pinus pinaster* seedlings (12- to 15-day-old), and divided to form small cell clusters (Faye and David, 1983).

2.3. *Regeneration of Plantlets*

Gymnosperm protoplasts isolated from non-embryogenic tissues cultured *in vitro* divide to form callus, but in no case has plant regeneration been reported. In contrast, protoplasts isolated from embryogenic suspension cultures of *Picea glauca* (Attree *et al.*, 1989b) and *Larix* × *eurolepis* (Klimaszewska, 1989) regenerated to form embryogenic cultures from which plantlets have been produced. In both cases plantlets were acclimatized and placed in the greenhouse. Von Aderkas (1992) reported the production of plants from tissue regenerated from haploid protoplasts of *Larix decidua* embryogenic cultures. In addition, regeneration of cotyledonary somatic embryos has been reported from protoplast-derived cultures of *Pinus caribaea* derived from embryogenic suspension cultures (Lainé and David, 1990). In all cases, maturation was achieved by transferring regenerated somatic embryos from liquid suspension culture to agar-solidified medium containing 8–12 μM abscisic acid. It was noticed that the culture line that best regenerated to plantlets often displayed the lowest protoplast yield and plating efficiency (Attree *et al.*, 1989b).

3. Protoplasts as a Tool for Germplasm Improvement

Protoplasts can be used as a means for germplasm improvement in several ways; by *in vitro* cell variation, by novel hybrid production via somatic cell fusion, by genetic transformation, and for true-to-type propagation.

3.1. *In Vitro* Cell Variation

In vitro cell variation describes the occurrence of variant plants produced through tissue culture. *In vitro* cell variation can arise through the specific use of mutagens such as ethyl methanesulphonate or through somaclonal variation (Larkin and Scowcroft, 1981), a comparatively benign process attributable to chromosomal rearrangements. Either process can be seen as having potential for germplasm improvement. Variant plants can be recovered following the use of selection pressure, such as growth in the presence of a herbicide or high salt concentration. It is a common observation, however, that expression of tolerance to selection pressure during callus culture is not necessarily correlated to tolerance at the whole plant level. Passage of the tolerance trait through one or more sexual cycles is needed to differentiate between genetic and epigenetic variation (variation at the biochemical or physiological level). The single-cell origin of protoplasts provides an especially useful attribute, the potential for avoidance of chimeral regenerants.

One of the more problematic aspects of somaclonal variation occurs in situations when true-to-type clonal propagation is required, including the multiplication of genetically modified tissues produced by any of the methods discussed below. In such situations it is advisable to minimize callus proliferation. As described in Section 2.3, plant regeneration from gymnosperm protoplast cultures has been achieved via somatic embryogenesis, this is a conservative developmental process which could be expected to eliminate variants. There are no examples of variant gymnosperm plants resulting from *in vitro* cell variation.

3.2. *Somatic* (*Parasexual*) *Hybridization*

The fusion of two or more protoplasts and the combination of their cytoplasms may lead to the production of homo- and hetero-karyons. In its simplest state such hybrids would possess two or more complete sets of cytoplasms and nuclei, or unequal amounts from each donor. Protoplast fusion has the potential for producing indiscriminate cell hybrids, generally with little chance for survival to the plant level. Greater likelihood of producing hybrid cells and improved chances for plant regeneration, can be achieved by fusion between protoplasts from haploid cells (e.g., see Sections 2.2.3 and 2.3) or by cybrid production. The latter approach has been used extensively with incompatible but closely related *Citrus* species (Vardi *et al.*, 1987). It involves asymmetric fusion between protoplasts in which nuclear divisions have been arrested, and those in which cellular metabolism has been arrested transiently. General methods for somatic hybridization of plant protoplasts have been reviewed elsewhere (see Gleba and Sytnik, 1984; Vasil, 1984).

Despite the availability of plant regeneration systems from protoplasts of gymnosperm somatic embryos, no examples have been reported of gymnosperm plants resulting from somatic hybridization. However, Kirsten *et al.* (1987) demonstrated electrofusion with cotyledon protoplasts of *Picea abies* and *Pinus sylvestris* and reported inter- and intra-specific fusion products. Intraspecific fusion of *Picea abies* protoplasts with PEG has also been observed (Ivanova, 1986). In neither of these two cases was division of the fusion product observed.

3.3. *Genetic Transformation*

Genetic transformation using protoplasts permits specific genetic changes in cells in short periods of time. For example, electroporation facilitates uptake of DNA into plant protoplasts permitting transient expression of foreign genes and eventually stable transformation (e.g., in *Zea mays* (Fromm *et. al.*, 1986) and in *Glycine max* (Dhir *et al.*, 1992)). Electroporation is a useful technique for rapid evaluation of the functionality of gene constructs.

Transient gene expression was observed in electroporated protoplasts of *Pseudotsuga menziesii* using the luciferase reporter gene (Gupta *et al.*, 1988) and in protoplasts of *Picea glauca* using chloramphenicol acetyltransferase (*cat*) and β-glucuronidase (*gus*) reporter genes introduced by electroporation (*cat* gene, Bekkaoui *et al.*, 1988) or by PEG (*gus* gene, Wilson *et al.*, 1989) (Table 2). Subsequently, expression of several reporter genes has been demonstrated for electroporated protoplasts of conifers (Table 2). The *gus* reporter gene was found to show a high level of background compared to the *cat* reporter gene (Bekkaoui *et al.*, 1988; Charest *et al.*, 1991). This GUS-like activity was found to be culture-line dependent (Bekkaoui *et al.*, 1990) and was due to a smaller molecule than GUS protein (Bekkaoui *et al.*, 1988). When *gus*, *nptII* (neomycin phosphotransferase) and *cat* were compared as reporter genes in *Larix* × *eurolepis*, the *cat* reporter was found to be the most useful, being more sensitive or with less background (Charest *et al.*, 1991). Using the *cat* reporter gene, it was found that different genotypes of

Table 2. Transient gene expression in gymnosperms protoplasts.

Species	Protoplast Source	Reporter Gene	Promoter	Gene Delivery System	Reference
Larix × *eurolepis*	Embryogenic suspension	*cat*, *gus*, *nptII*	35S, NOS, wound inducible	Electroporation	Charest *et al.*, 1991
Picea glauca	Embryogenic suspension	*cat*, *gus*	35S	Electroporation	Bekkaoui *et al.*, 1988, 1990
	Embryogenic suspension	*cat*, *gus*	35S	PEG	Wilson *et al.*, 1989
	Embryogenic suspension	*cat*	ADH	Electroporation	Good *et al.*, 1990
	Embryogenic suspension	*gus*	35S, 35SAMV, Ca_2, Ca_2AMV	Electroporation	Datla *et al.*, 1993
Picea mariana	Embryogenic suspension	*cat*, *gus*	35S, Ca_2	Electroporation	Tautorus *et al.*, 1989; Bekkaoui *et al.*, 1990
Pinus banksiana	Cell suspension	*cat*, *gus*	35S, Ca_2	Electroporation	Tautorus *et al.*, 1989; Bekkaoui *et al.*, 1990
Pinus taeda	Embryogenic suspension	Luciferase	35S	Electroporation, PEG	Gupta *et al.*, 1988
Pseudotseuga menziesii	Embryogenic suspension	Luciferase	35S	Electroporation, PEG	Gupta *et al.*, 1988

the same species varied in their responses (Tautorus *et al.*, 1989; Bekkaoui *et al.*, 1990).

The type of the promoter also has a significant effect on the level of transient gene expression. In contrast to results obtained with *Larix* × *eurolepis* (Charest *et al.*, 1991), the 35S promoter and the nopaline synthase (NOS) promoter gave similar levels of expression in *Picea glauca* (Bekkaoui *et al.* 1990). The duplicated 35S promoter (Ca_2, Kay *et al.*, 1987; Bekkaoui *et al.* 1990) or an addition of an alfalfa mosaic virus leader sequence enhanced the response several fold as compared to the 35S alone. The level of *gus* transient gene expression increased in the following order of promoters: 35S > Ca_2 > AMV-35S > AMV-Ca_2 (AMV-alfalfa mosaic virus). The AMV-Ca_2 produced a 20 fold higher expression compared to the 35S promoter alone (Datla *et al.*, 1993). The response to these different types of constructs in a gymnosperm was comparable to tobacco (Datla *et al.*, 1993). When using a wound-inducible promoter from the potato proteinase inhibitor IIK, no transient gene expression was obtained in electroporated protoplasts of *Larix* × *eurolepis* (Charest *et al.*, 1991).

Other factors affecting the level of transient gene expression in protoplasts include plasmid DNA concentration (Tautorus *et al.*, 1989; Bekkaoui *et al.*, 1990; Charest *et al.*, 1991), and a heat shock treatment (Tautorus *et al.*, 1989). The use of PEG in conjunction with electroporation was found to be beneficial in increasing transient gene expression of *Pseudotsuga menziesii* and *Pinus lambertiana* protoplasts (Gupta *et al.*, 1988) but remained without effect in *Larix* × *eurolepsis* (Charest *et al.*, 1991).

Indirect transformation techniques also have potential for application with gymnosperm protoplasts. *Agrobacterium* co-cultivation with protoplasts undergoing cell wall synthesis has resulted in bacterial attachment (Tsang *et al.*, unpublished observations). This treatment, and subsequent steps involving use of antibiotics resulted, however, in stress that is not conducive to cell and plant regeneration. Co-cultivation at subsequent stages in cell regeneration from protoplasts may have greater probability of success. The sensitivity of gymnosperm tissues to antibiotics during selection resulted however in abandoning this method since it is wrought with difficulties. Microinjection has potential in avoiding some of the stress, since the isolation of transgenic cells does not require selection processes based on antibiotics. The recovery of microcolonies from individually injected protoplasts and subsequent regeneration to plantlets is still problematic, however, and is a limitation to the extensive use of microinjection as a means for routine transformation of gymnosperm protoplasts.

To date, there have been no reports of successful stable incorporation of chimeric genes using a protoplast culture system in gymnosperms.

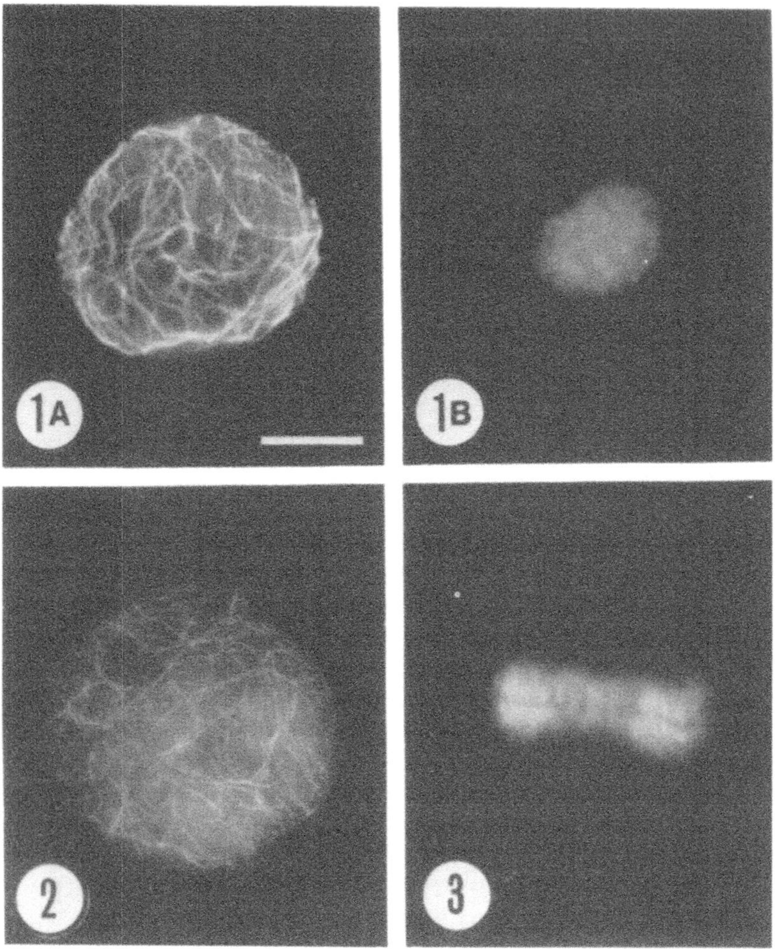

Plate 1. Figs. 1 to 3 illustrate the cytoskeleton of freshly isolated suspension-derived protoplasts of jack pine. Protoplasts were simultaneously stained for MTs by immunofluorescence and for DNA by Hoescht 33258 using the procedure described previously (Tautorus, 1990; Tautorus *et al.*, 1992).

(1a) Fluorescence micrograph showing freshly isolated uninucleate protoplasts containing a network of randomly arranged microtubules. Bar = 10 um.

(1b) Same material stained with Hoescht 33258 showing fusion nucleus. Magnification as per Fig. 1a.

(2) Fluorescence micrograph showing a freshly isolated binucleate protoplast containing a network of randomly arranged microtubles. Magnification as per Fig. 1a.

(3) Fluorescence micrograph showing a freshly isolated protoplast with a phragmoplast perserved from a cell in cytokinesis. Magnification as per Fig. 1a.

4. Protoplasts as a Tool for Cell Biology and Biochemistry Studies

For over 20 years protoplasts have been used in cell biology investigations (Bajaj, 1989; Cocking, 1972; Galun, 1981; Fowke and Constabel, 1985). Research has included the study of cell organelles for structural and metabolic investigations, cell wall regeneration, cell division, cytodifferentiation, membrane fusion, virology, membrane transport and metabolism. However, these investigations have rarely involved the specific use of gymnosperm protoplasts.

4.1. *Cell Biology*

4.1.1. *Cell Wall Regeneration and Cell Division*

Cortical microtubules (MTs) in protoplasts have been implicated in cell wall formation, cell shape determination and cell growth during interphase (Simmonds, 1991), and in controlling division and regeneration (Dijak and Simmonds, 1988). Studies using indirect immunofluorescence with cultured protoplasts has provided important information regarding cortical MT patterns. Immunofluorescence techniques are particularly useful as they provide a rapid means for visualizing the overall 3-dimensional organization of MTs in a cell (Fowke *et al.*, 1984). With *Picea glauca*, Fowke *et al.* (1990) developed immunofluorescence methods for examining the pattern of MTs in freshly isolated protoplasts, regenerating cells, and somatic embryos derived from the cultured protoplasts. Both uninucleate and multinucleate isolated protoplasts contained extensive networks of randomly arranged cortical MTs, and established parallel cortical MTs during cell wall formation. Dividing cells were characterized by pre-prophase bands of MTs, spindle MTs focused at the poles and a typical phragmoplast at telophase.

Protoplast isolation protocols which led to the disruption of MTs may prevent subsequent regeneration to microcolonies (Hahne and Hoffman, 1984; Lee *et al.*, 1989). However, suspension-derived protoplasts of *Pinus banksiana* which are highly recalcitrant to culture (Tautorus *et al.*, 1990b) showed a well developed cortical cytoskeleton after isolation (Figs. 1–3). *P. banksiana* protoplasts had a highly branched, dense, interconnecting network of fine and broad MT strands (Tautorus, 1990). Therefore, although a well developed cortical MT-network is a requirement for protoplast division, its presence does not guarantee division (Simmonds, 1991).

4.1.2. *Organelle Isolation*

Plant protoplasts provide an important system for understanding the structure, chemistry, and function of cell organelles (Fowke and Constabel, 1985). One of the major advantages of using protoplasts is that organelles can be isolated without the harsh mechanical methods necessary for disrupting plant cells (Galun, 1981). Employing angiosperm protoplasts, various cell components that have been isolated are plasmalemma, chloroplasts, mitochondria, vacuoles, nuclei, and chromosomes (see Galun, 1981; Fowke and Constabel, 1985). With gymnosperms, only chloroplasts have been isolated from protoplasts. Martinović *et al.* (1985) isolated chloroplasts from protoplasts derived from cotyledons of *Pinus nigra*, though only 38% of these chloroplasts were intact. Sandberg *et al.* (1990) isolated chloroplasts from cotyledons of *Pinus sylvestris* using a membrane filtration apparatus. The chloroplasts were used to study endogenous indole acetic acid (IAA) synthesis. Data indicated that IAA was synthesized in the cytosol and that approximately 30–40% was integrated into the chloroplasts. The physiological role of IAA in this organelle remains to be determined.

4.1.3. *Embryogenesis*

The single cell origin of a regenerable protoplast provides a useful system for studying embryogenesis. Gupta and Durzan (1987), using acetocarmine staining that reacts with the neocytoplasm, observed that the origin of the proembryo in *Pinus taeda* is from smaller dense embryonal protoplasts. With Evan's blue coloration, these authors also showed that the suspensor cell originated by asymmetric division of the embryonal cell. Klimaszweska (1989) in *Larix* × *eurolepis* and Attree *et al.* (1989b) in *Picea glauca* observed an asymmetric division of the protoplasts after 1 day in culture. In the same species, the formation of a 6–8 celled cluster at day 4 was also observed, with the appearance of suspensor cells at day 8.

4.2. *Cell Metabolism*

Because protoplasts from the same tissue are a relatively uniform material and are directly exposed to the culture media, they can provide valuable data for biochemical investigations.

David *et al.* (1989) compared two suspension culture lines derived from *Pinus pinaster* cotyledons, one of them had been obtained through protoplast culture. Growth kinetics for the protoplast-derived suspension culture showed faster growth in comparison with the cotyledon-derived suspension culture. The authors observed differences in the protein quantity and quality between the two suspension cultures, the protoplast-derived suspension culture had higher amounts of protein than the organ-derived cells. Using two-dimensional electrophoretic separation, two peptides were characteristic of the protoplast-derived cells and three other peptides were characteristic of the organ-derived cells. The authors suggest that these differences may be a result of variation in gene expression, or the result of selection pressure associated with protoplast culture.

The induction of the expression of the *cat* marker gene linked to monocot and dicot alcohol dehydrogenase (ADH) promoters was studied in protoplasts of an embryogenic suspension line of *Picea glauca* (Good *et al.*, 1990). Protoplasts were capable of anaerobically regulating a maize ADH reporter construct, however constructs with dicot promoters (*Pisum sativum* and *Ara-*

bidopsis) were not transiently expressed. Further studies in this area may provide interesting information on comparative gene expression among diverse plant groups and the evolution of regulatory genes.

The uptake of impermeable macromolecules (calcein and dextrans) into *Picea glauca* protoplasts was performed by electroporation (exponential decay wave form) or using polyethylene glycol 3350 (PEG) treatments (Bekkaoui and Dunstan, 1989). The electroporation technique gave better results than PEG treatments in terms of percentage of macromolecules taken into protoplasts and protoplast viability. There was also a difference of uptake depending on the size of the molecules. This type of approach can be used for the study of protoplast membrane pores during electroporation (Mehrle *et al.*, 1985), endogenous enzyme activities or probing intracellular metabolism by introducing antibodies (Maccaronne *et al.*, 1992) and in preliminary studies directed at DNA uptake (Bekkaoui and Dunstan, 1989).

In addition to these studies, protoplasts from gymnosperms have been used to study auxin metabolism in *Pinus sylvestris* (Sandberg and Hällgren, 1985; Sandberg *et al.*, 1990) and to study elicitors of fungal origin in *Pinus elliottii* (Lesney, 1990).

5. Conclusion and Prospects

Regeneration from gymnosperm protoplasts was not possible six years ago. Since then, regeneration to an embryo stage has been achieved in eight species and to a plantlet stage in three species (Table 1). This progress was possible because of the establishment of somatic embryogenesis in gymnosperms (for review see Tautorus *et al.*, 1991). However, plant regeneration from protoplasts is still not possible in many cases, for example with species such as *Pinus banksiana* or when an explant (e.g. needles) from mature trees is used as starting material. In both cases somatic embryogenesis has not been achieved.

Although direct DNA uptake by protoplasts has been a valuable tool in transformation studies (Table 2), stable transformation has not yet been obtained. Recently, stable transformation of gymnosperms was achieved using particle bombardment of somatic embryos (Robertson *et al.*, 1992; Bommineni *et al.*, 1993; Ellis *et al.*, 1993), this technique may be preferable to the use of protoplasts. Nevertheless, protoplasts have the advantage of being a more homogeneous cell material and can still be used for gene expression and regulation studies (Good *et al.*, 1990).

For studies directed toward cell biology and cell metabolism, protoplasts from many of the materials noted in the text (see Section 2.2) can be used. Where regeneration to plantlets is required, however, it is advisable to obtain protoplasts from cultures with embryogenic competence. Following the procedure outlined in the text (e.g., Section 2 and described more fully in the literature that has been cited), it should be possible to develop regenerable protoplast cultures reproducibly. In addition to using embryogenic cultures, it is advisable to compare at least two culture lines in protoplast culture or transient expression studies, because of the wide range of responses due to genotype (Attree *et al.*, 1987; Bekkaoui *et al.*, 1987, 1990; Tautorus *et al.*, 1989). These methods are not greatly different from generalized procedures of protoplast culture used with other plant groups. (for review see Vasil, 1984; Bajaj, 1989).

With the progress made with protoplasts from haploid plants of gymnosperms (Von Aderkas, 1992), it will be of an interest to perform somatic hybridization between haploid (n) protoplasts to achieve diploid hybrids (2n). This would be especially so if one or both of the donors could be genetically transformed.

Protoplasts have been widely used in angiosperms to study several new fundamental aspects, e.g., study of membrane channels (Kim *et al.*, 1993), study of environmental stresses (Rengel and Elliot, 1992) or use as a model to study plant microbe interactions (Pladys *et al.*, 1991). Due to recent progress, it will be even more feasible to perform such studies with gymnosperm protoplasts.

References

Attree, S.M., F. Bekkaoui, D.I. Dunstan and L.C. Fowke, 1987. Regeneration of somatic embryos from protoplasts isolated from an embryogenic suspension culture of white spruce (*Picea glauca*). Plant Cell Rep. 6: 480–483.

Attree, S.M., D.I. Dunstan and L.C. Fowke, 1989a. Initiation of embryogenic callus and suspension cultures, and improved regeneration of protoplasts, of white spruce (*Picea glauca*). Can. J. Bot. 67: 1790–1795.

Attree, S.M., D.I. Dunstan and L.C. Fowke, 1989b. Plantlet regeneration from embryogenic protoplasts of white spruce (*Picea glauca*). Bio/Tech. 7: 1060–1062.

Bajaj, Y.P.S., 1989. Biotechnology in Agriculture and Forestry, Vols. 8 and 9. Springer-Verlag, Berlin.

Becwar, M.R., T.L. Noland and S.R. Wann, 1987. Somatic embryo development and plant regeneration from embryogenic Norway spruce callus. Tech. Assoc. Pulp Paper Ind. J. 70: 155–160.

Bekkaoui, F., P.K. Saxena, S.M. Attree, L.C., Fowke and D.I. Dunstan, 1987. The isolation and culture of protoplasts from an embryogenic cell suspension culture of *Picea glauca* (Moench) Voss. Plant Cell Rep. 6: 476–479.

Bekkaoui, F., M. Pilon, E. Lainé, D.S.S. Raju, W.L. Crosby and D.I. Dunstan, 1988. Transient gene expression in electroporated *Picea glauca* protoplasts. Plant Cell Rep. 7: 481–484.

Bekkaoui, F. and D.I. Dunstan, 1989. Permeabilization of *Picea glauca* protoplasts to macromolecules. Can. J. For. Res. 19: 1316–1321.

Bekkaoui, F., R.S.S. Datla, M. Pilon, T.E. Tautorus, W.L. Crosby and D.I. Dunstan, 1990. The effects of promoter on transient expression in conifer cell lines. Theor. Appl. Genet. 79: 353–359.

Bommineni, V.R., R.N. Chibbar, R.S.S. Datla and E.W.T. Tsang, 1993. Transformation of white spruce (*Picea glauca*) somatic embryos by microprojectile bombardment. Plant Cell Rep. (in press).

Bonga, J.M., P. von Aderkas and D. James, 1988. Potential application of haploid culture of

tree species. In: J.W. Hanover and D.E. Keathley (Eds.), Genetic Manipulation of Woody Plants, pp. 57–77. Plenum Press, New York.
Brown, C.L. and R.H. Lawrence, 1968. Culture of pine callus on a defined medium. Forest. Sci. 14: 62–64.
Chalupa, V., 1985. Somatic embryogenesis and plantlet regeneration from cultured immature and mature embryos of *Picea abies* (L.) Karst. Communi. Inst. For. Cech. 14: 57–63.
Charest, P.J., Y. Devantier, C. Ward, C. Jones, U. Schaffer and K.K. Klimaszewska, 1991. Transient expression of foreign chimeric genes in the gymnosperm hybrid larch following electroporation. Can. J. Bot. 69: 1731–1736.
Cheng, T.-Y., 1977. Factors affecting adventitious bud formation of cotyledon culture of Douglas fir. Plant Sci. Lett. 9: 179–187.
Cocking, E.D., 1972. Plant cell protoplasts, isolation and development. Ann. Rev. Plant Physiol. 23: 29–50.
Cutler, A.J., M. Saleem and H. Wang, 1991. Cereal protoplast recalcitrance. In Vitro Cell Dev. Biol. 27P: 104–111.
Datla R.S.S., F. Bekkaoui, J. Hammerlindl, G. Pilate, D. Dunstan and W.L. Crosby, 1993. Improved high-level constitutive foreign gene expression in plants using an AMV RNA4 untranslated leader sequence. Plant Sci. (in press).
David, A., 1987. Conifer protoplasts. In: J.M. Bonga and D.J. Durzan (Eds), Cell and tissue culture in forestry, Vol. 2, pp. 2–15. Martinus Nijhoff Publishers, Dordrecht.
David, A. and H. David, 1979. Isolation and callus formation from cotyledon protoplasts of pine (*Pinus pinaster*). Z. Pflanzenphysiol. Bd. 94: 173–177.
David, H., A. David and T. Mateille 1981. Isolation and culture of protoplasts of two gymnosperms: *Pinus pinaster* and *Biota orientalis*. In: M. Boulay (Ed.), Proc. Intl. Colloq. *in vitro* Culture of Tissues of Forest Trees, pp. 339–347. AFOCEL, Nangis.
David, H., A. David and T. Mateille, 1982. Evaluation of parameters affecting the yield, viability and cell division of *Pinus pinaster* protoplasts. Physiol. Plant 56: 108–113.
David, H., E. Jarlet and A. David, 1984. Effects of nitrogen source, calcium concentration and osmotic stress on protoplasts and protoplast-derived cell cultures of *Pinus pinaster* cotyledons. Physiol. Plant 61: 477–482.
David, H., M.T. de Boucaud, J.M. Gaultier and A. David, 1986. Sustained division of protoplast-derived cells from primary leaves of *Pinus pinaster*, factors affecting growth and nuclear DNA content. Tree Physiol. 1: 21–30.
David, H., C. Laigneau and A. David, 1989. Growth and soluble proteins of cell cultures derived from explants and protoplasts of *Pinus pinaster* cotyledons. Tree Physiol. 5: 497–506.
Dhir, S.K., S. Dhir, M.A. Savka, F. Belanger, A.L. Kriz, S.K. Ferrand and J.M. Widholm, 1992. Regeneration of transgenic soybean (*Glycine max*) plants from electroporation protoplasts. Plant Physiol. 99: 81–88.
Dijak, M. and D.H. Simmonds, 1988. Microtubule organization during early direct embryogenesis from mesophyll protoplasts of *Medicago sativa* L. Plant Sci. 58: 183–191.
Duhoux, E., 1980. Protoplast isolation of gymnosperm pollen. Z. Pflanzenphysiol. Bd. 99: 207–214.
Dunstan, D.I., F. Bekkaoui, M. Pilon, L.C. Fowke and S.R. Abrams, 1988. Effects of abscisic acid and analogues on the maturation of white spruce (*Picea glauca*) embryos. Plant Sci. 58: 77–84.
Egertsdotter, U. and S. von Arnold, 1993. Classification of embryogenic cell-lines of *Picea abies* as regards protoplast isolation and culture. J. Plant Physiol. 141: 222–229.
Ellis, D.D., D.E. McCabe, S. McInnis, R. Ramachandran, D.R. Russell, K.M. Wallace, B.J. Martinell, D.R. Roberts, K.F. Raffa and B.H. McCown, 1993. Stable transformation of *Picea glauca* by particle acceleration. Bio/Tech. 11: 84–89.
Faye, M. and A. David, 1983. Isolation and culture of gymnosperm root protoplasts (*Pinus pinaster*). Physiol. Plant. 59: 359–362.
Fowke, L.C. and F. Constabel, 1985. Plant protoplasts. CRC Press Inc., Boca Raton, FL.
Fowke, L.C., D. Simmonds, P. van der Valk and G. Setterfield, 1984. Immunofluorescence

techniques for studies of plant microtubules. In: I.K. Vasil (Ed.), Cell Culture and Somatic Cell Genetics of Plants, Vol. 1, pp. 785–794. Academic Press, New York.

Fowke, L.C., S.M. Attree, H. Wang and D.I. Dunstan, 1990. Microtubule organization and cell division in embryogenic protoplast cultures of white spruce (*Picea glauca*). Protoplasma 158: 86–94.

Franclet, A., M. Boulay, F. Bekkaoui, Y. Fouret, N. Walker and B. Martouzet-Verchoore, 1987. Rejuvenation. In: J.M. Bonga and D.J. Durzan (Eds.), Cell and Tissue Culture in Forestry, Vol. 1, pp. 232–248. Martinus Nijhoff Publishers, Dordrecht.

Frearson, E.M., J.B. Power and E.C. Cocking, 1973. The isolation, culture and regeneration of petunia leaf protoplasts. Dev. Biol. 33: 130–137.

Fromm, M.E., L.P. Taylor and V. Walbot, 1986. Stable transformation of maize after gene transfer by electroporation. Nature 319: 791–793.

Gaff, D.F. and O. Okang'o-Ogola, 1971. The use of non-permeating pigments for testing the survival of cells. J. Exp. Bot. 22: 756–758.

Galun, E., 1981. Plant protoplasts as physiologycal tools. Ann. Rev. Plant Physiol. 32: 237–266.

Gamborg, O.L., R.A. Miller and K. Ojima, 1968. Nutrient requirements of suspension cultures of soybean root cells. Exp. Cell. Res. 50: 151–158.

Gautheret, R.J., 1959. La Culture des Tissus Végétaux. Techniques et Réalisations. Masson, Paris.

Gleba, Y.Y. and K.M. Sytnik, 1984. Protoplast Fusion. Springer-Verlag, Berlin.

Good, A.G., F. Bekkaoui, G. Pilate, D.I. Dunstan and W.L. Crosby, 1990. Anaerobic induction in conifers: expression of endogenous and chimeric anaerobically-induced genes. Physiol. Plant 78: 441–446.

Gupta, P.K. and D.J. Durzan, 1985. Shoot multiplication from mature trees of Douglas-fir (*Pseudotsuga menziesii*) and sugar pine (*Pinus lambertiana*). Plant Cell Rep. 4: 177–179.

Gupta, P.K. and D.J. Durzan, 1986. Isolation and cell regeneration of protoplasts from sugar pine (*Pinus lambertiana*). Plant Cell Rep. 5: 346–348.

Gupta, P.K. and D.J. Durzan, 1987. Somatic embryos from protoplasts of loblolly pine proembryonal cells. Bio/Tech. 5: 710–712.

Gupta, P.K., A.M. Dandekar and D.J. Durzan, 1988. Somatic proembryo formation and transient expression of a luciferase gene in Douglas fir and loblolly pine protoplasts. Plant Sci. 58: 85–92.

Hahne, G and F. Hoffman, 1984. Dimethyl sulfoxide can initiate cell divisions of arrested callus protoplasts by promoting cortical microtubule assembly. Proc. Natl. Acad. Sci. USA 81: 5449–5453.

Hakman, I. and S. von Arnold, 1983. Isolation and culture of protoplasts from cell suspensions of *Pinus contorta* Dougl. ex Loud. Plant Cell Rep. 2: 92–94.

Hakman, I. and Fowke L.C., 1987. An embryogenic cell suspension culture of *Picea glauca* (White spruce). Plant Cell Rep. 6: 20–22.

Hakman, I., S. von Arnold and A. Bengtsson, 1984. Cytofluorometric measurements of nuclear DNA in adventitious buds and shoots of *Picea abies* regenerated *in vitro*. Physiol. Plant 60: 321–325.

Hakman, I., L.C. Fowke, S. von Arnold and T. Eriksson, 1985. The development of somatic embryos in tissue culture initiated from immature embryos of *Picea abies* (Norway spruce). Plant Sci. 38: 53–59.

Hakman, I., S. von Arnold and H. Fellner-Feldegg, 1986. Isolation and DNA analysis of protoplasts from developing female gametophytes of *Picea abies* (Norway spruce). Can. J. Bot. 64: 108–112.

Hartmann, S., H. Lang and G. Reuther, 1992. Differentiation of somatic embryos from protoplasts isolated from embryogenic suspension cultures of *Abies alba* L. Plant Cell Rep. 11: 554–557.

Hohtola, A. and A-P. Kvist, 1991. Preparation of protoplasts from callus derived from buds of mature Scots pine and subsequent induction of cell proliferation. Tree Physiol. 8: 423–428.

Ivanova, E., 1986. Intraspecific fusion of the Norway spruce [*Picea abies* (L.) Karst.] protoplasts. Biologia 41: 841–846.

Kakoniova, D., I. Labudova and D. Liskova, 1987. Callus culture as substrate for the production of specific cell wall degrading enzyme for derived protoplast isolation. Biotech. Lett. 9: 721–724.

Kao, K.N. and M.R. Michayluk, 1975. Nutritional requirements for growth of *Vicia hajastana* cells and protoplasts at a very low population density. Planta 126: 105–110.

Kao, K.N., F. Constabel, M.R. Michayluk and O.L. Gamborg, 1974. Plant protoplast fusion and growth of intergenic hybrid cells. Planta 120: 215–224.

Kay, R., A. Chan, M. Daly and J. McPherson, 1987. Duplication of the CaMV 35S promoter sequences creates a strong enhancer for plant genes. Science 236: 1299–1302.

Kim, H.Y., G.G. Coté and R.C. Crain, 1993. Potassium channels in *Samanea saman* protoplasts controled by phytochrome and the biological clock. Science 260: 960–962.

Kirby, E.G., 1980. Factors affecting proliferation of protoplasts and cell cultures of Douglas fir. In: F. Sala, B. Parisa, R. Cella and O. Cifferi (Eds.), Plant Cell Cultures: Results and Perspectives, pp. 289–293. Elsevier/North Holland, Amsterdam.

Kirby, E.G., 1988. Recent advances in protoplast culture of horticultural crops: conifers. Sci. Hort. 37: 267–276.

Kirby, E.G. and T-Y. Cheng, 1979. Colony formation from protoplasts derived from Douglas-fir cotyledons. Plant Sci. Lett. 14: 145–154.

Kirby, E.G., M.A. Campbell and R.M. Penchel, 1989. Isolation and culture of protoplasts of forest tree species. In: Y.P.S Bajaj (Ed.), Biotechnology in Agriculture and Forestry, Vol. 8, pp. 262–274. Springer-Verlag, Berlin.

Kirsten, U., H.E. Jacob, M. Tesche and S. Kluge, 1987. First evidence of fusion of protoplasts from coniferous trees by electric field pulses. Studia biophysica 119: 85–87.

Klimaszewska, K., 1989. Recovery of somatic embryos and plantlets from protoplast cultures of *Larix* × *eurolepis*. Plant Cell Rep. 8: 440–444.

Lainé, E. and A. David, 1990. Somatic embryogenesis in immature embryos and protoplasts of *Pinus caribaea*. Plant Sci. 69: 215–224.

Lainé, E., H. David and A. David, 1988. Callus formation from cotyledon protoplasts of *Pinus oocarpa* and *Pinus patula*. Physiol. Plant 72: 374–378.

Lang, H. and H.W. Kohlenbach, 1989. Cell differentiation in protoplast cultures from embryogenic callus of *Abies alba* L. Plant Cell Rep. 8: 120–123.

Larkin, P.J. and W.R. Scowcroft, 1981. Somaclonal variation – a novel source of variability from cell culture for plant improvment. Theor. Appl. Genet. 60: 197–214.

Lee, N., H.Y. Wetzstein and C.H. Bornman, 1989. Cortical microtubule organization in *Vitis* protoplasts as affected by concentration of enzyme isolation medium and duration of incubation. Physiol. Plant. 77: 27–32.

Leinhos, V. and R.A. Savidge, 1993. Isolation of protoplasts from developing xylem of *Pinus banksiana* and *Pinus strobus*. Can. J. For. Res. 23: 343–348.

Lesney, M.S., 1990. Polycation-like behaviour of the chitosan on suspension-culture derived protoplasts of slash pine. Phytochemistry 29: 1123–1125.

Litvay, J.D., M.A. Johnson, D. Verma, D., Einspahr and K. Weyrauch, 1981. Conifer suspension culture medium development using analytical data from developing seeds. Institute Paper Chemistry Technical Paper Series #115, Appleton, WI.

Maccarrone, M., G.A. Veldnik and J.F.G. Vliegenthart, 1992. Inhibition of lipoxygenase activity in lentil protoplasts by monoclonal antibodies introduced into the cell via electroporation. Eur. J. Biochem. 205: 995–1001.

Martinović, B., M. Bogdanović and M. Vucković, 1985. Isolation of photoactive protoplasts from black pine cotyledons. Photosynthetica 19: 237–239.

Mehrle, W., U. Zimmermann and R. Hampp, 1985. Evidence for asymmetrical uptake of fluorescent dye through electropermeabilized membranes of *Avena* mesophyll protoplasts. FEBS Lett. 185: 89–94.

Menczel, L., F. Nagy, Z.R. Kiss and P. Maliga, 1981. Streptomycin resistant and sensitive somatic hybrids of *Nicotiana tabacum* + *Nicotiana knightiana*: correlation of resistance to *N. tabacum* plastids. Theor. Appl. Genet. 59: 101–118.

Murashige, T. and F. Skoog, 1962. A revised medium for rapid growth and bioassays with tobacco tissue cultures. Physiol. Plant 15: 473–497.

Patel, K.R., N.S. Shekhawat, G.P. Berlyn and T.A. Thorpe, 1984. Isolation and culture of protoplasts from cotyledons of *Pinus coulteri* D. Don. Plant Cell Tissue Organ Cult. 3: 85–90.

Pladys, D., L. Dimitrijevic and J. Rigaud, 1991. Localization of a protease in protoplast preparations in infected cells of French bean nodules. Plant Physiol. 97: 1174–1180.

Rengel, Z. and D.C. Elliot, 1992. Mechanism of aluminum inhibition of net $^{45}Ca^{+}$ uptake by *Amaranthus* protoplasts. Plant Physiol. 98: 632–638.

Robertson D., A.K. Weissinger, R. Ackley, S. Glover and R.R. Sederoff, 1992. Genetic transformation of Norway spruce (*Picea abies* (L.) Karst) using somatic embryo explants by microprojectile bombardment. Plant Mol. Biol. 19: 925–935.

Sandberg, G. and J.E. Hällgren, 1985. Catabolism of 3–indole acetic acid in protoplasts from etiolated seedlings of scot pine (*Pinus sylvestris* L.). Plant Cell Rep. 4: 100–104.

Sandberg, G, P. Gardeström, F. Sitbon and O. Olsson, 1990. Presence of indole-3–acetic acid in chloroplasts of *Nicotiana tabacum* and *Pinus sylvestris*. Planta 180: 562–568.

Simmonds, D.H., 1991. Microtubules in cultured plant protoplasts. Acta Bot. Neerl. 40: 183–195.

Sundberg, B., G. Sandberg and E. Jensen, 1985. Catabolism of indole-3–acetic acid to indole-3–methanol in a crude enzyme extract and in protoplasts from Scot pine (*Pinus sylvestris*). Physiol. Plant. 64: 438–444.

Tautorus, T.E., 1990. Tissue and cell culture studies of black spruce (*Picea mariana* Miller B.S.P.) and jack pine (*Pinus banksiana* Lambert). Ph.D. Thesis, Department of Biology, University of Saskatchewan, Saskatoon, Sask.

Tautorus, T.E., F. Bekkaoui, M. Pilon, R.S.S. Datla, W.L. Crosby, L.C. Fowke and D.I. Dunstan, 1989. Factors affecting transient gene expression in electroporated black spruce (*Picea mariana*) and jack pine (*Pinus banksiana*) protoplasts. Theor. Appl. Genet. 78: 531–536.

Tautorus, T.E., S.M. Attree, L.C. Fowke and D.I. Dunstan, 1990a. Somatic embryogenesis from immature and mature zygotic embryos, and embryo regeneration from protoplasts in black spruce (*Picea mariana* Mill.). Plant Sci. 67: 115–124.

Tautorus, T.E., L.C. Fowke, and D.I. Dunstan, 1990b. Comparative studies of protoplast development in jack pine (*Pinus banksiana*). Can. J. Bot. 68: 1774–1779.

Tautorus, T.E., L.C. Fowke and D.I. Dunstan, 1991. Somatic embryogenesis in conifers. Can. J. Bot. 69: 1873–1899.

Tautorus T.E., H. Wang, L.C. Fowke and D.I. Dunstan, 1992. Microtubule pattern and the occurence of pre-prophase bands in embryogenic cultures of black spruce (*Picea mariana* Mill.) and non-embryogenic cultures of jack pine (*Pinus banksiana* Lamb.) Plant Cell Rep. 11: 419–423.

Teasdale, R.D. and E. Rugini, 1983. Preparation of viable protoplasts from suspension-cultured loblolly pine (*Pinus taeda*) cells and subsequent regeneration to callus. Plant Cell Tissue Organ Cult. 2: 253-261.

Vardi, V., A. Breiman and E. Galun, 1987. *Citrus* cybrids: production by donor-recepient protoplast-fusion and verification by mitochondrial-DNA restriction profiles. Theor. Appl. Genet. 75: 51–58.

Vasil, I.K., 1984. Cell Culture and Somatic Cell Genetics of Plants, Vol. 1. Academic Press, Orlando, FL.

Vasil, I.K., 1988. Progress in the regeneration and genetic manipulation of ceral crops. Bio/Tech. 6: 397–402.

Verma, D.C., J.D. Litvay and M.A. Johnson, 1982. Media development for cell suspensions of conifers. In: Y. Yamada (Ed.), Proc. of the 5th International Congress Plant Tissue and Cell Culture, pp. 59–60. IAPTC, Tokyo.

Verma, D.C. and S.R. Wann, 1983. Isolation of high yields of viable protoplasts from quaking aspen seedlings and cultured loblolly pine cell suspensions. In: I. Potrykus, C.T. Harms, A. Hinnen, R. Hutter, P.J. King and R.D. Shillito (Eds.), Proc. 6th Intl. Protoplast Symp., pp. 10–11, Aug. 12–16, 1983, Basel. Birkhauser, Basel.

Von Aderkas, P. 1992. Embryogenesis from protoplasts of haploid European larch. Can. J. For. Res. 22: 397–402.

Von Arnold, S. and T. Eriksson, 1981. *In vitro* studies of adventitious shoot formation in *Pinus contorta*. Can. J. Bot. 59: 870–874.

Widholm, J.M., 1972. The use of fluorescein diacetate and phenosafranin for determining the viability of cultured cells. Stain Technol. 47: 189–194.

Wilson, S.M., T.A. Thorpe and M.M. Moloney, 1989. PEG-mediated expression of GUS and CAT genes in protoplasts from embryogenic suspension cultures of *Picea glauca*. Plant Cell Rep. 7: 704–707.

Winton, L.L., R.A. Parham and H.M. Kaustinen, 1975. Isolation of conifer protoplasts. Genet. Physiol. Notes #20: pp. 1–9. Inst. Pap. Chem., Appleton, WI.

10. Genetic Transformation of Angiosperms

Abhaya M. Dandekar

Contents

1. Introduction

A decade has elapsed since the first report appeared in the literature demonstrating the successful transformation of an Angiosperm (Fraley *et al.*, 1983). In the interim we have witnessed monumental advances in the expression of transgenes and the application of gene transfer technology. The focus of this review will be to evaluate the progress that has been made thus far and to identify areas where additional research could strengthen the emerging gene transfer and transgene technology and its applications. The term "transgenes" refers to any gene(s) that is stably incorporated via a process of gene transfer (genetic transformation) into the genome of plant cells. Plantlets regenerated from such cells are referred to as transgenic plants. This emerging technology has revolutionized our understanding of plant biology at the molecular level as well as paved the way for the first commercial application; the "Flavrsavr tomato" that came on sale in US markets in 1994.

Most woody angiosperms might well be considered to be one of the more intractable groups of experimental organisms, yet they are also some of the most valuable, both aesthetically and commercially. Because of their long generation time (Schuerman and Dandekar, 1991, 1993) manipulation of these plants for the introgression of new genetic information can be a difficult task for the laboratory researcher, breeder and grower alike. The advanced state of cell wall development in most woody angiosperm tissues can make

S. Jain, P. Gupta & R. Newton (eds.), Somatic Embryogenesis in Woody Plants, Vol. 1, 193–226.

cell culture, regeneration and biochemical analysis particularly demanding. Additionally, they tend to be especially rich in phenolic compounds which can react with proteins and nucleic acids during cell lysis and extraction procedures. However, the incentive for success certainly is the economic value of these plants and therefore, success can be particularly rewarding.

Another important limitation of woody angiosperm species is the introgression of specific traits. First, with generation times of years rather than months, an introgressive approach to the transfer of a single trait becomes temporally demanding. Second, genes can only be introgressed from sexually compatible donors. Third, most cultivars are not true-breeding; the levels of outcrossing which occur in such populations ensure a certain high degree of heterozygosity. For these reasons, production of commercial cultivars has resulted mostly from the large scale asexual propagation of chance seedlings and hybrids and clonal variants which showed some desirable commercial character. Such processes are somewhat random in generating a particular single-gene modification. In contrast, using gene transfer, a gene of interest can be inserted by one of several techniques (discussed below) into the genome of a single cell. The single cell can then be cultured and regenerated so that it develops into a plant, all cells of which carry the gene. This strategy is not limited by generation time but by regeneration time; that is, the time it takes to produce a plant from the single cell. Also, the phenomenon of detrimental linkage drag does not occur in transformation. However, there are many other obstacles that need to be overcome before one can obtain transgenic plants with commercial potential. This is discussed in detail in this review drawing upon the recent information published on the introduction of transgenes into different angiosperm species.

The critical barriers and challenges facing the transformation of most woody angiosperm are similar to that faced in the transformation of most herbaceous angiosperms. The problems, though similar, tend to be exacerbated in woody plants. Therefore, there is much that can be learnt from evaluating the emerging research on all angiosperms. In this review, therefore, we shall dwell not just upon the information obtained from woody plants but that obtained from all plants; most of which are much further along in the technology development process. Several basic issues listed below are critical to developing successful transformation protocols and the evaluation of transgenic Angiosperms using either the natural gene transfer system of *Agrobacterium* or direct DNA introduction techniques:

1) *Gene Transfer Process*: This pertains to the different factors that influence the efficiency of the method used for the introduction of transgenes into the target plant cells.
2) *Position and Frequency of Regeneration*: Position and frequency of cells in the target explant capable of regenerating transgenic plants.
3) *Genetic Selection and Screening*: Efficiency of genetic selection that would permit multiplication of a majority of the transformed cells.

4) *Transgene Expression*: Stability and location of the introduced DNA segment in the plant genome and its consequences on expression of the transgene(s).
5) *Field Introduction and Performance*: Field introduction and performance of transgenic plants arising from independent transformation events with respect to both transgenes and endogenous traits.

2. Gene Transfer Process

Gene transfer is the process by which DNA is introduced into plant cells or tissues. The DNA can come from virtually any source confirming the "infinite germplasm concept". Today gene transfer methodology has become part of an essential technology to manipulate plants for both scientific and commercial purposes. Transgenic plants, the products of this technology, are useful for dissecting the mechanism(s) of plant gene regulation. This technology is also useful in identifying and evaluating agriculturally useful traits (genes) as well as for their introduction into commercially valuable crops. The list of plants that have been successfully transformed to produce transgenic plants is long (Tables 1–4) encompassing a wide range of Angiosperms. Three techniques stand out and are responsible for the majority of success and these are; 1) *Agrobacterium*-mediated transformation (Table 1), 2) DNA coated particle bombardment (DNA coated microprojectiles, Table 2) and 3) electroporation (Table 3).

2.1. Agrobacterium-*Mediated Transformation Process*

This gene transfer technique takes advantage of the naturally evolved mechanisms of DNA transfer present in the common soil pathogen *Agrobacterium tumefaciens*. This bacterium is the causative agent of "crown gall" a tumorous growth that occurs when wounded tissues are infected by this bacterium usually at the plant-soil interface. *Agrobacterium rhizogenes* another closely related member to the above organism is also capable of DNA transfer and is the causative agent of "hairy root" disease that results in the formation of adventitious roots at the site of infection. Much is known today of the mechanism of tumorigenesis and rhizogenesis and how this information can be applied to genetically transform plants of interest (for reviews see Gelvin, 1990; Hooykaas and Schilperoort, 1992; Klee and Rogers, 1989; Tempe and Casse-Delbart, 1989; Zambryski, 1988, 1992; Zambryski *et al.*, 1989).

2.1.1. *Optimizing the* Agrobacterium-*Mediated Gene Transfer Process*
There are several factors that appear to be important and should be considered when using *Agrobacterium*. The genetics of the strain of *Agrobacterium* used is one such factor. It is well known that most dicot plants are

Table 1. Listing of Angiosperm species that have been transformed with *Agrobacterium tumefaciens* or *Agrobacterium rhizogenes* and transgenic plants regenerated.

Species	Common Name	Reference
Actinidia deliciosa	kiwi	Uematsu *et al.*, 1991
Allocasuarina verticillata		Phelep *et al.*, 1991
Anagallis arvensis	pimprenel	Mugnier, 1988
Antirrhinum majus	snapdragon	Handa, 1992
Apium graveolens	celery	Catlin *et al.*, 1988
Arabidopsis thaliana	mouse ear cress	Lloyd *et al.*, 1986; Pavingerova and Ondrej, 1986
Armoracia lapathifolia	horseradish	Noda *et al.*, 1987
Asparagus officinalis	asparagus	Bytebier *et al.*, 1987
Atropa belladonna		Saito *et al.*, 1992
Beta vulgaris	sugarbeet	Lindsey and Gallois, 1990
Brassica carinata		Narasimhulu *et al.*, 1992
Brassica juncea	India mustard	Barfield and Pua, 1991
Brassica napus	oilseed rape	Fry *et al.*, 1987; Pua *et al.*, 1987; Guerche *et al.*, 1987a
Brassica oleracea	cauliflower	Srivastava *et al.*, 1988; David and Tempe, 1988
	cabbage	Berthomieu and Jouanin, 1992
	broccoli	Hosoki *et al.*, 1991
Carica papaya	papaya	Fitch *et al.*, 1993
Carya illinoensis	pecan	McGranahan *et al.*, 1993
Catharanthus roseus		Brillanceau *et al.*, 1989
Cichorium intybus	chicory	Vermeulen *et al.*, 1992
Citrus jambhiri	rough lemon	Vardi *et al.*, 1990
Citrus sinensis	orange	Hidaka *et al.*, 1990
Citrus senensis × *Poncirus trifoliata*	Carrizo citrange	Moore *et al.*, 1992
Convolvulsus arvensis	morning glory	Tepfer, 1984
Cucumis melo	muskmelon	Fang and Grumet, 1990
Cucumis sativus	cucumber	Chee, 1990; Trulson *et al.*, 1986
Datura innoxia		Sangwan *et al.*, 1991
Daucus carota	carrot	Scott and Draper, 1987; Chilton *et al.*, 1982
Dendranthema indicum	chrysanthemum	Ledger *et al.*, 1991
Dianthus caryophyllus	carnation	Lu *et al.*, 1991
Fagopyrum esculentum	buckwheat	Miljus-Djukic *et al.*, 1992
Fragaria anannassa	strawberry	James *et al.*, 1990; Nehra *et al.*, 1990
Foeniculum vulgare	fennel	Mugnier, 1988
Glycine max	soybean	Hinchee *et al.*, 1988
Grossypium hirsutum	cotton	Firoozabady *et al.*, 1987; Umbeck *et al.*, 1987
Helianthus annus	sunflower	Everett *et al.*, 1987
Ipomoea batatas	sweet potato	Al-Juboory and Skirvin, 1991
Juglans regia	walnut	McGranahan *et al.*, 1988; Dandekar *et al.*, 1989
Kalanchoe laciniata		Jia *et al.*, 1989
Lactuca sativa	lettuce	Michelmore *et al.*, 1987
Linum usitatissimum	flax	Basiran *et al.*, 1987
Lotus corniculatus	bird's-foot trefoil	Otten and Schell, 1986

Table 1. Continued.

Species	Common Name	Reference
Lycopersicon esculentum	tomato	McCormick *et al.*, 1986; Shahin *et al.*, 1986b
Malus pumila	apple	James *et al.*, 1989
Medicago arborea		Damiani and Arcioni, 1991
Medicago sativa	alfalfa	Shahin *et al.*, 1986a; Golds *et al.*, 1991
Medicago trunculata	barrel medic	Thomas *et al.*, 1992
Medicago varia	alfalfa	Deak *et al.*, 1986
Nicotiana bigelovii		Schoelz *et al.*, 1991
Nicotiana clevlandii		Tavazza *et al.*, 1988
Nicotiana glauca		An *et al.*, 1986; Taylor *et al.*, 1985
Nicotiana hesperis		Waltonn and Belshaw, 1988
Nicotiana plumbaginifolia		Horsch *et al.*, 1984; De Block *et al.*, 1984; Jouanin *et al.*, 1987
Nicotiana rustica		An *et al.*, 1986
Nicotiana tabacum	tobacco	An *et al.*, 1986; Tepfer, 1984
Onobrychis viciifolia		Golds *et al.*, 1991
Petunia hybrida	petunia	Fraley *et al.*, 1985; Ondrej and Biskova, 1986
Pisum sativum	pea	De Kathen and Jacobsen, 1990; Puonti-Kaerlas *et al.*, 1990
Populus alba × *P. gradidentata*	aspen	Fillatti *et al.*, 1987
Populus alba × *P. tremula*	aspen	De Block, 1990
Populus trichocarpa × *P. deltoides*	poplar	De Block, 1990; Pythoud *et al.*, 1987
Prunus armeniaca	apricot	Machado *et al.*, 1992
Prunus domestica	plum	Mante *et al.*, 1991
Prunus persica	peach	Smigocki and Hammerschlag, 1991
Ribes nigrum	black currant	Graham and McNicol, 1991
Rubus spp.	brambles	Graham *et al.*, 1990
Solanum dulcamara		McInnes *et al.*, 1991
Solanum integrifolium		Rotino *et al.*, 1992
Solanum melongena	eggplant	Guri and Sink, 1988
Solanum muricatum	pepino	Atkinson and Gardner, 1991
Solanum nigrum	nightshade	Wei *et al.*, 1985
Solanum tuberosum	potato	An *et al.*, 1986; Shahin and Simpson, 1986; Ondrej *et al.*, 1989; Visser *et al.*, 1989
Stylosanthes humilis	Towsville stylo	Manners, 1988
Trifolium repens	white clover	White and Greenwood, 1987
Vicia narbonensis		Pickardt *et al.*, 1991
Vigna aconitifolia	moth bean	Eapen *et al.*, 1987
Vitis rupestris	grapevine	Mullins *et al.*, 1990
Zea mays	corn	Gould *et al.*, 1991

Table 2. Listing of Angiosperm species that have been transformed by DNA coated particle bombardment and transgenic plants regenerated.

Species	Common Name	Reference
Arabidopsis thaliana	mouse ear cress	Seki *et al.*, 1991
Carica papaya	papaya	Fitch *et al.*, 1990
Glycine max	soybean	McCabe *et al.*, 1988
Grossypium hirsutum	cotton	Finer and McMullen, 1990
Hordeum vulgare	barley	Wan and Lemaux, 1994
Nicotiana plumbaginifolia		Horth *et al.*, 1987
Nicotiana tabacum	tobacco	Klein *et al.*, 1988
Oryza sativa	rice	Christou *et al.*, 1991
Populus alba × *P. gradidentata*	aspen	McCown *et al.*, 1991
Saccharum spp.	sugar cane	Bower and Birch, 1992
Triticum aestivum	wheat	Vasil *et al.*, 1992; Weeks *et al.*, 1993
Vaccinium macrocarpon	cranberry	Serres *et al.*, 1992
Vigna aconitifolia	moth bean	Kohler *et al.*, 1987
Zea mays	corn	Gordon-Kamm *et al.*, 1990

Table 3. Listing of Angiosperm species that have been transformed by electroporation and transgenic plants regenerated.

Species	Common Name	Reference
Brassica napus	oilseed rape	Guerche *et al.*, 1987b
Dactylis glomerata	orchardgrass	Horn *et al.*, 1988
Fragaria anannassa	strawberry	Nyman and Wallin, 1992
Nicotiana tabacum	tobacco	Riggs and Bates, 1986
Oryza sativa	rice	Toriyama *et al.*, 1988; Zhang and Wu, 1988
Zea mays	corn	Rhodes *et al.*, 1988; D'Halluin *et al.*, 1992

susceptible to *Agrobacterium* as has been catalogued in the survey published by De Cleene and De Ley (1976, 1981). However, resistance of the target plant tissues to *Agrobacterium* could be an important factor influencing the virulence of this organism and ultimately reducing the efficiency of plant transformation when one attempts using this pathogen as a vector. There is a growing body of evidence that indicates wide variations in the virulence, depending upon the type of host plant used for most of the widely used strains of *Agrobacterium*. This appears to be of some significance in certain plant species such as soybean (Owens and Cress, 1985), strawberry (Uratsu *et al.*, 1990), *Zea mays* (Jarchow *et al.*, 1991) and a range of woody plants (Martin, 1987; Dandekar *et al.*, 1988, 1990; Morris *et al.*, 1989; Morris and Morris, 1990). Efforts to improve the efficiency of *Agrobacterium*-mediated transformation have centered on either identifying superior strains or increas-

Table 4. Listing of transgenic Angiosperm species that have been obtained by other methods of DNA transfer.

Species	Common Name	Reference
Direct DNA uptake:		
Dactylis glomerata	orchardgrass	Horn *et al.*, 1988
Festuca arundinacea	tall fescue	Wang *et al.*, 1992
Nicotiana tabacum	tobacco	Paszkowski *et al.*, 1984
Oryza sativa	rice	Zhang and Wu, 1988
Liposome mediated:		
Nicotiana tabacum	tobacco	Deshayes *et al.*, 1985
Ultrasonication:		
Nicotiana tabacum	tobacco	Zhang *et al.*, 1991
Injection of DNA into reproductive organs:		
Brassica napus	oilseed rape	Neuhaus *et al.*, 1987
Oryza sativa	rice	Luo and Wu, 1988
Secale cereale	rye	De la Pena *et al.*, 1987

ing the degree of virulence by optimizing the growth of the *Agrobacterium* strain used or that of the target plant material such that the physiological conditions are optimal for the induction of virulence. The super virulent *A. tumefaciens* strain A281 causes numerous large tumors (Hood *et al.*, 1984) and the cause of this phenotype has been localized to a segment of the Ti plasmid pTiBo542 (Hood *et al.*, 1986; Jin *et al.*, 1987). Incorporation of this segment into vectors offers the potential of increasing transformation efficiency (Pythoud *et al.*, 1987). Additional copies of the genes *virA*, *B*, and *G* greatly stimulate virulence of *A. tumefaciens* on apple (Dandekar *et al.*, 1990). Zyprian and Kado (1990) report increased efficiency of transformation using a binary vector system for *Agrobacterium* which includes virulence genes carried on a high-copy number plasmid.

Exogenously supplied growth regulators, notably auxins, have been shown to enhance hairy root induction (Ryder *et al.*, 1985) and tumor induction (El Khalifa and Lippincott, 1967; Hrouda and Ondrej, 1983) by *Agrobacterium*. Extended preculture of flax hypocotyls on growth regulator-containing medium before inoculation with the *Agrobacterium* vector resulted in higher frequencies of transgenic plant recovery (McHughen *et al.*, 1989). Interestingly, *A. tumefaciens* has been found to produce auxin and cytokinin, and this production of growth regulators has been implicated in virulence and host range of the organism (reviewed by Morris, 1986; Melchers *et al.*, 1990). Also, some compounds which promote cell-division in tobacco have been shown to induce virulence genes in *A. tumefaciens* (Hess *et al.*, 1991; Teutonico *et al.*, 1991).

Environmental factors like pH, temperature and osmotic conditions strongly influence the expression and induction of virulences genes (Alt-Mörbe *et al.*, 1988, 1989; Vernade *et al.*, 1988; Godwin *et al.*, 1991; Turk *et al.*, 1991). The most direct effects on virulence induction are mediated by the presence of phenolic compounds that seem to influence virulence gene expression (Bolton *et al.*, 1986; Melchers *et al.*, 1989). Compounds such as acetosyringone (3′,5′-dimethoxy-4′-hydroxyacetophenone), sinapinic acid, coniferyl alcohol, caffeic acid, ethyl ferrulate, methylsyringic acid are known inducers of virulence genes in *Agrobacterium* (reviewed by Kado, 1991). The virulence induction is also influenced by the presence of other compounds like monosaccharides (via the *chvE* gene; Ankenbauer and Nester, 1990; Cangelosi *et al.*, 1990; Huang *et al.*, 1990; Shimoda *et al.*, 1990) and opines (Veluthambi *et al.*, 1989). Osmoprotective compounds like betaine and proline were shown to synergistically enhance the effect of phenolic compounds like acetosyringone (James *et al.*, 1993; Vernade *et al.*, 1988). Betaine has been shown to increase the expression of several virulence genes in *Agrobacterium* (Vernade *et al.*, 1988). Proline or betaine may help the bacteria to adapt to rapid changes in pH and osmotic pressure caused by the proximity of wounded plant cells thus increasing the transformation efficiency (James *et al.*, 1993).

Agrobacterium has been shown to demonstrate tissue/cell selectivity (Castle and Morris, 1990; Colby *et al.*, 1991) and may preferentially transform metabolically active and/or actively dividing cells (Braun and Mandle, 1948; Braun, 1952; Chriqui *et al.*, 1988). Increasing the number of such cells in the explant should improve transformation efficiency. Examples of this would be the use of very young tissue, the use of tissue treated with growth regulators (McHughen *et al.*, 1989) or the use of pre-wounded tissue (Braun and Mandle, 1948). This latter application would need to be balanced against the transformation-decreasing effects of wound healing (Braun and Mandle, 1948; Davis *et al.*, 1991). Note that the media used during the cocultivation period in many *Agrobacterium*-mediated transformation protocols contain growth regulators to maintain viability of the explants; this may improve transformation rates as well. As mentioned earlier, some cell-division promoting compounds induce virulence of *Agrobacterium*, suggesting a connection between the bacterium's gene regulation and its apparent cell-selectivity.

2.2. *DNA Coated Particle Bombardment* (*DNA Coated Microprojectiles*)

This technique was developed by Sanford and coworkers and involves coating biologically active DNA onto small tungsten or gold particles (1 to 5 microns in size) and accelerating these in a gun such that they are bombarded into plant tissue at high velocity (Klein *et al.*, 1987, 1992; Sanford, 1988, 1990; Christou, 1992). The particles penetrate the plant cell wall and lodge themselves within the cell nucleus where the DNA is liberated resulting in transformation of the individual plant cells in an explant. The physical nature of particle bombardment can potentially overcome many of the biological barriers associated with other transformation methods such as the host-range specificity of *Agrobacterium* or regeneration of complete plants from protoplasts. Indeed, the introduction of DNA into organized, morphogenic tissues such as seeds, embryos or meristems has enabled the successful transformation and regeneration of several cereal crop plants recalcitrant to *Agrobacterium*-mediated procedure like wheat, rice, corn and barley and some recalcitrant dicot species like soybean (Table 2).

The particle guns used in biolistic transformation cause a certain amount of damage to the treated plant tissue which reduces the number of stable transformation events. Transformation can be maximized by re-configuring the target tissue and varying the distance between the target and the stopping plate (Armstrong and Hinchee, 1990). Guns which use chemical explosions must shield the plant tissue from the resulting heated vapors. The "spread" of the impacting pellets on the explant is another concern, especially when a small area (e.g., a meristem) is the target. Results from biolistic transformation experiments are becoming more reproducible as more researchers use this technique, however, many of the variables in the process are yet unaccounted for (Sanford, 1990).

2.3. *Electroporation*

Electroporation is another method for direct DNA delivery into plant cells and has been used for the transformation of recalcitrant species like monocots difficult to transform with *Agrobacterium*. The technique is a process whereby very short pulses of electricity are used to reversibly permeabilize the lipid bilayer of cell membranes (Fromm *et al.*, 1985, 1987). Basically, the electrical discharge enables the diffusion of macromolecules such as nucleic acids through an otherwise impermeable plasmalemma. Because the plant cell wall will not allow the efficient diffusion of most transgene constructs, protoplasts must be prepared. This requirement presents a major obstacle for many applications since protocols describing the regeneration of protoplasts into complete plants do not exist. Furthermore, the necessary extended culture period introduces the added risk of genotypic variation in recovered regenerants. Nevertheless, the method has been used to successfully to transform such important crops as rice, corn and wheat (Table 3).

Recently, a potentially important variation of this technique was reported (Dekeyser *et al.*, 1990; D'Halluin *et al.*, 1992). Electroporation was used to deliver DNA into intact, organized tissues of rice, corn, barley and wheat and transient expression of the introduced genes was examined (Dekeyser *et al.*, 1990). Transgenic plants were obtained using this procedure and the genes were found to be stably incorporated through three subsequent generations (D'Halluin *et al.*, 1992). This strategy has merit and could be used more frequently in future for stable transformation purposes because it circumvents the problems associated with plant regeneration from protoplasts.

2.4. *Other Methods of Plant Transformation*

Several additional methods have been used to successfully produce transgenic plants. Stable transformation of a number of species (see Table 4) has been achieved after permeabilizing cells with polyethylene glycol (Shillito *et al.*, 1985). As well, microinjection of DNA directly into the cell nucleus has been employed. However, as with electroporation, both of these techniques suffer from the drawback that protoplasts are utilized as the recipient host. Microinjection also has the added disadvantage of being labor intensive and tedious.

Intact reproductive tissues have also been injected with DNA resulting in the stable transformation of both rye and rice (see Table 4). DNA was introduced into the floral tillers of rye or pollen tubes of rice, allowing the successful recovery of transgenic seeds. The relatively small number of reports which utilized this approach suggests that it is not a feasible alternative for many situations; for example, Potrykus (1991) notes that many pollen tubes are probably plugged with callose and contain nucleases. Furthermore, this method is not appropriate for self-incompatible crops which are multiplied via clonal propagation.

Fusion of negatively charged, DNA-containing liposomes with plant protoplasts has also been employed as a method for transformation (Riggs and Bates, 1986). Only one report has appeared describing the recovery of a transgenic plant using this strategy, again implying that its general utility is probably limited. Potrykus (1991) noted that lipid-mediated transformation offers no technical advantage when compared with other methods of direct DNA uptake which utilize protoplasts.

Recently, a novel method for direct DNA transfer based on ultrasound has been developed (reviewed in Joersbo and Brunstedt, 1992). Ultrasonication has been used to successfully transform protoplasts and, more importantly, intact, organized plant tissue (Joersbo and Brunstedt, 1990; Zhang *et al.*, 1991). Using tobacco leaf explants, Zhang and coworkers (1991) recovered transgenic plantlets at an amazing frequency of 22 percent. Ultrasonication offers exciting potential because it is less costly and simpler than other physical methods of DNA transfer to intact plant cells (e.g., particle bombardment) and, as the initial report implies, perhaps more efficient.

3. Regeneration of Transgenic Plants

Once a plant cell has incorporated the introduced DNA in a stable manner the next step is to regenerate a plant from the transformed cells. Position, frequency and scope of regeneration events are critical to the isolation of transgenic plants (Dandekar *et al.*, 1993). Most often the major limiting step in the isolation of transgenic plants is due to the lack of regeneration occurring from within the transformed cell populations. There is a lot of variability

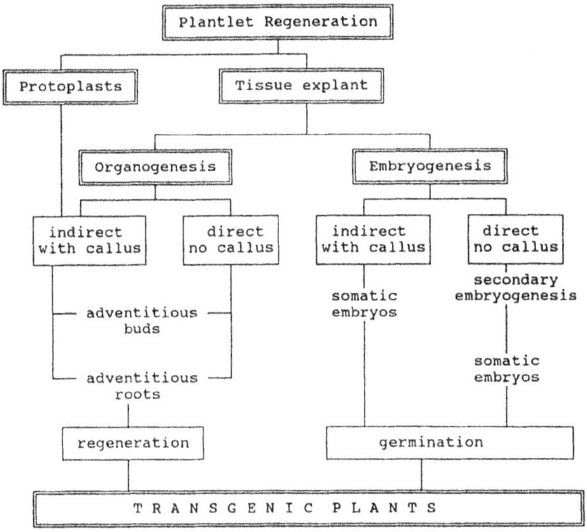

Figure 1. Pathways for the regeneration of Angiosperms.

in the frequency and scope of regeneration among different Angiosperm species as well as among different cultivars of any one species.

Two pathways of regeneration have been observed (Fig. 1) in most Angiosperms – organogenesis and somatic embryogenesis. Organogenesis involves the regeneration of adventitious shoots through the formation of a shoot meristem or the regeneration of adventitious roots through the formation of root meristems. The second pathway involves the formation of embryos or embryo-like structures from somatic tissues and is referred to as somatic embryogenesis. It has been suggested that somatic embryogenesis and organogenesis (adventitious bud formation) reflect different developmental events that are most likely mutually exclusive (Ammirato, 1985). Therefore, depending upon the conditions, totipotent cells can commit themselves to only one pathway such that they either undergo embryogenesis and become an embryo, a bipolar structure, or differentiate into a bud a monopolar structure. Somatic embryos closely resemble their zygotic counterparts in that they are bi-polar structures having both a shoot and a root meristem. The somatic embryos behave much like seeds and thus plants are obtained after germination of these embryos. In the case of adventitious bud formation shoots are regenerated and then are propagated and subjected to a hormone treatment with auxins to induce root formation. Organogenesis and somatic embryogenesis can take place either directly from the explant such as a leaf (direct mode, Fig. 1) or indirectly (indirect mode, Fig. 1) via the formation of callus (Williams and Maheswaran, 1986).

Somatic embryogenesis could be the fundamental process of regenerating many woody angiosperm plants for the purposes of transgene introduction. There are several excellent reviews that have appeared on the general subject of somatic embryogenesis in different woody plants (Attree and Fowke, 1993; Litz and Gray, 1992; Tulecke, 1987; Wann, 1988; Williams and Maheswaran, 1986). Induction, is the principal process that determines cell fate during embryogenesis. Different stimuli for example, hormonal, metabolic or environmental appear to be important factors for this induction process. These factors have been shown to stimulate totipotent cells triggering somatic embryogenesis. The key cellular event appears to be the unequal division of vacuolated cells, giving rise to embryogenic and parenchymatous daughter cells (Litz and Gray, 1992). It has been suggested that this stimulation of densely cytoplasmic cells from a population of rapidly dividing vacuolate parenchymatous cells is equivalent to induction (Litz and Gray, 1992). Differentiating cells can become delimited from the surrounding cells by the formation of thick cell walls as shown in *Citrus sinensis* (Button *et al.*, 1974). This isolation of somatic cells from the correlative influences of other cells has been thought to be essential for expression of totipotency (Steward *et al.*, 1964). The nature and distribution of fate determining factors in operation during embryogenesis has remained obscure until recently when it was shown that the cell wall could play a key role in determining cell fate (Berger *et al.*, 1994). Using laser microsurgery Berger *et al.* (1994), were able to demonstrate that the contact of one cell type with the wall of another caused a shift in developmental fate.

Perhaps the most critical question for the regeneration of transgenic plants is the number of cells that participate in organogenesis or somatic embryogenesis. If the origin is from a single cell then the recovery of transgenic plants could occur at a high frequency as long as the cells undergoing organogenesis or somatic embryogenesis are sensitive to *Agrobacterium* and the location or position of these cells is such that they come in contact with *Agrobacterium*. In many cases this is not so and the origin appears to be multicellular (D'Amato, 1985; Monacelli *et al.*, 1988) or several cell layers deep within the tissue (Colby *et al.*, 1991). If the origin involves more than one cell type then the formation of chimeras of transformed and nontransformed tissues is most likely to occur. Somaclonal variation (Larkin and Scowcroft, 1981) is always a distinct possibility in plants regenerated through a callus mode. Therefore, transgenic plants obtained through such a regeneration system must be extensively field tested to ensure that their genotype is identical to that of the parental material. Unfortunately culture conditions necessary for regeneration vary greatly from one plant species to another and in most instances between cultivars belonging to a single plant species. This is a major stumbling block for developing transgenic plants in any Angiosperm species. A multitude of variables have to be researched to develop a regeneration protocol for each system. Avenues to circumvent regeneration by directly transforming tissue that naturally regenerate e.g., meristems or embryos represent important areas of future innovation.

Table 5. Transgenes used for selection and identification of transformed cells.

Transgene Product	*Reference*
Antibiotic resistance:	
Kanamycin phosphotransferase (APH(3')II)	Herrera-Estrella *et al.*, 1983; Fraley *et al.*, 1983
Hygromycin phosphotransferase (HYG)	Waldron *et al.*, 1985
Gentamycin acetyltransferase (GENT)	Hayford *et al.*, 1988
Streptomycin phosphotransferase (STR)	Jones *et al.*, 1987
Resistance to bleomycin (BLE)	Hille *et al.*, 1986
Metabolic inhibitors:	
Dihydrofolate reductase (DHFR)	Herrera-Estrella *et al.*, 1983; Eichholtz *et al.*, 1987
EPSP synthase (EPSPS)	Comai *et al.*, 1985; Shah *et al.*, 1986
Acetolactate synthase (ALS)	Haughn and Somerville, 1986
Bromoxynil nitrilase (BNR)	Stalker *et al.*, 1988
Phosphinothricin acetyl transferase (PAT)	De Block *et al.*, 1987
Enzymatic activity:	
Nopaline synthase (NOS)	Zambryski *et al.*, 1983
Octopine synthase (OCS)	DeGreve *et al.*, 1982
Chloramphenicol acetyltransferase (CAT)	Herrera-Estrella *et al.*, 1983
β-glucuronidase (GUS)	Jefferson *et al.*, 1987
β-galactosidase (LAC)	Helmer *et al.*, 1984
Bacterial luciferase (LUX)	Koncz *et al.*, 1987
Firefly luciferase (LUC)	Ow *et al.*, 1986
Morphological:	
Tumour/root formation	Mártón *et al.*, 1979; Hernalsteens *et al.*, 1980
Agroinfection	Grimsley *et al.*, 1987

4. Genetic Selection

Genetic selection is the process of selecting preferentially for those cells that have been transformed by the incoming transgenes. A selective advantage can be conferred upon the transformed cells through the introduction of genes encoding antibiotic resistance or resistance to some metabolic inhibitor like an herbicide. In the presence of the antibiotic or herbicide the untransformed cells die whereas the transformed cells grow and multiply. If no form of genetic selection were used one would be faced with the option of screening every shoot that regenerated in a transformation experiment. In cases where the transformation frequency is high (i.e., the number of transformed cells or shoots arising from an explant), this would be feasible. However, for plant species that transformed at a low frequency this would become a laborious if not impossible task. Genetic selection is, therefore, an essential component of any plant transformation protocol and has been accomplished by using a variety of marker genes (Table 5).

Genes that encode resistance to an antibiotic are good examples of selectable marker genes. Resistance to the aminoglycoside antibiotic kanamycin has been used extensively in plant transformation experiments and is encoded by the aminoglycoside 3′ phosphotransferase gene [APH(3′)II] (Herrera-Estrella *et al.*, 1983),also referred to in the literature as neomycin phosphotransferase (NPTII). This enzyme inactivates kanamycin through phosphorylation. The bacterial transposable element Tn5 is the source of the gene encoding this enzyme that has been used in all vectors. Expression in plants has been obtained through the construction of chimeric genes where the structural sequences of the gene encoding APH(3′)II have been fused to 5′ and 3′ regulatory sequences that will enable expression in plant tissue.

Other antibiotic resistance genes have been used successfully for plant transformation and these include hygromycin resistance encoded by a hygromycin phosphotransferase gene (Waldron *et al.* 1985), gentamicin resistance encoded by an acetyltransferase gene (Hayford *et al.*, 1988), streptomycin resistance encoded by a phosphotransferase gene (Jones *et al.*, 1987) and resistance to bleomycin (Hille *et al.*, 1986). Metabolic inhibitors like trimethoprim that inhibit dihydrofolate reductase, a key enzyme in C1-metabolism, can also be used as an agent for selection by utilizing an altered gene which encodes an insensitive enzyme (Herrera-Estrella *et al.*, 1983; Eichholtz *et al.*, 1987).

Herbicides also represent a group of compounds that are potent metabolic inhibitors which usually act by affecting a particular step in the biosynthetic pathway of various amino acids. Herbicide resistance has been obtained through the incorporation and expression of an altered target that results in an insensitivity to the herbicide or through expression of an enzyme which detoxifies the inhibitor in much the same manner as the antibiotic resistance genes discussed above. Some examples of the former strategy are resistance to glyphosate ("Roundup", Monsanto Co.) encoded for by an altered EPSP synthase (Comai *et al.*, 1985; Shah *et al.*, 1986), and resistance to sulfonylurea encoded by a mutant form of acetolactate synthase (Haughn and Somerville, 1986). Examples of the latter approach include a bromoxynil nitrilase that detoxifies the herbicide bromoxinyl (Stalker *et al.*, 1988) and a phosphinothricin acetyl transferase (De Block *et al.*, 1987) that detoxifies the phosphinothricin herbicides like "bialaphos" and "ignite".

A key issue that concerns the use of any of the above selective agents relates to the expression of these gene products in the target tissue of the plant of interest. The gene product should be produced in sufficient quantities in the specific cell types that will undergo morphogenesis. Likewise, the concentration of the selective agent in those cells that are morphogenic must be sufficient to prevent the regeneration of "escapes" which do not harbor transgenes. For optimal selection, the pattern of expression should closely mimic the pattern of translocation of the selective agent in the target plant tissue.

A scorable marker gene allows the visual detection of transformed cells, tissues or plants through the expression of genes normally not found within

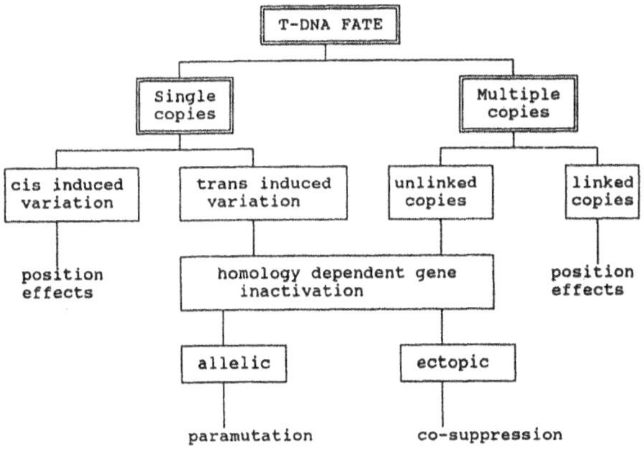

Figure 2. T-DNA fate and its effect upon the variation and inactivation of transgene expression.

the host plant's genome without necessarily imposing any form of selection. These genes usually encode an enzyme that is readily detectable through the use of chromogenic, fluorogenic, photon emitting or radioactive substrates. A good example of such a scorable marker is the *uid* A gene, also referred to as GUS (Jefferson *et al.*, 1987), which encodes β-glucuronidase, an enzyme that allows its detection using a variety of convenient methods (Jefferson, 1987). Scorable markers are of less utility with respect to recovering transgenic plants. Without a selectable marker gene, all the morphogenic cells within a given population may regenerate after a transformation procedure. Doubtless, the majority of regenerants will not carry the transgenes of interest. Consequently, a substantially higher number of plants will have to be screened in order to recover transformants. Nevertheless, this may be the only option available in instances where regeneration frequency is low or impossible in the presence of selective agents. Scorable marker genes are much more valuable as experimental tools employed to investigate, for example, expression patterns resulting from the use of a particular gene promoter.

5. Transgene Expression

The level of expression of a transgene in a transgenic plant is highly influenced by the surrounding DNA at the site of insertion (position effects) as well as by other homologous sequences (co-suppression) present in the genome of the transformed plant as outlined in Fig. 2. In most instances the interaction results in the inactivation of expression of the transgene (Matzke and Matzke, 1993). DNA methylation may be one of the key mechanisms responsible for the inactivation of transgene expression.

5.1. *DNA Methylation*

The methylation of the cytosine residue at the 5 position occurs post replicatively and is widely distributed both among eukaryotes and prokaryotes. In vertebrates DNA methylation occurs at the dinucleotide CpG and is involved in the regulation of gene expression (Cedar, 1988; Razin and Cedar, 1991). DNA methylation could well be one of the important epigenetic factors involved in the parental imprinting of genes that ultimately determine the pattern of gene activity during development (Holliday, 1989). Plant DNA is quite extensively methylated not only at the CpG dinucleotide but also at the trinucleotide CpXpG (where X = A, C or T; Gruenbaum *et al.*, 1981) where the total amount of 5-methyl cytosine (m5C) is as high as ~30% of total cytosines in some plant species (Hepburn *et al.*, 1987). In animals DNA methylation has been shown to inhibit transcription (Bird, 1992; Razin and Cedar, 1991) and can be reversed with 5-azacytidine (Jones, 1984). In plants dynamic changes in methylation patterns have been observed to correlate with expression of rRNA genes of pea (Watson *et al.*, 1987), maize storage protein genes (Spena *et al.*, 1983), transposition of the Ac element of maize (Schwartz, 1989), ripening specific genes in tomato fruit (Hadfield *et al.*, 1993) and chloroplast genes during ripening (Kobayashi *et al.*, 1990; Ngernprasirtsiri *et al.*, 1988). Profound effects have been observed with respect to methylation of T-DNA sequences and the expression of introduced genes (Amasino *et al.*, 1984; John and Amasino, 1989; Matzke and Matzke, 1991). Genes present on silent T-DNA have been activated by growth of the tissue on 5-azacytidine (Bochardt *et al.*, 1992; Hepburn *et al.*, 1983). The activity of the maize A1 gene under the control of the CaMV35S promoter in petunia was shown to be regulated by the amount of methylation of the promoter sequences and this was modulated by age and environmental influences in field grown plants (Meyer *et al.*, 1992). DNA methylation has also been implicated for some of the tissue culture-induced variation in plants (Kaeppler and Phillips, 1993).

5.2. *Position Effects*

Transgenes in a T-DNA may be present in one or more copies in the genome of the transformed plant. The fate of T-DNA with respect to expression of the transgenes is outlined in Fig. 2. Having a single copy of T-DNA in the transformed plant of interest is most desirable as it has the least amount of negative effects on expression. The only major complication is that of position effects where quantitative differences in the level of gene expression have been observed among individual transformants containing the same transgene (see Weising *et al.*, 1988). More recently a qualitative difference was noted when individual tobacco transformants containing the promoter of *ats*1A which encodes the light regulated small subunit of RUBISCO in *A. thaliana* fused to neomycin phosphotransferase II unexpectedly conferred expression in roots (De Almeida *et al.*, 1989). Moreover, the root specific expression varied widely among individual plants. Thus it appears that "positional effects" may not only be responsible for differences in the amount of expression, but also differences in tissue specificity directed by any particular promoter. It has been suggested that position effects are due to chromatin structure and/or the activity of endogenous promoters in the proximity of the foreign gene insertion site (Weising *et al.*, 1988). Methylation of DNA as explained above has also been shown to impair transcription in plants (John and Amasino, 1989; Ngernprasirtsiri, *et al.*, 1989) and this process may also contribute to some of the observed positional effects. These observations make thorough testing of individual transformants imperative before they are used for future purposes and perhaps may provide yet another level of complexity which can be exploited when producing transgenic plants to fulfill a particular need.

A few attempts have been made to try to correct this problem. The transfer of buffer DNA that normally flanks the gene does not appear to correct this problem. Position effects were still observed when the *rbc*S gene with 10 kb of 5′ and 13 kb of 3′ DNA that normally flanks the gene in the source plant was introduced into a heterologous host (Dean *et al.*, 1988). Recently a plant scaffold attachment region (SAR) was isolated and was shown to normalize transgene expression when the introduced transgene was flanked with this region (Breyne *et al.*, 1992). This scaffold attachment region was similar to the animal SARs and have probably been evolutionarily conserved among eukaryotes. Similar results were also obtained with a yeast ARS-1 element that contains a scaffold attachment region (Allen *et al.*, 1993). The presence of these elements flanking the transgene reduced the inhibitory effect on expression of transformants containing multiple copies as well as normalized the frequency distribution patterns of transgene expression from single copies more closely around a mean (Allen *et al.*, 1993; Breyne *et al.*, 1992).

5.3. *Homology Dependent Transgene Inactivation*

Homology dependent gene interactions can either occur between a gene and its allele on the homologous chromosome or between other duplicated or closely homologous gene sequences at unlinked loci. These have been referred to as "*trans*-interactions" and permit the understanding of natural gene interactions as well as the interaction of transgenes with the genome of the host plant (Dooner *et al.*, 1991; Matzke and Matzke, 1993). In the case of normal alleles these interactions can be either allelic or nonallelic/ ectopic (Fig. 2). A good example of an allelic interaction that influences gene expression is that of paramutation (Brink, 1973). Here the alteration in an allele persisted long after the *trans*-interaction and even after the alleles have segregated. There is a greater realization that the stable pairing or even transitory contact between alleles could leave some functional imprint that could have a profound effect on the expression pattern of genes in *Drosophila*

(Tartof and Henikoff, 1991) and animals (Monk, 1990). One could speculate that similar effects could result from the *trans*-interaction of a transgene and similar sequence present on the homologous chromosome in the genome of the host plant. In reality, however, allelic interactions have only been characterized for endogenous genes.

Ectopic or nonallelic interactions involve the interactions of homologous sequences at unlinked loci or present on non homologous chromosomes (Fig. 2). Such ectopic interactions were observed when tobacco plants were sequentially transformed with two T-DNAs that shared homologous regions that resulted in the inactivation of transgene expression from one of the T-DNAs (Matzke *et al.*, 1989; Matzke and Matzke, 1990, 1991). This suppression or inactivation of transgene expression could be correlated with the methylation of the promoter sequences (Matzke *et al.*, 1989). Another fine example of this phenomenon was the co-suppression of the anthocyanin pathway in petunia that were transformed with the gene encoding chalcone synthase (CHS) in an attempt to overproduce the color pigment (Napoli *et al.*, 1990). Similar results in the anthocyanin pathway were also obtained with the introduction of another gene in the same pathway encoding dihydroflavonol-4-reductase (DFR) in petunia (Van der Krol *et al.*, 1990). In both cases it was demonstrated that only the homologous genes were suppressed and that the suppression was not the result of the expression of the endogenous gene (Napoli *et al.*, 1990; Van der Krol *et al.*, 1990). This particular form of coordinate suppression of transgene and both alleles of the homologous endogenous gene and apparently no other gene in the plant is referred to as "co-suppression" (Jorgensen, 1990). It is clear from the above discussion that once a transgene enters in the nucleus its function will be influenced by other genes and that complex interactions between different loci can occur that will have a profound effect upon the expression of the transgene.

5.4. *Transcriptional Regulation of Expression of Transgenes*

Transcription in plants requires specific *cis* acting elements that will then interact with the corresponding *trans* acting factors in the target plant to correctly initiate transcription. Much is known about the mechanism and general principles of gene regulation in plants (Okamuro and Goldberg, 1989). Although much still remains to be learned about the mechanism of the initiation of transcription and the regulation of gene expression in plants, there are several excellent reviews on the subject (Benfey and Chua, 1989, 1990; Edwards and Coruzzi, 1990; Okamuro and Goldberg, 1989; Thompson and White, 1991; Weising *et al.*, 1988). As more genes are found in plants the choice of regulatory sequences to potentially regulate transgenes is increasing. The 35S promoter from cauliflower mosaic virus (CaMV) represents one of the most widely used promoters for the expression of transgenes in plants. Initial use of this promoter in chimeric genes showed that it conferred high-level expression in most cell types from virtually any species tested

(Weising *et al.*, 1988; Benfey and Chua, 1990). As a result, use of this promoter for biotechnology purposes is pervasive, and it has been intensively studied as a model for promoter function in plant cells (Benfey and Chua, 1990).

5.5. *Post-Transcriptional Regulation of Expression of Transgenes*

Gene expression is a complex, multi-step process which includes transcription, transcript processing and stability, mRNA stability and translatability and finally protein function and turnover. It is not surprising that regulation at any one of these levels could have a profound effect on the accumulation of the transgene product. There is a growing appreciation for the different cellular processes involved in the post-transcriptional regulation of transgenes. The various elements involved in the post-transcriptional regulation of genes in plants has been reviewed recently (Gallie, 1993). Part of the problem comes from the use of any gene from any source as a potential transgene for plants. Coding regions of genes from sources other than plants could have sequences that promote mRNA instability or protein turnover. It is now apparent that mRNA turnover in eukaryotes is a highly regulated process (Sachs, 1993). Highly conserved sequences have been identified that can cause rapid turnover of reporter transcripts in tobacco (Newman *et al.*, 1993).

Like other organisms plants have a preferred set of codons of the genetic code that are used in the coding regions of plant genes. Dicot plants prefer a set of 44 codons while monocots are more restrictive preferring a set of 38 codons (Campbell and Gowri, 1989). This preference for a particular set of codons was shown to be the key limiting factor in the expression of the genes encoding insecticidal crystal proteins of *Bacillus thuringiensis* in cotton (Perlak *et al.*, 1991b) and walnut (Dandekar *et al.*, 1994). Chemically synthesized genes with alterations in their codon usage demonstrated a much higher level of expression resulting in the mortality of both tobacco budworm and cotton bollworm, *Helicoverpa zea* (Boddie) (Perlak *et al.*, 1991a,b). Protein stability is another potential factor that needs attention and could be a rate limiting step for some transgene products. Protein degradation is quite a complex process and has been reviewed recently (Vierstra, 1993).

6. Field Introduction and Performance

The ultimate test for a transgenic plant is its performance in the field. The field trial not only permits one to determine the genetic stability and inheritance of the introduced trait but also the evaluation of other agricultural characteristics. There are many traits that are pertinent to field performance such as yield and quality. In instances where the transgenic plants have been produced by introducing the traits into existing commercial cultivars

Table 6. Number of field release permits issued by APHIS for the field testing of transgenic plants in the United States from 1988 to 1993.

Plant	No. of Permits	Plant	No. of Permits
Corn	125	Sugarbeet	3
Tomato	92	Apple	2
Soybean	87	Poplar	2
Potato	76	Papaya	1
Cotton	54	Plum	1
Tobacco	46	Chrysanthemum	1
Melon and squash	27	*Agrostis palustris*	1
Rapeseed	19	Allegheny Serviceberry	1
Alfalfa	13	Beets	1
Rice	6	Carrots	1
Cucumber	5	Peanut	1
Walnut	3	Petunia	1
Sunflower	3	Sorbus sp.	1
Lettuce	3		

Total number of permits issued = 576

Information courtesy of the Biotechnology, Biologics and Environmental Protection Biotechnology Permits Unit of the US Department of Agriculture Animal Plant Health Inspection Service (USDA/APHIS).

additional breeding may not be necessary. The emphasis of a field trial would be to evaluate the introduced trait, determine if any changes have occurred as a result of the transformation or regeneration process to other production traits. In cases where the useful genes have been introduced into relatives of commercial cultivars the field trial is vital to obtain transgenic plants that can become part of conventional breeding programs where crosses can be performed to introgress the transgenes into genetic backgrounds more appropriate for commercial production.

All field introductions require prior regulatory approval. In the United States the regulatory agency that oversees field trials is the United States Department of Agriculture's, Animal Plant Health Inspection Service (USDA/APHIS, or simply APHIS). Over 576 field releases have been approved by APHIS in the duration 1988 to 1993. These field trials involve 27 different plant commodities (Table 6). The list of different transgenic plant species actually tested in the field is much smaller than the list of transgenic plants that have been successfully obtained in the laboratory (Tables 1 to 4). The number of permits approved has been steadily increasing over the past several years since the first one was approved in 1987. The majority of applicants for these permits in the U.S. have been plant biotechnology companies or the commercial sector (83%), followed by Universities (12%) and the USDA (5%). About 83% of these permits can be accounted for by just 6 crops – corn, tomato, potato, cotton, soybean and tobacco, with the maximum number for corn. Calgene's "Flavr Savr" tomato will be the first genetically engineered plant product and is expected to enter the U.S. market place in 1994. It is not difficult to predict from Table 6 the crops that represent the future of genetically engineered products for the market place. The field release is the most significant test of gene transfer technology and is an essential component of a complete transformation program with commercial intentions.

7. Conclusions

Over the past decade that has elapsed since the first successful transformation experiments were performed in an angiosperm the different species where transgenic plants have been obtained has increased dramatically. The barriers to produce transgenic monocots are breaking down rapidly. It is inevitable that methods for gene transfer will be available for the introduction of transgenes into the tissues of most if not all angiosperm species. The challenges will be in two areas: 1) regeneration of transgenic plants from transformed cells and 2) regulation of transgene expression. In this review some of the challenges in these two areas have been outlined. It will be important that research occur in a wide variety of different angiosperm species as this would provide a diversity of research and options to overcome some of these difficulties.

Acknowledgements

I thank Henry Fisk and Peter Schuerman for their valuable input in this manuscript and Sandie Uratsu for her assistance.

References

Al-Juboory, K.H. and R.M. Skirvin, 1991. *In vitro* regeneration of *Agrobacterium*-transformed sweet potato (*Ipomoea batatas* L.). J. Plant Growth Regulator Society of America 19: 82–89.

Allen, G.C., G.E. Hall, Jr., J.C. Childs, A.K. Weissinger, S. Spiker and W.F. Thompson, 1993. Scaffold attachment regions increase reporter gene expression in stably transformed plant cells. Plant Cell 5: 603–613.

Alt-Mörbe, J., H. Kühlmann and J. Schröder, 1989. Differences in induction of Ti plasmid virulence genes virG and virD, and continued control of *vir*D expression by four external factors. Mol. Plant-Microbe Inter. 2: 301–308.

Alt-Mörbe, J., P. Neddermann, J. von Lintig, E.W. Weiler and J. Schröder, 1988. Temperature-sensitive step in Ti plasmid *vir*-region induction and correlation with cytokinin secretion by *Agrobacteria*. Mol. Gen. Genet. 213: 1–8.

Amasino, R.M., A.L.T. Powell and M.P. Gordon, 1984. Changes in T-DNA methylation and expression are associated with phenotypic variation and plant regeneration in a crown gall tumor line. Mol. Gen. Genet. 197: 437–446.

Ammirato, P.V., 1985. Patterns of development in culture. In: R.R. Henke, K.W. Hughes, M.J. Constantin and A. Hollaender (Eds.), Tissue Culture in Forestry and Agriculture, pp. 57–81. Plenum Press, New York.

Ankenbauer, R.G. and E.W. Nester, 1990. Sugar-mediated induction of *Agrobacterium tumefaciens* virulence genes: structural specificity and activities of monosaccharides. J. Bacteriol. 172:6442–6446.

An, G., B.D. Watson and C.C. Chiang, 1986. Transformation of tobacco, tomato, potato and *Arabidopsis thaliana* using a binary Ti vector system. Plant Physiol. 81: 301–305.

Armstrong, T.A. and M.A.W. Hinchee, 1990. Analysis of damage to plant tissue caused by particle gun transformation. In Vitro 26: 44A.

Atkinson, R.G. and R.C. Gardner, 1991. *Agrobacterium*-mediated transformation of pepino and regeneration of transgenic plants. Plant Cell Rep. 10: 208–212.

Attree, S.M. and L.C. Fowke, 1993. Embryogeny of gymnosperms: advances in synthetic seed technology of conifers. Plant Cell Tiss. Org. Cult. 35: 1–35.

Barfield, D.G. and E.-G. Pua, 1991. Gene transfer in plants of *Brassica juncea* using *Agrobacterium tumefaciens*-mediated transformation. Plant Cell Rep. 10: 308–314.

Basiran, N., P. Armitage, R.J. Scott and J. Draper, 1987. Genetic transformation of flax (*Linum usitatissimum*) by *Agrobacterium tumefaciens*: Regeneration of transformed shoots via a callus phase. Plant Cell Rep. 6: 396–399.

Benfey, P.N. and N.-H. Chua, 1989. Regulated genes in transgenic plants. Science 244: 174–181.

Benfey, P.N. and N.-H. Chua, 1990. The cauliflower mosaic virus 35S promoter: combinatorial regulation of transcription in plants. Science 250: 959–966.

Berger, F., A. Taylor and C. Brownlee, 1994. Cell fate determination by the cell wall in early Fucus development. Science 263: 1421–1423.

Berthomieu, P. and L. Jouanin, 1992. Transformation of rapid cycling cabbage (*Brassica oleracea* var. *capitata*) with *Agrobacterium rhizogenes*. Plant Cell Rep. 11: 334–338.

Bird, A., 1992. The essentials of DNA methylation. Cell 70: 5–8.

Bochardt, A., L. Hodal, G. Palmgren, O. Mattsson and F.T. Okkels, 1992. DNA methylation is involved in maintenance of an unusual expression pattern of an introduced gene. Plant Physiol. 99: 409–414.

Bolton, G.W., E.W. Nester and M.P. Gordon, 1986. Plant phenolic compounds induce expression of *A. tumefaciens* loci needed for virulence. Science 232:983–985.

Bower, P. and R.G. Birch, 1992. Transgenic sugarcane plants via microprojectile bombardment. Plant J. 2: 409–416.

Braun, A.C. and R.J. Mandle, 1948. Studies on the inactivation of the tumor-inducing principle in crown gall. Growth 12: 255–269.

Braun, A.C., 1952. Conditioning of the host cell as a factor in the transformation process in crown gall. Growth 16:65–74.

Breyne, P., M. Van Montagu, A. Depicker and G. Gheysen, 1992. Characterization of a plant scaffold attachment region in a DNA fragment that normalizes transgene expression in tobacco. Plant Cell 4: 463–471.

Brillanceau, M.H., C. David and J. Tempe, 1989. Genetic transformation of *Catharanthus roseus* G. Don by *Agrobacterium rhizogenes*. Plant Cell Rep. 8: 63–66.

Brink, R.A., 1973. Paramutation. Ann. Rev. Genet. 7: 129–152.

Button, J., J. Kochba and C.H. Bornman, 1974. Fine structure of and embryod development from embryogenic ovular callus of "Shamouti" orange (*Citrus sinensis* Osb.). J. Expt. Bot. 25: 446–457.

Bytebier, B., F. Deboeck, H. De Greve, M. Van Montagu and J.-P. Hernalsteens, 1987. T-DNA organization in tumor cultures and transgenic plants of monocotyledon *Asparagus officinalis* Proc. Natl. Acad. Sci. USA 84: 5345–5349.

Campbell, W.H. and G. Gowri, 1989. Codon usage in higher plants, green algae, and cyanobacteria. Plant Physiol. 92: 1–11.

Cangelosi, G.A., R.G. Ankenbauer and E.W. Nester, 1990. Sugars induce the *Agrobacterium* virulence genes through a periplasmic binding protein and a transmembrane signal protein. Proc. Natl. Acad. Sci. USA 87: 6708–6712.

Castle, L.A. and R.O. Morris, 1990. A method for the early detection of T-DNA transfer. Plant Mol. Biol. Rep. 8: 28–39.

Catlin, D., O. Ochoa, S. McCormick and C.F. Quiros, 1988. Celery transformation by *Agrobacterium tumefaciens*: cytological and genetic analysis of transgenic plants. Plant Cell Rep. 7: 100–103.

Cedar, H., 1988. DNA methylation and gene activity. Cell 53: 3–4.

Chee, P.P., 1990. Transformation of *Cucumis sativus* tissue by *Agrobacterium tumefaciens* and the regeneration of transformed plants. Plant Cell Rep. 9: 245–248.

Chilton, M.D., D.A. Tepfer, A. Petit, C. David, F. Casse-Delbart and J. Tempe, 1982. *Agrobacterium rhizogenes* injects T-DNA into the genomes of the host-plant root cells. Nature 295: 432–434.

Chriqui, D., C. David and S. Adam, 1988. Effect of the differentiated or dedifferentiated state of tobacco pith tissue on its behavior after inoculation with *Agrobacterium rhizogenes*. Plant Cell Rep. 7: 111–114.

Christou, P., 1992. Genetic transformation of crop plants using microprojectile bombardment. Plant J. 2: 275–281.

Christou, P., T.L. Ford and M. Kofron, 1991. Production of transgenic rice (*oryza sativa* L.) plants from agronomically important *indica* and *japonica* via electric discharge particle acceleration of exogenous DNA into immature zygotic embryos. Bio/Technology 9: 957–962.

Colby, S.M., A.M. Juncosa and C.P. Meredith, 1991. Cellular differences in *Agrobacterium* susceptibility and regenerative capacity restrict the development of transgenic grapevines. J. Am. Soc. Hort. Sci. 116: 356–361.

Comai, L., D. Facciotti, W.R. Hiatt, G. Thompson, R.E. Rose and D.E. Stalker, 1985. Expression in plants of a mutant *aro*A gene from *Salmonella typhimurium* confers tolerance to glyphosate. Nature 317: 741–744.

D'Amato, F., 1985. Cytogenetics of plant cell and tissue cultures and their regenerates. CRC Crit. Rev. Plant Sci. 3: 73–112.

Damiani, F. and S. Arcioni, 1991. Transformation of *Medicago arborea* L. with an *Agrobacterium rhizogenes* binary vector carrying the hygromycin resistance gene. Plant Cell Rep. 10: 300–303.

Dandekar, A.M., L.A. Martin and G.H. McGranahan, 1988. Genetic transformation and foreign gene expression in walnut tissue. J. Am. Soc. Hort. Sci. 113: 945–949.

Dandekar, A.M., G.H. McGranahan and D.J. James, 1993. Transgenic woody plants. In: S.D. Kung and R. Wu (Eds.), Transgenic Plants, Vol. 2, pp. 129–151. Academic Press, Inc. New York.

Dandekar, A.M., G.H. McGranahan, C.A. Leslie and S.L. Uratsu, 1989. *Agrobacterium*-mediated transformation of somatic embryos as a method for the production of transgenic plants. J. Tissue Cult. Meth. 12: 145–150.

Dandekar, A.M., S.L. Uratsu and N. Matsuta, 1990. Factors influencing virulence in *Agrobacterium*-mediated transformation of apple. Acta Hort. 280: 483–494.

Dandekar, A.M., G.H. McGranahan, P.V. Vail, S.L. Uratsu, C.A. Leslie, J.S. Tebbets and D.J. Hoffman, 1994. Low levels of expression of *cry*IA(c) sequences of *Bacillus thuringiensis* in transgenic walnut. Plant Sci. 96: 151–162.

David, C. and J. Tempe, 1988. Genetic transformation of cauliflower (*Brassica oleracea* L. var. *Botrytis*) by *Agrobacterium rhizogenes*. Plant Cell Rep. 7: 88–91.

Davis, M.E., A.R. Miller and R.D. Lineberger, 1991. Temporal competence for transformation of *Lycopersicon esculentum* (L. Mill.) cotyledons by *Agrobacterium tumefaciens*: relation to wound-healing and soluble plant factors. J. Expt. Bot. 42: 359–364.

Deak, M., G.B. Kiss, C. Konez and D. Dudits, 1986. Transformation of *Medicago* by *Agrobacterium* mediated gene transfer. Plant Cell Rep. 5: 97–100.

Dean, C., J. Jones, M. Favreau, P. Dunsmuir and J. Bedbrook, 1988. Influence of flanking

sequences on variability in expression levels of an introduced gene in transgenic tobacco plants. Nucleic Acids Res. 16: 9267–9283.
De Almeida, E.R.P., V. Gossele, C.G. Muller, J. Dockx, A. Reynaerts, J. Botterman, E. Krebbers and M.P. Timko, 1989. Transgenic expression of two marker genes under the control of an *Arbidopsis rbc*S promoter: sequences encoding the Rubisco transit peptide increase expression levels. Mol. Gen. Genet. 218: 78–86.
De Block, M., 1990. Factors influencing the tissue culture and the *Agrobacterium tumefaciens*-mediated transformation of hybrid Aspen and Poplar clones. Plant Physiol. 93: 1110–1116.
De Block, M., J. Botterman, M. Vanderwiele, J. Dockx, C. Thoen, V. Gossele, N. Rao Movva, C. Thompson, M. Van Montagu and J. Lee Mans, 1987. Engineering herbicide resistance in plants by expression of a detoxifying enzyme. EMBO J. 6: 2513–1518.
De Block, M., L. Herrera-Estrella, M. Van Montagu, J. Schell and P. Zambryski, 1984. Expression of foreign genes in regenerated plants and in their progeny. EMBO J. 3: 1681–1689.
De Cleene, M. and J. De Ley, 1976. The host range of crown gall. Bot. Rev. 42: 389–466.
De Cleene, M. and J. De Ley, 1981. The host range of infectious hairy root. Bot Rev. 47: 147–194.
De Kathen, A. and H.-J. Jacobsen, 1990. *Agrobacterium tumefaciens*-mediated transformation of *Pisum sativum* L. using binary and cointegrate vectors. Plant Cell Rep. 9: 276–279.
De Greve, H., J. Leemans, J.-P. Hernalsteens, L. Thia-Toong, M. DeBeuckeleer, L. Willmitzer, L. Otten, M. Van Montagu and J. Schell, 1982. Regeneration of normal and fertile plants that express octopine synthase, from tobacco crown galls after deletion of tumor-controlling functions. Nature 300: 752–755.
Dekeyser, R.A., B. Claes, R.M.U. De Rycke, M.E. Habets, M.C. Van Montagu and A.B. Caplan, 1990. Transient gene expression in intact and organized rice tissues. Plant Cell 2: 591–602.
De la Pena, A., H. Lorz and J. Schell, 1987. Transgenic rye plants obtained by injecting DNA into young floral tillers. Nature 325: 274–276.
Deshayes, A., L. Herrera-Estrella and M. Caboche, 1985. Liposome-mediated transformation of tobacco mesophyll protoplasts by an *Escherichia coli* plasmid. EMBO J. 4: 2731–2737.
D'Halluin, K., E. Bonne, M. Bossut, M. De Beuckeleer and J. Leemans, 1992. Transgenic maize plants by tissue electroporation. Plant Cell 4: 1495–1505.
Dooner, H.K., T.P. Robbins and R.A. Jorgensen, 1991. Genetic and developmental control of anthocyanin biosynthesis. Ann. Rev. Genet. 25: 173–199.
Eapen, S., F. Kohler, M. Gerdemann and O. Schieder, 1987. Cultivar dependence of transformation rates in moth bean after co-cultivation of protoplasts with *Agrobacterium tumefaciens*. Theor. Appl. Genet. 75: 207–210.
Edwards J.W. and G.M. Coruzzi, 1990. Cell-specific gene expression in plants. Annu. Rev. Genet. 24: 275–303.
Eichholtz, D.A., S.G. Rogers, R.B. Horsch, H.J. Klee, M. Hayford, N. Hoffmann, S. Braford, C. Fink, J. Flick, K. O'Connell and R. Fraley, 1987. Expression of mouse dihydrofolate reductase gene confers methotrexate resistance in transgenic plants. Somatic Cell Mol. Genet. 13: 67–76.
El Khalifa, M.D. and J.A. Lippincott, 1967. The influence of plant-growth factors on the initiation and growth of crown-gall tumours on primary pinto bean leaves. J. Exp. Bot. 19: 749–759.
Everett, N.P., K.E.P. Robinson and D. Mascarenhas, 1987. Genetic engineering of sunflower (*Helianthus annus* L.). Bio/Technology 5: 1201–1204.
Fang, G. and R. Grumet, 1990. *Agrobacterium tumefaciens* mediated transformation and regeneration of muskmelon plants. Plant Cell Rep. 9: 160–164.
Fillatti, J.J., J. Sellmer, B. McCown, B. Ilaissig and L. Comai, 1987. *Agrobacterium* mediated transformation and regeneration of *Populus*. Mol. Gen. Genet. 206: 192–199.
Finer, J.J. and M.D. McMullen, 1990. Transformation of cotton (*Gossypium hirsutum* L.) via particle bombardment. Plant Cell Rep. 8: 586–589.

Firoozabady, E., D.L. DeBoer, D.J. Merlo, E.L. Halk, L.N. Amerson, K.E. Rashka and E.E. Murray, 1987. Transformation of cotton (*Gossypium hirsutum* L.) by *Agrobacterium tumefaciens* and regeneration of transgenic plants. Plant Mol. Biol. 10: 105–116.

Fitch, M.M.M., R.M. Manshardt, D. Gonsalves, J.L. Slightom and J.C. Sanford, 1990. Stable transformation of papaya via microprojectile bombardment. Plant Cell Rep. 9: 189–194.

Fitch, M.M.M., R.M. Manshardt, D. Gonsalves and J.L. Slightom, 1993. Transgenic papaya plants from *Agrobacterium*-mediated transformation of somatic embryos. Plant Cell Rep. 12: 245–249.

Fraley, R.T., S.G. Rogers, R.B. Horsch, P.R. Sanders, J.S. Flick, S.P. Adams, M.L. Bittner, L.A. Brand, C.L. Fink, J.S. Fry, G.R. Galluppi, S.B. Goldberg, N.L. Hoffmann and S.C. Woo, 1983. Expression of bacterial genes in plant cells. Proc. Natl. Acad. Sci. USA 80: 4803–4807.

Fraley, R.T., S.G. Rogers, R.B. Horsch, D.A. Eichholtz, J.S. Flick, C.L. Fink, N.L. Hoffmann and P.R. Sanders, 1985. The SEV system: a new disarmed Ti plasmid vector system for plant transformation. Bio/Technology 3: 629–635.

Fromm, M., J. Callis, L.P. Taylor and V. Walbot, 1987. Electroporation of DNA and RNA into plant protoplasts. Methods Enzymol. 153: 351–366.

Fromm, M., L.P. Taylor and V. Walbot, 1985. Expression of genes transferred into monocot and dicot cells by electroporation. Proc Natl. Acad Sci. USA 82: 5824–5827.

Fry, J., A. Barnason and R.B. Horsch, 1987. Transformation of *Brassica napus* with *Agrobacterium tumefaciens* based vectors. Plant Cell Rep. 6: 321–325.

Gallie, D.R., 1993. Posttranscriptional regulation of gene expression in plants. Ann. Rev. Plant Physiol. Plant Mol. Biol. 44: 77–105.

Gelvin, S.B., 1990. Crown gall disease and hairy root disease. Plant Physiol. 92: 281–285.

Godwin, I., G. Todd, B. Ford-Lloyd and H.J. Newbury, 1991. The effects of acetosyringone and pH on agrobacterium-mediated transformation vary according to plant species. Plant Cell Rep. 9: 671–675.

Golds, T.J., J.Y. Lee, T. Husnain, T.K. Ghose and M.R. Davey, 1991. *Agrobacterium rhizogenes* mediated transformation of the forage legumes *Medicago sativa* and *Onobrychis viciifolia*. J. Exp. Bot. 42: 1147–1157.

Gordon-Kamm, W.J., M. Spencer, M.L. Mangano, T.R. Adams, R.J. Daines, W.G. Start, J.V. O'Brien, S.A. Chambers, W.R. Adams, Jr., N.G. Willetts, T.B. Rice, C.J. Mackey, R.W. Krueger, A.P. Kausch and P.G. Lemaux, 1990. Transformation of maize cells and regeneration of fertile transgenic plants. Plant Cell 2: 603–618.

Gould, J., M. Devey, O. Hasegawa, E.C. Ulian, G. Peterson and R.H. Smith, 1991. Transformation of *Zea mays* L. using *Agrobacterium tumefaciens* and the shoot apex. Plant Physiol. 95: 426–434.

Graham, J. and R.J. McNicol, 1991. Regeneration and transformation of *Ribes*. Plant Cell Tiss. Org. Cult. 24: 91–95.

Graham, J., R.J. McNicol and A. Kumar, 1990. Use of the GUS gene as a selectable marker for *Agrobacterium*-mediated transformation of *Rubus*. Plant Cell Tiss. Org. Cult. 20: 35–39.

Grimsley, N., T. Hohn, J.W. Davies and B. Hohn, 1987. *Agrobacterium*-mediated delivery of infectious maize streak virus into maize plants. Nature 325: 177–179.

Gruenbaum, Y., T. Naveh-Mary, H. Cedar and A. Razin, 1981. Sequence specificity of methylation in higher plant DNA. Nature 292: 860–862.

Guerche, P., M. Charbonnier, L. Jouanin, C. Tourneur, J. Paszkowski and G. Pelletier, 1987a. Direct gene transfer by electroporation in *Brassica napus*. Plant Sci. 52: 111–116.

Guerche, P., L. Jouanin, D. Tepfer and G. Pelletier, 1987b. Genetic transformation of oilseed rape (*Brassica napus*) by the Ri T-DNA of *Agrobacterium rhizogenes* and analysis of inheritance of the transformed phenotype. Mol. Gen. Genet. 206: 382–386.

Guri, A. and K.C. Sink, 1988. Agrobacterium transformation of eggplant. J. Plant Physiol. 133: 52–55.

Hadfield, K.A., A.M. Dandekar and R.J. Romani, 1993. Demethylation of ripening specific genes in tomato fruit. Plant Sci. 92: 13–18.

Handa, T., 1992. Genetic transformation of *Antirrhinum majus* L. and inheritance of altered phenotype induced by Ri T-DNA. Plant Sci. 81: 199–206.
Haughn, G. and C. Somerville, 1986. Sulfonylurea-resistance mutants of *Arabidopsis thaliana*. Mol. Gen. Genet. 204: 430–434.
Hayford, M., J. Medford, N. Hoffman, S.G. Rogers and H. Klee, 1988. Development of a plant transformation selection system based on expression of genes encoding gentamycin acetyltransferase. Plant Physiol. 86: 1216–1222.
Helmer, G., M. Casadaban, M. Bevan, L. Kayes and M.-D Chilton, 1984. A new chimeric gene as a marker for plant transformation: the expression of *Escherichia coli* β-galactosidase in sunflower and tobacco cells. Bio/Technology 2: 520–527.
Hepburn, A.G., L.E. Clark, L. Pearson and F. White, 1983. The role of cytosine methylation in the control of nopaline synthase gene expression in a plant tumor. J. Mol. Appl. Genet. 2: 315–329.
Hepburn, A., F. Belanger and J. Mattheis, 1987. DNA methylation in plants. Dev. Gen. 8: 475–493.
Hernalsteens,, J.P., F. VanViet, M. DeBeuckeleer, A. Depicker, G. Engler, M. Lemmers, M. Holsters, M. Van Montagu and J. Schell, 1980. The *Agrobacterium tumefaciens* Ti plasmid as a host vector system for introducing foreign DNA in plant cells. Nature 287: 654–656.
Herrera-Estrella, L., M. De Block, E. Messens, J.-P. Hernalsteens, M. Van Montagu and J. Schell, 1983. Chimeric genes as dominant selectable markers in plant cells. EMBO J. 2: 987–995.
Hess, K.M., M.W. Dudley, D.G. Lynn, R.D. Joerger and A.N. Binns, 1991. Mechanism of phenolic activation of *Agrobacterium* virulence genes: Development of a specific inhibitor of bacterial sensor/response systems. Proc. Natl. Acad. Sci. USA 88: 7854–7858.
Hidaka, T., M. Omura, M. Ugaki, M. Tomiyama, A. Kato, M. Ohshima and F. Motoyoshi, 1990. *Agrobacterium*-mediated transformation and regeneration of *Citrus* spp. from suspension cells. Jpn. J. Breed. 40: 199–207.
Hille, J., F. Verheggen, P. Roelvink, H. Franssen, A. Van Kammen and P. Zabel, 1986. Bleomycin resistance, a new dominant selectable marker for plant transformation. Plant Mol. Biol. 7: 171–176.
Hinchee, M.A.W., D.V. Connor-Ward, C.A. Newell, R.E. McDonnell, S.J. Sato, C.S. Gasser, D.A. Fischhoff, D.B. Re, R.T. Fraley and R.B. Horsch, 1988. Production of transgenic soybean plants using *Agrobacterium*-mediated DNA transfer. Bio/Technology 6: 915–922.
Holliday, R., 1989. A different kind of inheritance. Sci. Amer. 260(6): 60–73.
Hood, E.A., G.L. Helmer, R.T. Fraley and M.-D. Chilton, 1986. The hypervirulence of *Agrobacterium tumefaciens* A281 is encoded in a region of pTiBo542 outside T-DNA. J. Bacteriol. 168: 1291–1301.
Hood, E.A., G. Jen, L. Kayes, J. Kramer, R.T. Fraley and M.D. Chilton, 1984. Restriction endonuclease map of pTiBo542, a potential Ti plasmid vector for genetic engineering of plants. Bio/Technology 2: 702–709.
Hooykaas, P.J.J. and R.A. Schilperoort, 1992. *Agrobacterium* and plant genetic engineering. Plant Mol. Biol. 19: 15–38.
Horn, M.E., R.D. Shillito, B.V. Conger and C.T. Harms, 1988. Transgenic plants of orchardgrass (*Dactylis glomerata* L.) from protoplasts. Plant Cell Rep. 7: 469–472.
Horsch, R.T., R.T. Fraley, S.G. Rogers, P.R. Sanders, A. Lloyd and N. Hoffmann, 1984. Inheritance of functional foreign genes in plants. Science 223: 496–498.
Horth, M., I. Negrutiu, A. Burny, M. Van Montagu and L. Herrera-Estrella, 1987. Cloning of a *Nicotiana plumbaginifolia* protoplast-specific enhancer-like sequence. EMBO J. 6: 2525–2530.
Hosoki, T., T. Kigo and K. Shiraishi, 1991. Transformation and plant regeneration of broccoli (*Brassica oleracea* var. italica) mediated by *Agrobacterium rhizogenes*. J. Jpn. Soc. Hort. Sci. 60: 71–75.
Hrouda, M. and M. Ondrej, 1983. The effect of plant growth regulators on formation of crown gall tumors on potato tuber disks. Biol. Plant. (Praha) 25: 28–32.

Huang, M.-L.W., G.A. Cangelosi, W. Halperin and E.W. Nester, 1990. A chromosomal *Agrobacterium tumefaciens* gene required for effective plant signal transduction. J. Bacteriol. 172: 1814–1822.

James, D.J., A.J. Passey and D.J. Barbara, 1990. Agrobacterium-mediated transformation of the cultivated strawberry (*Fragaria* × *anannassa* Duch.) using desarmed binary vectors. Plant Sci. 69: 79–94.

James, D.J., A.J. Passey, D.J. Barbara and M. Bevan, 1989. Genetic transformation of apple (*Malus pumila* Mill.) using a disarmed Ti-binary vector. Plant Cell Rep. 7: 658–661.

James, D.J., S.L. Uratsu, J. Cheng, P. Negri, P. Viss and A.M. Dandekar, 1993. Conditions that induce Agrobacterium *Vir* genes also enhance apple cell transformation. Plant Cell Rep. 12: 559–563.

Jarchow, E., N.H. Grimsley and B. Hohn, 1991. *vir*F, the host-range-determining virulence gene of *Agrobacterium tumefaciens*, affects T-DNA transfer in *Zea mays*. Proc. Natl. Acad. Sci. USA 88: 10426–10430.

Jefferson, R.A., 1987. Assaying chimeric genes in plants: the GUS gene fusion system. Plant Mol. Biol. Rep. 5: 387–405.

Jefferson, R.A., T.A. Kavanagh and M.W. Bevan, 1987. GUS fusion: β-glucuronidase as a sensitive and versatile gene marker in higher plants. EMBO J. 6: 3901–3907.

Jia, S.-R., M.-Z. Yang, R. Ott and N.-H. Chua, 1989. High frequency transformation of *Kalanchoe laciniata*. Plant Cell Rep. 8: 336–340.

Jin, S., T. Komari, M.P. Gordon and E.W. Nester, 1987. Genes responsible for the supervirulence phenotype of *Agrobacterium tumefaciens* A281. J. Bacteriol. 169: 4417–4425.

Joersbo, M. and J. Brunstedt, 1992. Sonication: a new method for gene transfer to plants. Physiol. Plant. 85: 230–234.

Joersbo, M. and J. Brunstedt, 1990. Direct gene transfer to plant protoplasts by mild sonication. Plant Cell Rep. 9: 207–210.

John, M.C. and R.M. Amasino, 1989. Extensive changes in DNA methylation patterns accompany activation of a silent T-DNA *ipt* gene in *Agrobacterium tumefaciens*-transformed plant cells. Mol. Cell. Biol. 9: 4298–4303.

Jones, J.D.G., Z. Svab, E.C. Harper, C.D. Hurwitz and P. Maliga, 1987. A dominant nuclear streptomycin resistance marker for plant cell transformation. Mol. Gen. Genet. 210: 86–91.

Jones; P.A., 1984. Gene activation by 5–azacytidine. In: A. Razin, H. Cedar and A.D. Riggs (Eds.), DNA Methylation and Expression: Biochemistry and Biological Significance, pp. 165–187. Springer-Verlag, New York.

Jorgensen, R., 1990. Altered gene expression in plants due to *trans*-interactions between homologous genes. Trends Biotechnol. 8: 340–344.

Jouanin, L., F. Vilaine, J. Tourneur, V. Pautot, J.-F. Muller and M. Caboche, 1987. Transfer of a 4.3 kb fragment of the TL-DNA of *Agrobacterium rhizogenes* strain A4 confers the pRi transformed phenotype to regenerated plants. Plant Sci. 53: 53–63.

Kado, C.I., 1991. Molecular mechanisms of crown gall tumorigenesis. Critical Rev. Plant Sci. 10: 1–32.

Kaeppler, S.M. and R.L. Phillips, 1993. DNA methylation and tissue culture-induced variation in plants. In Vitro Cell. Dev. Biol. 29P: 125–130.

Klee, H.J. and S.G. Rogers, 1989. Plant gene vectors and genetic transformation systems based on the use of *Agrobacterium tumefaciens*. In: J. Schell and I.K. Vasil (Eds.), Cell Culture and Somatic Cell Genetics of Plants, Vol. 6, pp. 2–23. Academic Press, Inc., New York.

Klein, T.M., R. Arentzen, P.A. Lewis and S. Fitzpatrick-McElligott, 1992. Transformation of microbes, plants and animals by particle bombardment. Bio/Technology 10: 286–291.

Klein, T.M., E.C. Harper, Z. Svab, J.C. Sanford, M.E. Fromm and P. Maliga, 1988. Stable genetic transformation of intact *Nicotiana* cells by the particle bombardment process. Proc. Natl. Acad. Sci. USA 85: 8502–8505.

Klein, T.M., E.D. Wolf, R. Wu and J.C. Sanford, 1987. High-velocity microprojectiles for delivering nucleic acids into living cells. Nature 327: 70–73.

Kobayashi, H., J. Ngernprasirtsiri and T. Akazawa, 1990. Transcriptional regulation and DNA methylationin plastids during conversion of chloroplasts. EMBO J. 9: 307–313.

Kohler, F., C. Golz, S. Eapen, H. Kohn and O. Schieder, 1987. Stable transformation of moth bean *Vigna aconitifolia* via direct gene transfer. Plant Cell Rep. 6: 313–317.

Koncz, C., O. Olsson, W.H.R. Langridge, J. Schell and A.A. Szalay, 1987. Expression and functional assembly of bacterial luciferase in plants. Proc. Natl. Acad. Sci. USA 84: 131–135.

Larkin P.J. and W.R. Scowcroft, 1981. Somaclonal variation. A novel source of variability from cell cultures for plant improvement. Theor. Appl. Genet. 60: 197–214.

Ledger, S.E., S.C. Deroles and N.K. Given, 1991. Regeneration and *Agrobacterium*-mediated transformation of *chrysanthemum*. Plant Cell Rep. 10: 195–199.

Lindsey, K. and P. Gallois, 1990. Transformation of sugar beet (*Beta vulgaris*) by *Agrobacterium tumefaciens*. J. Exp. Bot. 41: 529–536.

Litz, R.E. and D.J. Gray, 1992. Organogenesis and somatic embryogensis. In: F.A. Hammerschlag and R.E. Litz (Eds.), Biotechnology of Perennial Fruit Crops. C.A.B., pp. 3–34. International, UK.

Lloyd, A.M., A.R. Barnason, S.G. Rogers, M.C. Byrne, R.T. Fraley and R.B. Horsch, 1986. Transformation of *Arabidopsis thaliana* with *Agrobacterium tumefaciens*. Science 234: 464–466.

Lu, C-Y., G. Nugent, T. Wardley-Richardson, S.F. Chandler, R. Young and M.J. Dalling, 1991. *Agrobacterium*-mediated transformation of carnation (*Dianthus caryophyllus* L.). Bio/Technology 9: 864–868.

Luo, Z.-X. and R. Wu, 1988. A simple method for the transformation of rice via the pollen-tube pathway. Plant Mol. Biol. Rep. 6: 165–174.

Machado, M.C., A. da C. Machado, V. Hanzer, H. Weiss, F. Regner, H. Steinkellner, D. Mattanovich, R. Plail, E. Knapp, B. Kalthoff and H. Katinger, 1992. Regeneration of transgenic plants of *Prunus armeniaca* containing the coat protein gene of Plum Pox Virus. Plant Cell Rep. 11: 25–29.

Manners, J.M., 1988. Transgenic plants of the tropical pasture legume *Stylosanthes humilis*. Plant Sci. 55: 61–68.

Mante, S., P.H. Morgens, R. Scorza, J.M. Cordts and A.M. Callahan, 1991. Agrobacterium-mediated transformation of plum (*Prunus domestica* L.) hypocotyl slices and regeneration of transgenic plants. Bio/Technology 9: 853–857.

Martin, L., 1987. Genetic transformation and foreign gene expression in tissue of different woody species, M.S. Thesis. University of California, Davis.

Márt6n, L., G.J. Wullems, L. Molendijk and R.A. Schilperoort, 1979. *In vitro* transformation of cultured cells from *Nicotiana tabacum* by *Agrobacterium tumefaciens*. Nature 277: 129–130.

Matzke, M.A., M. Primig, J. Trnovsky and A.J.M. Matzke, 1989. Reversible methylation and inactivation of marker genes in sequentially transformed tobacco plants. EMBO J. 8: 643–649.

Matzke, M.A. and A.J.M. Matzke, 1990. Gene interactions and epigenetic variation in transgenic plants. Dev. Genet. 11: 214–223.

Matzke, M.A. and A.J.M. Matzke, 1991. Differential inactivation and methylation of a transgene in plants by two supressor loci containing homologous sequences. Plant Mol. Biol. 16: 821–830.

Matzke, M. and A.J.M. Matzke, 1993. Genomic imprinting in plants: parental effects and trans-inactivation phenomena. Ann. Rev. Plant Physiol. Plant Mol. Biol. 44: 53–76.

McCabe, D.E., W.F. Swain, B.J. Martinell and P. Christou, 1988. Stable transformation of soybean (*Glycine max*) by particle acceleration. Bio/Technology 6: 923–926.

McCormick, S., J. Niedermeyer, J. Fry, A. Barnason, R.B. Horsch and R.T. Fraley, 1986. Leaf disc transformation of cultivated tomato (*L. esculentum*) using *Agrobacterium tumefaciens*. Plant Cell Rep. 5: 81–84.

McCown, B.H., D.E. McCabe, D.R. Russell, D.J. Robison, K.A. Barton and K.F. Raffa,

1991. Stable transformation of *Populus* and incorporation of pest resistance by electric discharge particle acceleration. Plant Cell Rep. 9: 590–594.

McGranahan, G.H., C.A. Leslie, A.M. Dandekar, S.L. Uratsu and I.E. Yates, 1993. Transformation of pecan and regeneration of transgenic plants. Plant Cell Rep. 12: 634–638.

McGranahan, G.H., C.A. Leslie, S.L. Uratsu, L.A. Martin and A.M. Dandekar, 1988. Agrobacterium-mediated transformation of walnut somatic embryos and regeneration of transgenic plants. Bio/Technology 6: 800–804.

McHughen, A., M. Jordan and G. Feist, 1989. A preculture period prior to *Agrobacterium* inoculation increases production of transgenic plants. J. Plant Physiol. 135: 245–248.

McInnes, E., A.J. Morgan, B.J. Mulligan and M.R. Davey, 1991. Phenotypic effects of isolated pRiA4 TL-DNA *rol* genes in the presence of intact TR-DNA in transgenic plants of *Solanum dulcamara* L. J. Exp. Bot. 42: 1279–1286.

Melchers, L.S., T.J.G. Regensburgtuink, R.B. Bourret, N.J.A. Sedee, R.A. Schilperoort and P.J.J. Hooykaas, 1989. Membrane topology and functional analysis of the sensory protein *virA* of *Agrobacterium tumefaciens*. EMBO J. 8: 1919–1925.

Melchers, L.S., M.J. Maroney, A. Dendulkras, D.V. Thompson, H.A.J. Van Vauren, R.A. Schilperoort and P.J.J. Hooykaas, 1990. Octopine and nopaline strains of *Agrobacterium tumefaciens* differ in virulence – molecular characterization of the *virF* locus. Plant Mol. Biol. 14: 249–259.

Meyer, P., F. Linn, I. Heidmann, H. Meyer, I. Niedenhof and H. Saedler, 1992. Endogenous and environmental factors influence 35S promoter methylation of maize A1 gene construct in transgenic petunia and its colour phenotype. Mol. Gen. Genet. 231: 345–352.

Michelmore, R., E. Marsh, S. Seely and B. Landry, 1987. Transformation of lettuce (*Lactuca sativa*) mediated by *Agrobacterium tumefaciens*. Plant Cell Rep. 6: 439–442.

Miljus-Djukic, J., M. Neskovic, S. Ninkovic and R. Crkvenjokov, 1992. *Agrobacterium*-mediated transformation and plant regeneration of buckwheat (*Fagopyrum esculentum* Moench.). Plant Cell Tiss. Org. Cult. 29: 101–108.

Monacelli, B., M.M. Altamura, G. Pasqua, M.G. Biasini and F. Sala, 1988. The histogensis of somaclones from tomato (*Lycopersicon esculentum* Mill.) cotyledons. Protoplasma 142: 156–163.

Monk, M., 1990. Variation in epigenetic inheritance. Trends Genet. 96: 110–114.

Moore, G.A., C.C. Jacono, J.L. Neidigh, S.D. Lawrence and K. Cline, 1992. *Agrobacterium*-mediated transformation of *Citrus* stem segments and regeneration of transgenic plants. Plant Cell Rep. 11: 238–242.

Morris, J.W. and R.O. Morris, 1990. Identification of an *Agrobacterium tumefaciens* virulence gene inducer from the pinaceous gymnosperm *Pseudotsuga menziesii*. Proc. Natl. Acad. Sci. USA 87: 3614-3618.

Morris, J.W., L.A. Castle and R.O. Morris, 1989. Efficacy of different *Agrobacterium tumefaciens* strains in transformation of pinaceous gymnosperms. Physiol. Mol. Plant Path. 34: 451–461.

Morris, R.O., 1986. Genes specifying auxin and cytokinin biosynthesis in phytopathogens. Ann. Rev. Plant Physiol. 37: 509–538.

Mugnier, J., 1988. Establishment of new hairy root lines by inoculation with *Agrobacterium rhizogenes*. Plant Cell Rep. 7: 9–12.

Mullins, M.G., F.C.A. Tang and D. Facciotti, 1990. *Agrobacterium*-mediated genetic transformation of grapevines: transgenic plants of *Vitis rupestris* scheele and buds of *Vitis vinifera* L. Bio/Technology 8: 1041–1045.

Napoli, C., C. Lemieux and R. Jorgensen, 1990. Introduction of a chimeric chalcone synthase gene into petunia results in reversible co-supression of homologous genes in *trans*. Plant Cell 2: 279–289.

Narasimhulu, S.B., P.B. Kirti, T. Mohapatra, S. Prakash and V.L. Chopra, 1992. Shoot regeneration in stem explants and its amenability to *Agrobacterium tumefaciens* mediated gene transfer in *Brassica carinata*. Plant Cell Rep. 11: 359–362.

Nehra, S.N., R.N. Chibbar, K.K. Kartha, R.S.S. Datla, W.L. Crosby and C. Stushnoff,

1990. Genetic transformation of strawberry by *Agrobacterium tumefaciens* using a leaf disk regeneration system. Plant Cell Rep. 9: 293–298.

Neuhaus, G., G. Spangenberg, O.M. Scheid and H.-G. Schweiger, 1987. Transgenic rapeseed plants obtained by the microinjection of DNA into microspore-derived embryoids. Theor. Appl. Genet. 75: 30–36.

Newman, T.C., M.O. Takagi, C.B. Taylor and P.J. Green, 1993. DST sequences, highly conserved among plant SAUR genes, target reporter transcripts for rapid decay in tobacco. Plant Cell 5: 701–714.

Ngernprasirtsiri, J., H. Kobayashi and T. Akazawa, 1988. DNA methylation occurred around lowly expressed genes of plastid DNA during tomato fruit development. Plant Physiol. 88: 16–20.

Ngernprasirtsiri, J., H. Kobayashi and T. Akazawa, 1989. Transcriptional regulation and DNA methylation of nuclear genes for photosynthesis in nongreen plant cells. Proc. Natl. Acad. Sci. USA 86: 7919–792.

Noda, T., N. Tanaka, Y. Mano, S. Nabeshima, H. Ohkawa and C. Matsui, 1987. Regeneration of horseradish hairy roots incited by *Agrobacterium rhizogenes* infection. Plant Cell Rep. 6: 283–286.

Nyman, M. and A. Wallin, 1992. Transient gene expression in strawberry (*Fragaria* × *ananassa Duch.*) protoplasts and the recovery of transgenic plants. 1992. Plant Cell Rep. 11: 105–108.

Okamuro, J.K. and R.B. Goldberg, 1989. Regulation of plant gene expresion: general principles. In: A. Marcus (Ed.), The Biochemistry of Plants, Vol. 15, pp. 1–82. Academic Press, Inc., New York.

Ondrej, M. and R. Biskova, 1986. Differentiation of *Petunia hybrida* tissues transformed by *Agrobacterium rhizogenes* and *Agrobacterium tumefaciens*. Biol. Plant 28: 152–155.

Ondrej, M., M. Hrouda and P. Kostrica, 1989. Potato transformation by *Agrobacterium rhizogenes* Ri plasmid. Biol. Plant 31: 312–314.

Otten, L. and J. Schell, 1986. Nodule-specific expression of a chimeric soybean leghaemoglobin gene in transgenic *Lotus corniculatus*. Nature 321: 669–674.

Ow, D.W., K.V. Wood, M. DeLuca, J.R. DeWet, D.R. Helinski and S.H. Howell, 1986. Transient and stable expression of the firefly luciferase gene in plant cells and transgenic plants. Science 234: 856–859.

Owens, L.D. and D.E. Cress, 1985. Genotypic variability of soybean response to *Agrobacterium* strains harboring the Ti or Ri plasmids. Plant Physiol. 77: 87–94.

Paszkowski, J., R.D. Shillito, M. Saul, V. Mandak, T. Hohn, B. Hohn and I. Potrykus, 1984. Direct gene transfer of plants. EMBO J. 3: 2717–2722.

Pavingerova, D. and M. Ondrej, 1986. Comparison of hairy root and crown gall tumors of *Arabidopsis thaliana*. Biol. Plant 28: 149–151.

Perlak, F.J., R.L. Fuchs, D.A. Dean, S.L. McPherson and D.A. Fischoff, 1991a. Modification of the coding sequence enhances plant expression of insect control protein genes. Proc. Natl. Acad. Sci. USA 88: 3324–3328.

Perlak, F.J., R.W. Deaton, T.A. Armstrong, R.L. Fuchs, S.R. Sims, J.T. Greenplate and D.A. Fischhoff, 1991b. Insect resistant cotton plants. Bio/Technology 8: 939–943.

Phelep, M., A. Petit, L. Martin, E. Duhoux and J. Tempe, 1991. Transformation and regeneration of a nitrogen-fixing tree, *Allocasuarina verticillata* Lam. Bio/Technology 9: 461–466.

Pickardt, T., M. Meixner, V. Schade and O. Schieder, 1991. Transformation of *Vicia narbonensis* via *Agrobacterium*-mediated gene transfer. Plant Cell Rep. 9: 535–538.

Potrykus, I., 1991. Gene transfer to plants: assessment of published approaches and results. Ann. Rev. Plant Physiol. Plant Mol. Biol. 42: 205–225.

Pua, E.-C., A. Mehra-Palta, F. Nagy and N.-H. Chua, 1987. Transgenic plants of *Brassica napus* L. Bio/Technology 5: 815–817.

Puonti-Kaerlas, J., T. Eriksson and P. Engstrom, 1990. Production of transgenic pea (*Pisum sativum* L.) plants by *Agrobacterium tumefaciens*-mediated gene transfer. Theor. Appl. Genet. 80: 246–252.

Pythoud, F., V.P. Sinkar, E.W. Nester and M.P. Gordon, 1987. Increased virulence of *Agrobac-*

terium rhizogenes conferred by the *vir* region of pTiBO542: Application to genetic engineering of poplar. Bio/Technology 5: 1323–1327.
Razin, A. and H. Cedar, 1991. DNA methylation and gene expression. Microbiol. Rev. 55: 451–458.
Rhodes, C.A., D.A. Pierce, I.J. Mettler, D. Mascarenhas and J.J. Detmer, 1988. Genetically transformed maize plants from protoplasts. Science 240: 204–207.
Riggs, C.D. and G.W. Bates, 1986. Stable transformation of tobacco by electroporation: evidence for plasmid concatenation. Proc. Natl. Acad. Sci. USA 83: 5602–5606.
Rotino, G.L., D. Perrone, P. Ajmone-Marsan and E. Lupotto, 1992. Transformation of *Solanum integrifolium* poir via *Agrobacterium tumefaciens*: Plant regeneration and progeny analysis. Plant Cell Rep. 11: 11–15.
Ryder, M.H., M.E. Tate and A. Kerr, 1985. Virulence properties of strains of *Agrobacterium* on the apical and basal surfaces of carrot root discs. Plant Physiol. 77: 215–221.
Sachs, A.B., 1993. Messenger RNA degradation in eukaryotes. Cell 74: 413–421.
Saito, K., M. Yamazaki, H. Anzai, K. Yoneyama and I. Murakoshi, 1992. Transgenic herbicide-resistant *Atropa belladonna* using an Ri binary vector and inheritance of the transgenic trait. Plant Cell Rep. 11: 219–224.
Sanford, J.C., 1988. The biolistic process. TIBTECH 6: 299–302.
Sanford, J.C., 1990. Biolistic plant transformation. Physiol. Plant 79: 206–209.
Sangwan, R.S., C. Ducrocq and B.S. Sangwan-Norreel, 1991. Effect of culture conditions on *Agrobacterium*-mediated transformation in datura. Plant Cell Rep. 10: 90–93.
Schoelz, J.E., K.-B. Goldgerg and J. Kiernan, 1991. Expression of cauliflower mosaic virus (CaMV) gene VI in transgenic *Nicotiana bigelovii* complements a strain of CaMV defective in long-distance movement in nontransformed *N. bigelovii*. Mol. Plant-Microbe Interact. 4: 350–355.
Schuerman P.L. and A.M. Dandekar, 1993. Transformation of Temperate Crop Species: Progress and Potentials. Sci. Hort. 55: 101–124.
Schuerman, P. and A.M. Dandekar, 1991. Potentials of woody plant transformation. In: B.B. Biswas and J.R. Harris (Eds.), Subcellular Biochemistry, Vol. 19, Plant Genetic Engineering, pp. 81–105. Plenum Publ., New York.
Schwartz, D., 1989. Gene-controlled cytosine demethylation in the promoter region of the Ac transposable element in maize. Proc. Natl. Acad. Sci. USA 86: 2789–2793.
Scott, R.J. and J. Draper, 1987. Transformation of carrot tissues derived from proembryogenic suspension cells: A useful model system for gene expression studies in plants. Plant Mol. Biol. 8: 265–274.
Seki, M., N. Shigemoto, Y. Komeda, J. Imamura, Y. Yamada and H. Morikawa, 1991. Transgenic *Arabidopsis thaliana* plants obtained by particle-bombardment-mediated transformation. Appl. Microbiol. Biotechnol. 36: 228–230.
Serres, R., E. Stang, D. McCabe, D. Russell, D. Mahr and B. McCown, 1992. Gene transfer using electric discharge particle bombardment and recovery of transformed cranberry plants. J. Amer. Soc. Hort. Sci. 117: 174–180.
Shah, D.M., R.B. Horsch, H.J. Klee, G.M. Kishore, J.A. Winter, N.E. Tumer, C.M. Hironaka, P.R. Sanders, C.S. Gasser, S. Aykent, N.R. Siegel, S.G. Rogers and R.T. Fraley, 1986. Engineering herbicide tolerance in transgenic plants. Science 233: 478–481.
Shahin, E.A. and R.B. Simpson, 1986. Gene transfer system for potato. HortSci. 21: 1199–1201.
Shahin, E.A., A. Spielmann, K. Sukhapinda, R.B. Simpson and M. Yashar, 1986a. Cell Biology and Molecular Genetics: Transformation of cultivated alfalfa using disarmed *Agrobacterium tumefaciens*. Crop Sci. 26: 1235–1239.
Shahin, E.A., K. Sukhapinda, R.B. Simpson and R. Spivey, 1986b. Transformation of cultivated tomato by a binary vector in *Agrobacterium rhizogenes*: transgenic plants with normal phenotypes harbor binary vector T-DNA, but no Ri-plasmid T-DNA. Theor. Appl. Genet. 72: 770–777.

Shillito, R.D., M.W. Saul, J. Paszkowski, M. Muller and I. Potrykus, 1985. High frequency direct gene transfer to plants. Bio/Technology 4: 1099–1103.

Shimoda, N, A. Toyoda-Yamamoto, J. Nagamine, S. Usami, M. Katayama, Y. Sakagami and Y. Machida, 1990. Control of expression of *Agrobacterium vir* genes by synergistic actions of phenolic signal molecules and monosaccharides. Proc. Natl. Acad. Sci. USA 87: 6684–6688.

Smigocki, A.C. and F.A. Hammerschlag, 1991. Regeneration of plants from peach embryo cells infected with a shoot mutant strain of *Agrobacterium*. J. Amer. Soc. Hort. Sci. 116: 1092–1097.

Spena, A., A. Viotti and V. Pirrotta, 1983. Two adjacent zein sequences: structure organization and tissue-specific restriction pattern. J. Mol. Biol. 169: 799–811.

Srivastava, V., A.S. Reddy and S. Guha-Mukherjee, 1988. Transformation and regeneration of *Brassica oleracea* mediated by an oncogenic *Agrobacterium tumefaciens*. Plant Cell Rep. 7: 504–507.

Stalker, D.M., K.E. McBride and L.D. Malyj, 1988. Herbicide resistance in transgenic plants expressing a bacterial detoxification gene. Science 242: 419–423.

Steward, F.C., M.O. Mapes, A.E. Kent and R.D. Holsten, 1964. Growth and development of cultured plant cells. Science 143: 20–27.

Tartof, K.D. and S. Henikoff, 1991. *Trans*-sensing effects from *Drosophila* to humans. Cell 65: 201–203.

Taylor, B., R. Amasino, F. White, E.W. Nester and M.P. Gordon, 1985. T-DNA analysis of plants regenerated from hairy root tumors. Mol. Gen. Genet. 201: 554–557.

Tavazza, R., R.J. Ordas, M. Tavazza, G. Ancora and E. Benvenuto, 1988. Genetic transformation of *Nicotiana clevelandii* using a Ti plasmid derived vector. J. Plant Physiol. 133: 640–644.

Tempe, J. and F. Casse-Delbart, 1989. Plant gene vectors and genetic transformation: *Agrobacterium* Ri plasmids. In: J. Schell and I.K. Vasil (Eds.), Cell Culture and Somatic Cell Genetics of Plants, Vol. 6, pp. 25–49. Academic Press, Inc., San Diego.

Tepfer, D., 1984. Transformation of several species of higher plants by Agrobacterium rhizogenes: Sexual transmission of the transformed genotype and phenotype. Cell 37: 959–967.

Teutonico, R.A., M.W. Dudley, J.D. Orr, D.G. Lynn and A.N. Binns, 1991. Activity and accumulation of cell division-promoting phenolics in tobacco tissue cultures. Plant Physiol. 97: 288–297.

Thomas, M.R., R.J. Rose and K.E. Nolan, 1992. Genetic transformation of *Medicago truncatula* using *Agrobacterium* with genetically modified Ri and disarmed Ti plasmids. Plant Cell Rep. 11: 113–117.

Thompson, W.F. and M.J. White, 1991. Physiological and molecular studies of light-regulated nuclear genes in higher plants. Annu. Rev. Plant Physiol. Plant Mol. Biol. 42: 423–466.

Toriyama, K., Y. Arimoto, H. Uchimiya and K. Hinata, 1988. Transgenic rice plants after direct gene transfer into protoplasts. Bio/Technology 6: 10–1024.

Trulson, A.J., R.B. Simpson and E.A. Shahin, 1986. Transformation of cucumber (*Cucumis sativus* L.) plants with *Agrobacterium rhizogenes*. Theor. Appl. Genet. 73: 11–15.

Tulecke, W., 1987. Somatic embryogensis in woody prennials. In: J.M. Bonga and D.J. Durzan (Eds.), Cell and Tissue Culture in Forestry. Vol. 2, pp. 61–91. Martinus Nijhoff Publishers, Dordrecht.

Turk, S.C.H.J., L.S. Melchers, H. den Dulk-Ras, A.J.G. Regensburg-Tuïnk and P.J.J. Hooykaas, 1991. Environmental conditions differentially affect *vir* gene induction in different *Agrobacterium* strains. Role of the *Vir*A sensor protein. Plant Mol. Biol. 16: 1051–1059.

Uematsu, C., M. Murase, H. Ichikawa and J. Imamura, 1991. *Agrobacterium*-mediated transformation and regeneration of kiwi fruit. Plant Cell Rep. 10: 286–290.

Umbeck, P., G. Johnson, K. Barton and W. Swain, 1987. Genetically transformed cotton (*Gossypium hirsutum* L.) plants. Bio/Technology 5: 263–266.

Uratsu, S.L., H. Ahmadi, R.S. Bringhurst and A.M. Dandekar, 1990. Relative virulence of *Agrobacterium* strains on strawberry (*Fragaria vesca*). Hort. Sci. 26: 196–199.

Van der Krol, A.R., L.A. Mur, M. Beld, J.N.M. Mol and A.R. Stuitje, 1990. Flavonoid genes in petunia: addition of a limited number of gene copies may lead to supression of gene expression. Plant Cell 2: 291–299.

Vardi, A., S. Bleichman and D. Aviv, 1990. Genetic transformation of *Citrus* protoplasts and regeneration of transgenic plants. Plant Sci. 69: 199–206.

Vasil, V., A.M. Castillo, M.E. Fromm and I.K. Vasil, 1992. Herbicide resistant fertile transgenic wheat plants obtained by microprojectile bombardment of regenerable embryogenic callus. Bio/Technology 10: 667–674.

Veluthambi, K., M. Krishnan, J.H. Gould, R.H. Smith and S.B. Gelvin, 1989. Opines stimulate induction of the *vir* genes of the *Agrobacterium tumefaciens* Ti plasmid. J. Bacteriol. 171: 3969–3703.

Vermeulen, A., H. Vaucheret, V. Pautot and Y. Chupeau, 1992. *Agrobacterium* mediated transfer of a mutant *Arabidopsis* acetolactate synthase gene confers resistance to chlorsulfuron in chicory (*Cichorium intybus* L.). Plant Cell Rep. 11: 243–247.

Vernade, D., A. Herrera-Estrella, K. Wang and M. Van Montagu, 1988. Glycine betaine allows enhanced induction of the *Agrobacterium tumefaciens vir* genes by acetosyringone at low pH. J. Bacteriol. 170: 5822–5829.

Vierstra, R.D., 1993. Protein degradation in plants. Annu. Rev. Plant Physiol. Plant Mol. Biol. 44: 385–410.

Visser, R.G.F., E. Jacobsen, B. Witholt and W.J. Feenstra, 1989. Efficient transformation of potato (*Solanum tuberosum* L.) using a binary vector in *Agrobacterium rhizogenes*. Theor. Appl. Genet. 78: 594–600.

Waldron, C., E.B. Murphy, J.L. Roberts, G.D. Gustafson, S.L. Armour and S.K. Malcom, 1985. Resistance to hygromycin-B. Plant Mol. Biol. 5: 103–108.

Waltonn, N. and N. Belshaw, 1988. The effect of cadaverine on the formation of anabasine from lyserine in hairy root cultures of *Nicotiana hesperis*. Plant Cell Rep. 7: 115–118.

Wan, Y. and P.G. Lemaux, 1994. Efficient production of fertile transgenic barley plants. J. Cell. Biol. Suppl. 18A: 103.

Wang, Z-Y., T. Takamizo, V.A. Iglesias, M. Osusky, J. Nagel, I. Potrykus and G. Spangenberg, 1992. Transgenic plants of tall fescue (*Festuca arundinacea* Schreb.) obtained by direct gene transfer to protoplasts. Bio/Technology 10: 691–696.

Wann, S.R., 1988. Somatic embryogensis in woody species. Hort. Rev. 10: 153–181.

Watson, J.C., L.S. Kaufman and W.F. Thompson, 1987. Developmental regulation of cytosine methylation in the nuclear ribosomal RNA genes of Pisum sativum. J. Mol. Biol. 193: 15–26.

Weeks. T.L., O.D. Anderson and A.E. Blechl, 1993. Rapid production of multiple independent lines of fertile transgenic wheat (*Triticum aestivum*). Plant Physiol. 102: 1077–1084.

Wei, Z., H. Kamada and H. Harada, 1985. Transformation of *Solanum nigrum* L. protoplasts by *Agrobacterium rhizogenes*. Plant Cell Rep. 5: 93–96.

Weising, K., J. Schell and G. Kahl, 1988. Foreign genes in plants: transfer, structure, expression, and applications. Annu. Rev. Genet. 22: 421–477.

White, D.W.R. and D. Greenwood, 1987. Transformation of the forage legume *Trifolium repens* L. using binary *Agrobacterium* vectors. Plant Mol. Biol. 8: 461–469.

Williams, E.G. and G. Maheswaran, 1986. Somatic embryogensis: Factors influencing coordinated behaviour of cells as an embryogenic group. Ann. Bot. 57: 443–462.

Zambryski, P., 1988. Basic processes underlying *Agrobacterium*-mediated DNA transfer to plant cells. Annu. Rev. Genet. 22: 1–30.

Zambryski, P.C., 1992. Chronicles from the *Agrobacterium*-plant cell DNA transfer story. Ann. Rev. Plant Physiol Plant Mol. Biol. 43: 465–490.

Zambryski, P., H. Joos, C. Genetello, J. Leemans, M. Van Montagu, and J. Schell, 1983. Ti plasmid vector for the introduction of DNA into plant cells without alteration of their normal regeneration capacity. EMBO J. 2: 2143–2150.

Zambryski, P., J. Tempe and J. Schell, 1989. Transfer and function of T-DNA genes from *Agrobacterium* Ti and Ri plasmids in plants. Cell 56: 193–201.

Zhang, L-J., L-M. Cheng, N. Xu, N.-M. Zhao, C.-G., Li, J. Yuan and S.-R. Jia, 1991. Efficient transformation of tobacco by ultrasonication. Bio/Technology 9: 996–997.
Zhang, W. and R. Wu, 1988. Efficient regeneration of transgenic plants from rice protoplasts and correctly regulated expression of the foreign gene in the plants. Theor. Appl. Genet. 76: 835–840.
Zyprian, E. and C.I. Kado, 1990. Agrobacterium-mediated plant transformation by novel mini-T vectors in conjunction with a high-copy *vir* region helper plasmid. Plant Mol. Biol. 15: 245–256.

11. Transformation of Gymnosperms

David D. Ellis

Contents

1. Introduction

Genetic improvement programs exist for only a handful of the genera within the gymnosperms and in contrast to the angiosperms, these programs are still in their infancy. Wide-scale selection from wild populations was done centuries ago for angiosperm trees, due mostly to their importance as fruit, nut, or oil crops. In contrast, with the exception of ornamental uses, selections for superior individuals within the gymnosperms has been limited to this century. Because of this and the long generation time required for tree breeding, few progenies greater than three generations exist. Genetic engineering offers a significant tool to breeders interested in improving gymnosperms because no other improvement strategies could have such a large potential impact in a relatively short period of time. Unfortunately, much of this potential remains theoretical as gene transfer systems for gymnosperms have only been successful in two gymnosperm genera, *Larix* (Huang *et al.*, 1991) and *Picea* (Ellis *et al.*, 1993). However recent successes with the regeneration of transformed conifers by new methods, such as pollen transformation (McCabe, personal communication) and by multiple research groups (Charest, personal communication; Tsang, personal communication; Smith, personal communication), offers hope that transformed plants will be regenerated from other commercially important gymnosperms. However, since virtually all the gene transfer work done in gymnosperms has been confined to a single family, the Pinaceae, this review focuses on research done in this family.

Foreign DNA has been expressed in virtually every conifer tissue thus far tested, including protoplasts (Bekkaoui *et al.*, 1990), zygotic and somatic embryos (Ellis *et al.*, 1993), embryogenic callus (Ellis *et al.*, 1991), seedling tissues (Huang *et al.*, 1991; Ellis *et al.*, 1991; Goldfarb *et al.*, 1991; Stomp *et al.*, 1991), needles, meristems, differentiating wood (Loopstra *et al.*, 1992)

S. Jain, P. Gupta & R. Newton (eds.), Somatic Embryogenesis in Woody Plants, Vol. 1, 227–251.

and pollen (McCabe, personal communication). *Agrobacterium* infection of conifers was reported in the mid 1930s (Smith 1935) and several studies have expanded on this work with the identification of numerous *Agrobacterium* strains infectious to several different gymnosperms (DeCleene and DeLay 1976; Sederoff *et al.*, 1986; Ellis *et al.*, 1989; Hood *et al.*, 1990; Stomp *et al.*, 1990; Morris *et al.*, 1989). Studies with both electroporation (Gupta *et al.*, 1988; Bekkaoui *et al.*, 1988; Tautorus *et al.*, 1989) and direct uptake of DNA (Wilson *et al.*, 1989) demonstrated the expression of marker genes in confer protoplasts. The use of particle acceleration greatly increased the ability to study transient gene expression and provided a tool to show that numerous bacterial, angiosperm, and gymnosperm promoters functioned in several different tissue types (Ellis *et al.*, 1991; Charest *et al.*, 1991, 1993; Duschane and Charest, 1992; Stomp *et al.*, 1991, Loopstra *et al.*, 1992; Newton *et al.*, 1992). From the above studies, it is clear that the placement of DNA into conifer cells is not the limiting factor to the successful transformation of this plant group.

The major limitation to the transformation of gymnosperms is the inability to regenerate whole plants from those single cells which contain and express the introduced DNA. In fact, for most of the gymnosperms little research has been done on the development of tissue culture regeneration systems. Even in the conifers, where considerable effort has been devoted to the development of tissue culture systems, matching these with existing gene transfer systems has been difficult. Clearly, targeting the DNA to those cells which are competent not only to regenerate but also to express the foreign DNA is crucial to success. Unfortunately, this is often not possible because those cells which are competent to regenerate have not been identified. Further, gene transfer technologies currently available are only capable of targeting the DNA in a broad sense. Despite this difficulty, in the few reports of the regeneration of transformed conifers, careful attention to the biological system into which the DNA was inserted has been crucial. In addition, the recent success of pollen transformation in *Picea glauca* (McCabe, personal communication) provides a means of introducing DNA into conifers which is independent of a tissue culture system.

Combined with the lack of adequate regeneration systems for gymnosperms, is a relatively high sensitivity to the antibiotics normally used for the selection of transformed cells. In *Picea glauca*, kanamycin, hygromycin and phosphinothricin were many times more toxic than in comparable tobacco tissue (Ellis *et al.*, 1989). This complicates the selection strategies because due to the slow growth of conifer tissue in culture, tissue often dies prior to the division and growth of those cells expressing the inserted DNA. In conifers, as with many other woody perennials, dying cells produce and/or release toxic compounds such as phenolics which are detrimental to surrounding cells. Therefore under selection regimes used for the regeneration of transformed angiosperms, the entire explant in conifers would die. The death of neighboring non-transformed cells could contribute to the death of the

Figure 1. Six month old transformed *Picea glauca* seedling in the greenhouse.

transformed cell. Therefore, a strategy of inhibiting growth but not killing cells has been successful. Matching a tissue culture system with a selection regime which allows the suppression of non-transformed tissue yet permits transformed tissue to actively divide, has been a limitation.

By surveying the literature, in this review I tried to 1) provide an overview of what has been done, and 2) provide insights into the limitations and progress of gene transfer in gymnosperms. Included is a list of published reports on the transfer and expression of genes in gymnosperms (Table 1). As previously mentioned very little work has been done with the transformation of non-coniferous gymnosperms and therefore the discussion is entirely devoted to the conifers. The gene transfer systems will not be described in detail as excellent reviews are available for each of the gene transfer methods discussed (Potrykus, 1991, Zambryski, 1992; Klein *et al.*, 1992; Christou, 1992). Since the success of genetic engineering in a long-lived perennial tree may be dependent on the long-term control of gene expression, special emphasis has been placed on studies looking at the regulation of gene expression, both in transient assays and when the DNA is stably incorporated into the genome. Finally, the future needs and potential applications of genetic engineering in conifers is discussed.

Table 1. Selected papers on the transformation of gymnosperms.

Species	Transformation Method	Gene Expression	Transgenic Plants	Promoter/ gene	References[a]
Abies alba	*A. tumefaciens*	galls	no	agro[b]	DeCleene and DeLey, 1976
A. cephalonica	*A. tumefaciens*	galls	no	agro	Smith, 1942
A. concolor	*A. tumefaciens*	galls	no	agro	Smith, 1942
A. firma	*A. tumefaciens*	galls	no	agro	Smith, 1942
A. holophylla	*A. tumefaciens*	galls	no	agro	Smith, 1942
A. nordmanniana	*A. tumefaciens*	galls	no	agro	DeCleene and DeLey, 1976
	A. tumefaciens	galls	no	agro	Clapham and Ekberg, 1986
A. procera	*A. tumefaciens*	galls	no	opines	Morris *et al.*, 1989
	A. rhizogenes	galls	no	opines	Morris *et al.*, 1989
Araucaria bidwillii	*A. tumefaciens*	galls	no	agro	Smith, 1936
Chamaecyparis lawsoniana	*A. tumefaciens*	galls	no	agro	Smith, 1935
Cunninghamia laceolata	*A. tumefaciens*	galls	no	agro	Smith, 1942
Cupressus spp.	*A. tumefaciens*	galls	no	agro	Smith, 1935, 1939
Juniperus spp.	*A. tumefaciens*	galls	no	agro	Smith, 1939
Larix decidua	*A. rhizogenes*	roots/galls	no	agro	Diner and Karnosky, 1987
	A. rhizogenes	stable	yes	opines	Huang *et al.*, 1991
	microprojectiles	trans	no	35S/gus	Duchesne *et al.*, 1993
L. × eurolepis	electroporation	trans	no	35S, nos/gus	Charest *et al.*, 1991
	microprojectile	trans	no	35S, Em, rbcS, nos, actin, arabin/gus	Duchesne and Charest, 1992
	microprojectile	trans	no	35S/gus	Duchesne *et al.*, 1993
	microprojectile	trans	no	Em, 2x35S/gus	Charest *et al.*, 1993
L. laricina	microprojectile	trans	no	4x35S/luc, ubiq/gus	Charest *et al.*, 1993
L. × leptolepis	microprojectile	trans	no	35S/gus	Duchesne *et al.*, 1993
Libocedrus decurrens	*A. tumefaciens*	galls	no	agro	Smith, 1935
	A. tumefaciens	galls	no	agro	Stomp *et al.*, 1988
	A. tumefaciens	galls	no	opines	Stomp *et al.*, 1990
Picea abies	*A. tumefaciens*	galls	no	agro	Clapham and Ekberg, 1986
	A. tumefaciens	galls	no	agro	Ahuja, 1988
	A. tumefaciens	galls	no	opines	Hood *et al.*, 1990

	A. tumefaciens	galls	no	agro	Clapham and Ekberg, 1986
	microprojectile	trans/stable	no	35S, Em/gus, npt	Robertson *et al.*, 1992
	microprojectile	trans	no	Dc8, 2x35S/gus	Newton *et al.*, 1992
	microprojectile	trans/stable	no	35S/gus	Bercetche *et al.*, 1993
P. engelmanii	*A. tumefaciens*	galls	no	opines	Ellis *et al.*, 1989
P. glauca	electroporation	trans	no	35S/cat	Bekkaoui *et al.*, 1988
	PEG	trans	no	35S/GUS, cat	Wilson *et al.*, 1989
	A. tumefaciens	galls	no	opines	Ellis *et al.*, 1989
	electroporated	trans	no	2x35S,nos, 35S/gus	Bekkaoui *et al.*, 1990
	microprojectile	trans	no	aux, HS, ADH, rbcS, PEP/gus	Ellis *et al.*, 1991
	microinjection	trans	no	35S/gus	Attree, unpublished
	microprojectile	trans/stable	yes	35S, Em, nos/gus, npt	Ellis, 1993
	microprojectile	trans	no	Em, 2x35S/gus	Charest *et al.*, 1993
	microprojectile	trans/stable		rbcs, ubiq/gus	Ellis, 1993
	microprojectile	trans/stable	yes	2x35S/gus	Bommineni, 1993
P. mariana	electroporation	trans	no	35S/cat	Tautorus *et al.*, 1989
	electroporation	trans	no	2x35S/cat	Bekkaoui *et al.*, 1990
	microprojectile	trans	no	35S, Em/gus	Duchesne and Charest, 1991
	microprojectile	trans	no	35S, Em, rbcS, nos, actin, arabin/gus	Duchesne and Charest, 1992
	microprojectile	trans	no	Em, 2x35S/gus	Charest *et al.*, 1993
	microprojectile	trans/stable	yes	2x35S/gus, npt	Charest *et al.*, 1993
P. rubens	microprojectile	trans	no	Em, 2x35S/gus	Charest *et al.*, 1993
P. sitchensis	*A. tumefaciens*	galls	no	opines	Ellis *et al.*, 1989
Pinus banksiana	electroporation	trans	no	nos, 35S/cat	Tautorus *et al.*, 1989
	electroporation	trans	no	35S/GUS, cat	Bekkaoui *et al.*, 1988
P. contorta	*A. rhizogenes*	roots	no	35S/gus	Lindroth *et al.*, 1993
P. eldarica	*A. tumefaciens*	galls	no	agro	Stomp *et al.*, 1988
	A. tumefaciens	galls	no	agro	Stomp *et al.*, 1990

Table 1. Continued.

Species	Transformation Method	Gene Expression	Transgenic Plants	Promoter/ gene	References[a]
P. echinata	*A. rhizogenes*	roots/galls	no	agro	Huang and Tauer, 1993
P. elliottii	*A. tumefaciens*	galls	no	agro	Stomp *et al.*, 1988
	A. tumefaciens	galls	no	opines	Stomp *et al.*, 1990
	A. rhizogenes	roots/galls	no	agro	Huang and Tauer, 1993
P. jeffreyi	*A. tumefaciens*	galls	no	agro	Stomp *et al.*, 1988
	A. tumefaciens	galls	no	agro	Stomp *et al.*, 1990
P. lambertiana	*A. tumefaciens*	galls	no	agro	Stomp *et al.*, 1988
	A. tumefaciens	galls	no	nos%npt	Loopstra *et al.*, 1990
P. nigra	*A. tumefaciens*	trans	no	35S/intron, gus	Ordas *et al.*, 1993
P. palustris	*A. rhizogenes*	galls	no	agro	Diner and Soliman, 1993
P. ponderosa	*A. tumefaciens*	galls	no	agro	Stomp *et al.*, 1988
	A. tumefaciens	galls	no	opines	Morris *et al.*, 1989
	A. rhizogenes	galls	no	agro	Morris *et al.*, 1989
P. radiata	*A. tumefaciens*	galls	no	agro	Stomp *et al.*, 1988
	A. tumefaciens	galls	no	opines	Stomp *et al.*, 1990
	microprojectile	trans	no	35S/gus	Campbell *et al.*, 1992
	electroporation	trans	no	35S/luc, gus	Campbell *et al.*, 1992
	microprojectile	trans	no	35S/gus	Gonzalez *et al.*, 1993
P. sylvestris	*A. tumefaciens*	galls	no	agro	Ahuja, 1988
	A. tumefaciens	galls	no	agro	Stomp *et al.*, 1988
	A. tumefaciens	galls	no	opines	Stomp *et al.*, 1990
P. taeda	*A. tumefaciens*	galls	no	opines	Sederoff *et al.*, 1986
	electroporation	trans	no	35S/luc	Gupta *et al.*, 1988
	A. tumefaciens	galls	no	agro	Stomp *et al.*, 1988
	A. tumefaciens	galls	no	opines	Stomp *et al.*, 1990
	microprojectile	trans	no	35S/gus	Stomp *et al.*, 1991
	microprojectile	trans	no	35S, Em/gus	Loopstra *et al.*, 1992
	A. rhizogenes	roots/galls	no	agro	Huang and Tauer, 1993
	microprojectile	trans	no	35S, Em, 3H-6, 14–A9/gus	Sargent *et al.*, 1993
P. virginiana	*A. tumefaciens*	galls	no	agro	Stomp *et al.*, 1988
	A. tumefaciens	galls	no	opines	Stomp *et al.*, 1990
Podocarpus elongatus	*A. tumefaciens*	galls	no	agro	Smith, 1942

P. macrophyllus	*A. tumefaciens*	galls	no	agro	Smith, 1942
Pseudotsuga menziesii	*A. tumefaciens*	galls	no	agro	Smith, 1935
	A. tumefaciens	galls	no	nos, 35S/npt	Dandekar *et al.*, 1987
	electroporation	trans	no	35S/luc	Gupta *et al.*, 1988
	A. tumefaciens	galls	no	agro	Stomp *et al.*, 1988
	A. tumefaciens	galls	no	35S/luc	Ellis *et al.*, 1989
	A. tumefaciens	galls	no	opines	Morris *et al.*, 1989
	A. rhizogenes	galls	no	opines	Morris *et al.*, 1989
	A. tumefaciens	galls	no	opines	Stomp *et al.*, 1990
	microprojectile	trans	no	35S/gus	Goldfarb *et al.*, 1991
Sciadopitys verticillata	*A. tumefaciens*	galls	no	agro	Smith, 1942
Sequia sempervirens	*A. tumefaciens*	galls	no	agro	Smith, 1942
Sequiadendron giganteum	*A. tumefaciens*	galls	no	agro	Smith, 1942
Taxus baccata	*A. tumefaciens*	galls	no	agro	Smith, 1942
	A. tumefaciens	galls	no	opines	Han *et al.*, 1993
T. brevifolia	*A. tumefaciens*	galls	no	agro	Smith, 1942
	A. tumefaciens	galls	no	opines	Han *et al.*, 1993
	A. rhizogenes	roots	no	opines	Plaut-Carasson *et al.*, 1993
T. cuspidata	microprojectiles	trans	no	35S/gus	Ellis, unpublished
	A. tumefaciens	galls	no	opines	Han *et al.*, 1993
T. × media	*A. tumefaciens*	galls	no	agro	Smith, 1942
	A. tumefaciens	galls	no	opines	Han *et al.*, 1993
Thuja occidentalis	*A. tumefaciens*	galls	no	agro	Smith, 1939
	A. tumefaciens	galls	no	agro	DeCleene and DeLey, 1976
	A. tumefaciens	galls	no	agro	Ellis, unpublished
T. orientalis	*A. tumefaciens*	galls	no	agro	Smith, 1939
T. plicata	*A. tumefaciens*	galls	no	agro	Smith, 1939
Torreya californica	*A. tumefaciens*	galls	no	agro	Smith, 1942
Tsuga heterophylla	*A. tumefaciens*	galls	no	opines	Morris *et al.*, 1989
	A. rhizogenes	roots	no	opines	Morris *et al.*, 1989

[a] References list only the first author.

[b] In cases where *Agrobacterium* was used and galls or other symptoms were observed the pomoters/genes expressed are denoted as agro meaning that we can assume the appropriate *Agrobacterium* T-DNA promoters/genes functioned.

agro – *Agrobacterium tumefaciens*; luc – luciferase; gus – β-glucuronidese; npt – neomycin phosphotransferase; nos – nopaline synthese; cat – chloramphenicol acetyltransferase; ADH – alcohol dehydrogenase; rbcS – the small subunit of ribulose biphosphate carboxylase; CaMV 35S – Cauliflower Thosaic Virus 35S promoter.

2. *Agrobacterium* Transformation

2.1. A. tumefaciens

Although it used to be believed that gymnosperms were not good hosts for *Agrobacterium*, all the work with the transformation of gymnosperms until the late 1980s was done with *Agrobacterium*. *A. tumefaciens* infection of gymnosperms was reported as early as the mid-1930s (Smith, 1934, 1935) and was extensively reviewed in 1976 (DeCleene and DeLey, 1976) (Table 1). These early studies found two classes of the gymnosperms very susceptible to *A. tumefaciens* infection, the Coniferopsida and the Taxopsida, although there was a large amount of variability between species. Since these studies were designed to investigate the host range of *Agrobacterium*, they tested widely divergent plant species, yet were limited in the number of *A. tumefaciens* strains tested. It is now known that the *Agrobacterium* strain used plays a large role in infectivity. The strain used by DeCleene and DeLey (1976), *A. tumefaciens* strain B6 (same as LMG187), was only a moderately infectious strain in *Picea* and *Pseudotsuga menziesii* (Ellis *et al.*, 1989).

Until 1986, all of the studies on *Agrobacterium* infection in gymnosperms relied solely on the formation of a crown gall to assess infectivity. Sederoff *et al.* (1986) tested 19 strains of *Agrobacterium* and found two that formed galls on *Pinus taeda* (Loblolly pine). What was crucial about this work, was that this was the first report that 1) gave frequencies of *Agrobacterium* infection in gymnosperms, and 2) demonstrated biochemically the expression of T-DNA genes in a gymnosperm, with the confirmation of opine production by tumor tissue. Interestingly, expression of the oncogenes could not be confirmed by hormone autonomous growth, which led the authors to speculate that the introduced DNA may have been modified and this accounted for the lack of expression of the genes for growth hormone synthesis. In the same year, Clapham and Ekberg (1986) reported the infection of *Abies nordmanniana* and *Picea abies* by *Agrobacterium* and confirmed the expression of the hormone biosynthetic genes (oncogenes) by hormone autonomous growth in tissue culture.

The first report of the transfer and expression of non-*Agrobacterium* DNA in a gymnosperm was by Dandekar *et al.* (1987). Using a binary vector, the *aph* gene driven by CaMV 35S was exprcssed in *Pseudotsuga menziesii*. Confirmation that the T-DNA was integrated into the conifer genome came in 1990, in *Picea abies* (Hood *et al.*, 1990) and *Pinus lambertiana* (Loopstra *et al.*, 1990). In the latter study, T-DNA mapping showed that both the number of T-DNA copies inserted and the sites of transfer of the border regions were similar to those in angiosperms.

Infection by *Agrobacterium* at a high frequency in conifers is dependent on the *Agrobacterium* strain used, the genotype of the individual inoculated and the developmental stage of the tree. Of 36 *A. tumefaciens* strains tested, galls with detectable opine production were only induced by 13 of the strains in *Picea glauca* seedings (Ellis *et al.*, 1989). *P. sitchensis*, *P. engelmanni*, and *Pseudotsuga menziesii* were inoculated with ten of these stains and a similar pattern of infection was noted, except in *Pseudotsuga menziesii* where the frequency of gall formation was up to 10 × higher than in the *Picea* spp. Further, opines were detected in all species except *Picea sitchensis* where arginine was also not detected, suggesting that this amino acid could be limiting in opine production.

The high frequency of gall formation in *Pseudotsuga menziesii* was also noted by Morris *et al.* (1989). In this study, 37 *A. tumefaciens* strains were tested against four different conifer genera and in addition to *Pseudotsuga menziesii*, *Abies procera* also showed a high frequency of infection. *Tsuga heterophylla* and *Pinus ponderosa* had the lowest infection frequencies. This lower frequency of infection in the genus *Pinus* has also been noted by Stomp *et al.* (1990). In addition to differences in infectivity between different *A. tumefaciens* strains, differences in infectivity of a single bacterium strain was found in different *Picea abies* (Clapham *et al.*, 1990) and *Pinus radiata* (Bergmann and Stomp, 1992) families. This highlights the importance of a broad based genetic screening program.

Data suggest that there might be some interaction between the chromosomal background of the *A. tumefaciens* and the origin of the Ti plasmid. Ellis *et al.* (1989) showed that infection frequency was higher in *Picea glauca* when a C58 chromosomal background (A136) was complemented with the Ti plasmid from Bo542 than when either had its native Ti plasmid and chromosomal complement. This phenomenon was clearly demonstrated by Morris *et al.* (1989), where the same C58 chromosomal background was used with Ti plasmids from 9 different *A. tumefaciens* strains. Considerable variation between the transconjugants was seen, suggesting that manipulation of *Agrobacterium* infection in conifers may be possible by optimizing the interaction between the Ti plasmid and the bacterial genome. However, since the nature of this interaction is unknown, the transformation frequency is not always enhanced by the introduction of a different Ti plasmid. Using *Pinus* as an example, in the three cases where the wild-type donor of the Ti plasmid was tested, two cases caused gall formation frequency to increase from 0 with the wild-type strain to 8 and 10% in the transconjugant. However in the third case, gall formation frequency dropped from 55% to 20%. Although some galls formed, the use of a Ri plasmid with an *A. tumefaciens* chromosomal background caused a large decrease in gall formation.

The frequency of *A. tumefaciens* infection is highly dependent on the developmental stage of the explant, with rapidly growing seedling tissue infected at a higher frequency than other tissues. In the three *Picea* spp. tested, gall formation was higher in the younger tissue than in the older tissue. In contrast, infection occurred more evenly evenly on old and new growth in *Pseudotsuga menziesii* (Ellis *et al.*, 1989; Table 2). This may be correlated to the higher frequency of gall formation in *P. menziesii*. This is in contrast to the general observation however that younger, actively growing

Table 2. Percent of galls forming on young vs. old[1] seedling tissue in four conifer species (*Picea glauca*, *P. engelmanii*, *P. sitchensis* and *Pseudotsuga menziesii*) in response to inoculation with 10 *A. tumefaciens* strains. Data summarized from Ellis *et al.* (1989).

Agro Strain	Tissue	*Picea glauca*	*Picea engelmanni*	*Picea sitchensis*	*Pseudotsuga menziesii*
Bo542	young	57	86	82	79
	old	43	14	18	21
W2/73	young	81	80	93	53
	old	19	20	7	47
A281	young	81	75	75	65
	old	19	25	25	35
A281/pEND4K	young	38	67	89	66
	old	62	33	11	34
C3/74	young	83	86	94	50
	old	17	14	6	50
K6/73	young	76	67	92	53
	old	24	33	8	47
I10/75	young	75	75	94	51
	old	25	25	6	49
B2/74	young	85	93	81	56
	old	15	7	19	44
40	young	86	89	93	62
	old	14	11	7	38
LMG 187	young	90	100	89	60
	old	10	0	11	40
Total	young	75	81	88	59
	old	25	19	12	41

[1] Young tissue was tissue which was actively elongating, old tissue was tissue which had formed prior to dormancy.

tissues are more susceptible to infection (Stomp *et al.*, 1988; Morris *et al.*, 1989). Whether the actively growing tissues respond differently to wounding, produce less detrimental secondary compounds, or are less sensitive to the hormonal stimulus of the T-DNA is not known.

It is also interesting to note that although embryonic material is the most responsive in culture, there have been no reports of infection of gymnosperm embryos with *Agrobacterium*. However, using a β-glucuronidase (*gus*) construct which does not express in *Agrobacterium*, expression of the *gus* gene has been detected in somatic embryos of *Picea glauca* (McInnis, personal communication). Since the detection of *Agrobacterium* infection is dependent on the formation of a gall, it is possible that infection occurs in embryos yet that the genes on the T-DNA are down regulated. The fact that embryos are highly responsive to externally applied phytohormones would infer that hormone receptivity is probably not the cause of the lack of a transformed phenotype.

There are no reports of the regeneration of conifer plants transformed by

A. tumefaciens and although callus derived from the galls has been cultured, long-term growth has been difficult to maintain. However, *A. tumefaciens* transformed *Taxus* callus has been cultured and shown to produce the anti-cancer compound taxol (Han *et al.*, 1993). The callus was derived from galls that formed from the cut ends of stem sections from mature trees within 3–4 weeks of inoculation. The frequency of gall formation was greatest (24% of the stems infected formed galls) with *A. tumefaciens* strain Bo542. Gall formation was also highly dependent on the *Taxus* species inoculated, with *T. baccata* forming significantly more galls than *T. brevifolia*. Long-term growth of these gall derived callus lines has been achieved and even after a year in culture, the callus lines still produce taxol (Han, personal communication).

For the production of a plant secondary metabolite, like taxol, differentiation of plants, or even organized structures are not necessary if rapidly growing callus produces the desired compound. In this case and for the production of other secondary compounds from gymnosperms, *A. tumefaciens* may provide a valuable tool for the insertion of genes which alter or enhance specific biosynthetic pathways of valuable metabolites. Unfortunately, many secondary metabolites are produced only by very specific cell or tissue types and therefore differentiation into whole plants is often required for the production of such metabolites.

2.2. A. rhizogenes

The induction of roots by *A. rhizogenes* infection in conifers is not a reliable indicator of infection, as gall formation may occur more frequently. In *Larix decidua* (European larch), swelling, blue needles and multiple buds were the most common pathological responses, although adventitious roots did form in a few instances (Diner and Karnosky, 1987). *A. rhizogenes* induced roots were also noted on *Tsuga heterophylla*, yet only galls formed on *Abies procera*, *Pseudotsuga menziesii* and *Pinus ponderosa* (Morris *et al.*, 1989). *A. rhizogenes* appears to be less virulent than *A. tumefaciens* on conifers as pathological responses were not observed in all conifers inoculated with *A. rhizogenes* (Diner and Karosky, 1987; Ellis *et al.*, 1989). As with *A. tumefaciens* there does appear to be a strong influence of the bacterial strain used, therefore the matching of strains to a particular species may be required for high frequency infection.

A. rhizogenes infection is also very dependent on the tissue inoculated (Diner and Karnosky, 1987). While no symptoms of infection were noted on cotyledons and radicles, swelling was noted on *Larix decidua* hypocotyls. In contrast, considerable symptoms, including adventitious root formation were noted from inoculation of the apex. Most important was the induction of multiple shoots at the inoculation site in the apex. Clearly, as with *A. tumefaciens*, infection is dependent on inoculating the proper tissue and at the proper developmental stage. The formation of these shoots has resulted in the development of a transformation system for the regeneration of transformed *Larix decidua* plants (Huang *et al.*, 1991).

For the regeneration of transformed *Larix decidua* shoots from *A. rhizogenes* infected tissue, seed was aseptically germinated for seven days and the seedling apex inoculated. The site of apex wounding was crucial for success and was done by making a small slice from the center of the apex through the hypocotyl at an area subtending a cotyledon. The seedling was then placed in tissue culture and adventitious shoots formed in 3–4 weeks at the cut surface. These shoots could then be excised and grown into plants. The frequency of transformed shoot regeneration was not reported but the method is reproducible. *Larix decidua* plants containing and expressing both a *Bacillus thuringiensis* (*B.t.*) endotoxin gene and the herbicide resistant *aroA* gene have been regenerated and tested positive for the expression of the respective resistances in the greenhouse (Huang, personal communication). These plants have been maintained in the greenhouse and no differences in growth have been noted between the transgenic trees and non-transformed controls. Current emphases with this transformation method is to expand the technique to other conifers, such as *Pinus taeda*.

A potential advantage of this transformation method is that it may be possible to by-pass an *in vitro* step in the transformation process. It is not known whether the *in vitro* germination after inoculation with *A. rhizogenes* is required. One possibility is that the in vitro environment offers increased humidity which might be needed for infection. If this is the case, elevated humidity chambers should provide a similar environment. Although no hormones were needed for *Larix decidua*, hormones may be needed for other genera to alter the phytohormone ratio such that buds and not roots or callus are formed. In this latter case a tissue culture system would be required.

A. rhizogenes hairy root cultures have been induced from *Taxus brevifolia* seedlings and very rapid growth rates have been reported. Hairy root cultures formed from 0.5% of the *T. brevifolia* embryos inoculated (Plaut-Carcasson *et al.*, 1993). These roots produce opines and DNA integration has been confirmed, yet taxane production has not been detected from these cultures (Plaut-Carcasson, personal communication). However, since taxol is present in the tips of roots from other *Taxus* spp. (Ellis, unpublished), one would expect taxol production from these cultures as well, unless the taxol in the roots was synthesized elsewhere in the plant, transported, and deposited in the roots. As in the case with *Taxus* infection by *A. tumefaciens*, the regeneration of a transformed plant may not be necessary if the root cultures could be manipulated to produce the desired secondary metabolite, such as taxol.

3. Electroporation and PEG-Mediated DNA Uptake

Although conifer protoplasts were isolated and cultured in the late 1970s (Kirby and Cheng, 1979; David and David, 1979), it was not until the development of an embryogenic system for conifers (Hakman and Von Arnold, 1985) that hope for the regeneration of whole plants from protoplasts became a reality (Gupta and Durzan, 1987; Attree *et al.*, 1989). Despite the demonstration that whole plants could be regenerated from protoplasts, regeneration frequencies were low. Therefore, gene transfer could be thought of as a numbers game, you need to get DNA expressed in as many cells as possible in order to have a chance of those few cells that divided and reinitiate new embryogenic callus lines contain the introduced DNA. Unfortunately, cell division has yet to be demonstrated from conifer protoplasts which contain foreign DNA.

Since the first embryogenic system in conifers was developed in *Picea*, much of the work was done in this genus. During the late 1980s considerable effort was devoted to the insertion of genes into conifer protoplasts by electroporation and polyethylene glycol (PEG)-mediated DNA uptake. These early studies proved valuable in that they demonstrated the transient expression of numerous marker genes (chloramphenicol acetyltransferase [CAT], β-glucuronidase [GUS], luciferase) and promoter constructs in conifers (Gupta *et al.*, 1988; Bekkaoui *et al.*, 1988, 1990; Wilson *et al.*, 1989; Tautorus *et al.*, 1989; Charest *et al.*, 1991). In electroporated conifer protoplasts GUS background levels posed some initial problems (Bekkaoui *et al.*, 1988), and based on increased sensitivity and ease of monitoring (Charest *et al.*, 1991), CAT was used as the marker gene of choice in virtually all the conifer electroporation studies.

One limitation to the use of protoplasts was that they were easily damaged and long-term expression studies of the introduced genes were not reported. Much of the information and experiments followed previous studies done in angiosperms that delineated the electroporation parameters, such as the voltage level and DNA concentration, which enhanced transient gene expression. As with most other gene transfer studies, variability from experiment to experiment and inconsistencies in results between species complicated the analysis of the data. For example, in *Picea glauca* (Bekkaoui *et al.*, 1988) and *Pinus banksiana* (Tautorus *et al.*, 1989), linearized DNA gave higher levels of expression than circular, while in *Picea mariana* circular DNA gave slightly higher gene expression. Whether the differences were significant between linear and circularized DNA is not known, as no indication of variation from run to run was indicated. With PEG-mediated DNA uptake, standard errors were reported and circular DNA was superior to linear DNA in *Picea glauca*, yet the expression level of the linear DNA was not above baseline (Wilson *et al.*, 1989).

As with *Agrobacterium*, the biological system, or the plant cell has a large influence on the level of gene expression following electroporation. Tautorus *et al.* (1989) tested four *Picea mariana* lines and three *Pinus banksiana* lines and found that the level of transient expression was cell line specific. Similar results were reported in *Picea glauca* (Bekkaoui *et al.*, 1990). It is not surprising that there were differences in the expression of introduced genes between cell lines as differences in growth rates, response to enzymatic digestion of the cells, protoplast survival, and division of protoplasts varies between cell lines (Attree *et al.*, 1989). Due to these factors, the differences in the level of expression could be due as much to the particular phase in the growth cycle that these cells were in during protoplast isolation as to true genetic differences between the cell lines.

In general, *Pinus banksiana* had a lower level of transient gene expression compared to either *Picea mariana* (Tautoris *et al.*, 1989) or *Picea glauca* (Bekkaoui *et al.*, 1990). As discussed above, since the cellular biology of the plant plays a key role in the expression of introduced genes, these differences could be due to a comparison of protoplasts isolated from embryogenic cells (*Picea*) with protoplasts from non-embryogenic cultures (*Pinus*). This is supported by the fact that protoplast viability of the *Pinus banksiana* protoplasts did not vary much and were actually higher at some voltages than the viability of *Picea glauca* protoplasts (Tautorus *et al.*, 1989). However, in contrast to *Picea*, transient expression of CAT in *Pinus banksiana* increased only slightly over the DNA levels tested. Promoters also influence gene expression and by using a double-CaMV 35S promoter, a relatively high level of transient gene expression was observed in electroporated *Pinus banksiana* protoplasts (Bekkaoui *et al.*, 1990).

4. Particle Acceleration

The first report of the use of microprojectiles for the introduction of DNA into plant cells (Klein *et al.*, 1987) clearly demonstrated the potential of particle acceleration as an important tool in the genetic engineering of plants. Using microprojectiles, virtually any tissue and any cell within that tissue can have DNA inserted into it. Tissues from conifers which have had DNA inserted into them by particle acceleration include suspension cells, embryogenic cultures, zygotic and somatic embryos, seedling parts, meristems, wood and pollen. Although this gene transfer technique is free from the constraints of actively growing tissue as with *Agrobacterium* or in protoplasts as with electroporation, it is limited by the same constraint as all gene transfer systems, the ability to regenerate a plant from that single cell which contains the foreign DNA. Again, the availability of a tissue culture system which complements the gene transfer system is the limitation to the regeneration of transformed plants. Despite this, there are four examples of the regeneration of transformed conifers using microprojectiles.

As with electroporation, most of the work with microprojectiles has been with elucidating factors important in transient gene expression. While no clear correlation has been established between the level of transient expression and the number of stable transformed lines which are recovered, it is clear that if no transient expression is detected, no stable transformants have been recovered. Therefore, the goal of many studies have been to optimize the gene transfer parameters to maximize transient expression.

While this approach has some merit, antibiotic selection for transformed cells relies on the transformed cell expressing the resistance gene. Parameters which maximize transient expression may not be the parameters which offer the highest chance of recovering a transformed cell line. For example, although the number of cells transiently expressing the introduced DNA can be greatly increased by decreasing the particle size used (Ellis, 1993), no increase in the regeneration of stable transformants has been observed. However, the same principle applies here as with electroporation, in that gene transfer can be thought of as a numbers game. The greater number of cells which express the inserted DNA, the greater the chances of hitting one that will regenerate. If this is true, then maximizing transient gene expression could be desirable.

Using the marker gene GUS, microprojectile bombardment provides a quick and easy way to quantify gene expression in any explant. GUS expressing cells can simply be counted using histochemical staining (Jefferson, 1987). In most cases, each blue GUS expressing area (spot) is comprised of more than one cell. Although it is probable that some of these spots represent GUS expression in greater than a single cell, the majority of the multicellular blue spots are thought to the expression from a single cell. The reaction responsible for the precipitation of the blue indigo dye relies on the cleavage of the x-gluc substrate by the GUS enzyme and the oxidative dimerization of the monomers. Therefore, the monomers could readily diffuse into neighboring cells until another monomer and oxygen are present. The single cell origin of the blue GUS spots is supported by the following observations: 1) the intensity of blue is greatest in the center and decreases as it radiates out, 2) the presence of lobes in some spots indicate multiple cell expression, and 3) single cells can be seen expressing GUS if the exposure to the substrate is minimized. Regardless of whether the GUS expressing spots are of single or multiple cell origin, a correlation between the number of GUS expressing spots and the level of GUS enzyme activity in *Picea glauca* zygotic embryos has been shown (Ellis *et al.*, 1991).

As with the other gene transfer methods, the explant plays an important role in the expression of the introduced genes (Ellis *et al.*, 1991). Further, the level of transient expression is influenced by the pretreatment regime of the explant. In *Pseudotsuga menziesii* cotyledons, pretreatment on a callus induction medium increased transient expression in a time dependent manner, up to seven days of pretreatment. Similarly, if the cotyledons were pretreated with a cytokinin pulse for bud induction, transient expression was also increased (Goldfarb *et al.*, 1991). Similar results were obtained with *Picea glauca* zygotic embryos, where pretreatment on a bud induction medium increased transient expression for up to seven days pretreatment. Further pretreatment did not significantly increase transient expression (Table 3). Because both non-pretreated and pretreated embryos were exposed to the same concentration of particles and DNA, this suggests that the cells in the untreated embryo or cotyledon are not competent to express

Table 3. The effect of pretreatment duration on a bud induction medium[a] on the transient expression of GUS[b] in *Picea glauca* zygotic embryos.

# of Days on the Bud Induction Medium	# of GUS Expressing Spots (+/− SE)
0	11.1 (1.0)
3	11.4 (3.8)
5	23.4 (8.5)
7	40.8 (6.8)
14	45.4 (4.1)
21	42.3 (6.1)

[a] The bud induction medium consisted of woody plant medium supplemented with 50 μM zeatin and 0.01 μM thidiazuron.
[b] Plasmid used was pTVBT41100 containing an enhanced CaMV 35S-GUS construct (Ellis *et al.*, 1993).

the introduced DNA. A pretreatment of the tissue allows the cells to achieve the competence to express the introduced DNA. However, the hormonal treatment alone is not responsible for this increased transient gene expression as a pretreatment does not enhance transient expression in seedling tissues which are not responsive to the hormones.

This has led to the idea that the induction of meristematic tissue is necessary and for a high level of transient expression (McCown *et al.*, 1991; Serres *et al.*, 1992; Ellis, 1993). In both *Pseudotsuga menziesii* cotyledons and *Picea glauca* embryos, the pretreatments induced meristematic tissues. In contrast, seedling tissue of *Picea glauca* is not competent to respond to the hormones and produce meristematic tissues, and therefore even with a hormonal pretreatment, does not transiently express GUS at a high level. Clearly, the physiological and developmental state of the cells play a role in the expression of introduced genes in plants. This may account for the relatively low level of GUS expression in *Pinus taeda* cotyledons (Stomp *et al.*, 1991).

Interestingly, embryogenic material which has a high percentage of meristematic cells, was the most variable of the *Picea glauca* tissues used. Although the GUS expression was confined to the meristematic head cells, supporting the theory of a need for meristematic cells for expression, the level of expression was highly variable. Further, based on target area, transient expression was lower than that of other explants (Ellis *et al.*, 1991). One reason for this variability may have been the moisture level on the surface of the embryogenic material. In contrast, the suspensor cells as well as non-organized cell clusters and not the pro-embryonal head cells expressed GUS in *Picea mariana* embryogenic material (Duchesne and Charest, 1991). It would be convenient if a meristematic state was all that was required for the competence to transiently express introduced genes because meristematic cells are generally competent to respond to a hormonal signal and differentiate into a plant.

A major contribution of particle bombardment for conifer genetic engin-

Table 4. The effect of different promoters on the transient (2d) and long-term (56d) expression of GUS in *Picea glauca* zygotic embryos. Portions of the data summarized from Ellis *et al.* (1991).

Promoter	# of GUS Spots at 2 Days (+/− SE)	Relative Promoter Strength Compared to CaMV 35S[a]	# of GUS Spots at 56 Days (+/− SE)
soybean saur	9.4 (1.8)	26	nd[b]
soybean heatshock			
plus heatshock[c]	29.0 (3.7)	80	3.0 (0.5)
minus heatshock	2.9 (0.7)	8	1.4 (0.7)
Arabidopsis rbcS	13.7 (2.8)	38	1.9 (0.6)
soybean rbcS	2.7 (0.5)	7	nd
larch rbcS	5.8 (1.2)	16	1.2 (0.4)
maize PEP	6.5 (0.7)	18	nd
CaMV 35S[d]	36.1 (4.9)	100	2.6 (1.3)
nopaline synthase	16.0 (3.2)	44	2.3 (1.3)
Arabidopsis ubiquitin	12.9 (1.9)	36	1.7 (0.6)
wheat Em	30.7 (4.3)	85	3.8 (0.7)

[a] Relative promoter strength is expressed as the relative percent of transient GUS spots (2 days) compared to CaMV 35S.
[b] No data.
[c] Tissue treated for one h at 42° followed by three hours at room temperature prior to placement into x-gluc.
[d] Enhanced 35S as described by Ellis *et al.* (1993).

eering has been the testing of numerous heterologous promoters. Of the promoters tested, the soybean heatshock inducible promotor is the only promoter to show inducible expression in a transient assay (Ellis *et al.*, 1991). All other promoters tested however have functioned in conifers, and based on transient assays, the relative strengths of these promoters in conifers can be compared. A list of the promoters expressed in conifers is included in Table 1 and the relative strength of select promoters in transient assays in *Picea glauca* embryos is listed in Table 4. In general, most of the promoters tested can be placed in the category of low transient expressors.. These promoters include a soybean auxin responsive saur and rbcS, maize PEP and ADH (Ellis *et al.*, 1991), CaMV 35S, tobacco rbcS, nos, rice actin, arabidopsis arabin (Duchsene and Charest, 1992), and carrot Dc8 (Newton *et al.*, 1992). Other promoters gave a consistently higher level of expression and these include an enhanced CaMV 35S, either double or with an alfalfa mosaic virus enhancer (Ellis *et al.*, 1991, 1993; Charest *et al.*, 1993) and wheat Em (Duchesne and Charest, 1992; Loopstra *et al.*, 1992; Ellis *et al.*, 1993; Charest *et al.*, 1993). Of the promoters tested in conifers, the wheat Em promoter from an ABA responsive gene (Marcotte *et al.*, 1989) has consistently given a high level of expression in all conifer tissues tested including embryogenic callus (Charest *et al.*, 1993), somatic embryos (Ellis *et al.*, 1993), zygotic embryos (Table 4) and differentiating wood (Loopstra *et al.*, 1992).

Although a majority of the work with particle acceleration has focused on the transient expression of genes, long-term expression plays a more important role in the recovery of transgenic cell lines. Unfortunately, few studies have looked at long-term gene expression. This is probably due to the rapid loss of expression within the first week following particle bombardment. With *Picea glauca* zygotic embryos, the initial level of transient expression was maintained for seven days after which time it decreased to a low baseline level of expression within 14 days. This baseline level of expression was then maintained for at least 8 weeks (Ellis *et al.*, 1992; Ellis, 1993). The only variation from this baseline was a highly reproducible but minor increase in expression about 3 weeks after particle bombardment. Several cultural treatments were tested to determine if the pattern of expression could be altered, and none have been successful in significantly changing this long-term expression pattern in *Picea glauca* zygotic embryos.

This expression pattern has been observed with a CaMV 35S promoter, a larch rbcS promoter, an arabidopsis ubiquitin promoter, the wheat Em promoter and a nos promoter (Ellis, 1993). Long-term gene expression has also been noted in *Pinus taeda* after 62 days (Stomp *et al.*, 1991), and in *Pinus radiata* suspension cultured cells after 6 weeks (Campbell *et al.*, 1992). For both the CaMV 35S and Em promotors, no expression was observed after 16 days in either *Larix* × *eurolepis* and *Picea mariana* (Duchesne and Charest 1992).

While expression in zygotic embryos decreased within the first two weeks, GUS expression in somatic embryos was maintained at the initial level for up to three weeks. As with the zygotic embryos, this long-term expression pattern was independent of the promoter used as both the CaMV 35S and Em promoters showed the same level of expression (Ellis *et al.*, 1993). Further, although the initial level of transient expression was different for the different developmental stages of the somatic embryos, the more mature stages showing a higher level of transient expression, the pattern of long-term expression was similar for all stages. Despite differences in the pattern of gene expression, it is interesting to note that the level of GUS expression 8 weeks after particle acceleration was similar for all explants and promoters tested (Table 4).

The identification of an explant which maintained a high level of gene expression for two or more weeks following particle acceleration was crucial in the development of a transformation system for *Picea glauca*. By combining a high sustained level of gene expression, a high regenerative potential and the ability to survive a sublethal selection regime, transformed embryogenic callus was induced from *Picea glauca* somatic embryos (Ellis *et al.*, 1993). Using cotyledon stage somatic embryos, the transformation frequency was 0.5–1.0%, in that one in 100–200 embryos exposed to particle acceleration initiated transformed embryogenic callus. Somatic embryos were exposed to particle acceleration, placed on embryogenic callus induction medium and selected with a sublethal level of kanamycin (5 μg/ml) two weeks after particle acceleration. Although this protocol is not optimized and the relative importance of each of the steps is unknown, careful following of the procedure has yielded transformed cell lines reproducibly over the past few years.

The important features of the procedure include the identification of the proper developmental somatic embryo stage such that the embryo will withstand all subsequent manipulations. Since particle acceleration is stressful to the explant, allowing the cells to recover prior to further manipulation is also important. Selection too soon always results in embryo death. After the cells have recovered from particle acceleration, the cells in the embryo have to be induced to divide before the selection pressure is applied. While this may not be important for rapidly dividing and highly responsive species such as tobacco, it is crucial for conifer species where cell divisions may not occur for weeks. After cell division has been initiated, a sublethal selection pressure can be applied so that non-transformed cells are suppressed but not killed. One drawback to this procedure is the high level of non-transformed embryogenic lines which are induced. In these cases GUS has proven to be 100% effective in screening *Picea glauca* transformants, as has the long-term exposure to a higher kanamycin concentration (10 μg/ml). Similar results have been obtained by two other independent labs (Tsang, personal communication; Charest, personal communication), demonstrating the reproducibility of this approach.

The expression of introduced genes in the whole plant is of particular interest. The expression of CaMV 35S in *Picea glauca* embryogenic cultures occurs in all cell types, while in somatic embryos and emblings, expression follows a similar pattern to that seen in angiosperms. In somatic embryos, expression is seen in the epidermis and phloem but is absent from the cortex and the pith. Studies currently underway are investigating expression patterns in regenerated plants (Fig. 1) and comparing these expression levels to those of a putative xylem specific promoter from pine (Loopstra, unpublished). Further, the expression of a *B.t.* endotoxin gene has been demonstrated and the effectiveness against spruce budworm has been confirmed. Additionally, transformed *Picea glauca* plants have been placed in the field this summer to initiate a long-term field study to monitor gene expression over a multi-year period.

One of the most exciting developments in conifer transformation is the success of pollen transformation in *Picea glauca* for two successive seasons (McCabe, personal communication). This gene transfer method for conifers may not rely heavily on the identification of nebulous developmental stages for DNA receptivity. DNA was introduced into pollen by particle acceleration and the pollen was subsequently used to pollinate receptive female strobili. The seed was allowed to mature and several transformed seedlings are now growing in the green house. Most importantly, since this gene transfer system does not rely on tissue culture, pollen transformation may provide a species independent transformation system for conifers. It will be intriguing to see if this gene transfer system will work with genera such as *Pinus* where pollination and fertilization occur in consecutive years. With accelerated breeding programs, the two year cycle can be significantly reduced and will likely not pose an insurmountable obstacle.

5. Future Directions and Needs

As mentioned throughout this review, regardless of the gene transfer method, DNA is placed into individual cells. Therefore, the ability to regenerate a whole plant from that single cell is required. It has been demonstrated that in conifers adventitious buds arise from subdermal cells and embryogenic cells differentiate from surface cells. Yet the specific cells in the embryo which respond to the hormonal stimulus and initiate organized division have not been identified. Currently, cell layers are targeted by the gene transfer systems, and serendipity plays a role in whether the proper cells are targeted. Obviously this system works, as a transformation system based on somatic embryos has been developed. However, better definition of these precursor cells would aid in the efficiency of gene transfer and in the recovery of transformed lines. Another approach would be to increase the number of cells which respond to the hormonal stimulus. This latter approach would include the development of more efficient tissue culture systems. Unfortunately, we know relatively little about the mechanisms controlling the ability of a cell to respond to a hormonal stimulus, thus enhancing the competence is difficult.

Fortunately, gene transfer systems have been developed and applied to conifers where transformation does not rely on a tissue culture system. In the case of pollen transformation the receptive cell is clearly identified and single cell targeting is possible. Using particle acceleration, a method for the transformation of soybean has been developed which involves introducing the DNA into the meristem and then inducing axillary bud formation (McCabe *et al.*, 1988). Such a system may also aid in alleviating the lack of adequate tissue culture systems for conifer transformation. Since most transformation systems in plants, including conifers, rely on a tissue culture system this is still the most logical avenue for success. Despite this, as work continues to progress with the more recalcitrant species, new innovative approaches will have to be developed.

Very little is known about the expression of native genes in gymnosperms, and the surface has just begun to be scratched with the expression of introduced genes. Clearly, heterologous promoters will play a major role in the regulation of the expression of introduced genes in conifers. Whether these promoters will function in the desired tissue specific or inducible manner is unknown. Transient assays have shown that heterologous promoters can be expressed but to date none have reliably demonstrated tissue specific expression. Proper expression of introduced genes will be important for everything from engineered pest resistance to the modification of quality properties such as lignin. Improper expression could place unneeded stresses on the tree and negate the benefit of the engineered improvement. Further, although the expression of an extra gene or two will probably not have a significant effect in the short-term, over the life of a tree, any extra drain on the energy balance of the tree could cause detrimental effects particularly during periods of high stress.

Since all of the gymnosperms which are being manipulated are wind pollinated, the control of gene spread will important. Elegant systems have been devised in other crops for the engineering of sterility. Similar approaches using antisense, ribozymes or ribonucleases regulated by flower specific promoters could be used for trees. Whether a tapetal specific promoter from an angiosperm would function in a gymnosperm is unknown, yet this is a logical starting point. Unfortunately, male sterility may not be adequate and total sterility might be a better goal. A strategy for the engineering of total sterility could hinge on the use of genes active during the transition of vegetative growth to flowering or early in flower formation. Several groups are currently looking for floral specific genes homologous to angiosperm sequences which might yield regulatory sequences valuable for such an approach.

The lack of genes that can be used to control important traits is also a limiting factor. While genes encoding for insect and herbicide resistance have been transferred into conifers, few other genes of economic importance have been isolated. Genes regulating disease resistance, wood and tree quality traits, resistance to drought and low nutrients and resistance to environmental insults will be needed.

Beyond answering basic scientific questions, the economic importance of transformed gymnosperms will depend on the ultimate use and added value of the trees. Related to this will be the deployment strategies used to outplant them. For example, with genetically engineered insect resistance, it is irresponsible to think of genetic engineering as a panacea. In this case it is another tool for forest managers to use in their arsenal of integrated pest management practices. Improper use of the genetically engineered tree could negate its value due to the evolution of resistant insect biotypes. However, with proper expression of the insecticidal genes, mixing of engineered trees with non-engineered trees and by providing enough non-engineered refuge for the insects to feed, selective pressures for the prevalence of resistant biotypes could be minimized.

6. Conclusions

Genetic engineering in gymnosperms has been confined to the conifers. Successful transformation systems have been established using both *Agrobacterium* and particle acceleration and genes encoding insect and herbicide resistance have been expressed in transformed plants. A field plot with a limited number of transformed *Picea glauca* has been established and gene expression over several years will be monitored. Despite these few successes, most of the conifers have yet to be stably transformed. Transient assays have demonstrated the expression of introduced genes but have been unable to do much more than confirm expression of different gene constructs. Research is needed to determine these promoters will function in a tissue-specific or inducible manner in conifers. Such studies require the testing of transformed plants. Further, genes encoding valuable traits for both horticultural and forest conifers need to be identified. The main limitation to the genetic engineering of most conifers is not gene transfer, it is the lack of adequate systems to regenerate plants from those cells which have had the DNA inserted into them. Progress with the embryogenesis of both *Pinus* spp. and *Pseudotsuga menziesii* offer hope that these systems could be coupled with gene transfer systems. The report of pollen transformation in *Picea glauca* provides yet the latest method which might expand genetic engineering into other gymnosperms.

Acknowledgements

I thank Pierre Charest, Kyung-Hawn Han, Yinghua Huang, Dennis McCabe, Stephanie McInnis, Yoke Plaut-Carcasson, and Ed Tsang for sharing unpublished results and allowing the inclusion of these results in this chapter. I also thank Rodney Serres for his critical reading of the manuscript and Jenni Rintamaki for technical assistance.

References

Ahuja, M.R., 1988. Gene transfer in forest trees. In: J.W. Hanover and D.E. Keathley (Eds.), Genetic Manipulation of Woody Plants, pp. 25–41. Plenum Publishing Corp., New York.

Attree, S.M., D.I. Dunstan and L.C. Fowke, 1989. Initiation of embryogenic callus and suspension cultures and improved regeneration of protoplasts of white spruce. Can. J. Bot. 67: 1790–1795.

Bekkaoui, F., M. Pilon, E. Laine, D.S.S. Raju, W.L. Crosby and D.I. Dunstan, 1988. Transient gene expression in electroporated *Picea glauca* protoplasts. Plant Cell Rep. 7: 481–484.

Bekkaoui, F., R.S.S. Datla, M. Pilon, T.E. Tautorus, W.L. Crosby and D.I. Dunstan, 1990. The effects of promoter on transient expression in conifer cell lines. Theor. Appl. Genet. 79: 353–359.

Bercetche, J., M. Dinant, M. Paques and R.F. Matagne, 1993. Genetic transformation of embryogenic tissues of *Picea abies* by microprojectile bombardment. Paper presented at the 5th International Workshop of the IUFRO Somatic Cell Genetics Working Party, Balsian, Spain, October 18–22, p. 51.

Bergmann, B.A. and A.-M. Stomp, 1992. Effect of plant genotype and growth rate on *Agrobacterium tumefaciens* mediated gall formation in *Pinus radiata*. In: Proceedings of the International Conifer Biotechnology Working Group Meeting, April 23–28, Raleigh, NC, p. 65.

Bommineni, V.R., R.S.S. Dalta and E.W.T. Tsang, 1993. Analysis of genetically transformed white spruce somatic embryos. In Vitro 29A: 66A.

Campbell, M.A., C.S. Kinlaw and D.B. Neale, 1992. Expression of luciferase and β-glucuronidase in Pinus radiata suspension cells using electroporation and particle bombardment. Can. J For. Res. 22: 2014.

Charest, P.J., Y. Devantier, C. Ward, C. Jones, U. Schaffer and K.K. Klimaszewska, 1991. Transient expression of foreign chimeric genes in the gymnosperm hybrid larch following electroporation. Can. J. Bot. 8: 1731–1736.

Charest, P.J., N. Calero, D. Lachance, R.S.S. Datla, L.C. Duchesne and E.W.T. Tsang, 1993. Microprojectile-DNA delivery in conifer species: factors affecting assessment of transient gene expression using the β-glucuronidase reporter gene. Plant Cell Rep. 12: 189–193.

Christou, P., 1992. Genetic transformation of crop plants using microprojectile bombardment. Plant J. 2: 275–281.

Clapham, D.H. and I. Ekberg, 1986. Induction of tumours by various strains of *Agrobacterium tumefaciens* on *Abies nordmanniana* and *Picea abies*. Scand. J. For. Res. 1: 435–437.

Clapham, D., I. Ekberg, G. Eriksson, E.E. Hood and L. Norell, 1990. Within-population variation in susceptibility to *Agrobacterium tumefaciens* A281 in *Picea abies* (L.) Karst. Theor. Appl. Genet. 79: 654–656.

Dandekar, A.M., P.K. Gupta, D.J. Durzan and V. Knauf, 1987. Transformation and foreign gene expression in micropropagated Douglas-fir (*Pseudotsuga menziesii*). Bio/Technology 5: 587–590.

David, A. and H. David, 1979. Isolation and callus formation from cotyledon protoplasts of pine. Z. Pflanzenphsiol. 94: 173–177.

DeCleene, M.D. and J.D. DeLey, 1976. The host range of crown gall. Bot. Rev. 42: 389–466.

Diner, A.M. and D.R. Karnosky, 1987. Differential responses of two conifers to *in vitro* inoculation with *Agrobacterium rhizogenes*. Eur. J. Plant. Path. 17: 211–216.

Diner, A.M. and K. Soliman, 1993. *Pinus palustris* transformation by *Agrobacterium rhizogenes*. In Vitro 29A: 86A.

Duchesne, L.C. and P.J. Charest, 1991. Transient expression of the β-glucuronidase gene in embryogenic callus of *Picea mariana* following microprojection. Plant Cell Rep. 10: 191–194.

Duchesne, L.C. and P.J. Charest, 1992. Effect of promoter sequence on transient expression of The β-glucruonidase gene in embryogenic calli of *Larix* × *eurolepis* and *Picea mariana* following microprojection. Can. J. Bot. 70: 175–180.

Duchesne, L.C., M.-A. Lelu, P. von Aderkas and P.J. Charest, 1993. Microprojectile-mediated DNA delivery in haploid and diploid embryogenic cells of *Larix* spp. Can. J. For. Res. 23: 312–316.

Ellis, D., D. Roberts, B. Sutton, W. Lazaroff, D. Webb and B. Flinn, 1989. Transformation of white spruce and other conifer species by *Agrobacterium tumefaciens*. Plant Cell Rep. 8: 16–20.

Ellis, D.D., D. McCabe, D. Russell, B. Martinell and B.H. McCown, 1991. Expression of inducible angiosperm promotors in a gymnosperm, *Picea glauca* (white spruce). Plant Mol. Biol. 17: 19–27.

Ellis, D.D., D.E. McCabe, D. Russell, B. McCown and B. Martinell, 1992. A transient assay for heterologous promoter activity in *Picea glauca*. In: M.R. Ahuja (Ed.), Woody Plant Biotechnology, pp. 283–294. Plenum Press, New York.

Ellis, D.D., D.E. McCabe, S. McInnis, R. Ramachandran, D.R. Russell, K.M. Wallace, B.J. Marinell, D.R. Roberts, K.R. Raffa and B.H. McCown, 1993. Stable transformation of *Picea glauca* by particle acceleration. Bio/Technology 11: 84–89.

Ellis, D.D., 1993. Transformation in *Picea*. In: Y.P.S. Bajaj (Ed.), Biotechnology in Forestry and Agriculture (in press).

Goldfarb, B., S.H. Strauss, G.T. Howe and J.B. Zaerr, 1991. Transient gene expression of microprojectile-introduced DNA in Douglas-fir cotyledons. Plant Cell Rep. 10: 517–521.

Gonzales, M.V., M. Rey, R.J. Ordas, G. Ancora and R. Tavazza, 1993. Transient gene expression in cultured radiata pine cotyledons. Paper presented at the 5th International Workshop of the IUFRO Somatic Cell Genetics Working Party, Balsian, Spain, October 18–22, p. 59.

Gupta, P.K. and D.J. Durzan, 1987. Somatic embryos from protoplasts of loblolly pine proembryonal cells. Bio/Technolgy 5: 710–712.

Gupta, P.K., A.M. Dandekar and D.J. Durzan, 1988. Somatic proembryo formation and transient expression of a luciferase gene in Douglas fir and Loblolly Pine protoplasts. Plant Sci. 58: 85–92.

Hakman, I. and S. von Arnold, 1985. Plantlet regeneration through somatic embryogenesisin *Picea abies*. J. Plant Physiol. 121: 149–158.

Han, K.-H., M.F. Gorden, M. Leper, H. Floss and S. Chilton, 1993. Genetic system of Taxus as an alternative system for taxol production. In: S. Scher and B.S. Schwarzschild (Eds.), Proceedings of the International Yew Resources Conference, March 12–13, Berkeley, CA, pp. 18–19.

Hood, E.E., D.H. Clapham, I. Ekberg and T. Johannson, 1990. T-DNA presence and opine production in tumors of *Picea abies* (L.) Karst induced by *Agrobacterium tumefaciens* A281. Plant Mol. Biol. 14: 111–117.

Huang, Y., A.M. Diner and D.F. Karnosky, 1991. *Agrobacterium rhizogenes*-mediated genetic transformation and regeneration of aconifer: *Larix decidua*. In Vitro Cell 27: 201–207.

Huang, Y and C.G. Tauer, 1993. Another tool for gene transfer in pine species: *Agrobacterium rhizogenes*. In Vitro 19A: 65A.

Jefferson, R.A., 1987. Assaying chimeric genes in plants: the GUS fusion system. Plant Mol. Biol. Rep. 5(4): 387–405.

Kirby, E.G. and T.Y. Cheng, 1979. Colony formation from protoplasts derived from PDouglas fir cotyledons. Plant Sci. Lett. 14: 145–154.

Klein, T.M., E.D. Wolf, R. Wu and J.C. Stanford, 1987. High velocity microprojectiles for delivering nulceic acids into living cells. Nature 327: 70–73.

Klein, T.M., R. Arentzen, P.A. Lewis and S. Fitzpatrick-McElligott, 1992. Transformation of microbes, plants and animals by particle bombardment. Bio/Technology 10: 286–291.

Lindroth, A., R. Gronroos and S. von Arnold, 1993. Transformation with *Agrobacterium* in lodgepole pine, *Pinus contorta*, and genes specifically expressed during root initiation. Paper presented at the 5th International Workshop of the IUFRO Somatic Cell Genetics Working Party, Balsian, Spain, October 18–22, p. 50.

Loopstra, C.A., A.M. Stomp and R.R. Sederoff, 1990. *Agrobacterium*-mediated DNA transfer in sugar pine. Plant Mol. Biol. 15: 1–9.

Loopstra, C.A., A.K. Weissinger and R.R. Sederoff, 1992. Transient gene expression in differentiating pine wood using microprojectile bombardment. Can. J. For. Res. 22: 993–996.

Marcotte, W.R., S.H. Russell and R.S. Quatrano, 1989. Abscisic-acidresponsive sequences from the Em gene of wheat. Plant Cell 1: 969–976.

McCabe, D.E., W.F. Swain, B.J. Martinell and P. Christou, 1988. Stable transformation of soybean by particle acceleration. Bio/Technology 6: 923–926.

McCown, B.H., D.E. McCabe, D.R. Russell, D.J. Robison, K.A. Barton and K.F. Raffa, 1991. Stable transformation of *Populus* and incorporation of pest resistance by electric discharge particle acceleration. Plant Cell Rep. 9: 590–594.

Morris, J.W., L.A. Castle and R.O. Morris, 1989. Efficacy of different *Agrobacterium tumefaciens* strains in transformation of pinaceous gymnosperms. Physiol. Mol. Plant. Path. 34: 451–461.

Newton, R.J., H.S. Yibrah, N. Dong, D.H. Clapham and S. von Arnold, 1992. Expression of an abscisic acid responsive promoter in *Picea abies* (L.) Karst. following bombardment from an electric discharge particle accelerator. Plant Cell Rep. 11: 188–191.

Ordas, R.J., M.A. Lopez, J.C. Pacheco, J.A. Manzanera, A. Bueno, J.A. Pardos and R. Rodriguez, 1993. Possibilities to use *A. tumefaciens* as gene transfer vector in *Pinus nigra*. Paper presented at the 5th International Workshop of the IUFRO Somatic Cell Genetics Working Party, Balsian, Spain, October 18–22, p. 52.

Plaut-Carcasson, Y.Y., L. Benkrima, M. Dawkins, N. Wheeler, A. Yanchuk and S. Misra, 1993. Taxol from *Agrobacterium*-transformed root cultures. In: S. Scher and B.S. Schwarzschild (Eds.), Proceedings of the International Yew Resources Conference, March 12–13, Berkeley, CA, p. 28.

Potrykus, I., 1991. Gene transfer to plants: assessment of published approaches and results. Ann. Rev. Plant Physiol. Plant Mol. Biol. 42: 205–225.

Robertson, D., A.K. Weissinger, R. Ackley, S. Glover and R.R. Sederoff, 1992. Genetic transformation of Norway spruce (*Picea abies* (L.) Karst) using somatic embryo explants by microprojectile bombardment. Plant Mol. Biol. 19: 925–935.

Sargent, W.A., R.J. Kodrzycki, L.W. Handley, A.P. Godbey, C.A. Loopstra and R.R. Sederoff, 1993. Biolistic parameters for transient expression of loblolly pine cultures. In Vitro 29A: 85A.

Sederoff, R., A.M. Stomp, W.S. Chilton and L.W. Moore, 1986. Gene transfer into Loblolly pine by *Agrobacterium tumefaciens*. Bio/Technology 4: 647–649.

Serres, R., E. Stang, D. McCabe, D. Russell, D. Mahr and B. McCown, 1992. Gene transfer using electric discharge particle bombardment and recovery of transformed cranberry plants. J. Am. Soc. Hort. 117: 174–180.

Smith, C.O., 1934. Crown gall on conifers. Phytopath. 25: 894.

Smith, C.O., 1935. Crown gall on the Sequoia. Phytopath. 26: 439–401.

Smith, C.O., 1939. Susceptibility of species of *Cupressaceae* to crown gall as determined my artificial inoculation. J. Agr. Res. 59: 919–925.

Smith, C.O., 1942. Crown gall on species of taxaceae, taxodiaceae, and pinaceae, as determined by artificial inoculations. Phytopath. 32: 1005–1009.

Stomp, A.M., C. Loopstra, R. Sederoff, S. Chilton, J. Fillatti, G. Dupper, P. Tedeschi and C. Kinlaw, 1988. Development of a DNA transfer system for pines. In: J.W. Hanover and D.E. Keathley (Eds.), Genetic Manipulation of Woody Plants, pp. 231–241. Plenum Press, New York.

Stomp, A.M., C. Loopstra, W.S. Chilton, R.R. Sederoff and L.W. Moore, 1990. Extended host range of *Agrobacterium tumefaciens* in the genus *Pinus*. Plant Physiol. 92: 1226–1232.

Stomp, A.M., A. Weissinger and R.R. Sederoff, 1991. Transient expression from microprojectile-mediated DNA transfer in *Pinus taeda*. Plant Cell Rep. 10: 187–190.

Tautorus, T.E., F. Bekkaoui, M. Pilon, R.S.S. Datla, W.L. Crosby, L.C. Fowke and D.I. Dunstan, 1989. Factors affecting transient gene expression in electroporated black spruce (*Picea mariana*) and jack pine (*Pinus banksiana*) protoplasts. Theor. Appl. Genet. 78: 531–536.

Wilson, S.M., T.A. Thorpe and M.M. Moloney, 1989. PEG-mediated expression of GUS and CAT genes in protoplasts from embryogenic suspension cultures of *Picea glauca*. Plant Cell Rep. 7: 704–707.

Zambryski, P.C., 1992. Chronicles from the *Agrobacterium*-plant cell DNA transfer story. Ann. Rev. Plant Physiol. Plant Mol. Biol. 43: 465–490.

12. Manufactured Seeds of Woody Plants

William C. Carlson and Jeffrey E. Hartle

Contents

Though I do not believe that a plant will spring up where no seed has been, I have great faith in a seed. Convince me that you have a seed there, and I am prepared for wonders.
Henry David Thoreau

1. Introduction

Redenbaugh (1986, 1993) defines a synthetic seed as a somatic embryo inside a coating, and as being directly analogous to a zygotic seed. There have been several names given such "seed" including artificial seed, synthetic seed, seed analog and somatic embryo seed. We believe that the term "manufactured" seed, as coined by our coworker James Dooley, reflects the nature of the construct in a more accurate way. The practical requirements of such seed are that they perform the basic functions of a botanic seed during the sowing and germination of a somatic embryo under field conditions. These basic functions include: protecting the embryo and surrounding nutritive matrix from mechanical damage, desiccation and microbial invasion; providing for an adequate supply of nutrients including carbon, gas exchange, and water to support germination; and physical properties that allow the germinating embryo to emerge normally from the seed under field conditions.

Many woody plants can be vegetatively propagated to allow genetic gain from maintaining traits that would be lost during meiotic recombination.

S. Jain, P. Gupta & R. Newton (eds.), Somatic Embryogenesis in Woody Plants, Vol. 1, 253–263.

Another substantial benefit in the production of vegetatively propagated woody plant crops is the opportunity to harvest uniform material. Increased raw material uniformity will open many new opportunities to make wood-based-manufacturing facilities more efficient.

In the future as genetic engineering becomes a functional part of genetic improvement programs the role of manufactured seed could be critical to success in at least two ways. First is the delivery of transgenic foundation stock. Secondly, if regulations mandate the use of sterile plants for field culture of transgenic crops then manufactured seed offer an economical method of delivery into the agronomic system.

Once a decision is made to vegetatively propagate a given woody plant, then the next decision is which method to use. Rooted cuttings can be used with many plants. There are commonly severe limitations to the number of propagules that can be produced per genotype. This is due to reduction in percentage of cuttings that root, and in some cases also reduced growth rate as the stock plant age increases. Also with many rooted cutting systems costs are limiting. Organogenesis offers another method of clonal multiplication, but costs are prohibitive because of the labor intensive nature of the process. Theoretically, somatic embryogenesis offers the potential for a low cost vegetative propagation method that allows a very large number of embryos to be propagated per genotype (not biologically limited) and allows cryopreservation of cultures while genotypes are being evaluated. Krikorian (1988), Bornman (1991), Mo (1993) and Redenbaugh (1993) point out that the quality of somatic embryos does not currently allow practical realization of this potential efficiency. Goebel-Tourand *et al.* (1993) demonstrated some of the consequences of poor somatic embryo morphology on normalcy during conversion. Once somatic embryos reach the quality necessary to produce zygotic-embryo-like vigor, then it will be possible to move forward toward commercialization. Manufactured seed offer the potential of delivering somatic embryos to the field or greenhouse utilizing standard agronomic, horticultural and forestry practices for seed sowing and crop culture. The technology will be very valuable at implementation but economical implementation demands major improvements in somatic embryo quality and in manufactured seed performance under normal agronomic conditions.

2. Historical Development of Manufactured Seed

Woody plant utilization of the manufactured seed concept arises out of earlier concepts applied to agronomic plants. There have been two schools of thought, desiccated and hydrated. We will, therefore, briefly summarize the historical derivation for the reader's reference. For a more complete discussion we refer the reader to Redenbaugh (1993).

Desiccated manufactured seed were first developed by Kitto and Janick (1986) at Purdue University. This seed consisted of desiccating an embryo

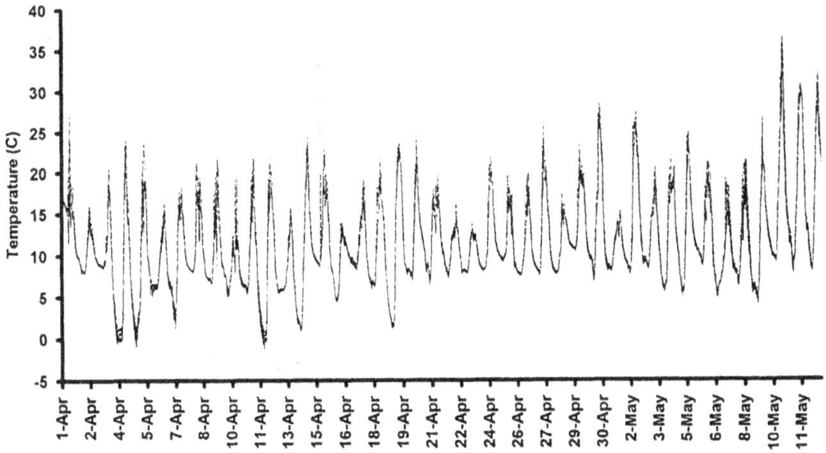

Figure 1. Temperature profile of the seed zone in a newly sown nursery bed in Weyerhaeuser Company's Mima Nursery near Olympia, WA. Thermisters were placed 6 mm under the soil surface, and temperatures represent maximum and minimums over each 30 min period.

coated with Polyox®. Germination, albeit at low levels, was achieved by placing the dried wafer in tissue culture media then plating on media moistened filter paper for germination. This approach has been followed by Gray at the University of Florida, and Attree and Fowke at the University of Saskatchewan at Saskatoon.

Hydrated manufactured seed were first developed by Redenbaugh, (1986). This method involves encapsulation of the embryo in a drop of sodium alginate followed by complexing with calcium ions to form a calcium alginate gel. Most of the published research on manufactured seed involves either this method used directly or with modifications. Communication of research and development of this alginate bead seed analog technology by Redenbaugh and associates galvanized the imagination of potential user groups and stimulated much research on a world-wide level.

3. Practical Considerations in the Design of Manufactured Seed

Natural seed have many functional features that facilitate their use in agriculture: 1) a hard seed coat that provides protection from mechanical damage, reduces the rate of drying and slows microbial invasion, 2) an endosperm, or in conifers a female gametophyte, that provides a controlled release of nutrients and an energy source, and physically constrains the germinating embryo such that elongation of the embryo promotes emergence of the radicle, then hypocotyl and finally the cotyledons from the seed, 3) an anatomically, physiologically and biochemically correct embryo which germinates and grows with substantial vigor, and 4), the capability to withstand wide variation of temperature (Fig. 1), soil moisture, oxygen availability, and microbial innoculum levels to yield high germination. Manufactured seed must mimic natural seed functions if they are to perform in the agricultural field in a manner that will allow economical application of the technology. Just as low vigor zygotic seed is often observed to undergo mortality from several secondary causal factors such as microbial invasion, manufactured seed must be vigorous to avoid a similar fate.

4. Current Status of Design of Manufactured Seed

4.1. *Manufactured Endosperm or Female Gametophyte*

Natural seed show a large variation within and among genera in the proportion of the total seed nutrient and energy reserves stored in the embryo. Some species require extra-embryo reserves from the endosperm/female gametophyte to germinate while others will germinate under ideal conditions without it. A simple test of germinability of excised embryos on sterile water agar will give an initial indication of whether the species in question will germinate without female gametophyte/endosperm, that is whether an artificial gametophyte (or artificial endosperm) is necessary. Fujii *et al.* (1992) demonstrated, for example, that naked alfalfa embryos placed in the field under styrofoam cups had a 25% conversion to autotrophic plants. Nature is extremely efficient, providing for controlled enzymatic cleavage of complex molecules to provide simpler molecules to the surfaces surrounding the embryo. In manufactured seed there are three choices in design of a system to supply nutrients: 1) limit nutrients to those carried in the embryo (i.e., no additional provided in the manufactured seed), 2) provide controlled release chemistry in the nutritive gel surrounding the embryo, or 3) provide a static supply of nutrients in the surrounding gel matrix but protect them from the leaching effects of water moving through the soil.

The Redenbaugh *et al.* (1986) manufactured seed as used on alfalfa and celery embryos usually utilize the first approach. No additional nutrients are supplied outside the somatic embryo. Embryo development protocols that induce the development of high levels of reserves in the somatic embryo increase manufactured seed germination (Fujii *et al.*, 1992). This approach is simple but it offers limited application, since most species either will not germinate without additional nutrients or have reduced vigor.

Kirin Brewing Company has pioneered the use of controlled release nutrients. Sanada *et al.* (1993) reported increased conversion frequencies of carrot and celery embryos encapsulated with microencapsulated sucrose. Microcapsules were 0.5 mm in diameter and were coated with ethylene-vinyl acetate copolymer and wax. They released sucrose over several weeks and stopped releasing at low temperatures. The microencapsulation technology was reported in Sakamoto *et al.* (1991a). This technology was patented in 1989 (Sakamoto *et al.*, 1991b). In the patent examples, germination of celery manufactured seed was improved from 30% without microcapsules to 90% with microcapsules when seed were germinated *in vitro* on a sugar free nutrient medium. Friend (1993) reported no improvement in conversion of alfalfa manufactured seed using sucrose microencapsulated with either cellulose acetate butyrate or gelatin. He suggested that low encapsulation efficiency could have caused the poor results apparently due to rapid release of sugar.

The third approach, involving freely available nutrients, has been used by many researchers beginning with Redenbaugh (1986), but the conditions of use define the result. Most of theses tests utilize the calcium alginate 4 mm diameter bead seed design. If the embryo is dependent on the surrounding gel for nutrients, then results were generally good only if the experiments were done with the seed sown on the surface of nutrient media, or in soil irrigated with such media. The results were generally poor if the seed were sown in a system where no additional nutrients were available. This is due to the fact that the volume of the 4 mm diameter alginate bead is inadequate to hold the required nutrient supply. This points out that it is critical to conceptualize use under agronomic conditions when determining the approach one will utilize. Germination on nutrient media can aid in defining some aspects of manufactured seed design, but the inference space of such experiments is very limited. For example, the effect of adding or subtracting a given nutrient can be studied in this way but the volume of the nutrient necessary to support germination in soil cannot.

Bapat and Rao (1988), and Bapat (1993) reported inclusion of nutrient salts in manufactured sandalwood seed which were beads of a composite of alginate and silica gel. Theoretically, silica gel should adsorb nutrients, acting to control their release to the embryo. In most experiments germination tests were carried out on nutrient media solidified with agar, thus nutrient depletion was not an issue. Where tap water was used, instead of nutrient media, germination was severely reduced. Similarly, Lulsdorf *et al.* (1993) added activated charcoal to alginate beads encapsulating interior spruce (*Picea glauca engelmannii* complex) and black spruce (*Picea mariana* Mill.). Such seed survived one month cold storage at 4°C on nutrient media, although results varied between species and between zygotic and somatic embryos. All germination was carried out on solid nutrient media plates, thus again nutrient supply was not an issue.

Mukunthakumar and Mathur (1992) obtained germination of male bamboo with encapsulation in 6% calcium alginate with MS salts with or without 3% sucrose. They coated seeds with paraffin oil to reduce microbial invasion and desiccation. Germination was 45% in initially sterile soil in covered pots under greenhouse conditions in spite of heavy microbial invasion of the alginate. The investigators attribute this success to the anatomy and disease-resistant properties of bamboo.

Figure 2. Weyerhaeuser Company manufactured Douglas-fir seed (Carlson *et al.*, 1992) in various stages of germination. In this case seeds were manufactured from excised zygotic embryos. (A) Newly emerging germinant in the "crook stage" of development, seeds were sown 6 mm under the soil surface. (B) Emergent germinant raising the seed coat above the soil. As with natural seed, manufactured seed sometimes emerge leaving the coat under the soil surface, and sometimes shed the coat as shown here. (C) Germinant has shed the seed coat and is free to grow. (D) A population of germinants from manufactured seed.

At Weyerhaeuser, Carlson and Hartle (unpublished data) anticipated the need to supply conifer embryos with nutrients over the entire, 6-week, "worst case", potential germination period. They measured the nutrient use of conifer embryos and cross checked the estimate from the dry weight gain over the germination period. They calculated that the weight of nutrients necessary to support a "worst case" germination period would be approximately 6 mg total salts including sugar. They tested several nutrient concentrations and determined that a gel volume of approximately 0.5 ml would support conifer embryos over the 6 week period. This volume provides both water and nutrients to support the germination process.

4.2. *Oxygen Supply in Hydrated Gels*

In many manufactured seed designs oxygen has been noted as limiting to germination. As gel volumes become larger this problem becomes more acute. Carlson *et al.* (1992) dramatically improved the performance of manufactured seed by making oxygen more available through the inclusion of oxygen carrier emulsions in the gel matrix. This increased the number of normal germinants four fold in alginate bead seeds germinated on agar media which included nutrient salts. Most importantly, it also makes possible the use of large manufactured gametophyte/endosperm gel volumes to support the embryo through a potentially long germination period in soil (Fig. 2).

4.3. *Manufactured Seed Coats*

Hydrated gels used for manufacturing analogs of female gametophyte/endosperm rapidly lose water to the ambient air or soil on sowing. Practical application of manufactured seed requires that this water loss be controlled. Similarly, mechanical damage from sowing equipment and entry of soil microbes must be reduced by a coating, if high vigor is to be expected under practical farming conditions.

Friend (1993) reviewed work done at SRI International in conjunction with Redenbaugh's group at Plant Genetics Inc. to develop hydrophobic coatings for alginate bead manufactured seed. Specifically, ethylene vinyl acetate copolymers of various types can be applied to the gel surface by spraying, dipping or immersion in a solution of the polymer in a solvent. Since the organic solvents suitable to the process are highly phytotoxic, it is very important to control the process carefully such that solvents do not penetrate to the depth of the embryo. If done carefully coatings can be applied without reducing the shoot emergence of alfalfa embryos from the alginate beads (Friend, 1993). Moisture loss can be controlled to less than 1% per day by such methods.

Natural seed have relatively hard seed coats that prevent rapid water loss and also protect the seed from mechanical damage during handling, mechanical sowing and in the soil. Carlson *et al.* (1992) developed a wax impregnated paper seed coat (Fig. 2). This coat has many of the favorable properties of a natural seed coat; it is harder than the coating mentioned above, and it is easily applied without exposing the manufactured gametophyte/endosperm or embryo to risk of organic solvent-induced damage.

4.4. *Physical Considerations in Germinant Emergence from Manufactured Seed*

Natural seed anatomy provides a direct exit for the root cap as the embryo begins to elongate during germination. As root, hypocotyl and cotyledon elongation occur, the root cap, then the root apex, then the hypocotyl and finally the cotyledons emerge along an anatomically defined route. Physical forces of elongation result in progressive shedding of the seed coat including any remaining nutritive tissues.

Hydrogel encapsulation can produce gels that are soft enough to allow penetration of the root cap and germinant emergence. Dehydration of the hydrogel aids in emergence. If the seed is coated with a seed coat that is hard enough to provide mechanical protection during mechanized sowing, there is a potential for reduced embryo emergence due to the radicle curving along the inside of the seed coat rather than emerging through it. Similarly, if the volume of gel is large enough to support the embryo through a several week germination period, then there can be problems with the enlarging cotyledons becoming trapped in the gel rather than shedding the seed coat.

Redenbaugh *et al.* (1993) reviewed a method for making self-breaking capsules as a modification of the alginate bead technology. Such capsules break when they are exposed to water after sowing. Masuda and Sakamoto (1993) have also filed a patent application for a self-breaking capsule. It was described as an interior-complexed alginic acid gel core. The core was coated with a hard shell described as a: polyamino acid, basic polysaccharide, bis biguanide, basic polymer containing an ion exchanger comprised of cellulose, and ion exchange resin, inorganic ion exchanger and polymer coagulate. This design also was described as breaking spontaneously upon sowing in the soil.

Self-breaking capsule designs would indeed allow the embryo to emerge easily but have the disadvantage that many of the benefits of manufactured seed are lost at the point of breakage, leaving the embryo to germinate in loose contact with the gel which would then readily desiccate.

To prevent the germinant from penetrating and becoming trapped in the gel, Carlson *et al.* (1992) enclosed the embryo in a cotyledon restraint cylinder. This cylinder orients the forces of the elongating embryo such that rapid germinant emergence is promoted (Fig. 2). Elongation of the radicle forces the root cap through a thin coating over the end of the cotyledon restraint cylinder. This design is analogous to natural seed in that the germinant does not shed the seed coat until nutrient reserves and protection from desiccation are no longer necessary after it has become autotrophic.

5. Woody Plant Manufactured Seed

Is the design of manufactured seed for a woody plant species really different from that of a crop plant? We believe that at the current state of knowledge there is little that can be noted as different. There is currently no ready-to-implement technology in the manufactured seed area in either woody or non-woody species. It is likely that there will be some adjustment in contents and volume of the artificial female gametophyte or endosperm as different species are considered, but we believe that there are not likely to be major differences in seed design.

6. Current Limitations to Implementation of Manufactured Seed

6.1. *Embryo Quality Must Improve*

Somatic embryo quality must improve to the point where embryos have the same characteristic vigor and morphology as zygotic embryos. There has been continuous improvement in embryo quality in many species, but more work is needed to achieve zygotic-like quality.

6.2. *Experimentation Must Be Carried out Under Appropriate Conditions*

Much current manufactured seed literature describes experiments in which embryos are encapsulated in alginate and germinated on nutrient media *in vitro*. These experiments do little to advance the state of the art. Experiments designed to evaluate the presence or concentration of components of an artificial female gametophyte/endosperm can be done on the surface of media, but one must realize that no knowledge of the volume required to support the embryo and related factors will be gained. Similarly when the whole seed structure is being tested, it should be tested when sown to a normal depth in soil.

We have previously stated that there is some variation among species in response of naked embryos to being placed on water agar in the total absence of supporting nutrients (organic or inorganic). We note as well that in our laboratory all species tested have responded with improved vigor when nutrients were supplied. High vigor is very important to the successful germination of natural seed in agricultural soils. If vigor is reduced either by seed quality factors or weather, then problems with soil microbes will increase. This will also be true for manufactured seed.

6.3. *Manufactured Seed as Analogs of Botanic Seed*

Manufactured seed have improved in our laboratory as they have become functionally more like natural seed. This has involved improving vigor through providing an adequate oxygen supply and by creating a cotyledon restraint system that allows nutrients to be available at the moist embryo-interface surface. The cotyledon restraint system also channels the forces of embryo growth to facilitate timely extraction of the germinant from the seed. Further work is needed in this area to achieve the predictable 95% germination and rapid emergence from agricultural soil common to natural seed. We believe that manufactured seed should perform as well as natural seed in order to be commercially viable. They must be capable of withstanding broad temperature ranges and moisture conditions ranging from short-term flooding to short-term dry periods. They must also be biodegradable, and must have a constituency and vigor that will ward off infestation. Several of these requirements cannot be fully tested and developed until large scale field experiments can be implemented. Large scale field tests will not be practical until germination rates > 80% in non-sterile soil freshly sampled from an agricultural field can be accomplished. Since traditional methods of methyl bromide/chloropicnin fumigation will probably no longer be available by the time of full implementation of this technology, seed should be designed to perform well in soil that has not been fumigated for several years. Seed must also be storable. Some storable desiccated seed designs have been proposed (Kitto and Janick, 1980, 1986; Attree and Fowke, 1993) but they are of more interest as methods of embryo desiccation than as seed since they require germination on nutrient media. Several tests have been done with hydrated or partially hydrated gel encapsulation and cool storage (e.g., Lulsdorf *et al.*, 1993; Redenbaugh and Fujii, 1988). Such techniques offer promise of manufactured seed storability. Manufacturing facilities, transport and on site handling will be much easier if manufactured seed are as storable as their natural counterparts. Many of these areas will require considerable research to reach practically implementable solutions.

7. Conclusions

Manufactured seed technology has improved substantially over the past decade. There is a high level of interest in the technology from agronomic, horticultural and forest industries. Manufactured seed technology could make it economically feasible to vegetatively propagate large numbers of plants originating as somatic embryos and sown as seed in agricultural fields with traditional equipment. This would be extremely valuable to several aspects of current genetics programs and could be critical to propagation of genetically engineered plants.

Both somatic embryo quality and manufactured seed design must improve prior to implementation of the technology. We believe that duplicating the chemical and physical attributes of natural seed that result in rapid emergence of normal germinants under field conditions will be the most productive way to approach research problems associated with operational use of manufactured seed.

References

Attree, S.M. and L.C. Fowke, 1993. Maturation, desiccation and encapsulation of gymnosperm somatic embryos. International Patent Application Number WO 93/11660, Filed December 19, 1991.

Bapat, V.A., 1993. Studies on synthetic seeds of sandlewood (*Santalum album* L.) and mulberry

(*Morus indica* L). In: K. Redenbaugh (Ed.), Synseeds: Applications of Synthetic Seeds to Crop Improvement, Chapter 21. CRC Press, Boca Raton, FL.

Bapat, V.A. and P.S. Rao, 1988. Sandlewood plantlets from synthetic seeds. Plant Cell Rep. 7: 434.

Bornman, C.H., 1991. Somatic embryo maturation is a critical phase in the development of a synthetic seed technology. Rev. Cytol. Veget. Bot. 14: 289–296.

Carlson, W.C., J.E. Hartle and B.K. Bower, 1992. Analogs of Botanic Seed. European Patent Application Number PCT/US91/07997, Filed in The United States October 26, 1990.

Friend, D.R., 1993. Hydrophobic coatings for synthetic seeds. In: K. Redenbaugh (Ed.), Synseeds: Applications of Synthetics Seeds to Crop Improvement, Chapter 4. CRC Press, Boca Raton, FL.

Fujii, J.A., D. Slade, J. Aguirre-Rascon and K. Redenbaugh, 1992. Field planting of alfalfa artificial seeds. In Vitro Cell. Dev. Biol. 28P: 73–80.

Goebel-Tourand, I., M. Mauro, L. Sossountzov, E. Miginiac and A. Deloire, 1993. Arrest of somatic embryo development in grapevine: histological characterization and the effect of ABA, BAP and zeatin in stimulating development. Plant Cell Tissue Organ Cult. 33: 91–103.

Kitto, S. and J. Janick, 1980. Water soluble resins as artificial seed coats. Hort. Sci. 15: 439 (Abstr.).

Kitto, S. and J. Janick, 1986. U.S. Patent 4,615,141, Filed August 14, 1984.

Krikorian, A.D., 1988. Plant tissue culture: Perceptions and realities. Proc. Indian Acad. Sci. (Plant Sci.) 98(6): 425–464.

Lulsdorf, M.M., T.E. Tautorus, S.I. Kikcio, T.D. Bethune and D.I. Dunstan, 1993. Germination of encapsulated embryos of interior spruce (*Picea gluca engelmannii* complex) and black spruce (*Picea mariana* Mill.). Plant Cell Rep. 12: 385–389.

Masuda, Y. and Y. Sakamoto, 1993. Self-rupturing artificial seed – comprises core of plant-propagating material, coat of alginic acid gel with complexing ions partly replaced and hard outer coat. Patent filed in Japan JP59102308. Listed in World Patent Index with abstract in English.

Mo, L.H., 1993. Somatic embryogenesis in Norway spruce (*Picea abies*). Res. Note 49. 1993. Dissertation, Department of Forest Genetics, Swedish University of Agricultural Sciences, Uppsala.

Mukunthakumar, S. and J. Mathur, 1992. Artificial seed production in the male bamboo, *Dendrocalamus strictus* L. Plant Science 87: 109–113.

Redenbaugh, K., 1986. Analogs of Botanic Seed. U.S. Patent 4,562,663, Filed October 12, 1982.

Redenbaugh, K., 1993. Introduction. In: K. Redenbaugh (Ed.), Synseeds: Application of Synthetic Seeds to Crop Improvement, Chapter 1. CRC Press, Boca Raton, FL.

Redenbaugh, M.K. and J.A. Fujii, 1988. Desiccated analogs of botanic seed. United States Patent 4,777,762, Filed December 24, 1986.

Redenbaugh, K., J. Fujii, D. Slade, P. Viss and M. Kossler, 1986. Synthetic seeds – Encapsulated somatic embryos. In: Agronomy – Adjusting to a Global Economy. Proc. American Society of Agronomy, Crop Science Society of America and Soil Science Society of America 78th Annual Meeting, New Orleans, LA, November 30 – December 5.

Redenbaugh, K., J.A. Fujii and D. Slade, 1993. Hydrated coatings for synthetic seeds. In: K. Redenbaugh (Ed.), Synseeds: Applications of Synthetic Seeds to Crop Improvement, Chapter 3. CRC Press, Boca Raton, FL.

Sakamoto, Y., N. Onushi, M. Hayashi, A. Okamoto, T. Mashiko and M. Sanada, 1991a. Artificial seed – Development of plant seed analog. Shokubutsu no kagaku chosei 26(2): 205–211.

Sakamoto, Y., S. Umeda and H. Ogishima, 1991b. Artificial seed comprising a sustained-release sugar granule. United States Patent 5,010,685, Filed May 1, 1989.

Sanada, M., Y. Sakamoto, M. Hayashi, T. Mashiko, A. Okamoto and N. Ohnishi, 1993. Celery and lettuce. In: K. Redenbaugh (Ed.), Synseeds: Applications of Synthetic Seeds to Crop Improvement, Chapter 17. CRC Press, Boca Raton, FL.

13. Scale-up of Embryogenic Plant Suspension Cultures in Bioreactors

T.E. Tautorus and D.I. Dunstan

Contents

1. Introduction

Somatic embryos of many woody species can be grown as suspensions in liquid medium (Tautorus *et al.*, 1991; Dunstan *et al.*, 1995). In this state cultures grow more rapidly than on agar-solidified medium, presenting opportunities for economical, large-scale propagation of plants. For laboratory-scale experimentation embryogenic suspension cultures are most commonly grown in 125 or 250 ml Erlenmeyer flasks on a gyratory shaker (Ammirato and Styer, 1985; Macek *et al.*, 1989; Denchev *et al.*, 1992). This procedure can be scaled-up in a limited fashion by employing more flasks of the same size or by increasing the size of each vessel. For example, assemblies of 1 L nipple flasks containing 100–200 ml medium have been used to culture embryogenic suspension cultures of conifers, e.g., *Pseudotsuga menziesii* (Durzan and Gupta, 1987), and *Picea abies* (Boulay *et al.*, 1988).

A more efficient large-scale culture of somatic embryos can be achieved through the use of bioreactors. These may be defined as vessels which are 1 l or larger, used to culture animal, plant, or microbial cells under controlled conditions of aeration, and agitation, and potentially other environmental parameters such as temperature. Often the term fermenter is used instead of bioreactor, especially in the literature referring to microorganisms, or with plant cells for production of secondary metabolites. The basic purpose of

S. Jain, P. Gupta & R. Newton (eds.), Somatic Embryogenesis in Woody Plants, Vol. 1, 265–292.

any bioreactor is to provide optimum growth conditions by regulating various chemical and physical environmental factors (Preil, 1991). The use of bioreactors offers the capability for rapid growth, maturation, transfer, and delivery of plant embryos (Ammirato and Styer, 1985). Although a variety of non-embryogenic plant cultures have been grown in an assortment of bioreactor types (for reviews see Scragg and Fowler, 1985; Panda *et al.*, 1989; Doran, 1993), there have been fewer reports of the use of bioreactors for the large-scale culture of somatic embryos.

The purpose of this chapter is to review the use of shake-flask and bioreactor technology for culture of somatic embryos. For comparative purposes, this review surveys somatic embryo cultures of woody and non-woody plants.

2. Shake-Flask Culture

Embryogenic tissues of many plant species have been cultured in liquid medium in shake-flasks; for examples of conifer species see Attree *et al.* (1991) and Tautorus *et al.* (1991). Using the following procedure most genotypes of *Picea glauca*, *Picea mariana*, and the species complex *Picea glauca-engelmannii* have been successfully established and maintained as embryogenic liquid suspension cultures in shake-flasks.

The inoculum for establishment of shake-flask cultures is rapidly growing embryogenic tissue, approximately 6–8 g fresh weight, transferred from agar-solidified plates to 50 ml of appropriate maintenance medium in 250 ml-capacity DeLong baffle flasks. Baffle ridges on the insides of each shake-flask help to break up the growing tissue, and help to maintain the tissue in suspension early in the establishment phase; eventually, flasks without baffles are used routinely. Agitation is effected by placing flasks on a gyratory shaker set at 100–150 rpm. Initially, the plant tissue is not subcultured for up to two weeks, to allow adequate growth. During the next 2–4 weeks of the establishment phase, it is generally adequate to divide each parent culture into 2–4 equal aliquots at each weekly transfer. Prior to subculture, suspension cultures should be dense and particulate when agitated. After cultures are established, they are usually subcultured by weekly transfer of 10–20 ml aliquots into 50 ml fresh medium. Alternatively, 2–4 g fresh weight of suspension culture tissue can be used. The exact conditions will relate to rate of growth of each culture and the intended use for the culture. Shake-flask culture is, therefore, a semicontinuous process, in that part of the culture is transferred to a fresh medium at regular intervals. The requirements for light, temperature, and nutrients may differ from those of the source tissue used to prepare the suspension culture, and may need to be re-examined periodically.

Shake-flask cultures can be used to determine growth kinetics, and nutritional requirements for growth and embryo yield. Accurate and rapid measurements of cell growth, and assessment of growth kinetics, are important for the development and optimization of methods for large-scale culture of plant material. Several parameters can be considered for use in these evaluations, including fresh and dry weight, settled and packed culture volume, embryo number, medium conductivity, osmolarity, and pH. Lulsdorf *et al.* (1992) determined for *Picea glauca-engelmannii* and *P. mariana* that the non-destructive parameters, conductivity and osmolarity, were closely correlated to culture growth and embryo yield. Krogstrup (1988) suggested the use of settled cell volume as a non-destructive quantitative measurement of growth of *Picea sitchensis* somatic embryos during shake-flask culture, though correlations with growth or embryo number were not reported. Some aspects of medium utilization can also be determined, e.g., by monitoring relative abundance of nutrients (e.g., carbohydrates, nitrogen compounds) in culture supernatants (Lulsdorf *et al.*, 1992). Similarly the metabolism of other compounds, such as abscisic acid (Dunstan *et al.*, 1992), can be conveniently monitored using shake-flask suspension cultures. Such investigations can provide insight into the fate of certain medium components.

The use of weekly subculture for maintenance of rapidly growing suspension cultures can raise questions as to long-term stability, and reliability, of the plant cells or tissues. Because each flask's culture is an independent unit, will there be flask-to-flask differences in e.g., inoculum, and aeration? Is there an inevitable selection for faster growing cell- or tissue-types? Will culture responses alter over time? Such questions are especially important when the use of shake-flask cultures is being considered for consistent production of somatic embryos. There are, indeed, indications that spruce somatic embryos can exhibit altered responses, e.g., during maintenance and during somatic embryo maturation, after extended periods in shake-flask culture. That this might be due to altered sensitivity to phytohormones is shown by the ability of some cultures to regain the response by suitable manipulative treatments, e.g., by auxin removal in the maintenance medium used immediately before treatment for spruce somatic embryo maturation (Dunstan *et al.*, 1993).

3. Bioreactor Culture

Bioreactors provide many advantages for the growth of plant cultures compared to shake-flasks, including increased working volume, maintenance of a nearly homogeneous culture, and control of the cultural and physical environment for optimum growth (Ammirato and Styer, 1985; Cazzulino *et al.*, 1991).

3.1. *Bioreactor Design*

Many configurations and sizes (1–75000 L) of vessels have now been used to grow plant cells. Choice of bioreactor design for a given culture may be

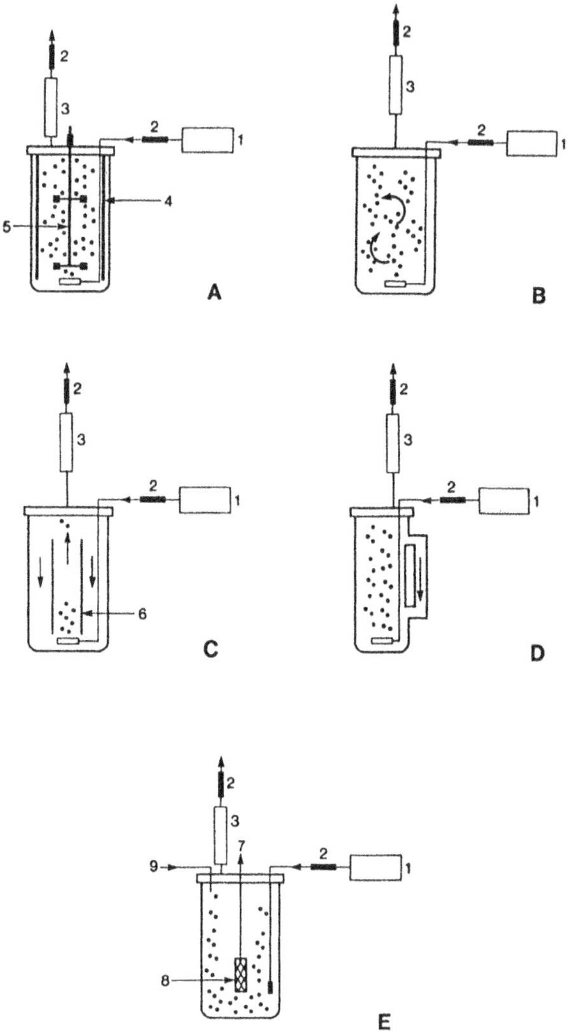

Figure 1. Schematic diagrams illustrating the various bioreactor configurations which have been successfully used with plant cell culture. (A) Mechanically-stirred. (B) Bubble column. (C) Airlift with internal loop. (D) Airlift with external loop. (E) Spin filter, Redrawn from Styer, 1985. Key: 1 = Air supply; 2 = Filter; 3 = Exit air condensor; 4 = Baffle; 5 = Agitator with flat bladed impellers; 6 = Draft tube; 7 = Spent medium out; 8 = Spin filter; 9 = Medium in.

determined largely through trial and error, and often depends upon several factors, including cell type and species, oxygen transfer and mixing requirements, shear rates, and the experimental objective (Styer, 1985; Treat *et al.*, 1989; Preil, 1991; Denchev *et al.*, 1992). Bioreactors for plant cell culture can be classified according to the type of agitation system used: mechanically (conventionally) stirred, or air flow agitation.

3.1.1. *Agitation by Mechanical-Stirring*

The stirred tank bioreactor, in which mixing and air dispersion are achieved by mechanical agitation, represents one of the most versatile bioreactor systems in use in industry at this time (Fig. 1A). Mechanically-stirred vessels can employ paddles, impellers, magnetic stir bars, or vibrating perforated plates (Cazzulino *et al.*, 1991). The effects of changes in number of impeller blades and their position, and the ratio between the diameter of the stirrer and the vessel have to be experimentally determined to ensure optimum conditions, especially for shear-sensitive species (Preil, 1991). Magnetic stirrers operate at higher revolutions per minute than impellers without shearing the cells (Styer, 1985). Stirred bioreactors are usually equipped with baffles to improve mixing within the vessel and to reduce vortexing.

A mechanically-stirred vessel is effective for mixing nutrient medium and cells, dispersing air bubbles (for improved gas transfer), and preventing the formation of large cell clumps. Dissolved oxygen concentration, temperature, and pH are generally easy to control in a mechanically-stirred bioreactor. Some disadvantages of mechanically-stirred vessels are that they may over-aerate plant cultures, and damage the cells through shear stress. Shear stress is generated in a stirred-bioreactor in a variety of ways, for example from the shear force at the impeller tip, from air bubbles coalescing and breaking at the liquid surface, from impact with other cells in the system, and from fluid flow, and turbulence caused by baffles (Leckie *et al.*, 1991).

3.1.2. *Agitation by Air-Flow*

Vessels that employ agitation by air-flow are either bubble column (Fig. 1B) or airlift type bioreactors (Fig. 1C,D). The bubble column bioreactor consists of a cylindrical vessel aerated from the bottom, where the air contacts gas, liquid, and cells. Low aeration rates cause a homogeneous or laminar flow (Preil, 1991). In such a system gas is dispersed pneumatically through a deep pool of liquid by means of nozzles or perforated plates. The bubble column bioreactor has not been extensively used for plant cell culture (Panda *et al.*, 1989).

Airlift bioreactors are of mechanically simple design since both the aeration and agitation of the medium are provided by compressed air entering the bottom of the vessel (Styer, 1985; Stuart *et al.*, 1987). An early design consisted of a rubber-stoppered 20 L carboy fitted with four tubes: air in, air out, nutrient medium in, and sample out (Tulecke and Nickell, 1959). This system was successfully used for culture of *Ginkgo*, *Ilex*, *Lolium* and *Rosa*. More recent airlift designs have used either i) an internal loop with draft tube (coaxially arranged cylinders) (Fig. 1C), or ii) an external outer loop (Fig. 1D) which is desgined to stabilize the liquid circulation patterns, producing a well-defined flow (Preil, 1991). The circulation of the liquid medium results from differences in gas content between the aerated and the non-aerated (downflow) regions. A more uniform mixing pattern is thus obtained in the airlift compared with the bubble column where a random flow pattern exists (Panda *et al.*, 1989). The main advantages of airlift bioreactors are: simplicity of construction, versatility, low operating costs, low and uniform shear stress (because of the use of hydrostatic agitation), and high mass-transfer and heat-transfer rates with low power input (Ammirato and Styer, 1985; Cazzulino *et al.*, 1991). Disadvantages include the development of dead zones inside the bioreactor, and insufficient mixing at high cell densities (Panda *et al.*, 1989; Doran, 1993).

Other bioreactor types used for plant cells include hollow fibre, photobioreactor (Lee and Bazin, 1990), hybrid bioreactor (Kim *et al.*, 1991a), recycling airlift (Townsley *et al.*, 1983), roller bottle (Hong *et al.*, 1989), and rotating drum (Panda *et al.*, 1989). Bioreactors have also been developed for hairy root culture (Hamill *et al.*, 1987; Toivonen *et al.*, 1990; Bisaria and Panda, 1991; Doran, 1993). For further discussion of bioreactor design the reader is referred to Margaritis and Wallace (1984), Scragg and Fowler (1985), Rokem (1988), Cazzulino *et al.* (1991), Taticek *et al.* (1991), and Doran (1993).

3.2. *Bioreactor Configurations Used with Somatic Embryo Cultures*

Bioreactor configurations that have been for somatic embryo culture are shown in Table 1. In most cases, relatively small airlift or mechanically-stirred vessels (less than 8 L) have been employed. A major limitation in many of the reports cited in Table 1 is the lack of specific information regarding bioreactor configuration, which will limit reproducibility of the reported experiments. Such information should include impeller size, their number and placement, the type and placement of sparger, and the size of baffles.

An early attempt to scale-up somatic embryogenesis was performed using *Daucus carota* cells in 20 L carboys, using an aeration tube extending to the bottom of the vessel (Backs-Hüsemann and Reinert, 1970). However, few embryos were formed. More recently, a magnetically-stirred bioreactor called a spin filter was successfully used for culture of *D. carota* somatic embryos (Fig. 1E) (Ammirato and Styer, 1985; Styer, 1985). The spin filter bioreactor resembles a mechanically-stirred vessel, except that the central shaft contains a filter housing. The filter is rotated by a magnetic stirrer that provides agitation of the medium without generating the shear observed with blade stirrers. The shaft is hollow and serves for removal of spent medium without loss of cells. This simplifies independent regulation of medium and cell flow. Preil (1991) suggested that short term exposures to phytohormones for induction of somatic embryogenesis can be performed more precisely with the spin filter bioreactor.

Styer (1985) and Ammirato and Styer (1985) used a two-stage spin-filter system with *D. carota* somatic embryos. Unfortunately, little data or methodology were reported. *D. carota* cell suspension cultures from shake-flasks were used to inoculate a first stage bioreactor, containing cell proliferation medium. Later, an aliquot of cells was transferred into a second stage bioreactor containing embryo development medium. This resulted in plantlet development by embryogenesis. Plants were successfully acclimatized in the greenhouse.

Euphorbia pulcherrima somatic embryos (Preil *et al.*, 1988) and *Medicago sativa* cells (Chen *et al.*, 1987) have been cultured in bioreactors which used vibration mixers. These produced their stirring effect by a vertical reciprocating motion of agitator shafts fitted with horizontally inserted discs. Conical holes in the discs caused an upward or downward stream (Preil, 1991). This mixing method prevented foam formation (a consequence of the absence of a vortex), inhibited cell aggregation, and produced a high degree of random cell distribution combined with very low shear forces.

3.3. *Growth Kinetics and Biomass Yields*

The growth of plant cells in bioreactors of any design can be characterized by the equations developed for microbial fermentations. Plant cells have been grown as batch, semicontinuous, and continuous (Chemostat, turbidostat) cultures depending upon the product. In batch cultures, the cells usually exhibit a lag phase, then enter the log phase of exponential growth, followed by a stationary phase (Styer, 1985).

Picea mariana and *Picea glauca-engelmannii* somatic embryos were cultured in 250 ml shake-flasks, 7 L mechanically-stirred bioreactors, and 7 L airlift bioreactors, in order to compare parameters for measuring growth under various cultivation conditions (Table 2). Both species showed little or no lag phase, a growth phase of approximately eight days, followed by a stationary phase from day eight when cultured in mechanically-stirred bioreactors (for example, see Fig. 2). Although specific growth rate (μ) differed with each culture system (Table 2), similar patterns of growth were observed. The biomass of cultures of each species, produced in the airlift bioreactor with 60 mM sucrose, was higher than that obtained with mechanically-stirred vessels, whereas the number of stage 1 somatic embryos (stage 1 = immature somatic embryo; the use in the text of somatic embryo stages for conifers is after von Arnold and Hakman, 1988; Hakman and von Arnold, 1988; Dunstan *et al.*, 1988; see also Tautorus *et al.*, 1991, for review) was similar in both types of vessel (Fig. 3). The number of somatic embryos per ml produced in either type of bioreactor was lower when compared to the number produced in 250 ml shake-flasks (Table 2). Nevertheless, each bioreactor would contain approximately 1.8×10^7 somatic embryos in 6 L of medium.

Stuart *et al.* (1987) reported that for *Medicago sativa* somatic embryos, airlift agitation gave slightly higher yields of embryos compared with mecha-

Table 1. Somatic embryos cultured in bioreactors.

Species	Vessel Type	Aeration	Culture Period	Embryos/ml	Plantlet Production	Reference
Daucus carota (carrot)	4 L mechanically-stirred	0.25 VVM (DO < 16%)	45–55 days	10	Yes	Kessell and Carr, 1972
		0.5 VVM (DO > 16%)	<1			
	Two-stage spin-filter	Not reported	25 days	Not reported	Yes	Ammirato and Styer, 1985; Styer, 1985
	3 L mechanically-stirred (Applikon, Holland)[1]	DO 10%	1.13 days; 2.18–25 days	170	Not reported	Jay *et al.*, 1992
		DO 100%		600		
Digitalis lanata (foxglove)	5 L airlift (LF2; ZWG Mytron, Germany)	0.5 VVM	28 days	Not reported	No (cardenolide synthesis)	Greideziak *et al.*, 1990
Euphorbia pulcherrima (poinsettia)	2 L vibration-stirred (Biostat M, B. Braun, Melsungen Germany)	DO 60%	4 weeks	100	Yes	Preil *et al.*, 1988
Medicago Sativa (alfalfa)	2 L airlift (LH Fermentation Ltd, Stoke Poges, Eng.)	0.5 VVM (DO 100%)	1–2 months	0	–	Chen *et al.*, 1987
	1 L hanging stir-bar (Virtis, NY, USA)	DO ≥ 13.5%	1–2 months	10	Yes	Chen *et al.*, 1987
	2 L mechanically-stirred (LSL Biolafitte, St. Germain-en-Laye, Fr.)	2 VVM (DO 21%)	14 days	0	–	Stuart *et al.*, 1987
		1.8 VVM (DO > 70%)	14 days	112	Yes	
	airlift	1.8 VVM (DO > 70%)	Not reported	228	Yes	Stuart *et al.*, 1987

Picea glauca-engelmannii (interior spruce)	7 L mechanically-stirred (Microferm, New Brunswick Sci., NJ, USA)	0.2–0.6 VVM	14 days	2278	Yes	Tautorus *et al.*, 1992, 1994
	7 L airlift	0.3–0.6 VVM	35 days	2698	Yes	Tautorus *et al.*, 1994
Picea mariana (black spruce)	7 L mechanically-stirred (New Brunswick)	0.2–0.6 VVM	14 days	3076	Yes	Tautorus *et al.*, 1992, 1994
	7 L airlift	0.3–0.6 VVM	18 days	2892	Yes	Tautorus *et al.*, 1994
Santalum album (sandalwood)	1) 7 L mechanically-stirred (New Brunswick)[1]	0.9 VVM	24 days	pro-globular stage	Yes	Bapat *et al.*, 1990
	2) 1 L mechanically-stirred		4 weeks	18		

[1] Two-stage process.
DO = dissolved oxygen; VVM = volume of air/volume of medium/minute.

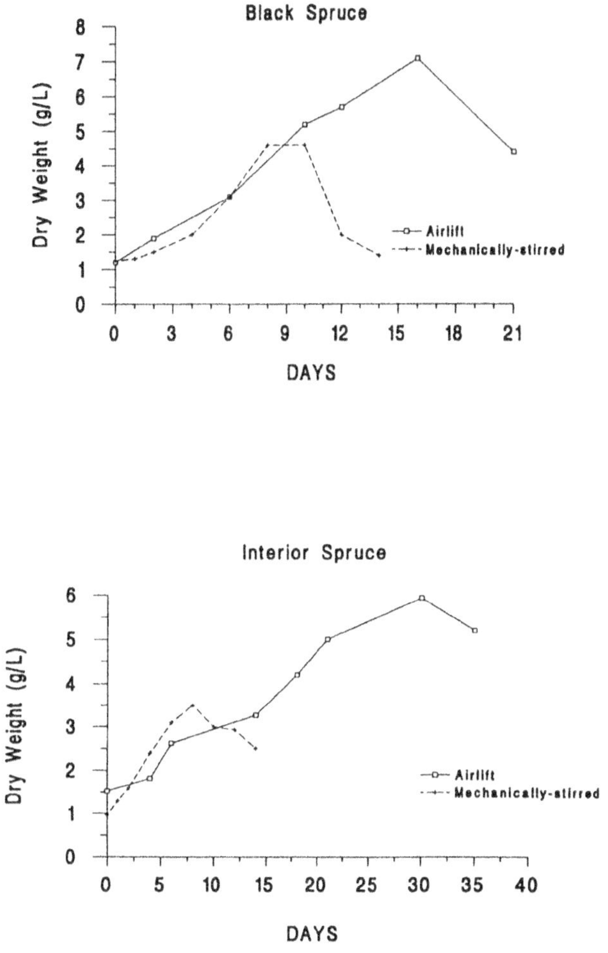

Figure 2. (*upper*) Growth of *Picea mariana* (black spruce) somatic embryo cultures in 7 L mechanically-stirred and airlift bioreactors with 60 mM sucrose. (*lower*) Growth of *Picea glauca-engelmanni* (interior spruce) somatic embryo cultures in 7 L mechanically-stirred and airlift bioreactors with 60 mM sucrose. Data taken from Tautorus *et al.*, 1992, and Tautorus *et al.*, 1994.

nically-stirred bioreactors and shake-flasks. In another study *M. sativa* somatic embryos were only produced in shake-flasks and mechanically-stirred vessels; airlift bioreactors produced no embryos (Chen *et al.*, 1987). With *Santalum album* somatic embryos, similar data for biomass and embryo yield were obtained when material was cultured in 250 or 500 ml shake-flasks, or 1 L mechanically-stirred bioreactors (Bapat *et al.*, 1990).

For comparison, growth rates and biomass levels of *Helianthus annuus*

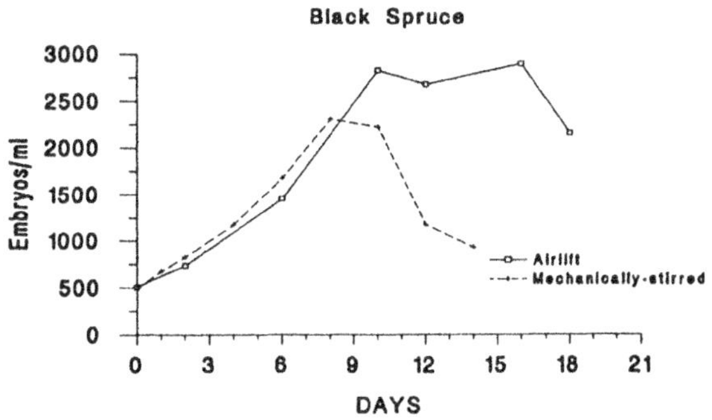

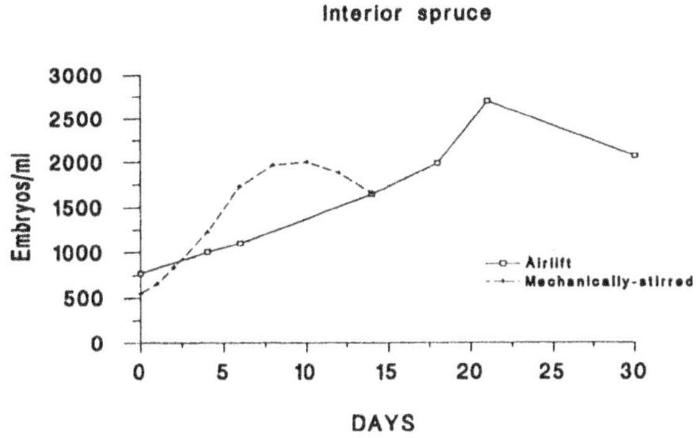

Figure 3. (*upper*) Production of stage 1 somatic embryos of *Picea mariana* (black spruce) in 7 L mechanically-stirred and airlift bioreactors with 60 mM sucrose. (*lower*) Production of stage 1 somatic embyros of *Picea glauca-engelmannii* (interior spruce) in 7 L mechanically-stirred and airlift bioreactors with 60 mM sucrose. Data taken from Tautorus *et al.*, 1992, and Tautorus *et al.*, 1994.

cells cultured in 250 ml shake flasks, 3 L mechanically-stirred bioreactors, and 80' L airlift bioreactors were very similar (Scragg, 1990). The growth rates of *Fragaria* cell suspension cultures were higher in roller bottle bioreactors without baffles than in shake-flasks, airlift bioreactors or mechanically-stirred bioreactors (Hong *et al.*, 1989).

3.4. *Effects of Medium Components*

The successful scale-up of any biological process requires knowledge of substrate requirements and their utilization. Analyses of changes in composi-

Table 2. Growth characteristics of *Picea mariana* (black spruce) and *Picea glauca-engelmannii* (interior spruce) somatic embryos cultured in 250 ml shake-flasks and 7 L mechanically-stirred or airlift bioreactors.[1]

	Black Spruce					Interior Spruce				
Vessel	Shake-flask	Mech.-stirred	Mech.-stirred	Mech.-stirred	Airlift	Shake-flask	Mech.-stirred	Mech.-stirred	Mech.-stirred	Airlift
Carbohydrate	30 mM sucrose	30 mM sucrose	60 mM sucrose	60 mM glucose	60 mM sucrose	30 mM sucrose	30 mM sucrose	60 mM sucrose	60 mM fructose	60 mM sucrose
Maximum dry weight (g/L)	5.8	4.6	6.3	6.3	7.1	4	3.5	4.3	5.6	5.9
Maximum no. embryos/ml	5000	2300	3076	2911	2892	3500	2000	2278	3235	2698
Lag phase length (days)	0	1	0	0	0	0	0	0–1	0	0
Growth phase length (days)	14	7	8–10	8	15	10	4–8	6–8	8	30
Specific growth rate per day (μ)	0.22	0.19	0.15	0.14	0.15	0.18	0.18	0.17	0.19	0.1

[1] Data are from Lulsdorf *et al.*, 1992; Tautorus *et al.*, 1992, 1994.

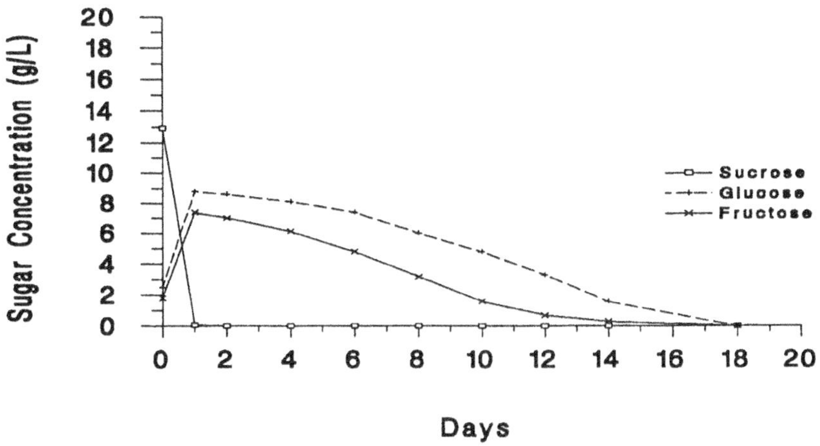

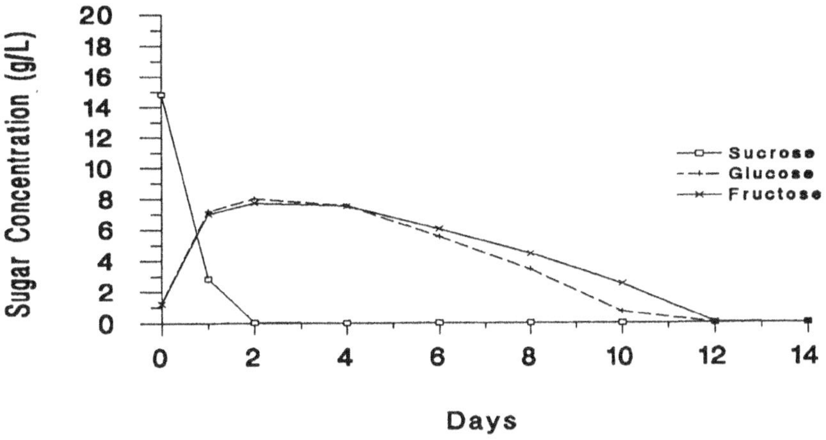

Figure 4. (*upper*) Carbohydrate utilization of *Picea glauca-engelmannii* (interior srpuce) somatic embryo cultures in 7 L mechanically-stirred bioreactors with 60 mM sucrose. (*lower*) Carbohydrate utilization of *Picea mariana* (black spruce) somatic embryo cultures in 7 L mechanically-stirred bioreactors with 60 mM sucrose. Data taken from Tautorus *et al.*, 1992, and Tautorus *et al.*, 1994.

tion of the liquid medium during suspension culture may give valuable information on factors which limit culture growth, allowing subsequent improvements in design and efficiency of bioreactor systems. Lulsdorf *et al.* (1992) found that somatic embryo suspension cultures of *Picea mariana* and *Picea glauca-engelmannii* grown in shake flasks became carbohydrate limited in medium with 30 mM sucrose. The effects of sucrose concentration in mechanically-stirred bioreactors were seen when somatic embryo suspension cultures of these two species were inoculated into medium with 30, 60 or 90 mM sucrose. For both species, the highest dry weight, and somatic embryo number were obtained with 60 mM sucrose after 10–14 days culture (Table 2) (Tautorus *et al.*, 1992). Both species rapidly hydrolysed sucrose into glucose and fructose, with each species showing a preference for a different monosaccharide (Fig. 4) (Tautorus *et al.*, 1994). For comparison, increased biomass production in shake-flask (Merillon *et al.*, 1984; Scragg *et al.*, 1990) and bioreactor (Scragg *et al.*, 1990) cultures of *Catharanthus roseus* occurred when sucrose was increased from 60 to 180 mM. Increased biomass of *Dioscorea deltoidea* suspension cultures was observed when the sucrose concentration was increased from 15 mM to 45, 90, 135, or 180 mM (Tal *et al.*, 1982). With *Betula* somatic embryos, 60 mM sucrose was not completely hydrolysed until day seven and there was no subsequent consumption of glucose and fructose during the first three weeks of shake-flask culture (Nuutila and Kauppinen, 1992). Carbohydrate hydrolysis is thought to be influenced by cell line, growth rate, and sucrose concentration (Taticek *et al.*, 1990).

Comparison of various carbohydrate sources on growth parameters of *Picea mariana* and *P. glauca-engelmannii* somatic embryo cultures has also been investigated (Tautorus *et al.*, 1994). For *P. glauca-engelmannii*, growth rate, biomass production, and the number of stage 1 embryos were improved when cultures were grown in mechanically-stirred bioreactors containing 60 mM fructose, as compared with 60 mM sucrose (refer to Table 2) (Tautorus *et al.*, 1994). Fructose was completely utilized by day 12. For *P. mariana* cultures, no substantial improvement in growth parameters was observed when grown in mechanically-stirred bioreactors containing 60 mM glucose, as compared to sucrose (Table 1). Liquid medium containing sucrose produced the highest biomass of *Solanum eleagnifolium*, as compared to glucose or fructose (Nigra *et al.*, 1990). Comparable growth rates and biomass of *Anchusa officinalis* (De-Eknamkul and Ellis, 1985), and *Dioscorea deltoidea* (Tal *et al.*, 1982) were produced in medium with sucrose, glucose, or fructose. Cultures of *Catharanthus roseus* grown in the presence of glucose gave a higher biomass yield at lower relative carbohydrate concentrations than cultures grown on sucrose, suggesting that sucrose was not utilized as effectively (Fowler, 1982b).

There is very little information concerning ion utilization by somatic embryos cultured in bioreactors. *Picea mariana* and *Picea glauca-engelmannii* somatic embryos showed similar ion (ammonium, nitrate, potassium) consumption patterns during culture in either an airlift or mechanically-stirred bioreactor (Tautorus *et al.*, 1994). Approximately 80% of the ammonium and 35% of the nitrate supplied were used by the cultures. Potassium remained in ample supply throughout the culture period. In many plant cell cultures, ammonium is preferentially used over nitrate (Dougall and Frazier, 1989). It is well established that a source of reduced nitrogen in addition to nitrate is beneficial for both embryo initiation and maturation (Ammirato, 1983a). Ammonium utilization contributes to cell organization and cell products associated with somatic embryogenesis (Durzan, 1987). In addition, potassium ions are essential for somatic embryogenesis (Ammirato, 1983a).

The low growth rates of plant cell cultures translate into long culture incubation times. Consequently, preventative measures against contamination are very important (Fowler, 1982a; Taticek *et al.*, 1991). The problem of contamination is increased by the use of complex plant cell culture nutrient media, which provide excellent growth environments for fungi and bacteria. No comprehensive data have been published on the effect of antibiotics on plant cell metabolism.

3.5. *Effect of PH*

Changes in medium pH can affect the relative availabilities of individual ions to plant cells, understanding how pH alters during culture is therefore an important consideration during the optimization of suspension cultures. During shake-flask and bioreactor culture with *Picea mariana* and *P. glauca-engelmannii* somatic embryos, inoculum pH was usually about pH 5.5. Within one day of culture the pH dropped to approximately pH 5.0, and then slowly increased to pH 6.5–6.8 by day 14 (stationary phase) (Lulsdorf *et al.*, 1992; Tautorus *et al.*, 1992). These changes in pH are common with most plant cells and are related to the balance between ammonium and nitrate uptake. The initial drop in pH is due to preferential uptake of ammonium at high ammonium levels, and the subsequent increase in pH is due to nitrate uptake after the ammonium has been depleted (Veliky and Rose, 1973; McDonald and Jackman, 1989; Treat *et al.*, 1989; Goodchild and Givan, 1990). It is suggested that the proton flux allows the cells to reduce internal fluctuations in pH.

Although many investigators have measured the pH during large-scale batch culture, little information has been reported on the effect on growth of fixed pH levels. Stuart *et al.* (1987) designed an experiment to estimate the effect of daily titration to pH 5.5 on cultures of *Medicago sativa* somatic embryos. Such pH adjustment did not result in a permanent modification to the medium pH. Rather, the pH of the medium was modified daily by the cells to a level near that of the non-titrated control vessel. However, in the control vessel 119 embryos/ml of medium were regenerated, whereas 147 embryos/ml were regenerated in the pH titrated vessel, illustrating the probable influence of pH on the regeneration processes. For comparison, maintenance of medium pH to pH 5.5 in bioreactor cultures of non-embryogenic alfalfa cells resulted in inhibition of growth (McDonald and Jackman, 1989). It was suggested that the controlled pH interferred with hydrogen ion requirements for nitrate and sugar transport.

3.6. *Effect of Dissolved Oxygen*

The level of aeration and the dissolved oxygen (DO) concentration in the medium has been shown to affect culture growth in bioreactors (Smart and Fowler, 1981; Preil *et al.*, 1988; Panda *et al.*, 1989; Smith *et al.*, 1990; Kim *et al.*, 1991b; Taticek *et al.*, 1991). In general, high aeration rates are detrimental to cell growth and result in reduced biomass yields. The range of useable aeration rates may be limited because volatilization of essential nutrients or product gases required for embryogenesis may occur (Cazzulino *et al.*, 1991). The volume of culture medium, the bioreactor design and the air flow are important considerations that will affect the rates of oxygen transfer, and these will vary among experimentors (Jay *et al.*, 1992). In addition, optimal oxygen requirements will vary among species and cell lines (Table 1). For example, Kessel and Carr (1972) first reported that *Daucus carota* somatic embryos required low oxygen levels (DO < 16%) for successful embryo formation. More recently however, Jay *et al.* (1992) studied the effects of DO on *D. carota* somatic embryos by means of a bio-processor, with which the DO was varied under constant agitation and air flow rate. Results showed that reduced oxygen concentration (10%) during the embryo differentiation phase resulted in a lower growth rate, reduced numbers of embryos, and reduced sugar uptake. Conflicting reports on the requirement for dissolved oxygen have also been published for *Medicago sativa* (Chen *et al.*, 1987; Stuart *et al.*, 1987).

Besides oxygen, carbon dioxide and ethylene have been reported to improve cell growth and metabolite synthesis in some plant cell cultures (Bisaria and Panda, 1991). In airlift bioreactor runs with *Thalictrum rugosum*, supplying exogenous carbon dioxide and ethylene to the culture reduced the inhibitory effects of high aeration, resulting in increased berberine synthesis (Kim *et al.*, 1991b). In addition, previous investigations on culture vessel environment have suggested that these gases influence both embryogenesis and organogenesis *in vitro* (Ammirato, 1983a; Kumar *et al.*, 1987, 1989). Thus the levels of these gases need to be evaluated during optimization of bioreactor culture conditions.

3.7. *Foaming*

Foaming is a common problem to bioreactor culture, and its severity is related to the rate of aeration, the composition of the medium and the bioreactor design (Fowler, 1982a; Taticek *et al.*, 1991). Cells become entrapped in the foam and gradually form a crust above the liquid level on the vessel walls, drive shafts, and probes. This affects the operation of the probes, the general stability of the culture, and complicates sampling. Cell stickiness appears to be caused by extracellular polysaccharides and is stimulated by high carbon levels (Scragg and Fowler, 1985). Many researchers have used silicone-based antifoam agents to reduce foaming, which in general have proved to be non-toxic for most species. However, foaming agents may cause reduced permeability of cells, affecting gas transfer rates into cells (Fowler, 1982a). Therefore, the use of antifoam agents should be viewed with caution. Few reports have indicated the use of mechanical foam breaking systems.

3.8. *Effect of Temperature and Light*

Although it is well established that temperature and light affect growth rate of plant cells, it is unfortunate that there are no reports on the effects of these environmental conditions during bioreactor culture of somatic embryos. Studies with a few plant cell suspension cultures suggest that the optimum temperature for growth is between 25–30°C but that individual species can vary in their requirements within this range (Scragg and Fowler, 1985).

3.9. *Immobilization*

Immobilization of plant cells, onto suitable carriers for cultivation in different types of bioreactors, has been extensively reviewed (see Hulst and Tramper, 1989; Panda *et al.*, 1989; Bisaria and Panda, 1991; Mazid, 1993). The advantages of immobilization compared with other culture systems include: the protection of cells from shear damage, control of cell aggregate size, increased production of secondary metabolites resulting from slower growth of cells and higher cell-cell contact, simplified product separation for extracellular compounds, and the possibility of maintaining high cell densities in the reactor which results in a high volumetric productivity and shorter culture periods. Common immobilization materials include calcium alginate, carrageenan, agar, agarose, chitosan, and foam particles. Many different types of bioreactor have been used in conjunction with cell immobilization, the most common being the stirred-tank (Mazid, 1993).

A recent development is the use of surface-immoblized plant cells (SIPC) in which cells are deposited onto a surface of fibrous material, e.g., short-fibre polyester, polypropylene geotextile (Archambault *et al.*, 1989). When compared with other immobilization techniques, this method allows better control over mass transfer, ease of scale-up, simplified medium changes, and reduced foaming. In addition, it is non-stressful to plant cells. This method was successfully used with 20 L airlift bioreactors in the culture of *Catharanthus roseus* (Archambault, 1991).

An example of the use of immobilization with somatic embryo cultures comes from *Medicago sativa*, with which somatic embryos were immobilized by permeation into cubes of polyurethane foam (Macek *et al.*, 1989). The formation of embryoids was visible after 10 days culture in MS medium. Further development of methods for immobilization of somatic embryos will be of considerable interest to those seeking to produce synthetic seeds (Macek *et al.*, 1989; Lulsdorf *et al.*, 1993). See also Section 4.1.1, Embryo Sorting.

4. Embryo Maturation, Plantlet Development and Automation

The ability to grow embryogenic suspension cultures and to produce embryos and "seedlings" from somatic embryos (emblings) on a large scale is essential if somatic embryogenesis is to be commercially used as a propagation tool (Styer, 1985; Parrott *et al.*, 1991). The production of emblings differs widely across genotypes, species, and culture systems. As seen in Table 1, somatic embryos of many species have been matured, with subsequent germination to emblings, following bioreactor culture. The majority of reports have limited their data to describing the successful recovery of plants, providing little or no information on somatic embryo maturation frequencies, and on germination and embling acclimatization successes.

4.1. *Embryo Maturation and Plantlet Development*

Conifer somatic embryo cultures can be transferred from suspension culture to agar-solidified medium containing abscisic acid (ABA), to achieve embryo maturation (Tautorus *et al.*, 1991). For example, Figs. 5 to 7 show stages in *Picea glauca-engelmannii* somatic embryo maturation and subsequent germination on agar-solidified medium, after culture multiplication for six days in a mechanically-stirred bioreactor. It has been possible to obtain development of somatic embryos using a liquid culture maturation phase (with or without a culture support system) e.g., *Picea abies* (Boulay *et al.*, 1988), *Picea glauca* (Attree *et al.*, 1994), *Picea mariana* (Tautorus, 1990), *Picea sitchensis* (Krogstrup, 1988), and *Pseudotsuga menziesii* (Durzan and Gupta, 1987). In most examples a support system was used to complete maturation and/or germination (e.g., cheesecloth, filter paper). Emblings of *Picea mariana* were obtained after six weeks incubation in liquid medium with 8 μM ABA without use of such a support system (Tautorus, 1990). However, continued development of the emblings on agar-solidified medium was poor.

Tautorus *et al.* (1994) recently compared the maturation response on agar-solidified medium of *Picea mariana* and *P. glauca-engelmannii* stage 1 somatic embryos derived from different maintenance regimes. Cultures were maintained either as suspensions in liquid medium in 250 ml or 500 ml shake-flasks, 7 L capacity airlift or mechanically-stirred bioreactors, or on agar-solidified medium. For both species the highest maturation frequency occurred with cultures grown in the airlift bioreactor (Fig. 8). Cultures maintained on agar-solidified medium produced the lowest number of cotyledonary somatic embryos (Tautorus *et al.*, 1994). These results suggest that maturation is significantly influenced by the type of vessel used during culture maintenance, thus emphasizing that this culture component should be optimized for the material being studied.

Ammirato (1983b) also showed that the maintenance condition used to grow *Carum carvi* somatic embryos greatly affected the pattern of embryo maturation. Tumble tubes, test tubes, and Erlenmeyer flasks each generated

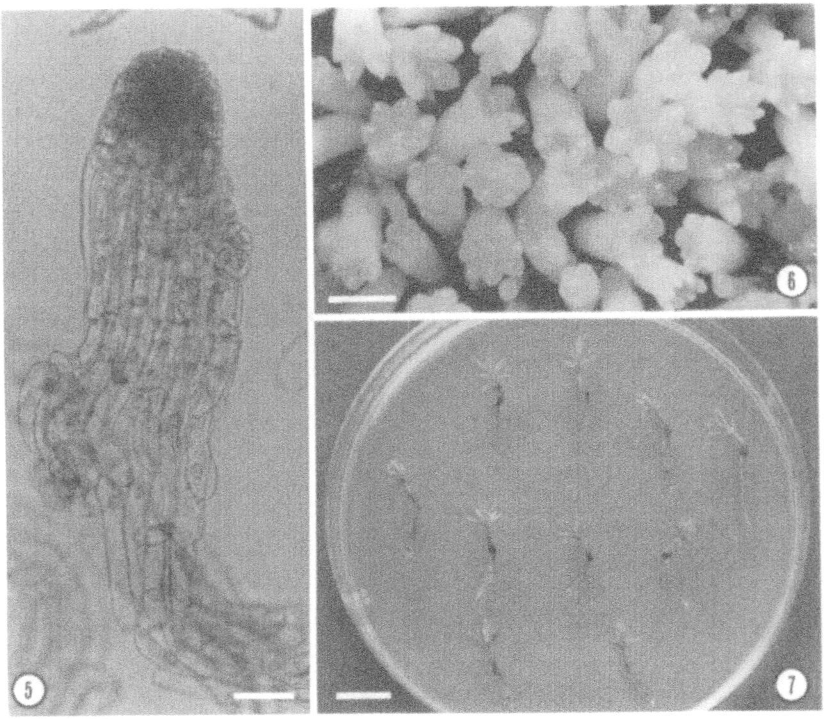

Figure 5–7. (5) Somatic embryo of *Picea glauca-engelmannii* (interior spruce) after 6 days culture in a 7 L mechanically-stirred bioreactor containing 60 mM sucrose. Bar = 100 μm. Reprinted from Tautorus *et al.* (1992).
6) Stage 3 somatic embryos of *Picea mariana* (black spruce) after 35 days incubation on medium containing 40 μM ABA. Cotyledonary stage embryos were derived from somatic embryos grown in a 7 L mechanically-stirred bioreactor for 6 days containing 60 mM sucrose. Bar = 100 μm. Reprinted from Tautorus *et al.* (1992).
7) Plantlets of *Picea mariana* (black spruce) after 6 weeks incubation on phytohormone-free medium, following 35 days on ABA medium. Plantlets were derived from somatic embryos grown in a 7 L mechanically-stirred bioreactor for 6 days containing 60 mM sucrose. Bar = 1 cm. Reprinted from Tautorus *et al.* (1992).

populations differing significantly in the frequency of normal and abnormal embryos. For *Medicago sativa* somatic embryos, germination and embling development was highest when cultures had been grown on agar-solidified medium, and lowest in shake-flasks and bioreactors (Stuart *et al.*, 1987). It was suggested that the poor performance of material maintained in liquid culture reflected the relative amount of time that had been spent working on optimizing embryo yield. The use of cryopreservation as a means of culture storage should be considered wherever possible to avoid such eventualities, as described for *Larix* × *eurolepis* (Klimaszewska *et al.*, 1992), *Picea*

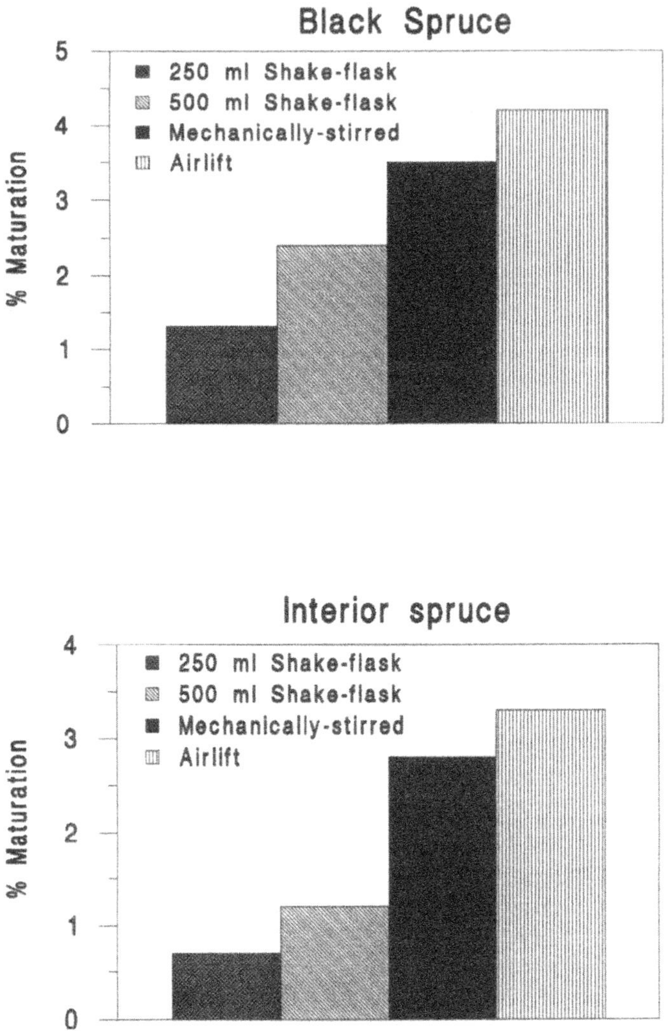

Figure 8. (*upper*) Maturation frequency of somatic embryos of *Picea mariana* (black spruce) derived from different maintenance regimes. (*lower*) Maturation frequency of somatic embryos of *Picea glauca-engelmannii* (interior spruce) derived from different maintenance regimes. Maturation frequency = (No. stage 3 embryos/No. stage 1 embryos) $\times$ 100.

abies (Gupta *et al.*, 1987; Bercetche *et al.*, 1990), *Picea glauca* (Kartha *et al.*, 1988) *Picea mariana* (Klimaszewska *et al.*, 1992) and *Pinus taeda* (Gupta *et al.*, 1987).

4.2. *Automation*

Conventional micropropagation techniques are labour-intensive, expensive, and require thousands of containers to produce a large number of plants (Preil, 1991; Doran, 1993; Giles and Morgan, 1987). A significant accomplishment towards commercialization of somatic embryogenesis will be the development of methods for somatic embryo maturation in bioreactor culture, amenable to automation. Suitable candidates to achieve this goal might include the automated Vitromatic system (Levin *et al.*, 1988) which supports the growth of both organogenic and embryogenic cultures. This liquid-based patented process integrates a bioreactor, and a bioprocessor which separates, sizes, and dispenses propagules (12,500/h) into culture vessels for growth and development. A transplanting machine placed plantlets in soil at a rate of 8000/h. According to Levin *et al.* (1988), the benefits of this system include: prevention of vitrification, lower contamination rates, savings in space, time and labor.

Weathers and Zobel (1992) recently reviewed the potential applications of aeroponics (i.e., culture of whole plants and/or tissues with their roots or the whole tissue fed by an air/water fog as opposed to immersion in water, soil, nutrient agar or other substrates) for bioreactor production of plants. It was suggested that aeroponic cultures permitted the control of root zone temperature, of nutrition, of moisture and of gaseous phases, providing an optimum environment. This method has been used successfully for production of *Capsicum frutescens* plants and production of shikonin from *Lithospermum* (Weathers and Zobel, 1992).

The nutrient mist bioreactor (NMB) developed by Weathers and Giles (1988) offers another approach using aeroponics. Plant tissues are situated on a liquid permeable mesh support (e.g., filter paper, rockwool) in the growth chamber. The tissues are sprayed by nutrient mists produced ultrasonically from the top or bottom, with excess media draining away continuously. Gas mixtures were used to modify humidity and circulate mist throughout the growth chamber. More than 13 different species of plants have been cultured in the NMB.

Aitken-Christie and Jones (1987) reported a novel method for *in vitro* shoot production of *Pinus radiata*. Shoots were cultured continuously as hedges for up to 18 months without subculturing, using liquid nutrient replenishment twice weekly. This method of growing shoots is suitable for species that are not sensitive to vitrification and can be cultured continuously in liquid (Aitken-Christie, 1991). The culture of shoots, roots, organs, and entire plantlets, in liquid medium using a variety of vessels, has been reported for many plant groups and species including ornamentals, herbs, (Park *et al.*, 1989), *Nicotiana*, fruits (Desamero *et al.*, 1993), woody species, and *Solanum* microtubers (for review see Aitken-Christie, 1991; Simonton *et al.*, 1991; Vasil, 1991; Ziv, 1991).

4.2.1. *Embryo Sorting*

In most plant cell suspension cultures, embryo induction occurs asynchronously with the resulting embryos varying in stage, shape, size, and morphology. In addition, with conifer species, many somatic embryos are attached to each other, with the result that the separation and selection of individual embryos (so-called "singulation") can be difficult and very labour-intensive (Gupta *et al.*, 1993). Automation will be essential to achieve economically feasible embryo-stage characterization, separation, and selection. There have been no developments of an integrated mechanized system reported for singulating and selecting conifer embryos, although the Vitromatic system (Levin *et al.*, 1988) described previously may have some applicability (Gupta *et al.*, 1991).

Several research groups have used size and density fractionation techniques to sort specific embryo stages, to synchronize development of the embryos, or to improve the embryo quality (Cazzulino *et al.*, 1991; Gupta *et al.*, 1991). Currently, computer image analysis systems are being developed to automatically count, sort, size, and grade somatic embryos (Cazzulino *et al.*, 1991, Harrell and Cantliffe, 1991). The basic hardware components include a video camera, TV monitor, and a computer with an imaging board. Cazzulino *et al.* (1990) utilized image analysis to estimate the number, and stage of development of carrot somatic embryos grown in suspension culture, this data was later used to verify a growth and developmental model. Harrell and Cantliffe (1991) recently designed a non-invasive machine vision system that could automatically quantify the developmental characteristics of *Ipomoea batatas* somatic embryos produced in an airlift bioreactor. This system could be envisaged in the eventual classification of bioreactor-grown spruce somatic embryos after maturation, into one of the following categories: stage 3 somatic embryos, stage 1 embryos, embryogenic tissue, and non-embryogenic tissue. The data obtained could be used to find out the effects of environmental changes on somatic embryo development and to help define appropriate bioreactor operating strategies (Cazzulino *et al.*, 1991). Selective harvest of mature embryos would also significantly improve development of technology for an automated synthetic seed production system.

5. Use of Somatic Embryos for Production of Metabolites

The mass culture of plant cells in bioreactors to supply phytochemicals has been proposed for some time. However, there is little information on the culture of somatic embryos for the production of secondary metabolites. High levels of secondary metabolites might be anticipated because of the observed effects of cellular organization on their yield (Taticek *et al.*, 1991; Doran, 1993). For example, when torpedo-stage somatic embryos in *Apium graveolens* suspension cultures were analyzed for production of flavour compounds, the levels produced by the embryo culture were comparable to that of the intact plant (Doran, 1993).

Merkle *et al.* (1990) suggested that somatic embryos may be an important source of lipids and seed storage proteins. Indeed the absence of seed tissues surrounding somatic embryos may be a significant advantage. For example, somatic embryos of *Theobroma cacao* accumulate lipids similar to those of zygotic embryos, including oleo-palmitostearin, the major ingredient of cocao butter. Such a system could theoretically produce a continuous supply of the metabolite throughout the year.

Somatic embryos of *Digitalis lanata* have been shown to synthesize cardiac glycosides (cardenolide) (Greideziak *et al.*, 1990). Batch cultures of *D. lanata* somatic embryos were successfully grown in 5 L gaslift bioreactors. Embryo development and synthesis of cardenolides in the embryos resembled that of cultures grown in shake-flasks. Both processes depended on the developmental stage of the somatic embryos at the time of inoculation, culture homogeneity, inoculum density, composition of the nutrient medium, irradiation intensity, and the composition of the gas mixtures used for agitation of the embryo suspension culture.

6. Conclusions

Large-scale culture of plant cells encounters problems of slow growth rates, shear damage, cell aggregation, poor mixing and inadequate mass transfer of oxygen. Any improvements will require analyses of the culture systems used, concurrent with more fundamental studies of plant cell metabolism. The conditions required for growth and development of cells into somatic embryos may be very different from those required for maximum production of metabolites, so the criteria for bioreactor design will vary accordingly.

Bioreactor technology provides the potential for producing large numbers of somatic embryos cheaply and efficiently (Ammirato and Styer, 1985; Styer, 1985; Tautorus *et al.*, 1991). Significant progress has been achieved in the development of bioreactor culture systems for somatic embryos of woody plants. To date, however, the application of this technology for large-scale plantlet production, on a level which is economically competitive with micropropagation, has not been shown for any species.

The mass culture of somatic embryos is impeded by both biological and technological difficulties. There is a general lack of knowledge about the physiological and developmental processes of somatic embryos in most species of horticultural importance (Preil, 1991). If all the steps of somatic embryo proliferation, development and maturation are to be carried out in the bioreactor, the conditions regulating these stages in liquid culture must be precisely defined (Stuart *et al.*, 1987; Parrott *et al.*, 1991). Understanding how to promote normal development of somatic embryos in liquid culture is a substantial challenge facing researchers, and a major effort is being made to define the requirements of somatic embryo liquid suspension cultures and their responses to different bioreactor conditions. This has proven to be a difficult and challenging area of bioengineering research (Doran, 1993).

There is also a need to develop reliable methods for direct delivery of propagules to the nursery greenhouse or field.

Reproducibility, one of the most important components of any industrial process, is still inadequate with most embryogenic propagation systems using liquid culture methods (Preil, 1991). It is well established that embryogenic tissues of some species are more amenable to suspension culture than others, and those which respond well in suspension culture will be more easily manipulated for large-scale production of somatic embryos (Merkle *et al.*, 1990). However, as shown by Stuart *et al.* (1987) with *Medicago*, some species which grow very well in suspension culture have proven to be more difficult during embryo maturation and plantlet development phases, when compared to the same tissues grown on semi-solid medium. It is not conclusive that a large bioreactor will provide the best conditions for embryo production and maturation, although even if these processes are less efficient there are likely to be economies-of-scale.

References

Aitken-Christie, J., 1991. Automation. In: P.C. Debergh and R.H. Zimmerman (Eds.), Micropropagation, pp. 363–388. Kluwer Academic Publishers, Dordrecht.

Aitken-Christie, J. and C. Jones, 1987. Towards automation: radiata pine shoot hedges *in vitro*. Plant Cell Tiss Org Cult. 8: 185–196.

Ammirato, P.V., 1983a. Embryogenesis. In: D.A. Evans, W.R. Sharp, P.V. Ammirato and Y. Yamada (Eds.), Handbook of Plant Cell Culture, Vol. 1, pp. 82–123. Macmillan Publishing Co., New York.

Ammirato, P.V., 1983b. The regulation of somatic embryo development in plant cell cultures: suspension culture techniques and hormone requirements. Bio/Tech. 3: 68–74.

Ammirato, P.V. and D.J. Styer, 1985. Strategies for large-scale manipulation of somatic embryos in suspension culture. In: M. Zaitlin, P. Day, A. Hollaender and C.M. Wilson (Eds.), Biotechnology in Plant Science – Relevance to Agriculture in the Eighties, pp. 161–178. Academic Press, Orlando.

Archambault, J., B. Volesky and W.G.W. Kurz, 1989. Surface immobilization of plant cells. Biotech. Bioeng. 33: 293–299.

Archambault, J. 1991. Large-scale (20 l) culture of surface-immobilized *Catharanthus roseus* cells. Enz. Microb. Tech. 13: 882–892.

Attree, S.M., D.I. Dunstan and L.C. Fowke, 1991. White spruce [*Picea glauca* (Moench) Voss] and black spruce [*Picea mariana* (Mill) B.S.P.]. In: Y.P.S. Bajaj (Ed.), Biotechnology in Agriculture and Forestry, Vol. 16, pp. 423–445. Springer-Verlag, Berlin.

Attree, S.M., M.K. Pomeroy and L.C. Fowke, 1994. Production of vigorous, desiccation tolerant white spruce (*Picea glauca* [Moench.] Voss.) synthetic seeds in a bioreactor. Plant Cell Rep. 13: 601–606.

Backs-Hüsemann, D. and J. Reinert, 1970. Embryo formation by isolated single cells from tissue cultures of *Daucus carota*. Protoplasma 70: 49–60.

Bapat, V.A., D.P. Fulzele, M.R. Heble and P.S. Rao, 1990. Production of sandalwood somatic embryos in bioreactors. Curr. Sci. 59: 746–748.

Bercetche, J., M. Galerne and J. Dereuddre, 1990. Efficient regeneration of plantlets from embryogenic callus of *Picea abies* (L) Karst after freezing in liquid nitrogen. Comptes Rendues de l'Academie des Sciences, Paris 310 (Série III): 357–363.

Bisaria, V. and A. Panda, 1991. Large-scale plant cell culture: methods, applications and products. Curr. Opinion in Biotech. 2: 370–374.

Boulay, M.P., P.K. Gupta, P. Krogstrup and D.J. Durzan, 1988. Development of somatic embryos from cell suspension cultures of Norway spruce (*Picea abies* Karst.). Plant Cell Rep. 7: 134–137.

Cazzulino, D.L., H. Pedersen, C.-K. Chin and D. Styer, 1990. Kinetics of carrot somatic embryo development in suspension culture. Biotech. Bioeng. 35: 781–786.

Cazzulino, D., H. Pedersen and C.-K. Chin, 1991. Bioreactors and image analysis for scale-up and plant propagation. In: I.K. Vasil (Ed.), Cell Culture and Somatic Cell Genetics of Plants, Vol. 8, pp. 147–177. Academic Press, San Diego, CA.

Chen, T.H.H., B.G. Thompson and D.F. Gerson, 1987. *In vitro* production of alfalfa somatic embryos in fermentation systems. J. Ferment. Technol. 65: 353–357.

De-Eknamkul, W. and B.E. Ellis, 1985. Effects of macronutrients on growth and rosmarinic acid formation in cell suspension cultures of *Anchusa officinalis*. Plant Cell Rep. 4: 46–49.

Denchev, P.D., A.I. Kuklin and A.H. Scragg, 1992. Somatic embryo production in bioreactors. J. Biotech. 26: 99–109.

Desamero, N.V., J.W. Adelberg, A. Hale, R.E. Young and B.B. Rhodes, 1993. Nutrient utilization in liquid/membrane system for watermelon micropropagation. Plant Cell Tiss Org Cult. 33: 265–271.

Doran, P.M., 1993. Design of reactors for plant cells and organs. In: A. Fiechter (Ed.), Bioprocess Design and Control, Vol. 48, pp. 116–169. Springer-Verlag, Berlin.

Dougall, D.K. and G.C. Frazier, 1989. Nutrient utilization during biomass and anthocyanin accumulation in suspension cultures of wild carrot cells. Plant Cell Tiss Org Cult. 18: 95–104.

Dunstan, D.I., F. Bekkaoui, M. Pilon, L.C. Fowke and S.R. Abrams, 1988. Effects of abscisic acid and analogues on the maturation of white spruce (*Picea glauca*) somatic embryos. Plant Sci. 58: 77–84.

Dunstan, D.I., C.A. Bock, G.D. Abrams and S.R. Abrams, 1992. Metabolism of (+)- and (−)-abscisic acid by somatic embryo suspension cultures of white spruce. Phytochem. 31: 1451–1454.

Dunstan, D.I., T.D. Bethune and C.A. Bock, 1993. Somatic embryo maturation from long-term suspension cultures of white spruce (*Picea glauca*). In Vitro Cell. Dev. Biol. Plant 29P: 109–112.

Dunstan, D.I., T.E. Tautorus and T.A. Thorpe, 1995. Somatic embryogenesis in woody plants. In: T.A. Thorpe (Ed.), *In Vitro* Embryogenesis in Plants. Kluwer Academic Publishers, Dordrecht, in press.

Durzan, D.J., 1987. Ammonia: Its analogues, metabolic products and site of action in somatic embryogenesis. In: J.M. Bonga and D.J. Durzan (Eds.), Cell and Tissue Culture in Forestry, Vol. 2, pp. 92–136. Martinus Nijhoff Publishers, Dordrecht.

Durzan, D.J. and P.K. Gupta, 1987. Somatic embryogenesis and polyembryogenesis in Douglas-fir cell suspension cultures. Plant Sci. 52: 229–235.

Fowler, M.W., 1982a. The large scale cultivation of plant cells. Prog. Ind. Microb. 16: 207–229.

Fowler, M.W., 1982b. Substrate utilization by plant-cell cultures. J. Chem. Tech. Biotechol. 32: 338–346.

Giles, K.L. and W.M. Morgan, 1987. Industrial-scale plant micropropagation. Trends Biotech. 5: 35–39.

Goodchild, J.A. and C.V. Givan, 1990. Influence of ammonium and extracellular pH on the amino and organic acid contents of suspension culture cells of *Acer pseudoplatanus*. Physiol. Plant. 78: 29–37.

Greideziak, N., B. Diettrich and M. Luckner, 1990. Batch cultures of somatic embryos of

Digitalis lanata in gaslift fermenters. Development and cardenolide accumulation. Planta Medica 56: 175–178.

Gupta, P.K., D.J. Durzan and B.J. Finkle, 1987. Somatic polyembryogenesis in embryogenic cell masses of *Picea abies* (Norway spruce) and *Pinus taeda* (loblolly pine) after thawing from liquid nitrogen. Can. J. For. Res. 17: 1130–1134.

Gupta, P.K., R. Timmis, G. Pullman, M. Yancey, M. Kreitinger, W. Carlson and C. Carpenter, 1991. Development of an embryogenic system for automated propagation of forest trees. In: I.K. Vasil (Ed.), Scale-Up and Automation in Plant Propagation, Vol. 8, pp. 75–93. Academic Press, San Diego, CA.

Gupta, P.K., G. Pullman, R. Timmis, M. Kreitinger, W.C. Carlson, J. Grob and E. Welty, 1993. Forestry in the 21st Century. The biotechnology of somatic embryogenesis. Bio/Tech. 11: 454–459.

Hakman, I. and S. von Arnold, 1988. Somatic embryogenesis and plant regeneration from suspension cultures of (*Picea glauca*) white spruce. Physiol. Plant. 72: 579–587.

Hamill, J.D., A.J. Parr, M.J.C. Rhodes, R.J. Robins and N.J. Walton, 1987. New routes to plant secondary products. Bio/Tech. 5: 800–804.

Harrell, R.C. and D.J. Cantliffe, 1991. Automated evaluation of somatic embryogenesis in sweet potato by machine vision. In: I.K. Vasil, (Ed.), Cell Culture and Somatic Cell Genetics of Plants, Vol. 8, pp. 179–195. Academic Press, San Diego, CA.

Hong, Y.C., T.P. Labuza and S.K. Harlander, 1989. Growth kinetics of strawberry cell suspension cultures in shake flask, airlift, stirred-jar, and roller bottle bioreactors. Biotech. Prog. 5: 137–143.

Hulst, A.C. and J. Tramper, 1989. Immobilized plant cells: a literature survey. Enz. Microb. Tech. 11: 546–558.

Jay, V., S. Genestier and J.-C. Courduroux, 1992. Bioreactor studies on the effect of dissolved oxygen concentrations on growth and differentiation of carrot (*Daucus carota* L.) cell cultures. Plant Cell Rep. 11: 605–608.

Kartha, K.K., L.C. Fowke, N.L. Leung, K.L. Caswell and I. Hakman, 1988. Induction of somatic embryos and plantlets from cryopreserved cell cultures of white spruce (*Picea glauca*). J. Plant Physiol. 132: 529–539.

Kessel, R.H.J. and A.H. Carr, 1972. The effect of dissolved oxygen concentration on growth and differentiation of carrot (*Daucus carota*) tissue. J. Exp. Bot. 23: 996–1007.

Kim, D-I, G.H. Cho, H. Pedersen and C.-K. Chin, 1991a. A hybrid bioreactor for high density cultivation of plant cell suspensions. Appl. Microb. Tech. 34: 726–729.

Kim, D.-I., H. Pedersen and C.-K. Chin, 1991b. Cultivation of *Thalictrum rugosum* cell suspension in an improved airlift bioreactor: stimulatory effect of carbon dioxide and ethylene on alkaloid production. Biotech. Bioeng. 38: 331–339.

Klimaszewska, K., C. Ward and W.M. Cheliak, 1992. Cryopreservation and plant regeneration from embryogenic cultures of larch (*Larix* × *eurolepis*) and black spruce (*Picea mariana*). J. Exp. Bot. 43: 73–79.

Krogstrup, P., 1988. Effect of culture densities on cell proliferation and regeneration from ebryogenic suspensions of *Picea sitchensis*. Plant Sci. 72: 115–123.

Kumar, P.P., D.M. Reid and T.A. Thorpe, 1987. The role of ethylene and carbon dioxide in differentiation of shoot buds in excised cotyledons of *Pinus radiata in vitro*. Physiol. Plant. 69: 244–252.

Kumar, P.P., R.W. Joy IV and T.A. Thorpe, 1989. Ethylene and carbon dioxide accumulation, and growth of cell suspension cultures of *Picea glauca* (white spruce). J. Plant Physiol. 135: 592–596.

Leckie, F., A.H. Scragg and K.C. Cliffe, 1991. Effect of bioreactor design and agitator speed on the growth and alkaloid accumulation by cultures of *Catharanthus roseus*. Enz. Microb. Tech. 13: 296–305.

Lee, E.T.Y. and M.J. Bazin, 1990. A laboratory scale air-lift helical photobioreactor to increase biomass output of photosynthetic algal cultures. New Phytol. 116: 331–335.

Levin, R., V. Gaba, B. Tal, S. Hirsch and D. DeNola, 1988. Automated plant tissue culture for mass propagation. Bio/Tech. 6: 1035–1040.
Lulsdorf, M.M., T.E. Tautorus, S.I. Kikcio and D.I. Dunstan, 1992. Growth parameters of embryogenic suspension cultures of interior spruce (*Picea glauca-engelmannii* complex) and black spruce (*Picea* mariana *Mill.*). Plant Sci. 82: 227–234.
Lulsdorf, M.M., T.E. Tautorus, S.I. Kikcio, T.D. Bethune and D.I. Dunstan, 1993. Germination of encapsulated embryos of interior spruce (*Picea glauca-engelmannii* complex) and black spruce (*Picea mariana* Mill.). Plant Cell Rep. 12: 385–389.
Macek, T., T. Vanek and P. Binarova, 1989. Biocatalyst obtained by embryo formation from immobilized plant cells. Planta Medica 55: 595–596.
Margaritis, M. and J.B. Wallace, 1984. Novel bioreactor systems and their applications. Bio/Tech. 2: 447–453.
Mazid, M.A., 1993. Biocatalysis and immobilized enzyme/cell bioreactors. Bio/Tech. 11: 690–695.
McDonald, K.A. and A.P. Jackman, 1989. Bioreactor studies of growth and nutrient utilization in alfalfa suspension cultures. Plant Cell Rep. 8: 455–458.
Merillon, J.M., M. Rideau and J.C. Chenieux, 1984. Influence of sucrose on levels of ajmalicine, serpentine, and tryptamine in *Catharanthus roseus* cells *in vitro*. Planta Medica 50: 497–501.
Merkle, S.A., W.A. Parrot and E.G. Williams, 1990. Applications of somatic embryogenesis and embryo cloning. In: S.S. Bhojwani (Ed.), Plant Tissue Culture: Applications and Limitations, Vol. 19, pp. 67–101. Elsevier, Amsterdam.
Nigra, H.M., M.A. Alvarez and A.M. Giulietti, 1990. Effect of carbon and nitrogen sources on growth and solasodine production in batch suspension cultures of *Solanum eleagnifolium* Cav. Plant Cell Tiss Org Cult. 21: 55–60.
Nuutila, A.M. and V. Kauppinen, 1992. Nutrient uptake and growth of an embryogenic and a non-embryogenic cell line of birch (*Betula pendula* Roth.) in suspension culture. Plant Cell Tiss Org Cult. 30: 7–13.
Panda, A.K., S. Misra, V.S. Bisaria and S.S. Bhojwani, 1989. Plant cell reactors – a perspective. Enz. Microb. Tech. 11: 386–397.
Park, J.M., W.-S. Hu and E.J. Staba, 1989. Cultivation of *Artemisia annua* L plantlets in a bioreactor containing a single carbon and a single nitrogen source. Biotech. Bioeng. 34: 1209–1213.
Parrott, W.A., S.A. Merkle and E.G. Williams, 1991. Somatic embryogenesis: potential for use in propagation and gene transfer systems. In: D.R. Murray (Ed.), Advanced Methods in Plant Breeding and Biotechnology, pp. 158–200. C.A.B. International, Wallingford.
Preil, W., 1991. Application of bioreactors in plant propagation. In: P.C. Debergh and R.H. Zimmerman (Eds.), Micropropagation, Technology and Application, pp. 425–445. Kluwer Academic Publishers, Dordrecht.
Preil, W., P. Florek, U. Wix and A. Beck, 1988. Towards mass propagation by use of bioreactors. Acta Hort. 226: 99–106.
Rokem, J.S., 1988. Conventionally stirred tank bioreactors. In: A. Mizrahi (Ed.), Upstream Processes: Equipment and Techniques, Vol. 7, pp. 49–78. Liss, New York.
Scragg, A.H., 1990. Large-scale cultivation of *Helianthus annuus* cell suspensions. Enz. Microb. Tech. 12: 82–85.
Scragg, A.H. and M.W. Fowler, 1985. The mass culture of plant cells. In: I.K. Vasil (Ed.), Cell Culture and Somatic Cell Genetics of Plants, Vol. 2, pp. 103–128. Academic Press, Orlando, FL.
Scragg, A.H., S. Ashton, A. York, P. Bond, G. Stepan-Sarkissian and D. Grey, 1990. Growth of *Catharanthus roseus* suspensions for maximum biomass and alkaloid accumulation. Enz. Microb. Tech. 12: 292–298.
Simonton, W., C. Robacker and S. Krueger, 1991. A programmable micropropagation apparatus using cycled liquid medium. Plant Cell Tiss Org Cult. 27: 211–218.
Smart, N.J. and M.W. Fowler, 1981. Effect of aeration on large-scale cultures of plant cells. Biotech. Lett. 3: 171–176.

Smith, J.M., S.W. Davison and G.F. Payne, 1990. Development of a strategy to control the dissolved concentrations of oxygen and carbon dioxide at constant shear in a plant cell bioreactor. Biotech. Bioeng. 35: 1088–1101.

Stuart, D.A., S.G. Strickland and K.A. Walker, 1987. Bioreactor production of alfalfa somatic embryos. HortSci. 22: 800–803.

Styer, D.J., 1985. Bioreactor technology for plant propagation. In: R.R. Henke, K.W. Hughes, M.J. Constantin, A. Hollaender and C.M. Wilson (Eds.), Tissue Culture in Forestry and Agriculture, pp. 117–130. Plenum Press, New York.

Tal, B., J. Gressel and I. Golberg, 1982. The effect of medium constituents on growth and diosgenin production by *Dioscorea deltoidea* cells grown in batch culture. Planta Medica 44: 111–115.

Taticek, R.A., M. Moo-Young and R.L. Legge, 1990. Effect of bioreactor configuration on substrate uptake by cell suspension cultures of the plant *Eschscholtzia californica*. Appl. Microb. Biotech. 33: 280–286.

Taticek, R.A., M. Moo-Young and R.L. Legge, 1991. The scale-up of plant cell culture: engineering considerations. Plant Cell Tiss Org Cult. 24: 139–158.

Tautorus, T.E., 1990. Tissue and cell culture studies of black spruce (*Picea mariana* Miller B.S.P.) and jack pine (*Pinus banksiana* Lambert). Ph.D. Thesis, University of Saskatchewan.

Tautorus, T.E., L.C. Fowke and D.I. Dunstan, 1991. Somatic embryogenesis in conifers. Can. J. Bot. 69: 1873–1899.

Tautorus, T.E., M.M. Lulsdorf, S.I. Kikcio and D.I. Dunstan, 1992. Bioreactor culture of *Picea mariana* Mill (black spruce) and the species complex *Picea glauca-engelmannii* (interior spruce) somatic embryos. Growth parameters. Appl. Microb. Biotech. 38: 46–51.

Tautorus, T.E., M.M. Lulsdorf, S.I. Kikcio and D.I. Dunstan, 1994. Nutrient utilization during bioreactor culture, and maturation of somatic embryo cultures of *Picea mariana* and *Picea glauca engelmannii*. In Vitro Cell. Dev. Biol. Plant. 308: 58–63.

Toivonen, L., M. Ojala and V. Kauppinen, 1990. Indole alkaloid production by hairy root cultures of *Catharanthus roseus*: growth kinetics and fermentation. Biotech. Lett. 12: 519–524.

Townsley, P.M., F. Webster, J.P. Kutney, P. Salisbury, G. Hewitt, N. Kawamura, L. Choi, T. Kurihara and G.G. Jacoli, 1983. The recycling air lift transfer fermentor for plant cells. Biotech. Lett. 5: 13–18.

Treat, W.J., C.R. Engler and E.J. Soltes, 1989. Culture of photomixotrophic soybean and pine in a modified fermentor using a novel impeller. Biotech. Bioeng. 34: 1191–1202.

Tulecke, W. and L.G. Nickell, 1959. Production of large amounts of plant tissue by submerged culture. Science 130: 863–864.

Vasil, I.K. 1991. Scale-Up and Automation in Plant Propagation. Cell Culture and Somatic Cell Genetics of Plants, Vol. 8, Academic Press, San Diego, CA.

Veliky, I.A. and D. Rose, 1973. Nitrate and ammonium as nitrogen nutrients for plant cell cultures. Can. J. Bot. 51: 1837–1844.

von Arnold, S. and I. Hakman, 1988. Regulation of somatic embryo development in *Picea abies* by abscisic acid (ABA). J. Plant Physiol. 132: 164–169.

Weathers, P.J.and K.L. Giles, 1988. Regeneration of plants using nutrient mist culture. In Vitro 24: 727–732.

Weathers, P.J. and R.W. Zobel, 1992. Aeroponics for the culture of organisms, tissues and cells. Biotech. Adv. 10: 93–115.

Ziv, M., 1991. Morphogenic patterns of plants micropropagated in liquid medium in shaken flasks or large-scale bioreactor cultures. Israel J. Bot. 40: 145–153.

14. Cryopreservation for Germplasm Collection in Woody Plants

Akira Sakai

Contents

1. Introduction

Tissue culture methods are commonly used for mass propagation of a wide variety of tree species. Meristem culture is utilized not only for clonal propagation, but also for the supply of virus-free plants. Furthermore, the progenies regenerated by *in vitro* culture of meristems have thus far, displayed a greater genetic stability compared to other methods of *in vitro* plant regeneration. A possible exception would be the stability attained through somatic embryogenesis. These unique attributes make plant meristems ideal candidates for the preservation of germplasm. Embryogenic cell cultures (Vasil, 1983) are also of considerable importance since they are currently a major source of totipotent cells and protoplasts for genetic transformation. The recent demonstration that embryogenic conifer cultures are amenable to cryopreservation creates an obvious application for long-term storage of conifer tissues for clonal forestry programs (Kartha *et al.*, 1988; Gupta *et al.*, 1987; Klimaszewska *et al.*, 1992). These *in vitro* practices will undoubtedly play an important role for large-scale production and improvement of fruit and forest trees, and eventually for increased production of biomass energy.

The technique of cryopreservation, in which cells and meristems are frozen under controlled conditions and stored in liquid nitrogen (LN_2), is a reliable method and has been an important tool for long-term preservation of germplasm. It requires minimum space, and maintenance, and is accompanied by few apparent genetic alterations. Availability or development of efficient and reliable protocols for cryopreservation and regeneration of plants from cryopreserved cells and meristems are the basic requirements to cryopreservation. This consideration obviously narrow the choice of the plant material

S. Jain, P. Gupta & R. Newton (eds.), Somatic Embryogenesis in Woody Plants, Vol. 1, 293–315.

currently available for cryopreservation unless research efforts are extended to include many other plant species, so far untreated. Cryopreservation is based on the non-injurious reduction and subsequent interruption of metabolic functions of biological materials by temperature reduction to the level of that of LN_2 (−196°C). Successful cryopreservation has generally been achieved by slow prefreezing to about −40°C prior to immersion in LN_2 in the presence of suitable cryoprotectants. This is the most widely used slow freezing method and requires controlled freezing equipment with complicated cryoprotective procedures. Thus, the development of a simple and reliable method for cryopreservation would allow for wider use of cryopreserved cells, meristems and embryos. Recently, some simple and reliable protocols such as simple freezing (Sakai *et al.*, 1991a), vitrification (Langis and Steponkus, 1990; Sakai *et al.*, 1990, 1991b) and alginate-coated dehydration (Dereuddre *et al.*, 1990) have been developed, and the number of cryopreserved woody species has been increased sharply over the last few years.

2. Conventional Slow Prefreezing

For maintaining viability after cooling to the temperature of LN_2, it is essential to avoid lethal intracellular freezing, which occurs during rapid cooling into LN_2 (Sakai and Yoshida, 1967). Thus, cells and meristems have to be sufficiently dehydrated to avoid intracellular freezing prior to immersion into LN_2 (Sakai, 1960). Based on the dehydration method prior to rapid cooling into LN_2, successful cryopreservation procedures at a practical cooling rate for cultured cells and meristems can be divided into four categories: 1) slow prefreezing, 2) simple freezing, 3) vitrification, 4) desiccation (Sakai, 1993).

The conventional prefreezing method is diagrammed in Fig. 1. Cultured cells and meristems are generally sensitive to freezing. Thus, material is pretreated with cryoprotectants to enable survival below −30°C. One effective and widely used cryoprotectant for cultured cells consists of a mixture of 5 to 10% dimethyl sulfoxide (DMSO) and 5 to 10% sugar or sorbitol (Table 1). To decrease toxicity, precooled cryoprotectants are gradually added to a cell suspension in an ice bath and subsequently held for 1 h at 0°C to allow equilibration. Preculture with a high concentration of sugar (Bertrand-Desbrunais *et al.*, 1988), or sorbitol (Chen et al., 1984; Kartha *et al.*, 1988; Klimaszewska *et al.*, 1992) improved the survival of less tolerant cells to freezing and dehydration (Table 1).

In the conventional slow prefreezing method (Fig. 1), cultured cells and meristems are slowly prefrozen in the presence of a suitable cryoprotectant at the rate of 0.3 to 0.5°C min^{-1} to about −40°C prior to being immersed into LN_2 (Sugawara and Sakai, 1974; Uemura and Sakai, 1980; Kobayashi *et al.*, 1990). Slow freezing to about −40°C results in a sufficient concentration of the unfrozen fraction of the suspending solution and cytosol to enable vitrification upon rapid cooling into LN_2. Survival of the vitrified cells de-

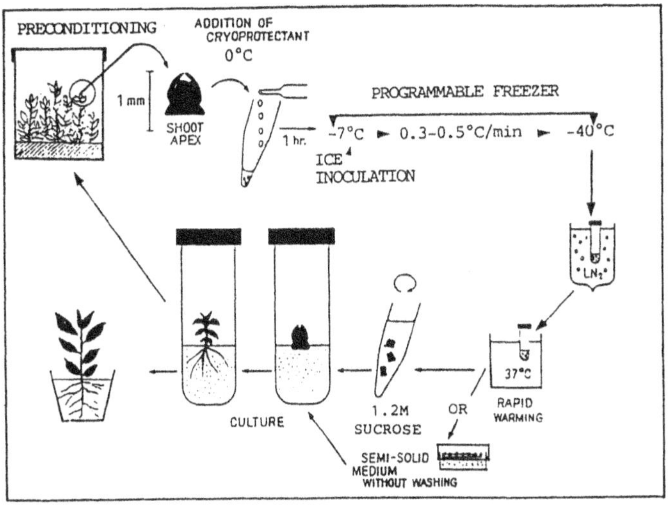

Figure 1. Procedure for cryopreservation by slow prefreezing method of apical meristems.

Table 1. Cryopreservation of cultured cells and meristems of woody plants by the slow prefreezing method and subsequent plant regeneration.

Species	Preculture	Cryoprotectant	Reference
Citrus sinensis (Navel orange) Nucellar cells	None	5% DMSO + 1.2 M suc	Kobayashi, 1990
Coffea arabica (Somatic embryo)	0.75 M suc, 1 day	5% DMSO + 0.75 M suc	Beltland-Desbrunais *et al.*, 1988
Larix xeurolepis (Larch)[a]	0.4 M suc, 1 day	0.4 M sorb + 10% DMSO	Klimaszewska *et al.*, 1992
Phoenix dactylifera (Date palm)	None	PGD	Ulich *et al.*, 1982
Picea abies (Norway spruce)[a]	None	PGD	Gupta *et al.*, 1987
Picea glauca (White spruce)	0.2 M sorb + 0.4 M sorb (each 1 day)	5% DMSO + 0.4 M sorb	Kartha *et al.*, 1988
Picea mariana (Black spruce)[a]	0.4 M sorb, 1 day	5% DMSO + 0.4 M sorb	Klimaszewska *et al.*, 1992
Pinus taeda (Loblolly pine)[a]	None	PGD	Gupta *et al.*, 1987
Pyrus 3 sp. 1 hyb (Pear) Meristem	Hardening, 5% DMSO – 2 days	PGD	Reed, 1990
Rubus 4 sp. 4 cvs Meristem	Hardening, 5% DMSO – 2 days	PGD	Reed, 1988, 1992
Vitis vinifera (Grape)[a]	None	5% DMSO + 0.25 M malt	Dussert *et al.*, 1991

[a] Embryogenic cells; PGD – PEG 10%, glucose 10%, DMSO 10% (each w/v%) (Finkle and Ulich, 1979); suc – sucrose; sor – sorbitol; malt – maltose; PEG – polyethylene glycol.

pends on the fate of the glassy cytoplasm during the warming process. When warming is slow enough to permit sufficient time for crystallization of the cytoplasm, low survival was observed due to crystallization during the warming process (Sakai *et al.*, 1991b). Thus, the vitrified cells following freeze-dehydration to −40°C must be warmed rapidly to produce a high level of survival. Particular attention has been given to the deleterious effects of washing with culture medium. Withers and King (1979) suggested that in addition to triggering injury by rapid deplasmolysis, post-thawing washing, especially hypotonic culture medium, may cause damage by removing vital solutes. Chen *et al.* (1984) achieved the regrowth of cryopreserved cells by their transfer without washing onto filter paper disc over nutrient medium solidified with agar. In our cryopreservation procedure, frozen and thawed cells or meristems were expelled into a 1.2 M sucrose solution dissolved in culture medium, and then dispensed onto a double layer filter paper disc over nutrient medium solidified with agar. After one day, the cells or meristems were transferred onto a fresh filter paper disc in a petri dish containing the same medium (Fig. 1).

Conventional slow freezing is still important for the cryopreservation of cultured cells. Successful cryopreservation procedures for cultured cells of woody plants are listed in Table 1. However, this method is time-consuming and laborious and requires controlled freezing equipment and complicated procedures.

3. Simple Freezing

Alternative cryogenic strategies reduce or eliminate the need for cellular dehydration during slow freezing by osmotically dehydrating cells or meristems prior to direct transfer to a freezer at −30°C (simple freezing method). With this method, cells are cryoprotected with a mixture of 2 M glycerol and 0.4 M sucrose at 25°C for 10 min and then spontaneously frozen by placing the 1.8 ml cryotube in a freezer at −30°C for 1 h prior to immersion into LN_2 (Sakai *et al.*, 1991a). The cell suspension nucleates spontaneously at about −15°C and then cooled to −30°C at the rate of about 2°C min^{-1} (Sakai *et al.*, 1991a). The cryoprotective effect in this method was compared for 2 or 3 M solutions of various solutes dissolved in culture medium supplemented with 0.4 M sucrose. The survival rate was highest for the cells treated with 2 M glycerol and 0.4 M sucrose at 25°C for 10 min (Table 2). All the cryoprotectants tested resulted in very low survival rates when the cells were frozen without sucrose.

With the simple freezing method, the effect of the terminal freezing temperature on survival prior to immersion into LN_2 on survival was examined (Nishizawa *et al.*, 1992). Cryoprotected cells were prefrozen by direct transfers at −15, −20, −30 and −40°C, respectively, and maintained there for 1 h prior to immersion in liquid nitrogen. Prefrozen cells at −30°C showed

Table 2. Effect of cryoprotectant on the survival of nucellar cells of navel orange cooled to −196°C by the simple freezing method.

Cryoprotectant	Survival (% ± S.E.)
2.0 M ethylene glycol	40.7 ± 1.6
3.0 M ethylene glycol	32.3 ± 1.9
2.0 M propylene glycol	66.1 ± 1.4
3.0 M propylene glycol	35.6 ± 1.8
2.0 M glycerol	91.2 ± 1.4
3.0 M glycerol	80.4 ± 1.5
2.0 M DMSO	63.4 ± 1.8
3.0 M DMSO	52.5 ± 1.6

Cryoprotection of the nucellar cells of navel orange was compared for various cryoprotectants dissolved in MT medium supplemented with 0.4 M sucrose at 25°C for 10 min. The cells were loaded into 0.5 ml straws and placed in a freezer at −30°C for 40 min prior to immersion in liquid nitrogen. The cell suspensions initiated freezing spontaneously at about −15°C and then cooled at about 2°C min^{-1} between −10 and −30°C. After rapid thawing in 40°C water, the cells were expelled into 2 ml of MT medium containing 1.2 M sucrose. In each treatment, two straws were used and about 100 cell clusters in each straw were examined for their viability by means of double FDA and phenosafranin (Sakai *et al.*, 1991a).

the highest survival (Fig. 2). By direct transfer from 25°C, the prefrozen cells at −40°C showed very low survival rate before or after immersion in LN_2. However, the prefrozen cells at −30°C which were additionally frozen at −40°C for 1 h showed a high survival. It was also observed that, these cells cooled slowly to −30 or −40°C at 0.5°C min^{-1} using a programmable freezer, showed nearly the same high rate of survival as those frozen in a freezer at −30°C. These results indicate that osmotic dehydration by a mixture containing 2 M glycerol and 0.4 M sucrose for 10 min at 25°C prior to freezing at −30°C resulted in high survival and simplification of the complex cryoprotective and freezing procedures (Sakai *et al.*, 1991a; Nishizawa *et al.*, 1992). This simple freezing method requires neither controlled freezing equipment nor ice inoculation to the extracellular medium. The temperature of −30°C corresponds to that of ordinary freezers. This method has been successfully applied to cultured cells such as orange, rice, carrot, asparagus, and meristemoids of white clover.

4. Vitrification

Vitrification refers to the physical process by which a highly concentrated cryoprotective solution supercools to very low temperatures and solidifies into metastable glass without undergoing crystallization (Fahy *et al.*, 1984). Recent work has focused on procedures that would eliminate the need for controlled freezing and enable cells and meristems to be cryopreserved by direct transfer into LN_2. Vitrification can also be achieved by direct immer-

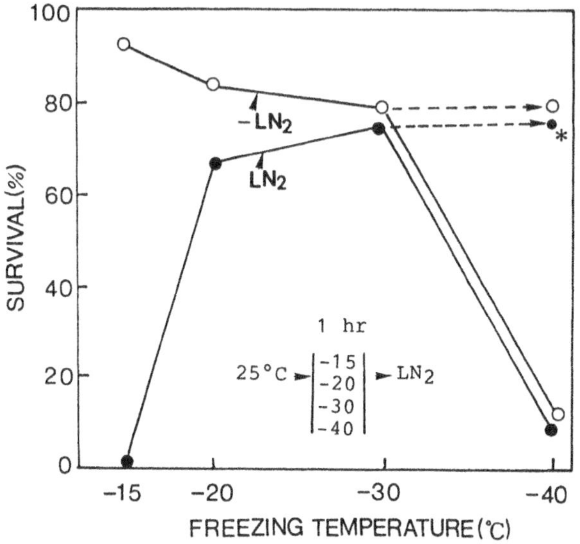

Figure 2. Effect of terminal freezing temperature on the survival of cells cooled to −196°C by the simple freezing method. Material:embryogenic cells of asparagus. The cells were cryoprotected with a mixture of 2 M glycerol and 0.4 M sucrose at 25°C for 10 min. Cryoprotected cells were frozen by direct transfer to freezers at −15, −20, −30 and −40°C, respectively. These cells frozen at each temperature for 1 hr were thawed rapidly before or after being immersed in liquid nitrogen. * Prefrozen cells were additionally frozen in a freezer at −40°C for 1 h prior to immersion in LN_2 (Nishizawa *et al.*, 1992).

sion into LN_2 without the freeze-concentration step by exposing the cells to extremely concentrated solutions of permeating and nonpermeating cryoprotectants. Such techniques are referred to as vitrification, distinct from conventional cryopreservation methods.

With vitrification, cells or meristems must be sufficiently dehydrated with a vitrification solution at about 25 or 0°C to avoid injury prior to immersion into LN_2. We used a glycerol-based, less toxic vitrification solution designated vitrification solution (Sakai *et al.*, 1991) (PVS2) which consists of 30% (w/v) glycerol, 15% (w/v) ethylene glycol and 15% (w/v) DMSO in the culture medium containing 0.4 M sucrose (pH 5.8) (Sakai *et al.*, 1990). This solution can be easily supercooled below −70°C on rapid cooling and becomes solidified into metastable glass at about −115°C. Upon subsequent slow warming in differential scanning calorimetory (DSC), the vitrified PVS2 expressed a glass transition (Tg) at about −115°C, with an exothermic devitrification (Td) (crystallization) at about −75°C and an endothermic melting (Tm) at about −36°C (Sakai *et al.*, 1990). The complete vitrification procedure involves: a) loading cells with an intermediate concentration of suitable cryoprotectants (cryoprotection); b) dehydrating cells by exposure to a concentrated vitrification solution (PVS2); c) placing cells in a 1.8 ml cryotube or loading into a 0.5 ml plastic straw followed by immersion into liquid nitrogen; d) rapid warming in a water bath at 40°C; e) unloading PVS2 (dilution) by transferring cells to culture medium containing 1.2 M sucrose.

4.1. *Survival of Nucellar Cells of Navel Orange Cryopreserved by Vitrification*

There are only a few reports on the applications of cryopreservation techniques to citrus. Marin and Duran-Vila (1988) reported that a small number of frozen embryos (3 to 5%) survived and developed into proliferating cultures that produced whole plants. However, the survival of somatic embryos did not refer to the survival of the whole embryos, but to recovery of proliferating structures from surviving cells by secondary embryogenesis. Kobayashi *et al.* (1987) obtained plants from protoplasts of navel orange nucellar callus and succeeded in producing somatic hybrid plants of *Rutaceae* by protoplast fusion (Kobayashi *et al.*, 1988). Thus, we attempted to apply the vitrification technique to nucellar callus of navel orange and succeeded in obtaining a high survival rate and subsequent regeneration of whole plants.

Two ml of a cell suspension from an 8-day-old culture containing about 0.2 ml packed cell volume of nucellar cells of navel orange was transferred to a 10 ml conical glass tube (110 × 15 mm) and allowed to settle. The supernatant was discarded and 4 ml of PVS2 at 25°C was added. The cell suspension was mixed and then centrifuged at 100 g for 20 s. The supernatant was discarded and 2 ml of fresh PVS2 was then added. The cells were treated with PVS2 at 25 or 0°C for various periods of time. Then, an aliquot of 0.2 or 0.4 ml of cell suspension was loaded into a 0.5 ml plastic straw or placed in a 1.8 ml cryotube. The straws or cryotubes were plunged into liquid nitrogen and held for a minimum of 30 min (Fig. 3). The mean cooling rate in the straw or the cryotube was about 1600°C min^{-1} or about 300°C min^{-1} in the range between −30 and −150°C, respectively. Cell suspensions in straws or cryotubes were warmed in a water bath at 25°C. The cell suspension was expelled into 2 ml of a diluting solution containing 1.2 M sucrose in the culture medium at 25°C and held for 10 min before the viability or growth capacity was assessed (Fig. 3).

The cells treated with PVS2 for various periods of time prior to immersion in LN_2 resulted in a time-dependent survival (Fig. 4). The highest survival was obtained for the cells treated with PVS2 for 3 min at 25°C. The vitrified cells treated with PVS2 for 1 to 5 min at 25°C still showed the high levels of survival (80 to 90%), but the survival rate decreased rapidly after a longer exposure. The survival of the cells treated with PVS2 at 25°C without cooling in LN_2 (treated control) attained levels approaching the vitrified cells. The vitrified cells treated with PVS2 at 0°C for 3 to 20 min showed high rates of survival (70–75%) after cooling to −196°C.

Vitrified and warmed cells were grown on filter paper discs over agar medium. Control and vitrified cells commenced growth within 4 days after

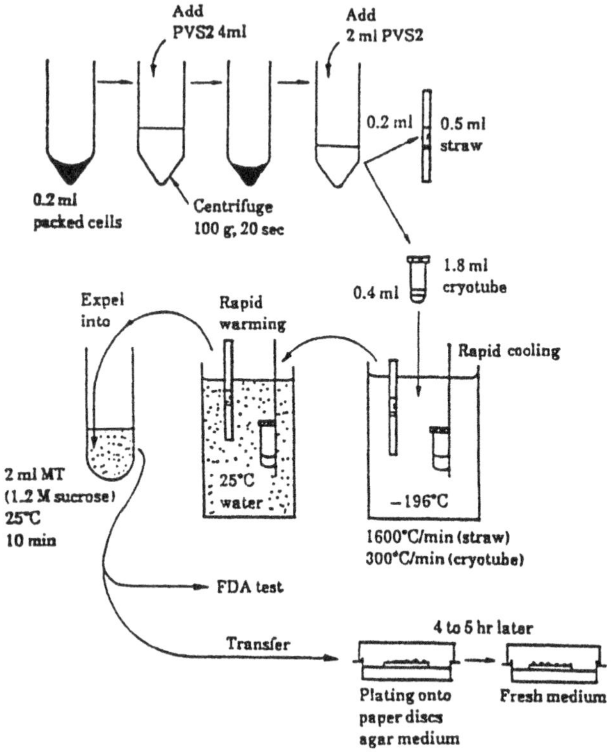

Figure 3. Vitrification procedure for nucellar cells of navel orange (Sakai *et al.*, 1991b).

plating, but the regrowth rate of the vitrified cells was lower than that of controls until 12 days after plating. Thereafter the growth pattern was similar to that of the controls (Fig. 5). Cotyledonary embryoids were produced from the cells within 2 to 3 months of culture and then developed into whole plants after 2 to 3 months (Fig. 6). The vitrified nucellar callus exhibited embryogenic potential identical with that of the nontreated controls, and plants regenerated from vitrified cells were morphologically uniform.

The vitrification procedure was successfully applied to the nucellar cells of three other citrus plants. High survival rates (above 90%) were obtained in these citrus plants that were cryopreserved for 320 days (Table 3). The survival of the nucellar cells of navel orange cooled to −196°C was compared with three different cryogenic protocols (Table 4). The highest survival was obtained from vitrified cells treated with PVS2 at 25°C for 3 min.

The warming rate of the nucellar cells was significant. Cells warmed rapidly at 300°C min^{-1} or above showed a high survival rate (about 90%) regardless of the cooling rate (Table 5). Conversely, most of the vitrified cells warmed slowly at 30°C min^{-1} or below showed very low survival rates. This obser-

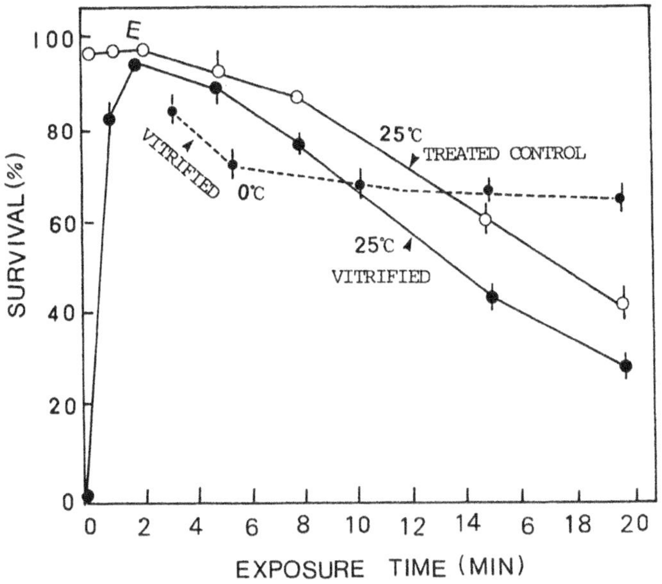

Figure 4. Effect of exposure time to PVS2 at 25 or 0°C on the survival of vitrified nucellar cells of navel orange. Cells treated with PVS2 at 25 or 0°C for various periods of time were loaded into a 0.5 ml plastic straw or placed in a 1.8 ml cryotube and then directly immersed into liquid nitrogen for 30 min. After rapid warming, the cell suspension was expelled into 2 ml of MT medium containing 1.2 M sucrose. Treated control: same as vitrified cells without cooling to −196°C (Sakai *et al.*, 1991b, modified).

vation can be attributed to the occurrence of intracellular freezing during the slow warming process. Thus, the time required for the temperature to rise from −80 to −40°C during which crystallization may occur was calculated based on the results presented in Table 5. As shown in Fig. 7, When cells were exposed to the temperature range −80 to −40°C within about 25 s (rates above 100°C min^{-1}), a high survival rate (about 90%) was expressed, in contrast, the survival rapidly decreased in cells warmed more slowly.

4.2. *Survival of Shoot Tips Cryopreserved by Vitrification*

This vitrification method was successfully applied to *in vitro*-grown shoot tips of certain woody plants. Excised shoot tips (about 2 mm long, 1.5 mm base diameter) from cold-hardened apple plantlets at 5°C for 3 weeks were precultured on a solidified agar medium supplemented with 0.7 M sucrose for 1 day at 5°C. Following preculture, shoot tips were treated with PVS2 for various periods of time at 25°C prior to immersion into LN_2 (Fig. 8). Shoot formation of vitrified shoot tips (2 mm in length) increased gradually with time of exposure to PVS2 and reached a maximum at about 80 min exposure. However, precultured meristems (0.5 to 1.0 mm in length) of white clover

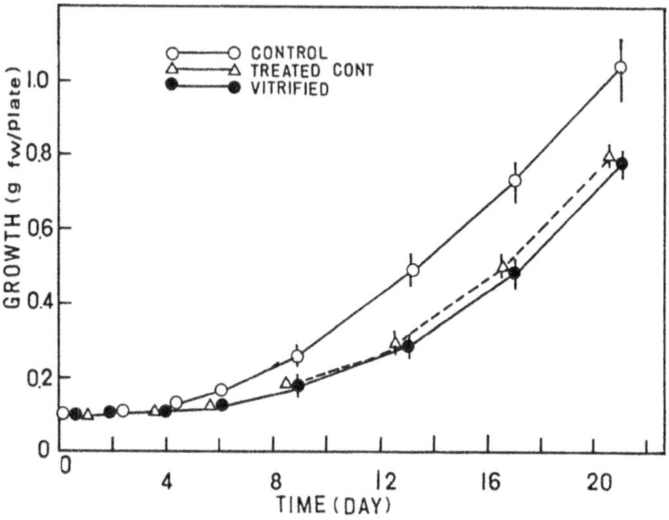

Figure 5. Recovery growth of vitrified-warmed cells of navel orange. Nucellar cells were treated with PVS2 at 25°C for 3 min prior to immersion in LN_2. After rapid warming, the cells were unloaded with 1.2 M sucrose and then cells (about 100 mg fresh weight) were plated onto a filter paper disc over agar medium. Control: without any treatment; Treated control: same as vitrified cells without cooling to −196°C (Sakai *et al.*, 1991b).

Figure 6. Plant regenerated from nucellar cells cooled to −196°C by vitrification, at about 1 month after transfer to soil in a pot. Bar represents 1 cm (Sakai *et al.*, 1991b).

Table 3. Survival rate of nucellar cells of four citrus plants cooled to −196°C by vitrification and stored for 320 days.

Species	Survival (%)
Citrus sinensis Osb. var. *brasiliensis* Tanaka (Navel orange)	90.0 ± 1.4
Citrus paradisi Macf. (Grapefruit)	96.0 ± 0.9
Citrus sudachi Hort. ex Shirai (Sudachi)	97.0 ± 2.0
Hybrid (Murcott' Tangor)	92.0 ± 1.4

Suspension culture nucellar cells treated with PVS2 at 25°C for 3 min were loaded into 0.5 ml straws and then immersed into LN_2. Values are means of duplicates (Sakai *et al.*, 1991b).

Table 4. Survival rate of nucellar cells of navel orange cooled to −196°C using different cryogenic protocols.

Cryogenic Protocol	Survival (% ± S.E.)
Conventional prefreezing method[1]	73.0 ± 1.4
Simple freezing method[2]	65.1 ± 1.6
Vitrification[3]	90.0 ± 1.1

[1] Cooled to −40°C at 0.3°C min^{-1} in the presence of 5% DMSO and 1.2 M sucrose (Kobayashi *et al.*, 1990).
[2] Cells were treated with a mixture of 2 M glycerol and 0.4 M sucrose at 25°C for 10 min and then directly transferred in a freezer at −30°C for 1 h prior to immersion in LN_2 (Sakai *et al.*, 1991a).
[3] Treated with PVS2 at 25°C for 3 min prior to immersion in LN_2 (Sakai *et al.*, 1991b).

Table 5. Effect of warming rate on the survival of the nucellar cells of navel orange cooled to −196°C by vitrification.

Containers	Cooling Rate (°C/min)	Warming Rate (Medium) (°C/min)	Survival (%)
0.5 ml straw	1.500	1.600 (25°C, water)	91.0 ± 0.8
	1.600	1.200 (0°C, water)	91.0 ± 0.8
	1.550	109 (50°C, air)	89.0 ± 0.9
	1.500	100 (25°C, air)	88.1 ± 1.2
1.8 ml cryotube	280	300 (25°C, water)	91.0 ± 1.2
	283	30 (25°C, air)	47.1 ± 1.4
	282	24 (0°C, air)	30.2 ± 1.4
	285	6 (−20°C, air)	23.9 ± 1.3

Nucellar cells treated with PVS2 at 25°C for 3 min were loaded into a 0.5 ml straw or placed in a 1.8 ml cryotube (0.4 ml cell suspension) prior to immersion in LN_2. Vitrified cells were warmed in a water bath or in a air at different temperatures, respectively. Cooling rate was calculated from the time required for the temperature to drop from −30 to −150°C. Warming rate was calculated from the time required for the temperature to rise from −80 to −40°C. Values are means of duplicates (Sakai *et al.*, 1991b).

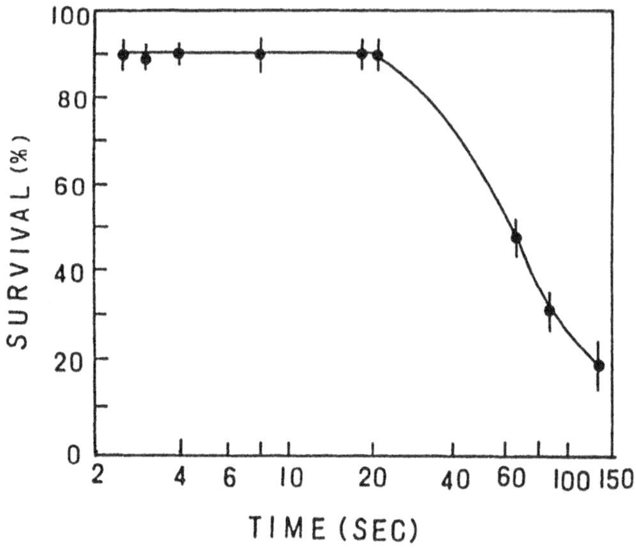

Figure 7. Effect of warming rate on the survival of nucellar cells cooled to −196°C by vitrification. Vitrified cells were warmed at various rates and time required for the temperature to rise from −80 to −40°C was calculated based on the results presented in Table 5. Vertical bars: SE (n = 2) (Sakai *et al.*, 1991b).

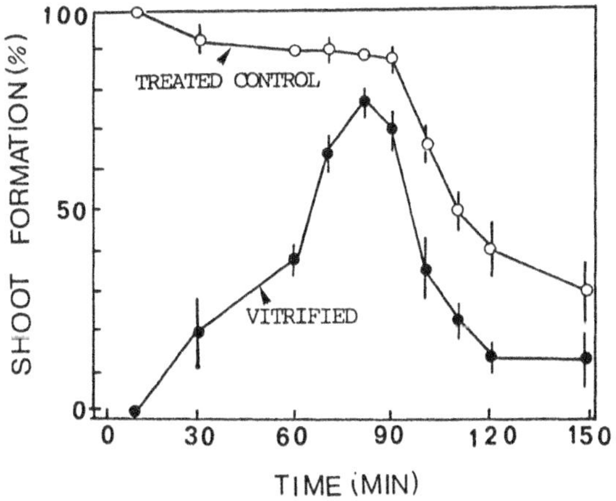

Figure 8. Effect of exposure time to PVS2 at 25°C on the shoot formation of apple shoot tips cooled to −196°C by vitrification. Material: *Malus domestica cv.* Fuji. Cold-hardended, precultured shoot tips (2 mm in length) were treated with PVS2 for various lengths of time at 25°C and then directly immersed in LN_2 for 1 day. Treated control: same as vitrified meristems, without cooling to −196°C (Niino *et al.*, 1992a).

Table 6. Effects of cold-hardening and preculturing on the shoot formation of apple shoot tips cool by vitrification.

Period of Preculture (Days)	Shoot Formation (% ± S.E.)	
	Hardening	Non-hardening
0	5.0 ± 2.9	0
1	60.0 ± 4.1	0
2	32.0 ± 7.5	0

Material: *Malus domestica* cv. Fuji. Cold-hardening: 3 weeks at 5°C (8h/day photoperiod); Preculturing: at 5°C on MS agar medium supplemented with 0.7 M sucrose; Shoot tips (2 mm in length) were treated with PVS2 for 80 min at 25°C and then directly immersed in LN_2. Shoot formation (%): percent of shoot tips produced shoots 40 days after plating (Niino *et al.*, 1992a).

(Yamada *et al.*, 1991) which were treated with PVS2 at 25°C for 5 min or at 0°C for 15 min prior to a plunge into LN_2 produced 80% plant regeneration. The incubation time in PVS2 may be associated with size of the excised meristems and appears to be species-specific. Recently we confirmed that the optimum exposure time for apple shoot tips of 0.5 to 1 mm in length was 20 to 25 min at 25°C.

Successfully vitrified apple shoot tips resumed growth in about 3 days after reculture, and started to develop shoots within two weeks without intermediary callus formation. Apple shoot tips, cooled to −196°C by vitrification, developed plantlets at 30 days after reculture. This vitrification method using PVS2 was successfully applied to *in vitro*-grown shoot tips of five apple species or cultivars, eight pear cultivars (Niino *et al.*, 1992a) and 13 mulberry species or cultivars including South East Asian cultivars (Niino *et al.*, 1992b), tea plants (Kuranuki and Sakai, 1992), Ribes (Reed, 1992) and many herbaceous plants.

Cold-hardening, preculturing and post-thaw handling of apical meristems from *in vitro* plantlets of woody plants are essential to successful cryopreservation by any other cryogenic protocol (Dereuddre *et al.*, 1990; Niino *et al.*, 1992a,b). *in vitro*-grown apple plantlets, preculturing at 5°C with 0.7 M sucrose for 1 day following cold-hardening at 5°C for 3 weeks improved the recovery rate of hardened shoot tips cooled to −196°C by vitrification (Table 6). Reed (1990, 1992) clearly demonstrated that cold-hardening significantly improved the recovery rates of apical meristems of all four pear cultivars including subtropical species cooled to −196°C by the slow freezing method.

It is particularly important that cryopreserved meristems produce plants identical to the nontreated phenotypes. Haskins and Kartha (1980) demonstrated that renewed growth was mainly originated from tissues other than that of the original meristem dome and that renewed growth led to differentiation and whole plant regeneration in frozen-thawed pea meristems treated with the conventional freezing method. Recovery of proliferated structures from surviving cells by secondary embryogenesis or adventitious buds was also reported (Towill, 1984). A callus phase prior to shoot formation is undesirable since callusing potentially increases the frequency of genetic variants. However, successfully vitrified and warmed meristems generally produced direct shoot formation in many materials tested in both herbaceous and woody plants (Yamada *et al.*, 1991; Niino *et al.*, 1992a,b; Schnabel-Preikstas *et al.*, 1992a,b,c). Recently, the vitrification procedure for cryopreservation has been applied to a wide range of cultured meristems of herbaceous plants. The vitrification technique appears to be a potential and ideal method for cryopreservation of meristems.

4.3. *Cryopreservation of Dehydration Sensitive Cultured Cells by Vitrification*

The vitrification method did not result in high rates of survival for the majority of cultured plant cells except for a limited number of dehydration tolerant cultured cells (Langis *et al.*, 1989; Uragami *et al.*, 1989; Sakai *et al.*, 1990, 1991b). We observed that direct exposure of less resistant cultured cells to a highly concentrated vitrification solution may lead to harmful effects due to osmotic stress or chemical toxicity. This limitation was almost completely overcome asparagus cultured cells by cryoprotective treatment (loading) with a mixture of 1.6 to 2 M glycerol and 0.4 M sucrose for 10 min at 25°C before the dehydration step (Table 7). Another approach for successful vitrification of dehydration sensitive cultured cells is the gradual addition of PVS2 at 0°C. The elaborate procedure for vitrification was more recently presented by Reinhoud *et al.* (1993). In the procedure, cells pretreated with 0.33 M mannitol or sucrose for 3 days were then treated with cold 20% PVS2 (20% solution of the stock PVS2) at 0°C for 5 min, and then gradually dehydrated by adding cold PVS2 with 1 min intervals at 0°C (Fig. 9). This procedure is currently being applied to one line of tobacco cells and has been successfully used for two lines of *Catharanthus roseus*. A two- or three-step loading procedure is less injurious than a one-step procedure for cells less tolerant to dehydration. The harmful effects due to dehydration can be alleviated or eliminated by reducing the concentration of PVS_2 and time of exposure for cell dehydration after which the cell suspension may become vitrified.

5. Drying Method

Another approach could involve extensive dehydration by air-drying. For example, cultured cells, somatic embryos (Senaratna *et al.*, 1990; Shimonishi *et al.*, 1991), lateral buds (Uragami et al., 1990) and apical meristems (Niino and Sakai, 1992) from *in vitro*-grown plantlets were cryopreserved after air-drying. However, the induction or modification of drought tolerance may be the main factor for successful cryopreservation in this approach. Alginate-coated shoot tips of *in vitro*-grown pear and potato plants were successfully

Table 7. Effect of loading on the survival of vitrified cells cooled to −196°C following dehydration with a vitrification solution.

Loading Solution	Survival (% ± S.E.)
Experiment 1	
Unloading	37.1 ± 1.3
2 M glycerol + 0.4 M sucrose	85.7 ± 0.6
1.7 M glycerol + 0.4 M sucrose	81.6 ± 1.3
1.2 M glycerol + 0.4 M sucrose	68.5 ± 1.2
2 M glycerol	74.3 ± 1.3
2 M EG + 0.4 M sucrose	51.6 ± 3.7
1.7 M EG + 0.4 M sucrose	53.4 ± 1.3
1.2 M EG + 0.4 M sucrose	47.0 ± 1.3
2 M EG	0.8 ± 1.5
Experiment 2	
2 M glycerol + 0.4 M sucrose	84.7 ± 0.8
1.7 M DMSO + 0.4 M sucrose	62.6 ± 2.0
1.2 M DMSO + 0.4 M sucrose	56.9 ± 1.7
2 M DMSO	0.6 ± 0.3
2 M prop. glycol + 0.4 M sucrose	7.0 ± 0.3
10% DMSO + 10% sucrose + 0.5 M glycerol	62.0 ± 1.5

Embryogenic cells of asparagus. Cells were loaded with each cryoprotectant for 10 min at 25°C and then dehydrated with PVS3 (5 g glycerol, 5 g sucrose and 3 g water) at 0°C for 20 min prior to immersion in LN_2. After rapid warming, cells were expelled into LS medium containing 1.2 M sucrose. The viability was compared with the controls by means of double staining with FDA and phenosafranin (Nishizawa *et al.*, 1993).

cryopreserved following air-drying (Dereuddre *et al.*, 1990; Fabre and Dereuddre, 1990). The encapsulating-dehydration technique is easy to handle and alleviates the dehydration process. In this method, resistance to dehydration and deep freezing was induced by preculturing encapsulated shoot tips in a medium enriched with sucrose before dehydration. In the encapsulation-dehydration technique, the sucrose molarity increased markedly during the drying process and reached or exceeded the saturation point of the sucrose solution resulting in a glass transition during cooling (Dereuddre et al., 1991a,b). The encapsulation-dehydration technique appears to be a practical method for cryopreservation of meristems and embryos. Recently, we have developed a modification of the encapsulation dehydration technique (Niino and Sakai, 1992). To induce dehydration tolerance, cold-hardened apple apical meristems were precultured before being embedded into alginate-coated beads by gradual daily transfer to media containing 0.1, 0.4, and 0.7 M sucrose (Table 8). The precultured shoot tips embedded into alginate-coated beads containing 0.5 M sucrose were treated in a medium supplemented with 1.0 M sucrose for 16 h at 5°C. The beads were dehydrated on sterile silica gel at 25°C before immersion in LN_2. Shoot formation of dehydrated alginate-coated *in-vitro* grown apple shoot tips before or after being cooled to −196°C is shown in Fig. 10. The alginate-coated shoot tips

VITRIFICATION OF PLANT CELL SUSPENSIONS
- Elavorate Dehydration Method-

MATERIAL:	* in rapid growth phase * cool 1 ml packed cells
TREATMENT 1:	* add 4 ml of cold 20%PVS2 for 5 min
TREATMENT 2: (dehydration)	* add 5 times 1 ml of PVS2 each time 1 min
	at 0°C
TREATMENT 3: (dehydration)	* centrifuge at 100 g * discard supernatant * repeat treatment 2
COOLING: (vitrify)	* fill 0.5 ml straws * immerse rapidly in liquid nitrogen
WARMING:	* water bath 40 °C * decontaminate (alcohol) * empty 6 straws in 5 ml 1.2 M sucrose, leave 20 min
RECOVERY:	* centrifuge, discard supernatant * plate on filter paper disks on medium with agarose * after 2 days, transfer upper disk to fresh plate

Figure 9. Elavorate procedure of cryopreservation by vitrification of plant cell suspensions (Reinhoud *et al.*, unpublished data, modified). PVS2 contains 30% (w/v) glycerol, 15% (w/v) ethylene glycol, 15% (w/v) DMSO and 0.4 M sucrose. 20% PVS2: 20% solution of the stock PVS2. Cells are precultured in a culture medium supplemented with 0.3 M sucrose or mannitol for 1–3 days.

of 37% water content produced a very low rate of shoot formation (15%) after immersion in LN_2, while prefreezing at −30°C following dehydration produced about 70% shoot formation (Fig. 10F). The modification method was successfully applied to four apple, one mulberry and three pear species or cultivars (Table 10) (Niino and Sakai 1992).

To eliminate the use of liquid nitrogen, dried alginate-coated shoot tips were transferred to −70, −135 and −196°C, respectively and stored for various lengths of time. Dried alginate-coated shoot tips stored at −135°C for at least 5 months showed little or no decrease in the rate of shoot development compared with those stored in liquid nitrogen (Table 9). However, little recovery was observed in dried alginate-coated shoot tips stored at −70°C for more than 2 days.

Successful cryopreservation of cultured cells and meristems of woody plants by vitrification or encapsulation dehydration techniques are summarized in Table 10.

Table 8. Effect of preculturing conditions on shoot formation of alginate-coated apple shoot tips cooled to −196°C after dehydration.

Preculturing Conditions: Duration, Sucrose Concentration (M)	Shoot Formation (% ± S.E.)
Nonprecultured	40 ± 10
0.1 M, 1 day	37 ± 9
0.4 M, 1 day	33 ± 3
0.4 M, 2 days	43 ± 3
0.7 M, 1 day	53 ± 9
0.7 M, 2 days	57 ± 3
0.1 M, 1 day and 0.4 M, 1 day	40 ± 10
0.4 M, 1 day; 0.7 M, 1 day	77 ± 10
0.1 M, 1 day; 0.4 M, 1 day; 0.7 M, 1 day	80 ± 3
Nonhardened and 0.1 M, 1 day; 0.4 M, 1 day; 0.7 M, 1 day	0

Material: *Malus domestica* cv. FUJI. Cold-hardening: 3 weeks at 5°C. Cold-hardened, precultured shoot tips were encapsulated in alginate-coated beads including 0.5 M sucrose and then treated in MS medium supplemented with 1 M sucrose at 5°C for 16 h prior to dehydration. The alginate-coated shoot tips were subjected to dehydration for 7 h at 25°C prior to immersion in LN_2. Shoot formation (%): percent of shoot tips that produced normal shoots 40 days after plating (Niino and
Sakai, 1992).

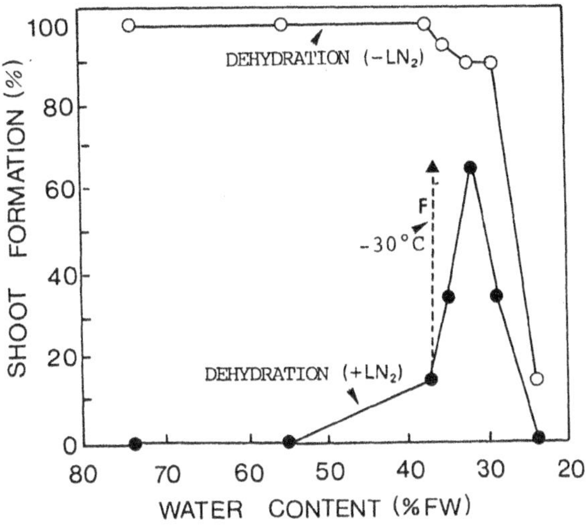

Figure 10. Shoot formation of dehydrated alginate-coated *in vitro* grown apple shoot tips before or after being cooled to 196°C. Material: *Malus domestica* cv. Fuji. Cold-hardened, precultured shoot tips in alginate-coated beads including 0.5 M sucrose was treated with 1.0 M sucrose at 5°C for 1 day. The beads were subjected to dehydration in Petri dishes containing 50 g dry silica gel held at 25°C for up to 24 h. Dehydrated beads were placed in a cryotube and transferred onto an agar MS medium without (−LN_2) or after (LN_2) cooling to −196°C. Beads were subjected to dehydration for 6 or 7 h (water content: about 33 or 37%) and were frozen at −30°C for 1 day prior to immersion in LN_2. F: The alginate-coated shoot tips of 37% water content were prefrozen at −30°C for 1 day before immersion into LN_2. Shoot formation (%): percent of shoot tips (n = 40) produced normal shoots 40 days after plating (Niino and Sakai, 1992).

Table 9. Shoot formation from dehydrated alginate-coated apple shoot tips at different temperatures directly or after cooling to 196°C.

Temperature and Duration	Shoot Formation (% ± S.E.)
196°C, 5 months	73 ± 3
135°C, 5 months[a]	77 ± 3
70°C, 3 days[a]	3 ± 3
196°C, 1 day and −135°C, 2 days	70 ± 0
196°C, 1 day and −70°C, 10 min	70 ± 6
196°C, 1 day and −70°C, 2 days	7 ± 3

[a] Direct transfer to −70°C or −135°C without cooling to −196°C.
Material: *Malus domestica* cv. Fuji. Cold-hardened and precultured shoot tips in alginate-coated beads including 0.5 M sucrose were treated in MS medium containing 1.0 M sucrose at 5°C for 16 h. The encapsulated shoot tips were dehydrated for 7 h (water content 33% FW base) before transfer to various temperatures (Niino and Sakai, 1992).

Table 10. Cryopreservation of cultured cells and meristems of woody plants by vitrification or encapsulation dehydration techniques and subsequent plant regeneration.

Species	Cryogenic Protocol	Reference
Citrus sinensis (Navel orange)	Vitrification	Sakai *et al.*, 1991b
Citrus paradisi (Grapefruit)	Vitrification	Sakai *et al.*, 1991b
Camellia sinensis (Tea plant)[a]	Vitrification	Kuranuki and Sakai, 1992
Elaeis guineensis (Oil palm)[b]	Desiccation	Dumet *et al.*, 1993
Malus 5 species or cvs (Apple)[a]	Vitrification	Niino *et al.*, 1992a
Malus 3 species or cvs[a]	Encapsulation-dehydration	Niino and Sakai, 1992
Morus 13 *species* or cvs (Mulberry)[a]	Vitrification	Niino *et al.*, 1992b
Morus bombysis 3 spec. or cvs[a]	Encapsulation-dehydration	Niino and Sakai, 1992
Pyrus 6 species or cvs (Pear)[a]	Vitrification	Niino *et al.*, 1992a
Pyrus communis[a]	Encapsulation-dehydration	Niino and Sakai, 1992
Ribes 3 species[a]	Vitrification, Encapsulation-dehydration	Reed, 1992

[a] *In vitro*-grown meristems.
[b] Somatic embryos.

6. Discussion

Vitrification can be achieved by direct immersion into LN_2 without going through the freeze-concentration step by exposing the cells and meristems to extremely concentrated solutions of cryoprotectants. Cryopreservation by the complete vitrification of cultured cells eliminates concern for the potentially damaging effects of intra- or extracellular crystallization (Rall, 1987). The vitrification technique does not require controlled freezing equipment or sophisticated, expensive apparatus. Thus, vitrification is a simple and attractive method for cryopreserving cells and meristems. However, vitrification procedures result in a high degree of osmotic stress. Thus, attention was given to potential injuries resulting from the osmotic excursions incurred during the loading, dehydration and unloading steps of the procedure. Many reports suggested that dehydration stress can be a major limitation during the excursion of the vitrification procedure (Langis and Steponkus, 1991). Injury to non-acclimated rye protoplasts resulting from severe cell dehydration was associated with several changes in the ultrastructure of the plasma membrane, including the formation of aparticulate domains in the plasma membrane, the formation of aparticulate lamellae subtending the plasma membranes, and lamellar-to-hexagonal$_{II}$ phase transitions in the plasma membrane. It was confirmed experimentally that loading with a cryoprotectant decreased the incidence of the dehydration-induced membrane ultrastractural changes (Pihakaski and Steponkus, 1987). With cold-acclimation, cells become tolerant to dehydration, and loading with cryoprotectants is not required to attain high levels of survival after either the dehydration step or cooling in LN_2. At least two possible reasons were suggested by Steponkus *et al.* (1989, 1990). During cold-acclimation the propensity for dehydration-induced lamellar-to-hexagonal phase transition is decreased because of alterations in the membrane lipid composition. Cold acclimation also results in the accumulation of endogeneous cryoprotectants such as sugars and sorbitol that increase the stability of membranes under conditions of severe dehydration (Ruldoph *et al.*, 1986; Crowe et al., 1987).

Glass fills in a tissue, and during dehydration may contribute to the prevention of additional tissue collapse, solute concentration, and pH alterations. Operationally, glass is expected to exhibit a lower water vapor pressure than the corresponding crystalline solid and thereby prevent further dehydration. As glass is exceedingly viscous and stops all chemical reactions that require molecular diffusion, its formation leads to dormancy and stability over time (Burke, 1986).

It is particularly important that cryopreserved cells produce plants identical with the nontreated phenotype. In citrus, plants regenerated from protoplasts through somatic embryogenesis also expressed phenotypic stability (Kobayashi, 1987). The nucellar callus developed from cryopreserved cells exhibited an embryogenic potential similar to that of the nontreated controls. Moreover, plants regenerated from cryopreserved cells were morphologically uniform. Further studies are necessary to confirm their phenotype by biochemical, cytological, genetic and morphological analysis. More recently, Kobayashi *et al.* (1990), confirmed that transformed nucellar cells of navel orange obtained by electroporation, were successfully cryopreserved by a vitrification method in LN_2 for 1 year. Southern blot analysis showed that the integrated neomycin-phosphotransferase II (NPTII) gene was stably maintained in the cryopreserved-regrown cells. No morphological differences were also observed between regenerated plants from cryopreserved cells and those from non-cryopreserved cells (personal communication).

It is necessary to expand the applicability of cryogenic protocols to a wide range of cultured cells and meristems of woody plants. To achieve this objective, it is essential to carry out basic studies on preconditioning for enhancing the freezing or dehydration tolerance and on basic studies on genetic stability of regenerated plants from cryopreserved cells and meristems.

7. Conclusion

To maintain viability of cultured cells and meristems, it is essential to avoid intracellular freezing during rapid cooling. Thus, cells and meristems have to be sufficiently dehydrated to avoid intracellular freezing prior to immersion in LN_2. Cryopreservation procedures can be divided into four categories based on the dehydration method prior to rapid cooling into LN_2. Recently, some simple and reliable protocols such as simple freezing, vitrification and alginate-coated dehydration have been developed, and the number of cryopreserved woody species has increased sharply over the last few years. Expanding the applicability of cryogenic protocols to a wide range of cultured cells and meristems is contingent on basic studies on preconditioning for enhancing freezing or dehydration tolerance.

References

Bertrand-Desbrunais, A., J. Fabre, F. Engelmann, J. Dereuddre and A. Charrier, 1988. Adventive embryogenesis recovery from coffee (*Coffea arabica* L.) somatic embryos after freezing in liquid nitrogen. C.R. Acad. Sci. Paris t. 307 Ser. III: 795–801.

Burke, M.J., 1986. The glassy state and survival of anhydrous biological systems. In: A.C. Leopold (Ed.), Membrane, Metabolism and Dry Organisms, pp. 358–364. Cornell Univ. Press, Ithaca, NY.

Chen, T.H.H., K.K. Kartha, N.L. Leung, W.G.M. Kurz, K.B. Chatson and F. Constabel, 1984. Cryopreservation of alkaloid-producing cell cultures of periwinkle (*Catharanthus roseus*). Plant Physiol. 75: 726–731.

Crowe, J.H., L.M. Crowe, J.F. Carpenter and C. Aurell Wistrom, 1987. Stabilization of dry phospholipid bilayers and proteins by sugars. Biochem. J. 242: 1–10.

Dereuddre, J., C. Scottez, Y. Arnaud and M. Duron, 1990. Resistance of alginate-coated axillary shoot tips of pear tree (*Pyrus communis* L. Beurre Hardy) *in vitro* plantlets to dehydration and subsequent freezing in liquid nitrogen: effects of previous cold hardening. C. R. Acad. Sci. Paris t. 310 Ser III 317–323.

Dereuddre, J., S. Blandin and N. Hassen, 1991a. Resistance of alginate-coated somatic embryos of carrot (*Daucus carota* L.) to desiccation and freezing in liquid nitrigen. 1: Effect of preculture. Cryo-Lett. 12: 125–134.

Dereuddre, J., N. Hassen, S. Blandin and M. Kaminski, 1991b. Resistance of alginate-coated somatic embryos of carrot (*Daucus carota* L.) to desiccation and freezing in liquid nitrogen: 2. Thermal analysis. Cryo-Lett. 12: 135–148.

Dumet, D., F. Engelmann, N. Chabrillange and Y. Duval, 1993. Cryopreservation of oil palm (*Elacis guineensis* Jacq.) somatic embryos involving a desiccation step. Plant Cell Rep. 12: 352–355.

Dussert, S., M.C. Mauro, A. Deloire, S. Hamon and F. Engelman. 1991. Cryopreservation

of grape embryogenic cell suspensions: 1. Influence of pretreatment, freezing and thawing conditions. Cryo-Lett. 12: 287–298.
Fabre, J. and J. Dereuddre, 1990. Encapsulation–dehydration: A new approach to cryopreservation of *Solanum* shoot tips. Cryo-Lett. 11: 413–426.
Fahy, G.M., D.R. MacFarlane, C.A. Angell and H.T. Meryman, 1984. Vitrification as an approach to cryopreservation. Cryobiol. 21: 407–426.
Finkle, B.J. and J.M. Ulich, 1979. Effects of cryoprotectants in combination on the survival of frozen sugarcane cells. Plant Physiol. 63:598–604.
Gupta, P.K., D.J. Durzan and B.J. Finkle, 1987. Somatic polyembryogenic cell masses of *Picea abies* (Norway spruce) and *Pinus taeda* (loblolly pine) after thawing from liquid nitrogen. Can. J. For. Res. 17: 1130–1134.
Haskins, R.H. and K.K. Kartha, 1980. Freeze-preservation of pea meristems: cell survival. Can. J. Bot. 58: 833–840.
Kartha, K.K.. L.C. Fowke, N.L. Leung, K.L. Caswell and I. Hawkman, 1988. Induction of somatic embryos and plantlets from cryopreserved cell cultures of white spruce (*Picea glauca*). J. Plant Physiol. 132: 529–539.
Klimaszewska, K., C. Ward and W.M. Cheliak, 1992. Cryopreservation and plant regeneration from embryogenic cultures of larch (*Larix* × *eurolepis*) and black spruce (*Picea mariana*). J. Exp. Bot. 43(246): 73–79.
Kobayashi, S., 1987. Uniformity of plants regenerated from orange (*Citrus sinensis* Osb.) protoplasts. Theor. Appl. Genet. 74: 10–14.
Kobayashi, S., T. Ohgawara, E. Ohgawara, I. Oiyama and S. Ishi, 1988. A somatic hybrid plant obtained by protoplast fusion between navel orange (*Citrus sinensis*) and satsuma mandarin (*C. unshiu*). Plant Cell Tiss. Org. Cult. 14: 63–69.
Kobayashi, S., A. Sakai and I. Oiyama, 1990. Cryopreservation in liquid nitrogen of cultured navel orange (*Citrus sinensis* Osb.) nucellar cells and subsequent plant regeneration. Plant Cell Tiss. Org. Cult. 23: 15–20.
Kuranuki, Y. and A. Sakai, 1992. Cryopreservation of shoot tips of *in vitro* tea plant by vitrification method. Jap. J. Breed. 42 (Suppl. 1): 330–331 (in Japanese).
Langis, R., B. Schnabel, E.D. Earle and P.L. Steponkus, 1989. Cryopreservation of *Brassica campestris* L. cell suspensions by vitrification. Cryo-Lett. 10: 421–428.
Langis, R. and P.L. Steponkus, 1990. Cryopreservation of rye protoplasts by vitrification. Plant Physiol. 92: 666–671.
Langis, R. and P.L. Steponkus, 1991. Vitrification of isolated rye protoplasts: protection against dehydration injury by ethylene glycol. Cryo-Lett. 12: 107–112.
Marin, M.L. and N. Duran-Vila, 1988. Survival of somatic embryos and recovery of plants of sweet orange (*Citrus sinensis* (L) Osb.) after immersion in liquid nitrogen. Plant Cell Tiss. Org. Cult. 14: 51–57.
Niino, T., A. Sakai and K. Nojiri, 1992a. Cryopreservation of *in vitro*-grown shoot tips of apple and pear by vitrification. Plant Cell Tiss. Org. Cult. 28: 261–266.
Niino, T., A. Sakai, S. Enomoto and S. Kato, 1992b. Cryopreservation of *in vitro*-grown shoot tips of mulberry by vitrification. Cryo-Lett. 13: 303–312.
Niino, T. and A. Sakai, 1992. Cryopreservation of alginate-coated *in vitro*-grown shoot tips of apple, pear and mulberry. Plant Sci. 87: 199–206.
Nishizawa, S., A. Sakai, Y. Amano and T. Matuzawa, 1992. Cryopreservation of asparagus (*Asparagus officinalis* L.) embryogenic suspension cells and subsequent plant regeneration by a simple freezing method. Cryo-Lett. 13: 379–388.
Nishizawa, S., A. Sakai, Y. Amano and T. Matuzawa, 1993. Cryopreservation of asparagus (*Asparagus officinalis* L.) embryogenic suspension cells and subsequent plant regeneration by vitrification method. Plant Sci. 88: 67–73.
Pihakaski, K. and P.L. Steponkus, 1987. Freeze-induced phase transitions in the plasma membrane of isolated protoplasts. Physiol. Plant. 69: 666–674.
Rall, W.F., 1987. Factors affecting the survival of mouse embryos cryopreserved by vitrification. Cryobiol. 24: 387–402.

Reed, B.M., 1988. Cold acclimation as a method to improve survival of cryopreserved *Rubus* meristems. Cryo-Lett. 9: 166–171.

Reed, B.M., 1990. Survival of *in vitro*-grown apical meristems of *Pyrus* following cryopreservation. HortSci. 25: 111–113.

Reed, B.M., 1992. Cryopreservation of *Ribes* apical meristems. Cryobiol. 29(6): 740.

Reinhoud, P.J., A. Uragami, A. Sakai and F. Van Iren, 1993. Vitrification of plant cell suspensions. In: J.G. Day and M.R. McLellan (Eds.), Methods in Molecular Biology, Cryopreservation and Freeze-Drying Protocols. Humana Press Inc., Clifton, NJ.

Rudolph, A.S., J.H. Crowe and L.M. Crowe, 1986. Effects of three stabilizing agents-proline, betaine, and trehalose on membrane phospholipids. Arch. Biochem. Biophys. 245: 134–143.

Sakai, A., 1960. Survival of the twig of woody plants at −196°C. Nature 185: 393–394.

Sakai, A. and S. Yoshida, 1967. Survival of plant tissue at super-low temperatures. VI. Effects of cooling and rewarming rates on survival. Plant Physiol. 42: 1695–1701.

Sakai, A., S. Kobayashi and I. Oiyama, 1990. Cryopreservation of nucellar cells of navel orange (*Citrus sinensis* Osb. var. *brasiliensis* Tanaka) by vitrification. Plant Cell Rep. 9: 30–33.

Sakai, A., S. Kobayashi and I. Oiyama, 1991a. Cryopreservation of nucellar cells of navel orange (*Citrus sinensis* Osb. var. *brasiliensis* Osb.) by a simple freezing method. Plant Sci. 74: 243–248.

Sakai, A., S. Kobayashi and I. Oiyama, 1991b. Survival by vitrification of nucellar cells of navel orange (*Citrus sinensis* var. *brasiliensis* Tanaka) cooled to −196°C. J. Plant Physiol. 137: 465–470.

Sakai, A., 1993. Cryogenic strategies for survival of plant cultured cells and meristems cooled to −196°C. In: JAICA (Ed.), Cryopreservation of Plant Genetic Resources Projects, pp. 5–21. Genetic Resources Projects, REF. 6, JAICA, Tokyo.

Schnabel-Preikstas, B., E.D. Earle and P.L. Steponkus, 1992a. Cryopreservation of potato shoot-tips by vitrification. Cryobiol. 29(6): 747.

Schnabel-Preikstas, B., E.D. Earle and P.L. Steponkus, 1992b. Cryopreservation of chrysanthemum shoot tips by vitrification. Cryobiol. 29(6): 747.

Schnabel-Preikstas, B., E.D. Earle and P.L. Steponkus, 1992c. Cryopreservation of sweet potato shoot-tips by vitrification. Cryobiol. 29(6): 738–739.

Senaratna, T., B.D. Mckersie and S.R. Bowley, 1990. Artificial seeds of alfalfa (*Medicago sativa* L.) induction of desiccation tolerance in somatic embryos. In Vitro Cell Dev. Biol. 26: 85–90.

Shimonishi, K., M. Ishikawa, S. Suzuki and K. Osawa, 1991. Cryopreservation of melon somatic embryos by desiccation method. Jap. J. Breed. 41: 347–351.

Steponkus, P.L. and D.V. Lynch, 1989. Freeze/thaw-induced destabilization of the plasma membrane and the effects of cold acclimation. J. Bioenerg. Biomembr. 21: 21–41.

Steponkus, P.L., D.V. Lynch and M. Uemura, 1990. The influence of cold acclimation on the lipid composition and cryobehaviour of the plasma membrane of isolated rye protoplasts. Phil. Trans. R. Soc. London B 326: 571–583.

Sugawara, Y. and A. Sakai, 1974. Survival of suspension-cultured sycamore cells cooled to the temperature of liquid nitrogen. Plant Physiol. 54: 772–774.

Towill, L.E., 1984. Survival of ultra-low temperatures of shoot-tips from *Solanum tuberosum* groups Andigena, Phureja, Stenotomum, and other tuber-bearing *Solanum* species. Cryo-Lett. 5: 319–326.

Uemura, M. and A. Sakai, 1980. Survival of carnation (*Dianthus caryophyllus* L.) shoot apices frozen to the temperature of liquid nitrogen. Plant Cell Physiol. 21: 85–94.

Ulich, J.M., B.J. Finkle and B.H. Tisserat, 1982. Effects of cryogenic treatment on plantlet production from frozen and unfrozen date palm callus. Plant Physiol. 69: 624–627.

Uragami, A., A. Sakai, M. Nagai and T.A. Takahashi, 1989. Survival of cultured cells and somatic embryos of *Asparagus officinalis* cryopreserved by vitrification. Plant Cell Rep. 8: 418–421.

Uragami, A., A. Sakai and M. Nagai, 1990. Cryopreservation of dried axially buds from plantlets of *Asparagus officinalis* L. grown *in vitro*. Plant Cell Rep. 9: 328–331.

Vasil, I.K., 1983. Regeneration of plants from single cells of cereals and grasses. In: P.F. Lurquin and A. Kleinhofs (Eds.), Genetic Engineering in Eukaryotes, pp. 233–252. Plenum, New York.

Yamada, T., A. Sakai, T. Matsumura and S. Higuchi, 1991. Cryopreservation of apical meristems of white clover (*Trifolium repens* L.) by vitrification. Plant Sci. 78: 81–87.

Withers, L.A. and P.J. King, 1979. A novel cryopreservation for the freeze preservation of cultured cells of *Zea mays* L. Plant Physiol. 64: 675–678.

15. The Biochemistry of Conifer Embryo Development: Amino Acids, Polyamines and Storage Proteins

Russell P. Feirer

Contents

1. Introduction

Much of the success in the initiation and further development of conifer somatic embryos can be attributed to optimization of culture conditions, including growth regulator types and levels, osmolarity, and choice of explant. Less attention has been given to optimization of other constituents of culture media (macro- and microelements). Knowledge of the basic biochemical environment associated with zygotic embryo development and germination may aid attempts at "rational medium design", just as our understanding of the basic morphological changes during the process has facilitated much of the success concerning the induction of somatic embryogenesis. The importance of the environments within and immediately surrounding the developing zygotic embryo have been recognized by many who have attempted to characterize the inorganic, organic and hormonal composition within developing seeds (see Raghavan, 1976, 1986; Simola and Santanen, 1990). Rational media design based upon information gained in the study of zygotic and *in vivo* systems has, and may continue to, facilitate/enhance the successful production of high quality somatic embryos from conifer cultures. The medium devised by Litvay *et al.* (1985), based upon analysis of the inorganic composition of developing conifer seeds and the studies of Teasdale *et al.* (1986) are examples of this approach. This chapter describes a continuing attempt to characterize the biochemical environment of the developing and germinating zygotic conifer embryo to aid in the further optimization of conifer culture media.

S. Jain, P. Gupta & R. Newton (eds.), Somatic Embryogenesis in Woody Plants, Vol. 1, 317–336.

2. Amino Acids

The choices of nitrogen source in culture media have resulted from studies of *in vivo* and zygotic systems. Reflecting knowledge of the biochemistry of nitrogen transport in plants, early studies identified glutamine as a compound which could be utilized by carrot (*Daucus carota*) cultures, with Wetherell and Dougall (1976) reporting that it could serve as the sole nitrogen source in culture media supporting somatic embryogenesis. Kamada and Harada (1984) continued this work with *D. carota*, measuring intracellular levels of free and protein-bound amino acids. Glutamine and glutamic acid were among the amino acids reaching the highest concentrations during somatic embryo formation. Rapid turnover of amino acids and proteins accompanied somatic embryo development, and it was suggested that glutamine may play a central role in these activities due to its ability to serve as a nitrogen donor/transport molecule in amino acid metabolism.

Changes in amino acid levels during zygotic embryo development/germination, as well as experimental studies testing the ability of organic nitrogen sources to support continued growth of excised zygotic embryos have also demonstrated the key role of the amides, glutamine and asparagine. While studying the formation and germination of zygotic cotton (*Gossypium hirsutum*) embryos, Capdevila and Dure (1977) found glutamine and asparagine in relatively high amounts. Glutamine was also reported to be the major source of nitrogen during *in vivo* development of pea (*Pisum sativum*) embryos (Murray and Cordova-Edwards, 1984). *In vitro* development of maize (*Zea mays*) kernels has been achieved using medium supplemented with amino acids. Cully *et al.* (1984) added the twenty protein amino acids to the culture medium, with glutamic acid and aspartic acids being in high concentrations, and arginine being added at relatively low levels. The amino acid additions have also been shown to be vitally important to the deposition of seed storage proteins in the seed endosperm. Protein content/deposition was dependent upon the levels of amino acids added to the medium (Singletary and Below, 1989).

Amino acid supplements have led to significant enhancement of somatic embryo initiation and further development in a number of non-conifer species including alfalfa (*Medicago sativa*) (Stuart and Strickland, 1984a,b) and *Zea mays* (Armstrong and Green, 1985). In a large scale screening of over twenty organic sources of nitrogen, including all of the protein amino acids, Stuart and Strickland (1984a) identified glutamine, arginine, alanine and proline as acceptable supplements to the culture media. Arginine was found to increase embryo size, while glutamine was responsible for the best conversion of somatic embryos into plantlets.

Glutamine also plays a central role in the nitrogen metabolism of conifer species. This amino acid is most responsible for the transport of nitrogen in pines, which was demonstrated by the classic work of Barnes and others. Analysis of xylem sap of *Pinus taeda* demonstrated the abundance of glutamine, which likely serves as the primary form of translocated nitrogen in this species. Glutamine accounted for over 80% of the total free amino acid pool in the sap of this as well as six other species of pine (Barnes, 1963). Glutamine synthesis was increased by application of nitrogen fertilizers, suggesting glutamine's role as a transport compound in this conifer (Barnes, 1962).

As demonstrated in *Daucus carota* and several other herbaceous species, organic sources of nitrogen have also been shown to be important in the initiation and growth of Douglas-fir (*Pseudotsuga menziesii*) callus (Kirby, 1982), whereas the addition of glutamine to the medium (50 mM) apparently increased the growth rate of the callus. Not unexpectedly, the addition of the ureides allantoin and allantoic acid, compounds important in the transport of nitrogen in leguminous plants, did not significantly affect growth of the cultures. Glutamine and casein hydrolysate markedly enhanced callus production by cultured megagametophytes of *Picea abies* (Simola and Honkanen, 1983). The concentration of inorganic nitrogen sources in the culture media, especially NH_4, and the ratio of NO_3/NH_4 affect morphogenesis in cultured conifer tissues. High levels of NH_4, equivalent to the level found in MS medium (Murashige and Skoog, 1962), were shown to inhibit adventitious shoot development in cultured white pine (*Pinus strobus*) embryos (Flinn *et al.*, 1986).

Casein hydrolysate and glutamine have been added to media used for the initiation and maintenance of embryogenic calli, and for the subsequent growth of the somatic embryos of a number of conifer species including *Pseudotsuga menziesii* (Durzan and Gupta, 1987), *Pinus lambertiana* (Gupta and Durzan, 1986), *Pinus taeda* (Gupta and Durzan, 1987), *Larix decidua* (Nagmani and Bonga, 1985) and *Picea glauca* (Lu and Thorpe, 1987). Supplementation or replacement of inorganic nitrogen sources, especially ammonium, with organic sources (amino acids) has become common in many media formulations. Simola and Santanen (1990) reported that the omission of casein hydrolysate (100 mg/l), arginine (0.25 mM) and glutamine (0.5 mM) led to less overall growth of white embryogenic *Picea abies* callus. Different metabolic requirements of chlorophylous tissue were suggested by the ability of this tissue to grow equally as well on a medium containing or lacking these organic nitrogen sources. In addition to the growth stimulation of white embryogenic callus, the organic nitrogen sources were also reported to be essential for further development of proembryos into more developed structures (Simola and Santanen, 1990).

While initiation of embryogenic *Picea abies* callus routinely takes place on a medium containing a mixture of amino acids present in low concentrations (glutamine, alanine, cysteine, arginine, leucine, phenylalanine, tyrosine, glycine) (Hakman *et al.*, 1985), development of somatic embryos into plantlets was favored by a medium containing higher levels of glutamine and alanine (400 and 100 mg/L, respectively) (Hakman and Von Arnold, 1985). Furthermore, the amino acid supplements that improved the culture response in non-conifer species, especially the addition of proline as reported by Stuart

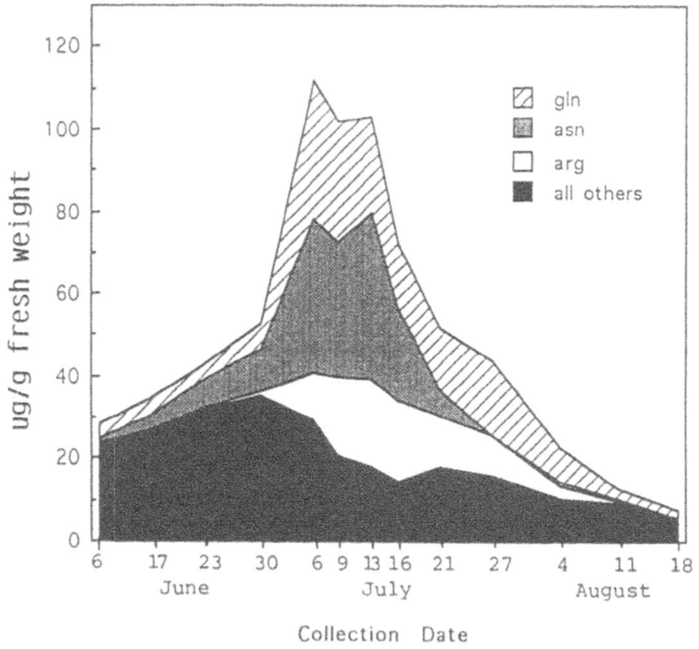

Figure 1. Free amino acid content (μg/g fresh weight) of developing *Pinus strobus* ovules. Values represent the means of triplicate analyses.

and Strickland (1984a,b), did not improve the rate of initiation of embryogenic callus from *Picea abies* explants (Von Arnold, 1987). Studies in other laboratories have also reported that the initiation was slightly better on a medium lacking amino acids (Jain *et al.*, 1988).

Free amino acid levels have been studied in developing zygotic embryos and seeds of several herbaceous species including *Gossypium hirsutm*, *Pisum sativum*, *Daucus carota* and *Lactuca sativa* (Capdevila and Dure, 1977; Murray and Cordova-Edwards, 1984; Kamada and Harada, 1984; Takeba, 1980). Similar studies of coniferous species are rarc in the literature, although at least one early report described changes in amino acid and sugar composition in developing pine embryos (*Pinus roxburghii*, Konar, 1958). This semi-quantitative study found the presence of amides during all stages of development, relatively low levels of aromatic amino acids and essentially no differences in the free amino acid content of developing gametophyte and embryos, although minor changes in amino acid pools were noted in the tissues over the course of development. Among the goals of our laboratory was to study and quantify changes in free amino acid levels during the course of zygotic embryo development in pines.

Table 1. Free amino acid content (µg/g f wt) of developing *P. strobus* ovules.

	June 6	June 17	June 23	June 30	July 6	July 9	July 13	July 16	July 21	July 27	Aug 4	Aug 11	Aug 18	Aug 26
Asp	1.2	0.9	0.9	1.9	3.7	3.7	3.4	3.3	3.2	2.1	1.1	0.7	0.4	0.7
Asn	0.9	4.1	6.5	9.9	37.3	32.7	40.5	22.5	6.3	0.3	0.9	t	t	2.5
Glu	4.6	4.4	3.6	5.7	7.4	7.3	7.4	6.6	8.7	8.8	4.9	3.9	2.1	1.1
Gln	3.9	3.9	3.9	6.6	40	29.6	23.4	15.5	15	17.9	8.6	2.6	1.8	1.2
Thr	0.5	0.7	1.1	1.7	0.9	0.6	0.6	0.3	0.2	0.2	t	t	t	t
Ser	1.2	2.2	3.2	7.2	4.1	1.4	1.6	0.8	1.2	0.9	0.6	0.7	0.3	0.3
Pro	8.9	8.3	9.5	3.2	3.6	0.2	0.3	0.2	0.3	0.6	0.7	1.6	1.5	0.5
Tyr	0.2	0.3	0.4	0.6	0.2	0.2	0.1	0.1	0.1	0.1	t	t	t	t
Phe	0.5	0.5	1.3	0.9	0.1	0.1	0.1	0.1	0.1	0.1	t	t	t	t
Trp	0.1	0.3	0.5	1.1	0.7	0.9	t	0.4	0.4	0.5	0.3	0.2	0.2	0.3
Gly	0.6	0.7	0.6	0.6	0.6	0.4	0.3	0.2	0.2	0.2	0.1	0.1	0.1	t
Ala	2.2	3.4	3.2	4.8	4.5	1.9	1.9	1.4	2.5	2.1	1.2	1.1	0.7	0.5
Val	0.3	0.8	1.8	1.8	0.8	0.7	0.4	0.2	0.3	0.2	0.1	0.1	t	t
Leu	0.5	0.9	1.4	1.7	0.4	0.3	0.2	0.2	0.2	0.1	0.1	0.1	0.1	0.1
Ile	0.4	0.8	1.2	1.6	0.2	0.2	0.1	0.1	0.2	0.1	0.1	0.1	0.1	0.1
Met	0.1	0.2	0.2	0.2	0.2	0.2	0.1	0.1	0.1	0.1	0.1	0.1	0.1	0.1
Cys	t	t	t	t	t	t	t	t	t	t	0.2	0.2	t	t
His	0.4	1.1	1.1	0.9	0.8	0.6	0.5	0.3	0.1	0.1	0.1	0.1	0.1	0.1
Lys	0.4	0.4	0.9	1.1	0.5	0.4	0.3	0.3	0.2	0.2	0.1	0.1	0.1	0.1
Arg	0.4	0.3	0.6	1.2	11.1	19.2	21.2	19.5	12.6	9.7	3.3	0.9	0.3	1.1
Total	28.7	34.7	42.5	53.1	112.3	101.6	103.3	72.4	52.1	44.6	22.8	12.8	7.7	10.3

3. Free Amino Acid and Polyamine Levels in Developing White Pine Seeds

Developing female cones were obtained from plantation-grown white pine (*Pinus strobus*) trees, approximately 20 years old, located near Waupaca, Wisconsin. Developing seeds were obtained from cones collected at various times throughout the summer. Seeds were excised from the cones within 4 h of collection, and stored on ice until use. Judged microscopically, fertilization of the *Pinus strobus* ovules was estimated to occur on or about June 30. This date corresponds to the time that amino acid and spermidine levels increased dramatically in the developing seed (Fig. 1, Table 1).

Ovules or seeds (embryo + gametophyte) were analyzed after removal of the seedcoats. Free amino acids were measured in 80% EtOH extracts of the tissues using a Beckman Model 119CL Amino Acid Analyzer employing lithium citrate buffers. Free polyamine levels were measured in 5% perchloric acid extracts of the tissues.

Dramatic changes in both absolute and relative proportions of several amino acids were observed in developing *Pinus strobus* ovules (Table 1, Fig. 1). While the total free amino acid pool was slowly expanding on a μg/g fresh weight basis before fertilization, the levels of several amino acids began to rise dramatically at the time of fertilization (judged to occur on or about June 30). Both of the amides glutamine and asparagine rose to high concentrations becoming the predominant free amino acids in the developing ovules, at times representing over 50% of the total amino acid pool. While the evidence is circumstantial, these data suggest that both glutamine and asparagine are used as transport compounds or intermediates in the nitrogen metabolism of developing pine embryos and female gametophytes. This apparent key role of glutamine and asparagine during zygotic embryo development is interesting in light of the studies of Verhagen and Wann (1989) concerning amino acid additions to culture medium of another conifer species Norway spruce (*Picea abies*). Successful initiation of embryogenic calli from mature embryos was carried out on a medium having very reduced levels of inorganic nitrogen, ammonium was absent from the medium, while having high levels of organic nitrogen in the form of glutamine and asparagine (1500 and 100 mg/L, respectively).

While tissues of many plants exhibit relatively high levels of single amides, conifers are somewhat unique in also having relatively high levels of arginine in the free amino acid pools (Durzan, 1968). Accompanying the dramatic elevations of the amides after fertilization, arginine was also found to constitute more of the amino acid pool of developing pine ovules (Fig. 1). From a "baseline" concentration of ~1 μg/g fresh weight, its level increased over twenty-fold during the course of embryo development (Table 1). While arginine occupies a key position in the urea cycle and is a nitrogen-rich amino acid, making it well suited as a transport compound, it may have a unique role in conifers as it is rarely found in such relative abundance in other plant species.

Most of the other amino acids exhibited minor fluctuations in concentrations over the course of seed development, although proline was notable as it was present in moderate levels in the tissues collected at the earliest stages of development and declined throughout most of the collection period (Table 1). Proline is known to accumulate in plant tissues exposed to both cold and drought stress (Durzan, 1968; Kramer and Kozlowski, 1979), suggesting that perhaps the relatively high levels of proline are related to the wide temperature flucuations observed during spring. The steadily declining levels of proline may also represent the utilization of much higher levels of proline which may have been present during the late winter months, although earlier tissue collections would be required to document this.

4. Polyamines

While early studies reported high levels of the amides (especially glutamine) present in xylem sap (Barnes, 1963), subsequent work demonstrated arginine to be the most abundant free amino acid in fertilized *Pinus taeda* seedling tissues, although the amides again comprised a large proportion of the amino acid pool (Pharis *et al.*, 1964). Similar findings were reported with *Abies grandis* (Carrow, 1973) and *Picea glauca* (Durzan, 1968). The data presented in this chapter also document high levels of free and protein-bound arginine in pine seeds. Arginine is especially interesting due to its high nitrogen composition, containing four nitrogens per molecule, and its being a substrate for the biosynthesis of polyamines such as putrescine, spermidine, and spermine (Smith, 1985). The involvement of polyamines in basic processes of plants such as growth, flowering, responses to growth regulators and stress, and environmental stimuli have been well documented (Evans and Malmberg, 1989; Slocum *et al.*, 1984; Smith, 1985). Polyamines have been shown to affect biochemical events including protein synthesis, DNA synthesis and conformation, membrane stability, etc. The most relevant consideration in the development of somatic embryos, however, is the role of polyamines during differentiation and embryogenesis. Polyamines appear to be essential during differentiation of tissues and embryo development of animals (Pegg and McCann, 1982) as well as plants. Polyamine levels have been shown to rise during development and germination of *Glycine max*, *Hordeum vulgare*, *Orzya sativa*, *Phaseolus vulgaris* and *Zea mays* seeds (Lin, 1984; Kyriakidis, 1983; Sen *et al.*, 1981; Bagni, 1970; Sepulveda and Jimenez, 1988). Studies have also documented active polyamine metabolism during somatic embryogenesis of carrot (Montague *et al.*, 1979), with subsequent work demonstrating the essential role of polyamines in *in vitro* development. Experimentally induced reductions in intracellular levels of polyamines were shown to block embryogenesis of carrot and other herbaceous species (Feirer *et al.*, 1984; Tiburcio *et al.*, 1988).

While more is known about the involvement of polyamines in the growth and development in herbaceous plants, the body of knowledge concerning conifer species is growing (see contribution by S. Minocha, this volume). As in many herbaceous species, polyamine levels have been shown to rise during woody plant growth as well. Elevations of polyamines have been observed during periods of intensive vegetative growth of *Picea abies*, with the highest levels of polyamines being present in buds, where meristematic activity is high (Konigshofer, 1989, 1991). Experimental manipulation of polyamine levels have also suggested their involvement and importance in woody plant systems. Reduction of intracellular polyamine levels with inhibitors of their biosynthesis resulted in significant reductions in the mitotic index of cells in the radicle of germinating *Acer saccharum* seeds (Walker *et al.*, 1985). Other published reports, however, have focused on the role of polyamines during *in vitro* growth and development of woody species.

Additions of polyamines to culture media affects the *in vitro* growth of woody plants. Ornithine, a precursor of polyamine biosynthesis, and putrescine stimulated cell divisions of cultured alder (*Alnus glutinosa*) and *Pinus oocarpa* and *P. patula* (Huhtinen *et al.*, 1982; Laine *et al.* 1988), although replacement of organic nitrogen sources with polyamines led to significantly reduced callus production in *Picea abies* (Simola and Honkanen, 1983). These results strongly indicate that polyamines have a function other than a major role in the nitrogen nutrition of plant cells, although their role in this capacity has not been suggested (Slocum *et al.*, 1984).

Studies of the role of polyamines in organized growth of woody plants *in vitro* have produced mixed results. Polyamines have been shown to rise during adventitious root formation in a number of woody species (*Prunus*, Biondi *et al.*, 1990), although exogenously added polyamines did not significantly affect this *in vitro* development. While successful inhibition of polyamine biosynthesis has been reported in cultured tissues of *Pinus radiata*, relationships between polyamines and growth/development in these tissues were not clear (Biondi *et al.*, 1988). Levels of both free and conjugated polyamines exhibited significant fluctuations during the growth of subcultured embryogenic *Picea abies* calli (Santanen and Simola, 1992). Spermidine was the polyamine presumed to be correlated to active embryogenesis as levels of this compound were significantly higher in embryogenic calli when compared to non-embryogenic calli. Clearly more work needs to be done in order to investigate the role of polyamines during the organized growth of woody plants, especially conifers. For this reason and based upon studies of amino acid metabolism in developing zygotic embryos (with special interest in high levels of free arginine), levels of free polyamines were determined during zygotic embryogenesis of *Pinus*. Since elevations of polyamines had been associated with the organized development in a number of plant species, including conifers, it was logical to extend those studies to include the measurement of polyamine levels during pine seed formation. Using the same tissues collected for the previously described amino acid analyses, free polyamines were measured in pine ovules during the course of zygotic embryo

Table 2. Amino acid composition of seed storage proteins (expressed as residues per 100 residues; or mole%).

	P. strobus	*P. taeda*	*P. taeda* (44*kD*)	*P. menziesii*
Asx*	8.2	8.4	9.6	9.1
Glx*	19.2	18.9	15.5	18.1
Thr	2.5	2.5	5.1	2.7
Ser	7.3	6.7	6.5	7.8
Pro	5.9	5.8	4.9	6.2
Tyr	3.6	3.5	2.8	2.2
Phe	2.4	2.4	4.2	2.7
Trp	nd	nd	nd	nd
Gly	7.2	7.2	11.4	7.5
Ala	5.8	5.4	5.8	6.2
Val	5.1	5.5	6.1	5.1
Leu	7.1	6.9	7.6	6.5
Ile	2.9	2.8	4.7	3.6
Met	2.1	1.9	1.9	1.2
Cys	nd	nd	nd	nd
His	1.4	1.6	3.1	2.1
Lys	1.1	1.2	2.1	2
Arg	18.1	19.1	8.5	14.5

* Asx and Glx represent the sum of aspartic acid + asparagine, and glutamic acid + glutamine, respectively, as the amides decompose during protein hydrolysis.

development. The polyamines were benzoylated and separated on a C18-reverse phase column using a Varian Model 5000 HPLC as previously described (Feirer *et al.*, 1984; Flores and Galston, 1982).

While putrescine and spermine exhibited unremarkable changes during the course of ovule development, a significant elevation of spermidine was observed between June 30 and July 6, just after fertilization (Fig. 2). Spermidine levels rose nearly fivefold from 59 to 258 nmoles/gram fresh weight and remained elevated throughout the collection and analysis period. It appears, then, that elevated levels of spermidine accompany zygotic embryo development in pines, while levels of the other polyamines were not significantly changed. Several hypotheses follow these observations, the first being the correlation of the increased activity of amino acid metabolism with that of polyamine synthesis. The data suggest that the elevated levels of free arginine in the developing tissues may serve as a substrate for the production of polyamines via direct decarboxylation of arginine by arginine decarboxylase or through the conversion of arginine to ornithine, which then leads to polyamine synthesis through the action of ornithine decarboxylase (Slocum *et al.*, 1984). The observation that spermidine exhibited larger elevations than putrescine or spermine suggests that spermidine is the polyamine most involved in pine ovule growth and development. This is similar to findings in several herbaceous plants in which spermidine was the polyamine most correlated to somatic embryogenesis (Feirer *et al.*, 1984).

Perhaps the most interesting observation remains that polyamines are elevated during ovule development in a coniferous species. Whether polyamine (spermidine) levels are partially responsible for the organized growth of the ovules or are simply a consequence of this development is not known. Their exact function/role remains unclear in pines just as in other plant species. Nonetheless, attention should be paid to polyamine levels in somatic embryo development if a goal is to recapitulate the biochemical events of zygotic embryo development.

5. Storage Proteins

Early attempts to optimize protocols for somatic embryogenesis of many species involved the study and manipulation of variables important during embryo initiation and early development. Once somatic embryogenesis has been achieved, however, the focus logically shifts to the further development and subsequent germination of the somatic embryos. While zygotic embryos may be dormant and metabolically inactive in dry seeds, the commencement of germination is associated with the mobilization of nutrients from seed reserves and a dramatic change in the biochemical environments surrounding the embryos. The germinating zygotic embryo, then, depends upon the seed reserves for nitrogen and carbon skeletons and the nature of these storage compounds becomes especially relevant for those attempting to reconstitute the environment within the germinating seed.

The appearance of seed storage proteins is one event in the development of zygotic embryos which has received a great deal of attention and study, owing probably to the importance of seed storage proteins in human and animal diets. These proteins are relatively abundant in the seed, which has made the isolation and study of these proteins relatively easy. A primary function of these proteins is to act as storage reserves, both nitrogen and carbon skeletons, utilized during the development and germination of the embryo. The literature concerning the biochemistry and molecular biology of the seed storage proteins from the major crop plants, such as *Zea mays*, *Glycine max*, *Hordeum vulgare*, *Pisum sativum*, etc., is vast (see reviews of: Daussant *et al.*, 1983; Higgins, 1984; Shannon and Chrispeels, 1986; Shotwell and Larkins, 1989). In contrast to the large body of work done on crop plants, little characterization of the seed storage proteins of conifers has been reported. One of the first studies of conifer systems found the storage proteins to be relatively insoluble and present as cytoplasmic crystaloids, comprising over 75% of the insoluble protein fraction (Gifford, 1988). Although the proteins isolated from 9 species of *Pinus* appeared to be similar, these crystaloid proteins were found to be immunologically distinct from the crystaloid proteins found in the angiosperm *Ricinus*.

Misra and Green have isolated and studied the seed storage proteins in *Picea* and *Pseudotsuga menziesii* (see chapter by Misra, this volume). As in pines, the storage proteins are present in protein bodies containing insoluble globoid and crystalloid proteins (Misra and Green, 1990; Green *et al.*, 1991). These proteins, present in both megagametophyte and embryonic tissues, had molecular weights of 42, 34.5–35 and 22.5–23 kD. In both *Picea abies* and *P. glauca* the proteins were found to accumulate very early during embryo development, detectable as early as three days after fertilization (Hakman *et al.*, 1990; Misra and Green, 1991). Misra and Green (1990) also found that most of the storage proteins had been depleted by the ninth day of germination, demonstrating their role as reserves utilized during the seed germination.

These proteins have been found in somatic embryos of several spruce species, including *Picea abies*, *P. glauca*, and *P. glauca engelmanni* complex (Cyr *et al.*, 1991; Flinn *et al.*, 1991; Hakman *et al.*, 1990; Joy *et al.*, 1991; Misra *et al.*, 1993) and in many systems have been found to undergo hydrolysis during germination, presumably to free amino acids. The accumulation of storage proteins and triglycerides is enhanced by, if not totally dependent upon, ABA and high osmolarity of the medium (Cyr *et al.*, 1991; Hakman *et al.*, 1990; Misra *et al.*, 1993). These same conditions are necessary for the successful conversion of somatic embryos into viable plantlets (Attree *et al.*, 1991; Roberts, 1991). It appears, then, that storage proteins play a key/major role in the development and late germination of somatic as well as zygotic conifer embryos.

During germination seed storage proteins are hydrolyzed to their component free amino acids which serve as nitrogen sources for the embryo/plantlet. Nitrogen sources used to support the late development and germination of somatic embryos might then be expected to reflect these forms of nitrogen delivered to the germinating zygotic embryos. Measurement of free amino acid levels during ovule development is important in our understanding of the biochemical changes accompanying zygotic embryo development, but perhaps more important is the role of seed storage proteins in later stages of growth. Seed storage proteins generally function to supply nitrogen to the germinating embryo, with much of the requirement for carbon skeletons being met by storage lipids also present in the seed. The amino acid composition of the seed storage proteins becomes an important consideration in our understanding of the nitrogen nutrition of the germinating plantlet, then, as they define the form of nitrogen supplied to the growing plantlet. Obviously, knowledge about the amino acids present in and released from the storage proteins may be useful in the formulation of "germination" or conversion media. To further characterize the proteins for a better understanding of their function as nutrient reserves in conifer seeds, storage proteins were isolated from mature loblolly pine (*Pinus taeda*), white pine (*Pinus strobus*) and Douglas-fir (*Pseudotsuga menziessi*) seeds. The isolated proteins were used for the determination of amino acid composition. In addition, individual storage proteins of *Pinus taeda* were resolved by SDS-PAGE and transferred to solid supports for both amino acid composition and sequence determination. Studies of the hydrolysis of storage proteins during the stratification of *Pseudotsuga menziessi* seeds were also conducted, revealing the expected loss of storage proteins and the concomitant elevation of the free amino acid pool available for use by the embryo axis.

6. Extraction and Amino Acid Composition of Seed Storage Proteins

After removal of seedcoat and underlying membranous tissues, the embryos were separated from the surrounding gametophytic tissues. The tissues, typically from one or two seeds, were homogenized in 800 μl of cold 50 mM HEPES (pH = 7.5) containing 1 mM PMSF (phenylmethylsulfonyl fluoride, Sigma Chemical Co.) in a Kontes microfuge tube with matching grinding rod. After centrifuging 5 min at full speed in Eppendorf microfuge (5°C), the supernatant was discarded. The remaining pellet was re-extracted with the HEPES buffer two more times to remove soluble proteins from the sample. The pellet was then homogenized in 200 μl 6M urea, 50 mM Tris (pH = 7.4), 2.5% mercaptoethanol, placed in a boiling water bath for 5 min and centrifuged 5 min in an Eppendorf microfuge at full speed. The supernatant, containing the dissolved storage proteins, was transferred to new microfuge tube. The storage proteins were precipitated with 80% acetone by adding 800 μl acetone to the 200 μl sample. After incubating on ice for 30 min, the tube was again centrifuged for 5 min at full speed in an Eppendorf centrifuge. The acetone was decanted and the pellet dried in a vacuum desicator. The pellet was then either used directly for analysis, dissolved in 0.1 M NaOH for Bradford protein analysis or dissolved in electrophoresis loading buffer.

SDS-PAGE analysis of proteins was carried out using 12% acrylamide or 7–20% acrylamide gradient gels, followed by staining with Coomassie Blue. Amino acid composition of the storage proteins was performed by hydrolysis of proteins in 6 M HCl at 105°C for 20 h and analysis using a Beckman Model 6300 Amino Acid Analyzer.

7. Free Amino Acid and Protein Content of Stratified Seeds

Mature seeds of Douglas-fir (*Pseudotsuga menziesii* (mirb.) Franco) were stratified by soaking in water for 48 h, drained, placed in polyethylene bags and stored at 5°C for 30 days. After removal of the seedcoat, stratified and unstratified seeds were homogenized and extracted 3 times with 80% ethanol. These extracts were summed, dried, and used for free amino acid analysis.

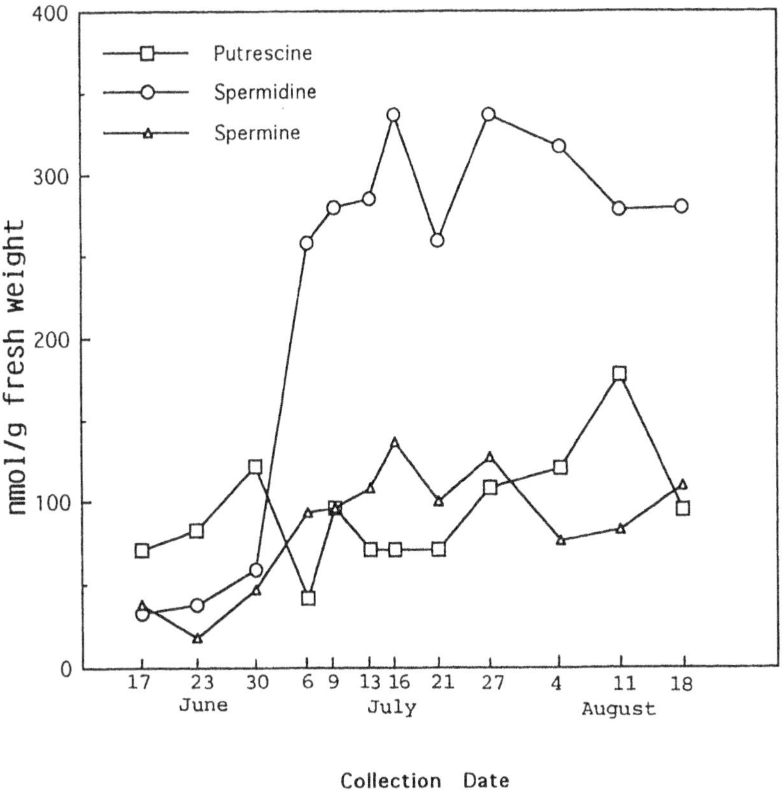

Figure 2. Free polyamines in developing *Pinus strobus* ovules. Values represent the means of triplicate analyses.

Total hydrolyzable protein in the tissues was estimated by hydrolysis of the proteins remaining in the insoluble pellet remaining after the ethanol extraction described above and subsequent quantification of the released amino acids.

Analysis of seed storage proteins isolated from three coniferous species revealed that they share some characteristics of the seed storage proteins isolated from crop species (Table 2). While the proportion of hydrophobic amino acids may be slightly higher in these conifer species, the levels of asx and glx (which include the amides asparagine and glutamine) and most other amino acids were similar to those found in the storage proteins of *Pisum*, *Zea* and *Triticum* (see Pernollet and Mosse, 1983). A striking difference between the proteins found in conifer seeds and the crop species, however, was the abundance of arginine in conifer proteins. While the mole% of arginine in the crop species ranges from less than 1% to just over 7% (Pernollet and Mosse, 1983), arginine comprised over 14% of the amino acid

Table 3. Protein composition in stratified and unstratified Douglas-fir seeds (mg hydrolyzable protein/gm dry wt).

	unstratified	stratified
gametophyte	624	389
embryo	264	287

residues in most of the conifer proteins analyzed (Table 2). This represents a very unique feature of conifer storage proteins, distinguishing them from the storage proteins of other plants.

Electrophoretic analysis of the storage proteins isolated from *P. strobus* and *P. taeda* revealed the existence of four principle storage proteins (data not shown). The number and apparent molecular weights of the proteins are similar to those found in other coniferous species (Gifford, 1988; Flinn *et al.*, 1991). The amino acid composition of one of the purified proteins, having a molecular weight of ~44 kD, was slightly different than that found in the total storage protein preparation (containing all four of the proteins). Specifically, the arginine content was less than one-half that present in the total protein pool (Table 2). At present this unexpected result can be explained by the hypothesis that one or more of the other three storage proteins must be especially rich in arginine.

Storage proteins are utilized during the germination of the embryo axis, with the mobilization of the storage proteins beginning in the female gametophyte tissue during seed stratification. While this hydrolysis of storage proteins in the gametophytic tissue was documented by a reduction in protein content, the level of protein in the embryo actually increased slightly during stratification (Table 3). The hydrolysis of storage proteins in the gametophytic tissue was accompanied by the expected rise in free amino acid levels. The free amino acid pool exhibited nearly a two-fold expansion in the gametophyte after stratification, with the transport of free amino acids to the embryo being suggested by an elevation of free amino acid levels in the stratified embryo as well (Table 4). Examination of the composition of the free amino acid pool in the stratified tissues shows that the amino acids released during the hydrolysis of the storage proteins closely reflects the amino acid content of the proteins, with arginine being dominant free amino acid. Aspartic acid and glutamic acid were also present in relative abundance. The high levels of aspartic and glutamic acids, compared to asparagine and glutamine, may reflect the loss of the amide nitrogen in the transamination reactions expected during the active metabolism of amino acids, although this is speculative.

Similar results have been reported with some, although not all plant species. During red light induced germination of lettuce (*Lactuca*) seeds, degradation of seeds storage proteins was followed by elevations in the levels of free amino acids, including arginine, glutamic and aspartic acids and the

Table 4. Free amino acid levels in *P. menziesii* seeds (μg/g dw).

	Gametophyte		Embryo	
	Unstratified	Stratified	Unstratified	Stratified
Asp	846	944	2557	3183
Asn	129	55	163	180
Asx				
Glu	399	645	1179	1235
Gln	44	415	101	222
Glx				
Thr	15	37	104	58
Ser	29	66	388	137
Pro	136	239	1133	1140
Tyr	179	285	215	270
Phe	4	22	51	24
Trp	242	359	trace	331
Gly	17	34	188	50
Ala	103	143	240	121
Val	23	34	108	
Leu	32	59	86	56
Ile	27	40	77	29
Met	17	23	52	61
Cys				
His	95	182	162	259
Lys	63	65	146	154
Arg	2241	4331	10974	14424
Total	4691	8172	18281	22233

amides (Takeba, 1980). In studies of the formation and germination of zygotic cotton (*Gossypium*) embryos, however, Capdevila and Dure (1977) noted that the free amino acid pool did not reflect the amino acid content of the endosperm and "conspicuously high levels of arginine were maintained through embryogenesis and germination". At times arginine represented over 37% of the free amino acid pool in the developing embryo, while comprising only 1.5% of the that in the endosperm.

The amino acid composition of the seed storage proteins and the amino acids released upon their hydrolysis during the mobilization of storage compounds during stratification and subsequent germination should be of special interest to those concerned with the optimization of conifer somatic embryogenesis protocols. Roberts *et al.* (1990) have utilized the appearance of storage proteins as an indicator of development of somatic embryos of interior spruce (*Picea glauca* × *engelmanni*), the appearance of the proteins being strongly correlated to the degree of maturation and likely success of normal development of the embryos into plantlets. Compared to the germination of zygotic embryos within the seed and embryos freed of surrounding megagametophyte tissues, though, somatic embryos were found to begin hydrolysis of their storage proteins sooner than those within the intact seed (Cyr *et al.*, 1991). These authors made an important comment, noting that storage proteins and triglycerides accumulate in somatic embryos, although most of the storage reserves in zygotic conifer seeds reside in the megagametophyte. With respect to *in vitro* systems, then, the composition of the germination medium takes on special importance as it must be a substitute for the megagametophyte to supply adequate amounts of nitrogen and carbon skeletons. It must be remembered that even a "high quality somatic embryo", having abundant protein and triglyceride reserves, depends in large part upon the nutrients supplied by the medium, just as the zygotic embryo is dependent upon the megagametophyte.

8. Conclusions

Based upon the arguments presented, it may be reasonable to assume that a culture medium formulated to support the development of somatic embryos or germination (conversion) of the embryos into plantlets should reflect the nutrients supplied to the developing and germinating zygotic embryos. This approach led to the formulation of a medium based upon the composition of inorganic nutrients found in developing conifer seeds (Litvay *et al.*, 1985; Teasdale *et al.*, 1986). This work should be extended, however, to include an evaluation of the organic compounds present in the culture media. From the analysis of the free amino acid pool during ovule development and storage proteins hydrolyzed during stratification/germination, it is clear that arginine and to a lessor extent the amides play a significant role in the nutrition of zygotic conifer embryos. Replacement of inorganic sources of nitrogen with these amino acids in culture medium designed to support development of somatic embryos may be justified if a goal is to duplicate the biochemical environment of the seed. While the addition of casein hydrolysate and glutamine to culture media have become routine, relatively few publications report results of additions of other amino acids supplements. In one of these few reports a combination of arginine and glutamine did little to promote normal development of *Picea glauca* somatic embryos into plantlets (Hakman and Von Arnold, 1988). The effects of these amino acids on the biochemical composition of the embryos, including storage proteins and lipids, was not reported in that study. Nonetheless, these supplements did not yield a normal or even better quality plantlet, which is the ultimate goal of any program. It is possible that additions of higher levels of these amino acids will find more success, however.

Acknowledgements

Portions of this work were carried out as dues-funded research at the Institute of Paper Chemistry, formerly of Appleton, WI. This work was also supported by a grant from the Hewlett Foundation Grant of Research Corporation to St. Norbert College. The technical assistance of D. Henning and C.D. Vaughan is also gratefully acknowledged.

References

Armstrong, C.L. and C.E. Green, 1985. Establishment and maintenance of friable, embryogenic maize callus and the involvement of L-proline. Planta 164: 207–214.

Attree, S.M., D. Moore, V.K. Sawhney, and L.C. Fowke, 1991. Enhanced maturation and desiccation tolerance of white spruce (*Picea glauca*) somatic embryos: Effects of a non-plasmolysing water stress and abscisic acid. Ann. Bot. 68: 519–525.

Bagni, N., 1970. Metabolic changes of polyamines during the germination of *Phaseolus vulgaris*. New Phytol. 69: 159–164.

Barnes, R.L., 1962. Glutamine synthesis and translocation in pine. Plant Physiol. 37: 323–326.

Barnes, R.L., 1963. Organic nitrogen compounds in tree xylem sap. For. Sci. 9: 98–102

Biondi, S., P. Torrigiani, A. Sansovini and N. Bagni, 1988. Inhibition of polyamine biosynthesis by dicyclohexylamine in cultured cotyledons of *Pinus radiata*. Physiol. Plant. 72: 471–476.

Biondi, S., T. Diaz, I. Iglesias, G. Gamberini and N. Bagni, 1990. Polyamines and ethylene in relation to adventitious root formation in *Prunus avium* shoot cultures. Physiol. Plant. 78: 474–483.

Capdevila, A.M. and L. Dure, 1977. Developmental biochemistry of cottonseed embryogenesis and germination. VIII. Free amino acid pool composition during cotyledon development. Plant Physiol. 59: 268–273.

Carrow. J.R., 1973. Free amino acids in Grand-fir needles and effects of different forms of foliar applied nitrogen. Can. J. For. Res. 3: 465–471.

Cully, D.E., B.G. Gengenbach, J.A. Smith, I. Rubenstein, J.A. Connelly and W.D. Park, 1984. Endosperm protein synthesis and L-[^{35}S]methionine incorporation in maize kernels cultured *in vitro*. Plant Physiol. 74: 389–394.

Cyr, D.R., F.B. Webster and D.R. Roberts, 1991. Biochemical events during germination and early growth of somatic embryos and seed of interior spruce (*Picea glauca engalmanni* complex). Seed Sci. Res. 1: 91–97.

Daussant, J., J. Mosse and J. Vaughan, 1983. Seed Proteins. Academic Press, New York.

Durzan, D.J., 1968. Nitrogen metabolism of *Picea glauca*. I. Seasonal changes of free amino acids in buds, shoot apices and leaves, and the metabolism of uniformly labelled ^{14}C-arginine by buds during the onset of dormancy. Can. J. Bot. 46: 909–919.

Durzan, D.J. and P.K. Gupta, 1987. Somatic embryogenesis and polyembryogenesis in Douglas-fir cell suspension cultures. Plant Sci. 52: 229–235.

Evans, P.T. and R.L. Malmberg, 1989. Do polyamines have roles in plant development? Ann. Rev. Plant Physiol. Plant Mol. Biol. 40: 235–269.

Feirer, R., G. Mignon and J. Litvay, 1984. Arginine decarboxylase and polyamines required for embryogenesis in the wild carrot. Science 223: 1433–1435.

Flinn, B.S., D.R. Roberts and E.P. Taylor, 1991. Evaluation of somatic embryos of interior spruce. Characterization and developmental regulation of storage proteins. Physiol. Plant 82: 624–632.

Flinn, B.S., D.T. Webb and W. Georgis, 1986. *In vitro* control of caulogenesis by growth regulators and media components in embryonic explants of eastern white pine (*Pinus strobus*). Can. J. Bot. 64: 1948–1956.

Flores, H. and A.W. Galston, 1982. Analysis of polyamines in higher plants by high performance liquid chromatography. Plant Physiol. 69: 701–706.

Gifford, D.J., 1988. An electrophoretic analysis of the seed proteins from *Pinus monticola* and eight other species of pine. Can. J. Bot. 66: 1808–1812.

Green, M.J., J.K. McLeod and S. Misra, 1991. Characterization of Douglas fir protein body composition by SDS-PAGE and electron microscopy. Plant Physiol. Biochem. 29: 49–55.

Gupta, P.K. and D.J. Durzan, 1986. Somatic polyembryogenesis from callus of mature sugar pine embryos. Biotechnology 4: 643–645.

Gupta, P.K. and D.J. Durzan, 1987. Biotechnology of somatic polyembryogenesis and plantlet regeneration in loblolly pine. Biotechnology 5: 147–151.

Hakman, I. and S. von Arnold, 1985. Plantlet regeneration through somatic embryogenesis in *Picea abies* (Norway spruce). J. Plant Physiol. 121: 149–158.

Hakman I., L.C. Fowke, S. von Arnold and T. Eriksson, 1985. The development of somatic embryos in tissue cultures initiated from immature embryos of *Picea abies* (Norway spruce). Plant Sci. 38: 53–59.

Hakman, I., P. Stabel, P. Engstrom and T. Eriksson, 1990. Storage protein accumulation during zygotic and somatic embryo development in *Picea abies* (Norway spruce). Physiol. Plant. 80: 441–445.

Hakman, I. and S. von Arnold, 1988. Somatic embryogenesis and plant regeneration from suspension cultures of *Picea glauca* (white spruce). Physiol. Plant. 72: 579–587.

Higgins, T.J.V., 1984. Synthesis and regulation of major proteins in seeds. Ann. Rev. Plant Physiol. 35: 191–221.

Huhtinen, O., J. Honkanen and L.K. Simola, 1982. Ornithine and putrescine-supported divisions and cell colony formation in leaf protoplasts of alders (*Alnus glutinosa* and *A. incana*). Plant Sci. Lett. 28: 3–9.

Jain, S.M., R.J. Newton and E.J. Soltes, 1988. Enhancement of somatic embryogenesis in Norway spruce (*Picea abies* L.). Theor. Appl. Genet. 76: 501–506.

Joy, R.W., E.C. Yeung, L. Kong and T.A. Thorpe, 1991. Development of white spruce somatic embryos: I. Storage product deposition. In Vitro Cell. Dev. Biol. 27P: 32–41.

Kamada, H. and H. Harada, 1984. Changes in endogenous amino acid compositions during somatic embryogenesis in *Daucus carota*. Plant Cell Physiol. 25: 27–38.

Kirby, E.G., 1982. The effects of organic nitrogen sources on growth of cell cultures of Douglas-fir. Physiol. Plant. 56: 114–117.

Konar, R.N., 1958. A quantitative survey of some nitrogenous substances and fats in the developing embryos and gametophytes of *Pinus roxburghii* sar. Phytomorphology 8: 174–176.

Konigshofer, H., 1989. Distribution and seasonal variation of polyamines in shoot axes of spruce (*Picea abies*). J. Plant Physiol. 137: 607–612.

Konigshofer, H., 1991. Seasonal changes in polyamine content in different parts of juvenile spruce trees (*Picea abies*). J. Plant Physiol. 134: 736–740.

Kramer, P.J. and T.T. Kozlowski, 1979. Physiology of Woody Plants. Academic Press, New York.

Kyriakidis, D.A, 1983. Effect of plant growth hormones and polyamines on ornithine decarboxylase activity during the germination of barley seeds. Physiol. Plant 57: 499–504.

Laine, E., H. David and A. David. 1988. Callus formation from cotyledon protoplasts of *Pinus oocarpa* and *Pinus patula*. Physiol. Plant. 72: 374–378.

Lin, P.P., 1984. Polyamine anabolism in germinating *Glycine max* (L.) seeds. Plant Physiol 76: 372–380.

Litvay, J.D, D.C. Verma and M.A. Johnson, 1985. Influence of a loblolly pine culture medium and its components on growth and somatic embryogenesis of the wild carrot. Plant Cell Rep. 4: 325–328.

Lu, C.Y. and T.A. Thorpe, 1987. Somatic embryogenesis and plantlet regeneration in cultured immature embryos of *Picea glauca*. J. Plant Physiol. 128: 297–302.

Misra, S. and M.J. Green, 1990. Developmental gene expression in conifer embryogenesis and germination. I. Seed proteins and protein body composition of mature embryo and the megagametophyte of white spruce (*Picea glauca*). Plant Sci. 78: 61–71.

Misra, S. and M.J. Green, 1991. Developmental gene expression in conifer embryogenesis and germination. II. Crystalloid protein synthesis in the developing embryo and megagametophyte of white spruce (*Picea glauca*). Plant Sci. 78: 61–71.

Misra, S., S.M. Attree, I. Leal and L.C. Fowke, 1993. Effect of abscisic acid, osmoticum, and desiccatio on synthesis of storge proteins during the development of white spruce somatic embryos. Ann. Bot. 71: 11–22.

Montague, M.J, T.A. Armstrong and E.G. Jaworski, 1979. Polyamine metabolism in embryogenic cells of *Daucus carota*. Changes in intracellular content and rates of synthesis. Plant Physiol. 63: 341–345.

Murashige, T. and F. Skoog, 1962. A revised medium for rapid growth and bioassays with tobacco tissue culture. Physiol. Plant. 15: 473–497.

Murray, D.R. and M. Cordova-Edwards, 1984. Amino acid and amide metabolism in the hulls and sseds of developing fruits of garden pea, *Pisum staivum*. New Phytol. 97: 243–252.

Nagmani, R. and J.M. Bonga, 1985. Embryogenesis in subcultured callus of *Larix decidua*. Can. J. For. Res. 15: 1088–1091.

Pegg, A.E. and P.P. McCann, 1982. Polyamine metabolism and function. Am. J. Physiol. 243: 212–221.

Pernollet J.C. and J. Mosse, 1983. Structure and location of legume and cereal seed storage proteins. In: J. Daussant, J. Mosse and J. Vaughan (Eds.), Seed Proteins, pp 155–191. Academic Press, New York.

Pharis, R.P., R.L. Barnes and A.W. Naylor, 1964. Effects of nitrogen level, calcium level and nitrogen source upon the growth and composition of *Pinus taeda* L. Physiol. Plant. 17: 560–572.

Raghavan, D., 1976. Experimental Embryogenesis in Vascular Plants. Academic Press, New York.

Raghavan, D., 1986. Embryogenesis in Angiosperms: A Developmental and Experimental Study. Cambridge University Press, Cambridge.

Roberts, D.R., 1991. Abscisic acid and mannitol promote early development, maturation and storage protein accumulation in somatic embryos of interior spruce. Physiol. Plant. 83: 247–254.

Roberts, D.R., S.S. Flinn, D.T. Webb, F.B. Webster and B.C.S. Sutton, 1990. Abscisic acid and indole-3–butyric acid regulation of maturation and accumulation of storage proteins in somatic embryos of interior spruce. Physiol. Plant. 78: 355–360.

Santanen, A. and L.K. Simola, 1992. Changes in polyamine metabolism during somatic embryogenesis in *Picea abies*. J. Plant Physiol. 140: 475–480.

Sen, K., M. Choudhuri and B. Ghosh, 1981. Changes in polyamine contents during developent and germination of rice seeds. Phytochemistry 20: 631–633.

Sepulveda, G. and E.S. Jimenez, 1988. Polyamine distribution among maize embryonic tissues and its relation to seed germination. Biochem. Biophys. Res. Comm. 153: 881–887.

Shannon, L.M. and M.J. Chrispeels, 1986. Molecular Biology of Seed Storage Proteins and Lectins. American Society of Plant Physiologists, Rockville, MD.

Shotwell, M.A. and B.A. Larkins, 1989. The biochemistry and molecular biology of seed storage proteins. In: P.K. Stumpf and E.E. Conn (Eds.), The Biochemistry of Plants, pp. 297–345. Academic Press, New York.

Simola, L.K. and J. Honkanen, 1983. Organogenesis and fine structure in megagametophyte callus lines of *Picea abies*. Physiol. Plant. 59: 551–561.

Simola, L.K. and A. Santanen, 1990. Improvement of nutreint medium for growth and embryogenesis of megagametophyte and embryo callus lines of *Picea abies*. Physiol. Plant 80: 27–35.

Singletary, G.W and F.E. Below, 1989. Growth and composition of maize kernels cultured in vitro with varying supplies of carbon and nitrogen. Plant Physiol. 89: 341–346.

Slocum, R., R. Kaur-Sawhney and A.W. Galston, 1984. The physiology and biochemistry of polyamines in plants. Arch. Biochem. Biophys. 235: 283–303.

Smith, T.A., 1985. Polyamines. Ann. Rev. Plant Physiol. 36: 117–143.

Stuart, D.A. and S.G. Strickland, 1984a. Somatic embryogenesis from cell cultures of *Medicago sativa*. I. The role of amino acid additions to the regeneration medium. Plant Sci. 34: 165–174.

Stuart, D.A. and S.G. Strickland, 1984b. Somatic embryogenesis from cell cultures of *Medicago sativa*. II. The interaction of amino acids with ammonium. Plant Sci. 34: 175–181.

Takeba, G., 1980. Phytochrome-mediated accumulation of free amino acids in embryonic axes of New York lettuce seeds. Plant Cell Physiol. 21: 1651–1656.

Teasdale, R.D., P.A. Dawson and W.H. Woolhouse, 1986. Mineral nutrient requirements of a loblolly pine (*Pinus taeda*) cell suspension Culture. Evaluation of a medium formulated from seed composition data. Plant Physiol. 82: 942–945.

Tiburcio, A.F., R. Kaur-Sawhney and A.W. Galston, 1988. Polyamine biosynthesis during vegetative and floral bud differentiation in thin layer tobacco tissue cultures. Plant Cell Physiol. 29: 1241–1249.

Verhagen, S.A. and S.R. Wann, 1989. Norway spruce somatic embryogenesis: high-frequency initiation from light-cultured embryos. Plant Cell. Tiss. Org. Cult. 16: 103–111.

Von Arnold, S., 1987. Improved efficiency of somatic embryogenesis in mature embryos of *Picea abies*(L.) Karst. J. Plant Physiol. 128: 233–244.

Walker, M.A., D.R. Roberts, C.Y. Shih and E.B. Dumbroff, 1985. A requirement for polyamines during the cell division phase of radicle emergence in seeds of *Acer saccharum*. Plant Cell Physiol. 26: 967–971.

Wetherell, D.F. and D.K. Dougall, 1976. Sources of nitrogen supporting growth and embryogenesis in cultured wild carrot tissue. Physiol. Plant. 37: 97–103.

16. Somatic Embryogenesis and Polyamines in Woody Plants

Rakesh Minocha, Subhash C. Minocha and Liisa Kaarina Simola

Contents

1. Introduction

The formation of whole plants from cultured cells is interesting not only because of its applications for mass propagation but also as a prime example of the process of controlled development and differentiation in plants. Cultures capable of producing somatic embryos with high frequency provide ideal experimental systems to study and understand the biochemical basis of hormonal regulation of the developmental process. This knowledge should lead to the planning of media and various treatments that allow us to obtain whole plants from single cells in recalcitrant plant species.

In contrast to zygotic embryos which mature under the restrictive environment of endospermic tissue and show a high degree of commitment at very early stages of development (Bhojwani and Razdan, 1983; Goldberg *et al.*, 1989), somatic embryos are highly plastic and much less committed to maturity (Carman, 1990). Their development is easily disturbed by changes in medium constituents, physical environment, density of cultures, and other exogenously applied chemical factors.

Whereas a large amount of literature has accumulated on the morphogenetic and biochemical events underlying embryogenic development in cell cultures of herbaceous dicotyledons as exemplified by carrot (Street, 1977; Sharp *et al.*, 1980; Ammirato, 1984; Raghavan, 1986; Minocha, 1988; Carman, 1990; De Jong *et al.*, 1993), only limited information is available for the development of zygotic or somatic embryos in woody plants. A few laboratories have attempted to characterize the metabolic status of conifer cell and tissue cultures undergoing somatic embryogenesis or adventitious

S. Jain, P. Gupta & R. Newton (eds.), Somatic Embryogenesis in Woody Plants, Vol. 1, 337–359.

shoot regeneration (Simola and Honkanen, 1983; Grey *et al.*, 1987; Durzan, 1987; Wann *et al.*, 1987; Hakman and von Arnold, 1988; Kumar and Thorpe, 1989; Simola and Santanen, 1990; Tautorus *et al.*, 1991). These studies have focused mostly on the metabolism of carbohydrates and nitrogen. Somatic embryos of *Picea abies* cultured under appropriate conditions demonstrate ultrastructural and chemical similarities (Hakman, 1993) to zygotic conifer embryos (Simola, 1974a,b, 1975). Changes in proteins and mRNA during somatic and zygotic embryogenesis in *Picea glauca* have recently been reported by Misra and Green (1990, 1991).

2. Polyamines

Polyamines are naturally occurring aliphatic amines that carry a net positive charge at physiological pH. The most prevalent polyamines in higher plants are spermidine, spermine, and their precursor putrescine. Other less frequently observed polyamines include cadaverine, norspermidine, norspermine, homospermidine and homospermine. Their ubiquitous occurrence in all living organisms and in all cell types has led to the speculation that polyamines are essential for cellular growth and differentiation (Smith, 1985; Evans and Malmberg, 1989; Slocum and Flores, 1991). Polyamines have been found to exist in free, conjugated, and bound forms.

Polyamine biosynthesis and cellular concentrations can be modulated by various plant growth regulators (Evans and Malmberg, 1989; Rastogi and Davies, 1991, and references therein). Likewise, the modulation of cellular polyamines by exogenous polyamines or polyamine biosynthetic inhibitors can alter the endogenous levels of plant growth regulators (Roberts *et al.*, 1984). A possible function of these compounds as growth regulators or as secondary messengers in plant cells has been discussed (Slocum *et al.*, 1984; Smith, 1985; Galston and Kaur-Sawhney, 1987; Phillips *et al.*, 1987; Evans and Malmberg, 1989).

The variety of physiological responses in which a role for polyamines has been suggested is indeed large. Obviously, a single mode of action cannot be envisioned. Undoubtedly, the polybasic nature of these compounds is essential to the many metabolic activities they help to regulate, but their mechanism of action remains in question. It is believed that polyamines accomplish many of their functions by binding with negatively charged sites on various macromolecules (see Bachrach and Heimer, 1989a,b; Slocum and Flores, 1991, and references therein).

In plants, polyamines have been implicated in the regulation of light-induced growth responses, embryogenesis, organogenesis, pollen formation, flower development, fertilization, fruit development and senescence (Slocum *et al.*, 1984; Minocha, 1988; Evans and Malmberg, 1989; Slocum and Flores, 1991). In addition, a role for polyamines has also been proposed in the response of plants to stress from both biotic and abiotic factors (Young and

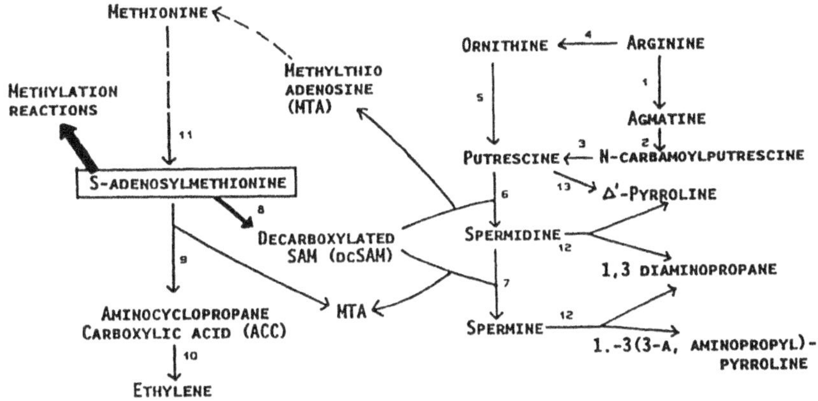

Figure 1. Combined pathway for biosynthesis of polyamines and ethylene in plants. The enzymes involved are: (1) arginine decarboxylase (ADC); (2) agmatine iminohydrolase; (3) N-carbamoylputrescine amidohydrolase; (4) arginase; (5) ornithine decarboxylase (ODC); (6) spermidine synthase; (7) spermine synthase; (8) SAM decarboxylase (SAMDC); (9) ACC synthase; (10) ethylene forming enzyme; (11) S-adenosylmethionine synthase; (12) polyamine oxidase (PAO); (13) diamine oxidase (DAO).

Galston, 1983; Flores and Galston, 1984; Villanueva *et al.*, 1987; Krishnamurthy and Bhagwat, 1989; Wang and Zug, 1989; Slocum and Flores, 1991; Tenter and Wild, 1991; Minocha *et al.*, 1992; Prediari *et al.*, 1993).

Ornithine decarboxylase (ODC) (EC 4.1.1.17) is a ubiquitous enzyme that catalyzes the conversion of L-ornithine to the diamine putrescine (Fig. 1). Ornithine decarboxylation is the only *de novo* pathway to putrescine synthesis in animals. In higher plants as well as prokaryotes and some fungi, there is a second pathway for putrescine synthesis (Slocum *et al.*, 1984; Smith, 1985; Pegg, 1986; Khan and Minocha, 1989a,b). In this pathway, the decarboxylation of L-arginine by arginine decarboxylase (ADC; EC 4.1.1.19) leads to the formation of putrescine, via agmatine and N-carbamoylputrescine intermediates. The distribution of these enzymes in different tissues and in different plants is regulated in a developmental and tissue-specific manner. For example, proliferating suspension cultures of carrot possess only ADC while developing green mature somatic embryos contain both ODC and ADC, the former being predominant (Robie and Minocha, 1989).

Spermidine and spermine are synthesized by the sequential addition of an aminopropyl group to putrescine (Fig. 1). The aminopropyl moeity is donated to putrescine by decarboxylated S-adenosylmethionine (dcSAM) (Torget *et al.*, 1979; Cohen *et al.*, 1981; Pegg, 1986). In addition to its role as the major biological methylating agent and in the biosynthesis of polyamines, SAM is also the precursor for ethylene biosynthesis in plants. Decarboxylation of SAM by SAM decarboxylase is nonreversible, committing SAM to the polyamine biosynthetic pathway. Spermidine and spermine synthases are assumed to be separate enzymes, though the mechanisms of their action may be identical (Smith, 1985; Pegg, 1986).

Many other polyamines have been reported in plants whose synthesis and potential functions were reviewed by Phillips and Kuehn (1991). Polyamines can also serve as precursors for secondary metabolites including anabasine, nicotine, atropine and unusual cinnamic acid amide conjugates. Some of the conjugates appear to be involved in resistance to pathogen infections, while others apparently are involved in plant reproduction (Smith *et al.*, 1983; Flores and Martin-Tanguy, 1991). Total concentrations of cellular polyamines are obviously influenced by the concentration of these secondary metabolites. Techniques are now available for isolation and quantitation of both perchloric acid (PCA) soluble and insoluble (i.e., free vs. conjugated or covalently bound) polyamines to help in the accurate determination of endogenous polyamine levels (Minocha *et al.*, 1990; Smith, 1991a). This is particularly important in view of the possibility that polyamine titers are regulated in part by interconversion between free polyamines and their conjugates.

A number of compounds have been synthesized that act as potent inhibitors of the polyamine biosynthetic enzymes (Pegg, 1986; McCann *et al.*, 1987). These compounds have proven useful in elucidating the importance of polyamines in cellular activities, and in understanding the regulation of polyamine biosynthesis. The suicide inhibitors, DL-α-difluoromethylornithine (DFMO) and DL-α-difluoromethylarginine (DFMA) specifically inhibit ODC and ADC, respectively. In addition, D-ornithine and monofluoromethylornithine have also been used to inhibit ODC activity.

Inhibition of cell growth with these compounds seems to be cytostatic rather than cytotoxic as the exogenous application of polyamines generally reverses the effect of the ODC and ADC inhibitors. Despite a specific and almost universal inhibition of ODC activity by DFMO in animals, this compound does not function as an effective inhibitor for all prokaryotic and plant ODCs. For example, DFMO does not inhibit ODC activity of certain cultivars of *Solanum tuberosum*, *Avena sativa*, *Pisum sativum*, *Amaranthus* sp., *Daucus carota*, *Catharanthus roseus*, and a few cultivars of *Nicotiana tabacum* (Galston, 1983; Flores and Galston, 1984; Slocum and Galston, 1985; Robie and Minocha, 1989; Minocha *et al.*, 1991a) when tested *in vitro*. In general, DFMA has been highly effective as a suicide inhibitor of ADC in plants.

Another widely used inhibitor of polyamine biosynthesis is MGBG (methylglyoxal bis(guanylhydrazone)), a competitive and reversible inhibitor of SAMDC that generally inhibits spermidine and spermine biosynthesis. In some instances, MGBG is not specific for SAMDC inhibition (Pegg, 1986) and may also cause stabilization of this enzyme (Hiatt *et al.*, 1986; Malmberg and Hiatt, 1989).

Polyamines are oxidatively deaminated by the enzymes diamine oxidases (DAOs) and polyamine oxidases (PAOs) (Federico and Angelini, 1991). Plant diamine oxidases are most active with putrescine and cadaverine, though enzymes with broad specificities also occur in some species. Polyamine oxidases are known to occur mostly in monocots, especially cereals (Smith, 1991b). Only recently have these enzymes been detected in dicots (Bagga *et al.*, 1991) and conifers (Santanen and Simola, 1994). Unlike the diamine oxidases, the PAOs are specific for aliphatic polyamines. It is unclear why di- and polyamine oxidases occur so erratically throughout the plant kingdom. Some plants, such as those in the Solanaceae, apparently do not contain amine oxidase activity. The ultimate fate of polyamines in plants is thus quite complicated and not fully understood.

3. Polyamines in Conifers and Hardwoods: General Comments

Cellular levels of polyamines in different tissues of a number of woody plant species have been reported (Table 1). It is obvious from the published work that the major polyamines (putrescine, spermidine, and spermine) found in woody plants are the same as in the herbaceous plants. There are no published reports containing any information on the occurrence of rare polyamines, possibly because only a few species of woody plants have been studied. Like the situation with herbaceous plants, different species and different tissues of the same plant differ widely in having either putrescine or spermidine as the predominant polyamine. Both ODC and ADC have been observed in woody plants, their distribution also being tissue and species dependent as in herbaceous plants (R. Minocha and S.C. Minocha, unpublished data). While the information is scant, both conjugated and bound forms of polyamines have been reported in woody plants.

Some polyamines and their combinations have been shown to have favorable effects on protoplast and callus cultures of woody plants. Ornithine (25 μM) and putrescine (50 and 100 μM) were found to stimulate cell division and colony formation in mesophyll protoplasts of *Alnus glutinosa* and *A. incana* (Huhtinen *et al.*, 1982/83). When the effect of polyamines was tested on initiation of megagametophyte callus of *Picea abies*, it was found that these compounds could not replace casein hydrolysate and glutamine in most cases, but a combination of three polyamines (putrescine 0.25, spermidine 0.1, and spermine 0.025 mM) favored the development of roots (Simola and Honkanen, 1983). The microspore callus cultures of this species were able to grow in the dark if the nutrient medium contained putrescine (0.1 mM). The effects of spermidine and spermine on growth were similar but root differentiation was stimulated in blue, red, and fluorescent light by spermine only (Simola and Huhtinen, 1986).

Table 1. Amount of polyamines in different woody plants.

Species	Putrescine	Spermidine	Spermine	Reference
Larix decidua				
callus	52.8	48.5	14.6	Minocha *et al.*, unpublished data
Picea abies				
needles, clone 14*	50	27	16	Dohmen *et al.*, 1990
needles, clone 11*	65	38	11	
callus	228.7	52.7	0.0	Minocha *et al.*, 1993
zygotic embryos	209.4	728.1	221.9	Santanen and Simola, unpublished data
megagametophyte	25.6	131.0	21.3	Santanen and Simola, unpublished data
Picea rubens				
needles	112.6	173.2	7.0	Minocha *et al.*, unpublished data
callus	227.3	65.0	10.3	Minocha *et al.*, 1993
roots	109.4	57.6	0.0	Minocha *et al.*, unpublished data
suspension	139.7	354.3	78.5	Minocha *et al.*, unpublished data
Pinus radiata				
cotyledon (ng/cot)	130 (0d) 40 (3d)	105 (0d) 80 (3d)	105 (0d) 70 (3d)	Biondi *et al.*, 1986
cotyledon (μmol/gFW)*	1.05 (0d) 0.4 (3d)	2.9 (0d) 1.1 (3d)	0.2 (0d) 0.25 (3d)	Kumar and Thorpe, 1989
Citrus aurantium leaves	41	90*	75	Kushad and Yelenosky, 1987
Citrus jambhiri	28	90*	14	
Citrus sinensis	29	70*	210	
Hevea brasiliensis				
callus (40 d)	4060	322	270	El Hadrami and D'Auzac, 1992
Mangifera indica (μmol/gFW)				
nucellus-monoembryonic	9.3	0.6	0.0	Litz and Schaffer, 1987
callus	382.3	2.2	0.4	
somatic embryos	8.6	0.0	0.0	
zygotic embryos	0.5	1.4	0.2	
Populus nigra × *maximowiczii*				
suspension	2039	386	45	Sun and Minocha, unpublished data
Populus tremuloides				
callus	5.3	40.0	6.3	Minocha *et al.*, unpublished data
Prunus avium leaf*	110	45	–	Biondi *et al.*, 1990
shoot 63	37	traces		
*Pyrus communis**				
shoot cultures (2 wk)	460	120	19	Prediari *et al.*, 1993

The polyamine levels are expressed as nmol/gFW unless otherwise stated (d = days).
* Indicates estimated amount from published figures.

4. Polyamines and Somatic Embryogenesis: Background

In addition to a multitude of potential functions of polyamines in plants, it has been suggested that polyamines play a critical role in morphogenesis in plant cell cultures (Minocha, 1988; Galston and Flores, 1991). With respect to somatic embryogenesis, the role of polyamines has been studied extensively in carrot (Montague *et al.*, 1978, 1979; Feirer *et al.*, 1984; Minocha, 1988; Mengoli *et al.*, 1989; Robie and Minocha, 1989; Minocha *et al.*, 1991b,c). The following statements summarize the relationship between polyamines and somatic embryogenesis in this tissue: 1) presence of auxin in the medium completely suppresses somatic embryogenesis; 2) increased biosynthesis of polyamines precedes the development of somatic embryos upon removal of auxin; 3) inhibitors of polyamine biosynthesis (e.g., DFMA and MGBG) strongly inhibit the development of somatic embryos. Studies from our laboratory (Robie and Minocha, 1989; Khan and Minocha 1991; Minocha *et al.*, 1991b,c; Nissen and Minocha, 1993; D.R. Bastola and S.C. Minocha, unpublished data) further show that: 1) cells grown in the presence of auxin contain less polyamines than those grown in an auxin-free medium; 2) DFMO, in a manner unrelated to its effects on ODC, promotes the biosynthesis of polyamines, inhibits ethylene production and promotes somatic embryogenesis in the absence of auxin, and, most importantly, allows the development of somatic embryos even in the presence of auxin; 3) the promotory effects of DFMO cannot be mimicked by exogenous addition of polyamines or other analogs; 4) transgenic cell lines of carrot that overexpress a mouse ODC cDNA and thus produce considerably higher levels of putrescine (as compared to control cells), also show a stimulation of somatic embryogenesis in the auxin-free medium, and produce somatic embryos even in the presence of otherwise inhibitory concentrations of 2,4-D. Thus, stimulation of putrescine biosynthesis achieved either through treatment with DFMO (Robie and Minocha, 1989) or through the overexpression of ODC (D.R. Bastola and S.C. Minocha, unpublished data) have similar consequences, i.e., a partial reversal of the inhibitory effect of auxin. Additional research is needed on mechanism(s) by which increased cellular levels of putrescine induce the development of somatic embryos.

Parallel studies with several other herbaceous angiosperms also have demonstrated the inhibition of somatic embryogenesis by inhibitors of polyamine biosynthesis (Meijer and Simmonds, 1988; Minocha, 1988; Galston and Flores, 1991). These results indicate that a continued biosynthesis of polyamines is an essential aspect of the metabolism during the differentiation and development of somatic embryos in plants.

In contrast to the herbaceous angiosperms, however, there have been only a few studies on polyamines in relation to the process of embryogenesis in woody plants, especially trees. Litz and Schaffer (1987) were the first to study changes in cellular polyamine content during the development of somatic and adventitious embryos in a tree species (*Mangifera indica*). Biondi *et al.* (1986,

1988) have reported changes in polyamine metabolism in the cotyledons of *Pinus radiata* cultured on shoot-forming and non shoot-forming media. Santanen and Simola (1992) compared the free, bound, and conjugated polyamines in non-embryogenic callus cultures with the differentiating embryogenic callus cultures of *Picea abies* grown on ABA-containing maturation medium. Minocha *et al.* (1993) have studied the metabolism of free polyamines in *Picea abies* and *Picea rubens* pro-embryogenic callus cultures grown on the proliferation medium containing auxin and cytokinin, and on the maturation medium containing ABA and IBA. The effects of exogenous addition of polyamines and their biosynthetic inhibitors on free polyamine levels and growth rate in the differentiating embryogenic cultures of rubber tree (*Hevea brasiliensis*) were reported by El Hadrami and D'Auzac (1992). All of these studies, like those on carrot, were done with tissues that were producing somatic embryos but also contained a large proportion of non-differentiating cells. However, a detailed analysis of polyamines in different stages of the developing somatic or zygotic embryos is not currently available, though, as discussed by Feirer, Chapter 15 of this book, preliminary studies have been undertaken in *Pinus strobus*. Santanen and Simola (1994) also reported on the putrescine and spermidine catabolism in non-embryogenic and embryogenic callus lines and late stages of developing embryos of *Picea abies*. The following is a summary of the available literature on polyamines and somatic embryogenesis and morphogenesis in tissue cultures of woody plants, and how these results compare with the reported data on herbaceous angiosperms.

5. Polyamines and Somatic Embryogenesis in Woody Plants

Somatic embryogenesis in conifers is known to be quite different from that in carrot and other angiosperm species. In most angiosperms, somatic embryogenesis entails a rapid growth phase for several days during which embryos and plantlets are formed that continue to grow or "germinate" without an intervening period of dormancy (maturation). In most conifers, on the other hand, tissue on the proliferation medium grows fast and produces only early stage embryos (Tautorus *et al.*, 1991). Maturation of these embryos is similar to the maturation of zygotic embryos. Maturation medium for conifers commonly contains ABA in contrast to the medium used for somatic embryogenesis in angiosperms (Hakman and Fowke, 1987a,b; Boulay *et al.*, 1988; Hakman and von Arnold, 1988; Misra and Green, 1990, 1991; Simola and Santanen, 1990). These mature embryos must be transferred to a fresh medium lacking ABA and other growth hormones in order to initiate germination.

As described elsewhere in this book, in most conifers, the pro-embryogenic tissue maintained on the proliferation medium consists of numerous organized meristematic clusters attached to long suspensor-like cells. Pro-embryogenic callus produces visible globular embryos on maturation medium within 10 to 15 days in both *Picea abies* and *Picea rubens*. Embryos with greenish or yellowish green cotyledons are visible in large quantities within 4 to 8 weeks. Multiple embryos attached to the same group of loosely organized suspensor-like cells are commonly observed. Since embryogenesis is not synchronous in these cultures, embryos at different stages of development can be found at any one time on the same piece of tissue. Embryogenic cultures maintained on maturation medium increase in dry as well as wet weight mostly due to the developing embryos. However, the increase in weight of these cultures always is less than that of the pro-embryogenic tissue maintained on the proliferation medium.

Differentiating embryogenic cultures of *Picea abies* contained significantly higher levels of polyamines than non-embryogenic cultures (Fig. 3). These observations are similar to those reported for carrot and egg plant (*Solanum melongena*) tissues (Montague *et al.*, 1978; Fobert and Webb, 1987; Robie and Minocha, 1989). Differentiating embryogenic tissues of *Picea abies* and *Picea rubens*, maintained on the maturation medium, had lower levels of polyamines compared to the same tissues grown on the proliferation medium (Fig. 2). The decline of polyamines on the maturation medium may partly be related to the effects of ABA itself on this tissue and not to embryogenesis since non-embryogenic tissue grown on ABA containing medium also showed this decline, though at a slower rate (Santanen and Simola, 1992). Either putrescine or spermidine was the predominant polyamine in embryogenic tissue depending on how long the tissue had been on a particular medium, the developmental stage of the tissue, and whether the cultures were grown in suspension or on solid media. Spermine was present in minute quantities at all times tested. In general, embryogenic cultures had a higher concentration of free putrescine than spermidine both in *Picea abies* and *Hevea brasiliensis* (El Hadrami and D'Auzac, 1992; Minocha *et al.*, 1993). By contrast, Litz and Schaffer (1987) had found a much lower concentration of polyamines in the somatic as well as zygotic embryos compared to nucellus tissue and callus of *Mangifera indica* (Table 1).

5.1. *Putrescine*

The cellular content of free putrescine increased between 2 to 12 days after subculture of the pro-embryogenic tissue of two different tissue lines (genotypes) of *Picea abies* on the proliferation medium (Fig. 2). In contrast, there was a reverse trend for changes in free putrescine content during the same period in embryogenic cultures grown on the maturation medium. A similar profile of changes also was seen in *Picea rubens* cultures grown on maturation medium (Fig. 2). Free putrescine was always higher in tissues grown on the proliferation medium compared to those grown on the maturation medium (for details see Minocha *et al.*, 1993). In *Picea abies* the level of conjugated putrescine was found to be high in the ABA-containing medium at the

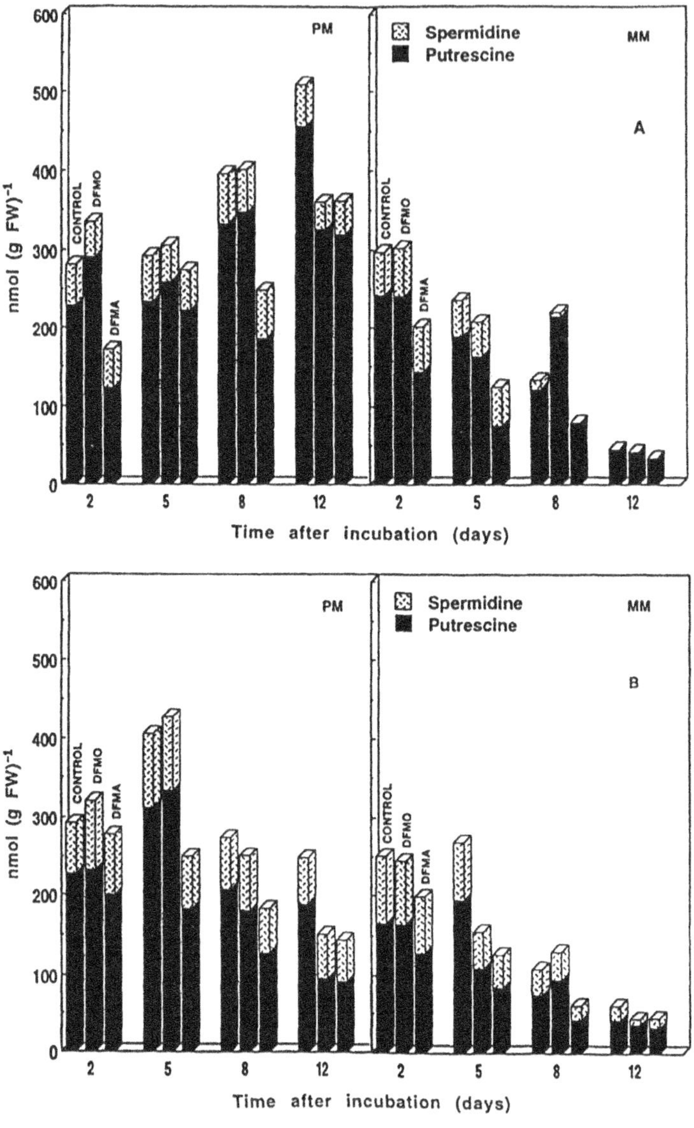

Figure 2. Effect of 2.0 mM DFMO and 0.1 mM DFMA on cellular levels of putrescine and spermidine in (A) *Picea abies* and (B) *Picea rubens* tissues grown on proliferation (PM) or maturation (MM) medium for different lengths of time. Values are mean ± SE of 4 replicates for *Picea abies* and 3 replicates for *Picea rubens*. (Modified from Minocha *et al.*, 1993.)

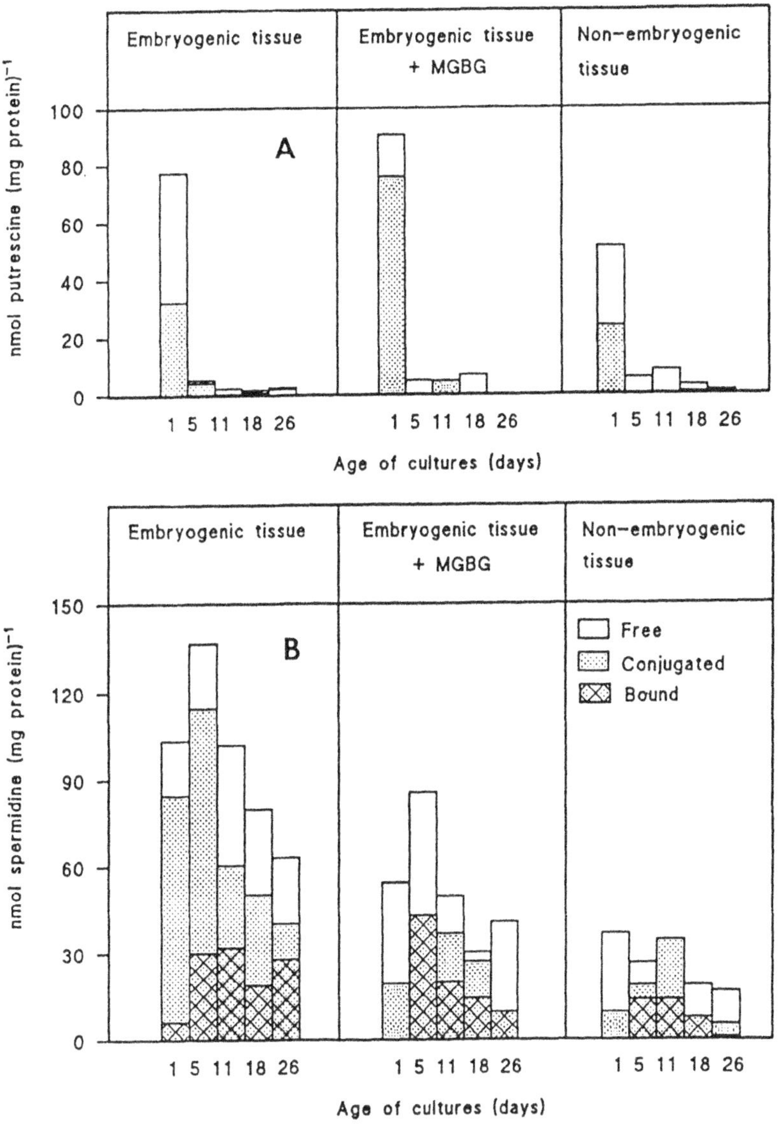

Figure 3. Changes in (A) putrescine and (B) spermidine content in embryogenic and non-embryogenic tissues of *P. abies* on ABA (+MgBg) containing medium. (Modified from Santanen and Simola, 1994.)

beginning of experiment in embryogenic as well as non-embryogenic cultures (Fig. 3).

Covalently bound putrescine was not found in *Picea abies* embryogenic tissue (Santanen and Simola, 1992). With few exceptions, DFMO (2.0 mM) had little effect on free putrescine level for the first 8 days of subculture in any experiment (Fig. 2). Occasionally, free putrescine level was higher in DFMO treated cells than control cells. This effect of DFMO is similar to the situation with carrot (Robie and Minocha, 1989; Nissen and Minocha, 1993). By contrast, DFMA was a strong inhibitor of free putrescine levels at all times tested (Fig. 2). DFMO has been shown to have stimulatory effects on embryogenesis in the case of carrot and maize (*Zea mays*) cell cultures (Mengoli *et al.* 1989; Robie and Minocha, 1989; Nissen and Minocha, 1993; Torné *et al.* 1993). However, in *Hevea brasiliensis* and in *Medicago sativa*, DFMO inhibited the embryogenesis process (Meijer and Simmonds, 1988; El Hadrami and D'Auzac, 1992). This may be due to the different pathways for putrescine biosynthesis in different tissues. Unfortunately, the distribution of ODC and ADC in these tissues is not known.

In a study of the changes in cellular polyamines during development of shoot forming (+BA) and non shoot-forming (−BA) callus from *Pinus radiata* cotyledons, Kumar and Thorpe (1989) observed a decline in putrescine levels in both types of cultures during the 21-day culture period; the sharpest decline occurred during the first 3 days. There was a somewhat faster decline in the non shoot-forming tissue than in the shoot-forming tissue. These results are similar to those of Biondi *et al.* (1986) with the same tissue. In their studies, however, an earlier decline (1 to 3 days) was followed by a steady increase for up to 10 days.

Marked differences in polyamine levels were observed in embryos and calli originating from different tissues of different varieties of *Mangifera indica* with either monoembryonic or polyembryonic adventitious embryos (Litz and Schaffer, 1987). The nucellar calli produced from monoembryonic varieties contained significantly more polyamines than those from polyembryonic varieties. Also, in contrast to the observations with *Picea abies* (Santanen and Simola, 1992), the non-embryogenic calli contained higher levels of polyamines than the embryogenic calli. Putrescine was the predominant polyamine in all varieties and in all tissues except in the zygotic embryos, where low levels of all polyamines were detected. Cellular polyamine levels in somatic embryos were comparable to those in the nucellus and were much lower (10 to 50 fold) than those in the callus. Exogenous supply of polyamines generally had no effect on the initiation of callus or somatic embryos.

In *Hevea brasiliensis*, the levels of putrescine increased with the time of culture up to 40 days, thereafter showing a decline up to 70 days (El Hadrami and D'Auzac, 1992). The addition of putrescine or arginine into the medium increased somatic embryogenesis potential of the calli while both DFMO and DFMA caused a reduction in somatic embryogenesis. While DFMA inhibited cellular putrescine in this tissue, there was a significant increase in putrescine in the presence of DFMO. Surprisingly, MGBG, which inhibited both spermidine and spermine synthesis, also caused a reduction in cellular putrescine. On the other hand, spermidine as well as spermine levels were higher in the presence of DFMA. No direct analysis of the enzyme activities for ODC, ADC, or SAMDC was reported.

Although MGBG (0.1 mM) inhibited free putrescine levels in differentiating embryogenic cultures of *Picea abies*, it showed a simultaneous increase in conjugated putrescine (day 1, Fig. 3). El Hadrami and D'Auzac (1992) also observed a similar decrease in free putrescine in embryogenic cultures of *Hevea brasiliensis*. However, free putrescine levels more than doubled within 4 days of subcultures on auxin-free medium containing MGBG in differentiating carrot cultures (Minocha *et al.*, 1991c).

5.2. *Spermidine*

In *Picea abies*, free spermidine levels did not change appreciably in the proembryogenic tissue during 12 days of culture on the proliferation medium (Fig. 2). In *Picea rubens* tissue, on the other hand, there was a slight increase in spermidine on day 5 under similar conditions. However, a decline in free spermidine levels was observed with time in cultures of both the species grown on the maturation medium (Fig. 2). These results are apparently different from the findings of Santanen and Simola (1992) who observed an increase in free spermidine levels at day 11, concomitant with a decrease in conjugated spermidine (Fig. 3). It should be pointed out, however, that the two laboratories have reported polyamine levels differently: whereas Minocha *et al.* (1993) expressed polyamine levels as nmol g^{-1}FW, Santanen and Simola (1992) expressed these as nmol mg^{-1} protein.

Similar to its effects on putrescine, DFMO usually had no effect on free spermidine levels in either *Picea abies* or *Picea rubens* tissues. DFMA inhibited spermidine only in differentiating embryogenic tissue on the maturation medium after 5–8 days of culture (Fig. 2). MGBG strongly decreased conjugated spermidine level in differentiating embryogenic tissues (Fig. 3).

Pinus radiata cotyledons grown on media with or without BA showed a sharp decline in cellular spermidine during the first 3 days of culture, after which only slight changes (mostly an increase) were seen during the next 18 days (Kumar and Thorpe, 1989). There were no significant differences in response to the presence of BA in the medium. These results are in apparent contrast to the data of Biondi *et al.* (1986), who observed a sharp rise in spermidine in the cotyledons by day 3 following a decline during the first 2 days. Again, it should be noted that the data reported by Kumar and Thorpe (1989) and Biondi *et al.* (1986) are in different units, i.e., μmol/gFW in the former and ng/cotyledon in the latter. Dicyclohexylamine (DCHA, 1.0 mM) inhibited the growth rate of cotyledons and caused a severe decline in cellular spermidine levels both in the presence and the absence of BA. The formation of adventitious buds in the presence of BA was only partially inhibited by DCHA.

Spermidine levels increased within 20 days of culture in *Hevea brasiliensis*, remaining high for the next 70 days (El Hadrami and D'Auzac, 1992). Both DFMA and DFMA + DFMO treatments caused an increase in spermidine on day 40 while DFMO alone caused a decrease of spermidine at this time.

No effects of inhibitors were seen on days 60 and 70. Exogenous supply of spermidine enhanced the embryogenic potential of this tissue when used alone and in combination with other polyamines. Spermidine was also able to reverse the inhibitory effect of MGBG on somatic embryogenesis.

Cellular spermidine levels were higher in the non-embryogenic callus obtained from nucellus tissue of a monoembryonic mango as compared to the embryogenic callus as well as the somatic and zygotic embryos (Litz and Schaffer, 1987). However, in embryogenic callus raised from the nucellus of a polyembryonic variety, the cellular spermidine was higher than that in the non-embryogenic callus, somatic embryos and adventitious embryos.

6. Spermine

Spermine levels always were low in both *Picea rubens* and *Picea abies* tissues (Minocha *et al.*, 1993). Neither DFMO nor DFMA had an effect on the production of cellular spermine. In contrast to putrescine and spermidine, both of which decreased by 50 to 60 percent during the first 3 days of culture of *Pinus radiata* cotyledons, spermine levels increased by 60 percent during this period in the presence of BA (Kumar and Thorpe, 1989). No such increase in spermine was seen in the absence of BA. In earlier studies with the same tissue, Biondi *et al.* (1986) observed a decline in spermine during the first 2 days. This was followed by a steady rise between days 3 and 10. This increase in spermine was completely prevented by DCHA, presumably through its inhibition of spermidine biosynthesis.

In *Hevea brasiliensis*, however, spermine levels were comparable to those of spermidine and showed a peak at day 40, the same time as for putrescine and spermidine (El Hadrami and D'Auzac, 1992). DFMA caused an increase in spermine, while DFMO caused a slight decrease. MGBG inhibited cellular spermine as expected.

7. ODC, ADC, and SAMDC Activities

As is the case with several plants like *Daucus carota*, *Catharanthus roseus*, and *Nicotiana* sp. (Tiburcio *et al.*, 1987; Robie and Minocha, 1989; Minocha *et al.*, 1991a), ADC seems to be the dominant pathway for putrescine production in embryogenic tissues of both *Picea abies* and *Picea rubens* (Minocha *et al.*, 1993). In both these species, DFMA inhibits putrescine production presumably through the inhibition of ADC activity (Fig. 4). Whereas ADC activity did not show specific trends during the 12 days of growth of the proembryogenic tissue kept on the proliferation medium, it increased with time in tissues grown on the maturation medium for both species. In line with its lack of effect on putrescine production, DFMO either slightly promoted or had no effect on ADC activity. While no enzyme assays were reported for

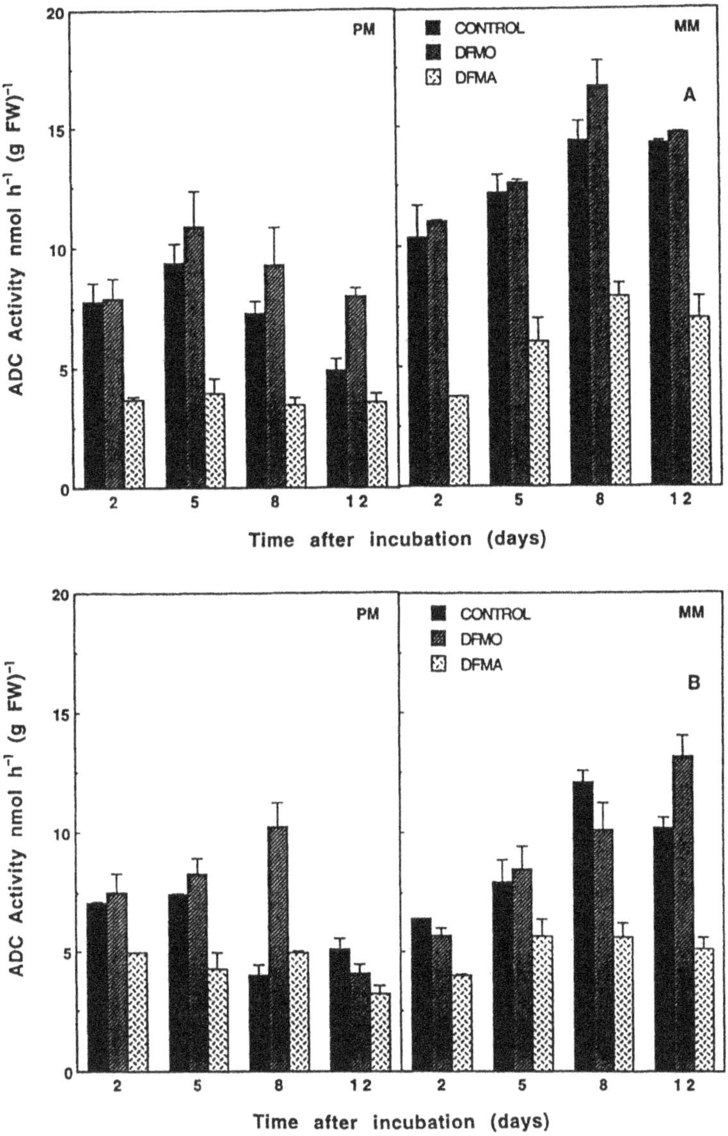

Figure 4. Effect of 2.0 mM DFMO and 0.1 mM DFMA on cellular levels of arginine decarboxylase (ADC) activity in (A) *Picea abies* and (B) *Picea rubens* tissues grown on proliferation (PM) or maturation (MM) medium for different lengths of time. Values are mean ± SE of 4 replicates for *Picea abies* and 3 replicates for *Picea rubens*. (Minocha *et al.*, unpublished data.)

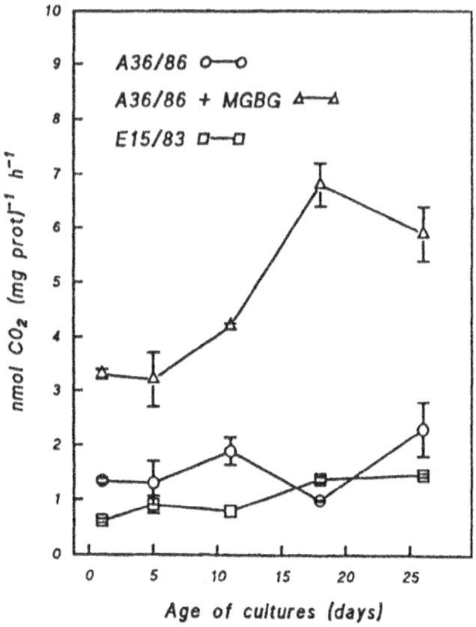

Figure 5. SAMDC activities in callus cultures of *P. abies* during 26 days of culture on ABA-containing medium A36/38 embryogenic callus, E15/83 non-embryogenic callus. Values are mean ± SE of three replicates. (Santanen and Simola, 1992.)

Hevea brasiliensis tissue by El Hadrami and D'Auzac (1992), their results on the effects of DFMO and DFMA indicate that ADC also is the main enzyme responsible for putrescine biosynthesis in this tissue. In contrast, in *Solanum melongena* and other solanaceous species, ODC seems to play a major role in putrescine biosynthesis (Slocum and Galston, 1985; Fobert and Webb, 1987). In the case of carrot embryogenic suspensions, ADC was the predominant enzyme during the first 10 days after culture and its inhibition by DFMA inhibited both putrescine levels and somatic embryogenesis. However, by day 12 when mature green embryos had appeared in these cultures, ODC was detected and showed a steady increase thereafter, accompanying the growth of these embryos (Robie and Minocha, 1989).

The activity of SAMDC also increased with time in embryogenic cultures of *Picea abies* grown on the maturation medium (Fig. 5). The specific activity of SAMDC was slightly higher in embryogenic cultures than in non-embryogenic cultures except for day 18. It is difficult to explain the increase in ADC and SAMDC activities in tissue grown on the maturation medium where overall polyamine levels as well as growth rate were lower than in the pro-embryogenic callus cultures on the proliferation medium. A further complication arises from the fact that somatic embryos constitute only a small proportion of the total tissue mass. While the somatic embryos are showing active growth, the remainder of the tissue mass is not. Since the enzyme and polyamine analysis were done on total tissue, the observed results are probably representative of the slow-growing tissue. Studies are underway to separately analyze enzyme activities in the somatic embryos and the surrounding tissue.

In the cotyledons of *Pinus radiata*, the activity of SAMDC increased sharply within the first 2 days of culture. This was followed by a steady decline during the next 8 days (Biondi *et al.*, 1988). This change coincided with change in the rate of conversion of ^{14}C-putrescine into spermidine and spermine. However, dicyclohexylamine, which inhibited the production of spermidine in this tissue, had little effect on measurable SAMDC activity during the entire 10-day culture period. No enzyme activities have been reported for *Hevea* or *Mangifera*.

8. Polyamine Degradation and Interconversions

The uptake and metabolism of putrescine has been studied in *Pinus radiata* cotyledons grown on shoot forming and non shoot-forming media (Kumar and Thorpe, 1989). Most of the ^{14}C-putrescine was metabolized into γ-aminobutyric acid (GABA), aspartic acid and glutamic acid. A similar fast metabolic conversion of ^{14}C-putrescine into GABA was reported in other higher plants (Flores and Filner, 1985). Only a small proportion (2 to 9 percent) of ^{14}C appeared in spermidine and spermine. The conversion of ^{14}C-putrescine into spermidine and spermine increased by 4 to 5 fold in the non shoot-forming cotyledons on days 3 and 10. However, overall catabolism of putrescine was highest in the shoot forming cotyledons where only 17 percent of the label was still in putrescine after 3 days as compared to 33 percent in the non shoot-forming cotyledons at the same time. A similar slow conversion of putrescine into spermidine and spermine in the same tissue was also reported earlier by Biondi *et al.* (1988). In their studies, a peak of incorporation of ^{14}C-putrescine into spermidine and spermine was observed on day 2 in the presence of BA and on day 5 in the absence of BA. Dicyclohexylamine inhibited the conversion of putrescine into both spermidine and spermine and also inhibited the uptake of putrescine. However, the activity of SAMDC in the tissue was not significantly affected by the presence of DCHA in the medium. Biondi *et al.* (1986) had earlier shown that treatment of cotyledons with DCHA prevented the accumulation of spermidine and spermine in the tissue during the 10-day culture period.

Catabolism of putrescine and spermidine in embryogenic and non-embryogenic tissues of *Picea abies* was studied by Santanen and Simola (1994) using (1,4–^{14}C)-putrescine and (1,-4–^{14}C)-spermidine as substrates. Except for day 1, both putrescine oxidation and spermidine oxidation rates increased with time of subculture. Activity was highest toward the end of growth period and also in later stages of embryo development (Fig. 6). Because both putrescine and aminopropylpyrroline were formed by the degradation of spermidine in non-embryogenic tissue (Santanen and Simola, 1994), it was suggested that there might be two separate routes for spermidine oxidation

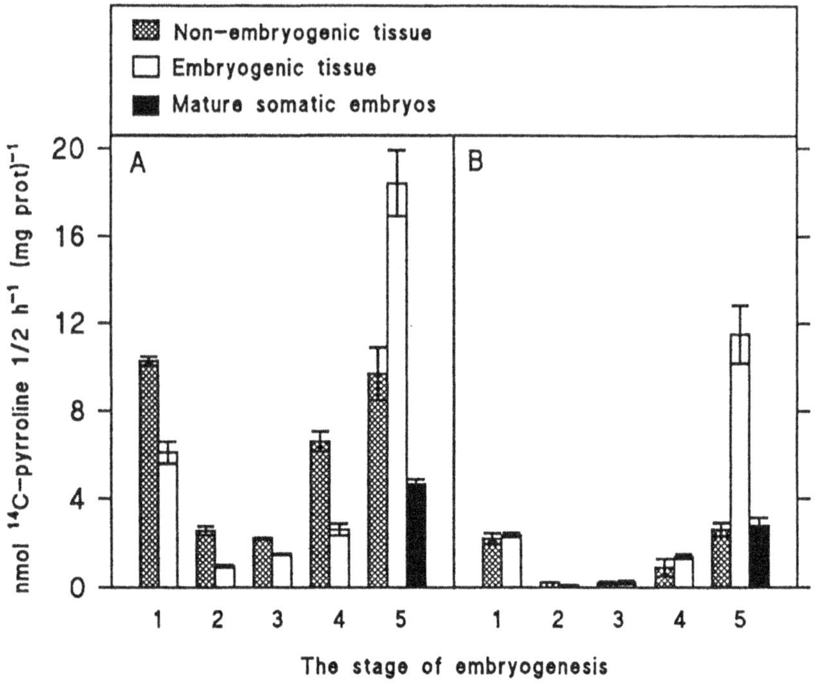

Figure 6. Activities of putrescine and spermidine oxidative enzymes in embryogenic and non-embryogenic tissues of *P. abies* (pH 8) during 5 stages of somatic embryogenesis in embryogenic tissue. At developmental stage 5, enzyme activity in somatic embryos and non-differentiated tissue were assayed separately. (A) DAO activity; (B) PAO activity. Values are mean ± SE of three replicates. (Modified from Santanen and Simola, 1994.)

in spruce tissue; one to aminopropylpyrroline by DAO and a second to putrescine and 3-aminopropanal by PAO, as reported earlier in fungal cells (Kobayashi and Horikoshi, 1982; Bagni and Pistocchi, 1992). In general, putrescine showed about a 5 times higher rate of degradation as compared to spermidine in both types of cultures. Maturing embryos showed much lower rates of putrescine and spermidine oxidation as compared to the non-differentiating embryogenic tissue growing on the same maturation medium (Fig. 6). In nonembryogenic tissue, oxidation of putrescine and spermidine by DAO and PAO proceeded via the formation of pyrroline intermediate, since pyrroline dehydrogenase activity also was observed in these cultures. Both DAO and PAO activities were inhibited in *Picea abies* by 1.0 mM aminoguanidine. Whereas the highest level of spermidine oxidation activity was located in cell wall fraction, the location of putrescine oxidation activity varied between cell wall, supernatant fraction and the homogenates (Santanen and Simola, 1994). The rather high polyamine degradation activity in aging embryogenic tissue (non-differentiated cells growing on maturation medium, Fig. 6) is not consistent with observations on some other tissues. The dissimilar nitrogen metabolism, developmental stage and physiological state of the tissue may explain why some tissues have low activity (ageing leaves, Kaur-Sawhney *et al.*, 1981) while others have high activity (intensively dividing tuber cells of *Helianthus tuberosus*, Torrigiani *et al.*, 1989).

9. Conclusions

The paucity of information on the metabolism of polyamines in cell cultures of woody plants makes it difficult to draw conclusions on their role in somatic embryogenesis. Further, the fact that different laboratories have reported results in different units (see Table 1) makes it nearly impossible to compare the results of these studies. However, analysis of results of the published studies with herbaceous plants leaves no doubt as to the importance of polyamine metabolism in the development of somatic embryos. In-depth studies are needed with well-established embryogenic cultures of conifers and hardwoods to establish the role of polyamines in the process of somatic embryogenesis in these species. Nevertheless, manipulation of polyamine biosynthetic pathways by use of inhibitors and transgenic techniques has demonstrated that somatic embryogenesis can be affected through polyamine metabolism in a number of unrelated species. This should lead to greater success in obtaining somatic embryogenesis in recalcitrant woody species, and enhancing it in those that show somatic embryogenesis at a low frequency.

References

Ammirato, P.V., 1984. Induction, maintenance, and manipulation of development in embryogenic cell suspension cultures. In: I.K. Vasil (Ed.), Cell Culture and Somatic Cell Genetics of Plants, Vol. 1, pp. 139–151. Academic Press, New York.

Bachrach, U. and Y.M. Heimer (Eds.), 1989a. The Physiology of Polyamines, Vol. 1. CRC Press, Boca Raton, FL.

Bachrach, U. and Y.M. Heimer (Eds.), 1989b. The Physiology of Polyamines, Vol. 2. CRC Press, Boca Raton, FL.

Bagga, S., A. Dharma, G.C. Phillips and G.D. Kuehn, 1991. Evidence for the occurrence of polyamine oxidase in the dicotyledonous plant *Medicago sativa* L. (alfalfa). Plant Cell Rep. 10: 550–554.

Bagni, N. and R. Pistocchi, 1992. Polyamine metabolism and compartmentation in plant cells. In: K. Mengel and J. Bilbeam (Eds.), Nitrogen Metabolism of Plants, pp. 229–248. Oxford Science Publications, Oxford, UK.

Bhojwani, S.S. and M.K. Razdan, 1983. Plant Tissue Culture: Theory and Practice, pp. 91–112; 199–236. Elsevier, New York.

Biondi, S., N. Bagni and A. Sansovini, 1986. Dicyclohexylamine uptake and effects on polyamine content in cultured cotyledons of radiata pine. Physiol. Plant. 66: 41–45.

Biondi, S., T. Diaz, I. Iglesias, G. Gamberini and N. Bagni, 1990. Polyamines and ethylene in relation to adventitious root formation in *Prunus avium* shoot cultures. Physiol. Plant. 78: 475–483.

Biondi, S., P. Torrigiani, A. Sansovini and N. Bagni, 1988. Inhibition of polyamine biosynthesis by dicyclohexylamine in cultured cotyledons of *Pinus radiata*. Physiol. Plant. 72: 471–476.

Boulay, M.P., P.K. Gupta, P. Krogstrup and D.J. Durzan, 1988. Development of somatic embryos from cell suspension cultures of Norway spruce (*Picea abies* Karst.). Plant Cell Rep. 7: 134–137.

Carman, J.G., 1990. Embryogenic cells in plant tissue cultures: Occurrence and behavior. In Vitro Cell Dev. Biol. 26: 746–753.

Cohen, S.S., R. Ballint and R.K. Sindhu, 1981. The synthesis of polyamines from methionine in intact and disrupted leaf protoplasts of virus-infected chinese cabbage. Plant Physiol. 68: 1150–1155.

De Jong, A.J., E.D.L. Schmidt and S.C. DeVries, 1993. Early events in higher plant embryogenesis. Plant Mol. Biol. 22: 367–377.

Dohmen, G.P., A. Koppers and C. Langebartels, 1990. Biochemical response of Norway spruce (*Picea abies* (L.) Karst.) towards 14 month exposure to ozone and acid mist: Effects on amino acid, glutathione and polyamine titers. Environ. Pollut. 64: 375–383.

Durzan, D.J., 1987. Physiological states and metabolic phenotypes in embryonic development. In: J.M. Bonga and D.J. Durzan (Eds.), Cell and Tissue Culture in Forestry, Vol. 2, pp. 405–439. Martinus Nijhoff Publishers, Dordrecht.

El Hadrami, I. and J. D'Auzac, 1992. Effects of polyamine biosynthetic inhibitors on somatic embryogenesis and cellular polyamines in *Hevea brasiliensis*. J. Plant Physiol. 140: 33–36.

Evans, P.T. and R.L. Malmberg, 1989. Do polyamines have roles in plant development? Annu. Rev. Plant Physiol. Plant Mol. Biol. 40: 235–269.

Federico, R. and R. Angelini, 1991. Polyamine catabolism in plants. In: R.D. Slocum and H.E. Flores (Eds.), Biochemistry and Physiology of Polyamines in Plants, pp. 41–56. CRC Press, Boca Raton, FL.

Feirer, R.P., G. Mignon and J.D. Litvay, 1984. Arginine decarboxylase and polyamines required for embryogenesis in the wild carrot. Science 223: 1433–1435.

Flores, H.E. and P. Filner, 1985. Polyamine catabolism in higher plants: characterization of pyrroline dehydrogenase. Plant Growth Reg. 3: 277–291.

Flores, H.E. and A.W. Galston, 1984. Osmotic stress-induced polyamine accumulation in cereal leaves. Plant Physiol. 75: 102–109.

Flores, H.E. and J. Martin-Tanguy, 1991. Polyamines and plant secondary metabolites. In: R.D. Slocum and H.E. Flores (Eds.), Biochemistry and Physiology of Polyamines in Plants, pp. 57–76. CRC Press, Boca Raton, FL.

Fobert, P.R. and D.T. Webb, 1987. Effects of polyamines, polyamine precursors, and polyamine biosynthetic inhibitors on somatic embryogenesis from eggplant (*Solanum melongena*) cotyledons. Can. J. Bot. 66: 1734–1742.

Galston, A.W., 1983. Polyamines as modulators of plant development. Bioscience 33: 382–388.

Galston, A.W. and H.E. Flores, 1991. Polyamines and plant morphogenesis. In: R.D. Slocum and H.E. Flores (Eds.), Biochemistry and Physiology of Polyamines in Plants, pp. 175–186. CRC Press, Boca Raton, FL.

Galston, A.W. and R. Kaur-Sawhney, 1987. Polyamines as endogenous growth regulators. In: P.J. Davies (Ed.), Plant Hormones and Their Role in Plant Growth and Development, pp. 280–295. Kluwer Academic Publishers, Dordrecht.

Goldberg, R.B., S.J. Baker and L. Perez-Grau, 1989. Regulation of gene expression during plant embryogenesis. Cell 56: 149–160.

Grey, D., G. Stepan-Sarkisian and M.W. Fowler, 1987. Biochemistry of forest tree species in culture. In: J.M. Bonga and D.J. Durzan (Eds.), Cell and Tissue Culture in Forestry, Vol. 2, pp. 31–60. Martinus Nijhoff Publishers, Dordrecht.

Hakman, I., 1993. Embryology in Norway spruce (*Picea abies*): An analysis of the composition of seed storage proteins and deposition of storage reserves during seed development and somatic embryogenesis. Physiol. Plant. 87: 148–159.

Hakman, I. and L.C. Fowke, 1987a. An embryogenic cell suspension culture of *Picea glauca* (white spruce). Plant Cell Rep. 6: 20–22.

Hakman, I. and L.C. Fowke, 1987b. Somatic embryogenesis in *Picea glauca* (white spruce) and *Picea mariana* (black spruce). Can. J. Bot. 65: 656–659.

Hakman, I. and S. von Arnold, 1988. Somatic embryogenesis and plant regeneration from suspension cultures of *Picea glauca* (white spruce). Physiol. Plant. 72: 579–587.

Hiatt, A.C., J. McIndoo and R.L. Malmberg, 1986. Regulation of polyamine biosynthesis in tobacco. J. Biol. Chem. 261: 1293–1298.

Huhtinen, O., J. Honkanen and L.K. Simola, 1982/83. Ornithine- and putrescine-supported divisions and cell colony formation in leaf protoplasts of alders (*Alnus glutinosa* and *A. incana*). Plant Sci. Lett. 28: 3–9.

Kaur-Sawhney, R., H.E. Flores and A.W. Galston, 1981. Polyamine oxidase in oat leaves: A cell wall-localized enzyme. Plant Physiol. 68: 494–498.

Khan, A.J. and S.C. Minocha, 1989a. Biosynthetic arginine decarboxylase in phytopathogenic fungi. Life Sci. 44: 1215–1222.

Khan, A.J. and S.C. Minocha, 1989b. Polyamine biosynthetic enzymes and the effect of their inhibition on the growth of some phytopathogenic fungi. Plant Cell Physiol. 30: 655–663.

Khan, A.J. and S.C. Minocha, 1991. Polyamines and somatic embryogenesis in carrot. II. The effects of cyclohexylammonium phosphate. J. Plant Physiol. 137: 446–452.

Kobayashi, Y. and K. Horikoshi, 1982. Purification and characterization of extracellular polyamine oxidase produced by *Penicillium* sp. Biochim. Biophys. Acta. 705: 133–138.

Krishnamurthy, R. and K.A. Bhagwat. 1989. Polyamines as modulators of salt tolerance in rice cultivars. Plant Physiol. 91: 500–504.

Kumar, P.P. and T.A. Thorpe, 1989. Putrescine metabolism in excised cotyledons of *Pinus radiata* cultured *in vitro*. Physiol. Plant. 76: 521–526.

Kushad, M.M. and G. Yelenosky, 1987. Evaluation of polyamine and proline levels during low temperature acclimation of citrus. Plant Physiol. 84: 692–695.

Litz, R.E. and B. Schaffer, 1987. Polyamines in adventitious and somatic embryogenesis in mango (*Mangifera indica* L.). J. Plant Physiol. 128: 251–258.

Malmberg, R.L. and A.C. Hiatt, 1989. Polyamines in plant mutants. In: U. Bachrach and Y.M. Heimer (Eds.), The Physiology of Polyamines, Vol. 2, pp. 148–159. CRC Press, Boca Raton, FL.

McCann, P.P., A.E. Pegg and A. Sjoerdsma (Eds.), 1987. Inhibition of Polyamine Metabolism, pp. 1–371. Academic Press, New York.

Meijer, E.G.M. and J. Simmonds, 1988. Polyamine levels in relation to growth and somatic embryogenesis in tissue cultures of *Medicago sativa* L. (alfalfa). J. Exp. Bot. 39: 787–794.

Mengoli, M., N. Bagni, G. Luccarini, V. Nuti Ronchi and D. Serafini-Fracassini, 1989. *Daucus carota* cell cultures: Polyamines and effect of polyamine biosynthesis inhibitors in the pre-embryogenic phase and different embryo stages. J. Plant Physiol. 134: 389–394.

Minocha, R., H. Kvaalen, S.C. Minocha and S. Long, 1993. Polyamines in embryogenic cultures of Norway spruce (*Picea abies*) and red spruce (*Picea rubens*). Tree Physiol. 13: 365–377.

Minocha, R., S.C. Minocha, A. Komamine and W.C. Shortle, 1991a. Regulation of DNA synthesis and cell division by polyamines in *Catharanthus roseus* suspension cultures. Plant Cell Rep. 10: 126–130.

Minocha, R., S.C. Minocha, S. Long and W.C. Shortle, 1992. Effects of aluminum on DNA synthesis, cellular polyamines, polyamine biosynthetic enzymes, and inorganic ions in cell suspension cultures of a woody plant, *Catharanthus roseus*. Physiol. Plant. 85: 417–424.

Minocha, S.C., 1988. Relationship between polyamine and ethylene biosynthesis in plants and its significance for morphogenesis in cell cultures. In: V. Zappia and A.E. Pegg (Eds.), Progress in Polyamine Research, pp. 601–606. Plenum Publishing Corporation, New York.

Minocha, S.C., R. Minocha and A. Komamine, 1991b. Effects of polyamine biosynthesis inhibitors on s-adenosylmethionine synthetase and s-adenosylmethionine decarboxylase activities in carrot cell cultures. Plant Physiol. Biochem. 29: 231–237.

Minocha, S.C., R. Minocha and C.A. Robie, 1990. High-performance liquid chromatographic method for the determination of dansyl-polyamines. J. Chromatogr. 511: 177–183.

Minocha, S.C., N.S. Papa, A.J. Khan and A.I. Samuelsen, 1991c. Polyamines and somatic embryogenesis in carrot III. Effects of methylglyoxal bis(guanylhydrazone). Plant Cell Physiol. 32: 395–402.

Misra, S. and M.J. Green, 1990. Developmental gene expression in conifer embryogenesis and germination. I. Seed proteins and protein body composition of mature embryo and the megagametophyte of white spruce (*Picea glauca* [Moench] Voss.). Plant Sci. 68: 163–173.

Misra, S. and M.J. Green, 1991. Developmental gene expression in conifer embryogenesis and germination II. Crystalloid protein synthesis in the developing embryo and megagametophyte of white spruce (*Picea glauca* (Moench) Voss). Plant Sci. 78: 61–71.

Montague, M.J., T.A. Armstrong and E.G. Jaworski, 1979. Polyamine metabolism in embryogenic cells of *Daucus carota* L. II. Changes in arginine decarboxylase activity. Plant Physiol. 63: 341–345.

Montague, M.J., J.W. Koppenbrink and E.G. Jaworski, 1978. Polyamine metabolism in embryogenic cells of *Daucus carota*. I. Changes in intracellular content and rates of synthesis . Plant Physiol. 62: 430–433.

Nissen, P. and S.C. Minocha, 1993. Inhibition by 2,4-D of somatic embryogenesis in carrot as explored by its reversal by difluoromethylornithine. Physiol. Plant. 89: 673–680.

Pegg, A.E., 1986. Recent advances in the biochemistry of polyamines in eukaryotes. Biochem. J. 234: 249–262.

Phillips, G. and G. Kuehn, 1991. Uncommon polyamines in plants and other natural sources. In: R.D. Slocum and H.E. Flores (Eds.), Biochemistry and Physiology of Polyamines in Plants, pp. 121–136. CRC Press, Boca Raton, FL.

Phillips, R., M.C. Press and A. Eason, 1987. Polyamines in relation to cell division and xylogenesis in cultured explants of *Helianthus tuberosus*: Lack of evidence for growth regulatory action. J. Expt. Bot. 38: 164–172.

Prediari, S., D.T. Krizek, C.Y. Wang, R.M. Mirecki and R.H. Zimmerman, 1993. Influence of UV-B radiation on developmental changes, ethylene, CO_2 flux and polyamines in cv. Doyenne d'Hiver pear shoots grown *in vitro*. Physiol. Plant. 87: 109–117.

Raghavan, V., 1986. Somatic embryogenesis. Embryogenesis in Angiosperms, pp. 115–151. Cambridge Univ. Press, Cambridge, UK.

Rastogi, R. and P.J. Davies, 1991. The effects of light and plant growth regulators on polyamine metabolism in higher plants. In: R.D. Slocum and H.E. Flores (Eds.), Biochemistry and Physiology of Polyamines in Plants, pp. 187–199. CRC Press, Boca Raton, FL.

Roberts, D.R., M.A. Walker, J.E. Thompson and E.B. Dumbroff, 1984. The effects of inhibitors of polyamine and ethylene biosynthesis on senescence, ethylene production and polyamine levels in cut carnation flowers. Plant Cell Physiol. 25: 315–322.

Robie, C.A. and S.C. Minocha, 1989. Polyamines and somatic embryogenesis in carrot. I. The effects of difluoromethylornithine and difluoromethylarginine. Plant Sci. 65: 45–54.

Santanen, A. and L.K. Simola, 1992. Changes in polyamine metabolism during somatic embryogenesis in *Picea abies*. J. Plant Physiol. 140: 475–480.

Santanen, A. and L.K. Simola, 1994. Catabolism of putrescine and spermidine in embryogenic and non-embryogenic callus lines of *Picea abies*. Physiol. Plant. 90: 125–129.

Sharp, W.R., M.R. Sondhal, L.S. Caldas and S.B. Maraffa, 1980. The physiology of *in vitro* asexual embryogenesis. Hort. Rev. 2: 268–310.

Simola, L.K., 1974a. The ultrastructure of dry and germinating seeds of *Pinus sylvestris* L. Acta Bot. Fenn. 103: 1–31.

Simola, L.K., 1974b. Subcellular organization of cotyledons of *Picea abies* during germination. Port. Acta Biol., Ser. A, 14: 413–428.

Simola, L.K., 1975. Changes in the subcellular organization of endosperm and radicle cells of *Picea abies* during germination. Z. Pflanzenphysiol. 78: 41–51.

Simola, L.K. and J. Honkanen, 1983. Organogenesis and fine structure in megagametophyte callus lines of *Picea abies*. Physiol. Plant. 59: 551–561.

Simola, L.K. and O. Huhtinen, 1986. Growth, differentiation, and ultrastructure of microspore callus of *Picea abies* as affected by nitrogenous supplements and light. N.Z. J. For. Sci. 16: 357–368.

Simola, L.K. and A. Santanen, 1990. Improvement of nutrient medium for growth and embryogenesis of megagametophyte and embryo callus lines of *Picea abies*. Physiol. Plant. 80: 27–35.

Slocum, R.D. and H.E. Flores, 1991. Biochemistry and Physiology of Polyamines in Plants, pp. 1–264. CRC Press, Boca Raton, FL.

Slocum, R.D. and A.W. Galston, 1985. *In vivo* inhibition of polyamine biosynthesis and growth in tobacco ovary tissue. Plant Cell Physiol. 26: 1519–1526.

Slocum, R.D., R. Kaur-Sawhney and A.W. Galston, 1984. The physiology and biochemistry of polyamines in plants. Arch. Biochem. Biophys. 235: 283–303.

Smith, M.A., 1991a. Chromatographic methods for the identification and quantitation of polyamines. In: R.D. Slocum and H.E. Flores (Eds.), Biochemistry and Physiology of Polyamines in Plants, pp. 229–242. CRC Press, Boca Raton, FL.

Smith, T.A., 1985. Polyamines. Ann. Rev. Plant Physiol. 36: 117–143.

Smith, T.A., 1991b. A historical perspective on research in plant polyamine biology. In: R.D. Slocum and H.E. Flores (Eds.), Biochemistry and Physiology of Polyamines in Plants, pp. 1–22. CRC Press, Boca Raton, FL.

Smith, T.A., J. Negrel and C. Bird, 1983. The cinnamic acid amides of the di- and polyamines. Adv. Polyamine Res. 4: 347–369.

Street, H.E., 1977. Embryogenesis and chemically induced organogenesis. In: W.R. Sharp, P.O. Larsen, E.F. Paddock and V. Raghavan (Eds.), Plant Cell and Tissue Culture Principles and Applications, pp. 123–153. Ohio State University Press, Columbus, OH.

Tautorus, T.E., L.C. Fowke and D.I. Dunstan, 1991. Somatic embryogenesis in conifers. Can. J. Bot. 69: 1873–1899.

Tenter, M. and A. Wild, 1991. Investigations on the polyamine content of spruce needles relative to the occurrence of novel forest decline. J. Plant Physiol. 137: 647–654.

Tiburcio, A.F., R. Kaur-Sawhney and A.W. Galston, 1987. Effect of polyamine biosynthetic inhibitors on alkaloids and organogenesis in tobacco callus cultures. Plant Cell Tiss. Org. Cult. 9: 111–120.

Torget, R., L. Lapi and S.S. Cohen, 1979. Synthesis and accumulation of polyamines and S-adenosylmethionine in chinese cabbage infected by turnip yellow mosaic virus. Biochem. Biophys. Res. Commun. 87(4): 1132–1139.

Torné, J.M., I. Claparols, X. Figueras and M. Santos, 1993. Effect of DL-alpha-difluoromethylornithine pretreatments in maize callus differentiation. Plant Cell Physiol. 34: 371–374.

Torrigiani, P., D. Serafini-Fracassini and A. Fara, 1989. Diamine oxidase activity in different physiological stages of *Helianthus tuberosus* tuber. Plant Physiol. 89: 69–73.

Villanueva, V.R., M. Mardon, Th.M. Le Goff and F. Moncelon, 1987. Development of a multicomponent analysis system. Application and preliminary results of a comparative study of cellular metabolism in healthy and damaged *Picea* trees from polluted areas. J. Chromatogr. 393: 97–105.

Wang, C.Y. and L.J. Zug, 1989. Effect of low-oxygen storage on chilling injury and polyamines in zucchini squash. Sci. Hort. 39: 1–7.

Wann, S.R., M.A. Johnson, T.L. Noland and J.A. Carlson, 1987. Biochemical differences between embryogenic and nonembryogenic callus of *Picea abies* (L.) Karst. Plant Cell Rep. 6: 39–42.

Young, N.D. and A.W. Galston, 1983. Putrescine and acid stress: Induction of arginine decarboxylase activity and putrescine accumulation by low pH. Plant Physiol. 71: 767–771.

17. An Evaluation of Somaclonal Variation during Somatic Embryogenesis

L.L. Deverno

Contents

1. Introduction

The process of woody plant tissue culture and subsequent regeneration of plantlets should result, theoretically, in the production of clones phenotypically and genetically identical to the original material. However, the passage of cells through *in vitro* culture induces variation in many plant species, a phenomenon termed somaclonal variation (Larkin and Scowcroft, 1981). Somaclonal variation manifested as phenotypic changes and biochemical alterations arising from tissue culture has been documented (Schaeffer, 1982; Armstrong and Phillips, 1988). Changes in methylation and chromosome number and structure have also occurred during *in vitro* culture (Edallo *et al.*, 1981; Prat, 1983; Evans and Sharp, 1983; Muller *et al.*, 1990; Grisvard *et al.*, 1990; Lewis-Smith *et al.*, 1990). However, as dedifferentiation and organogenesis occur, the level of DNA variation appears to decrease; most regenerated plants are genetically similar to controls (Brown *et al.*, 1991). *In vitro* culture conditions appear to affect the stability of plant genomes, with different plant species and genotypes responding in various ways. Somaclonal variation may, therefore, provide a valuable source of genetic variation for plant breeding, but it is undesirable when genetic stability is required for production of cloned plantlets.

Typically, most mutations are deleterious; therefore, somaclonal variation will tend to result in decreased vigour, yield, and performance characteristics (Jackson and Dale, 1989). Somaclonal variation occurs at a frequency of about one out of every 100 regenerated plants (Evans and Sharp, 1983) in contrast to the expected rate of 10^{-7}–10^{-9} mutations per nucleotide pair per generation. Some undesirable mutations will not be recognized immediately.

S. Jain, P. Gupta & R. Newton (eds.), Somatic Embryogenesis in Woody Plants, Vol. 1, 361–377.

These include recessive mutations, expressed only during certain developmental stages or under specific environmental conditions, or those that result in small changes in quantitative traits (DeKlerk, 1990).

Many potential practical applications of somatic embryogenesis for woody plants will never be realized unless selected genotypes remain stable during culture and regenerate true-to-type. The purpose of this chapter is to review current literature that describes genetic variability which occurs in somatic embryogenic cell culture, as well as somaclonal variation in plantlets regenerated from somatic embryogenic cells of various woody plants. An initial discussion will introduce the subjects of somaclonal variation in plants and genetic variation arising during *in vitro* cell and callus culture. Tissue culture genetic instability in somatic embryogenic cells and somaclonal variation in regenerated angiosperm and gymnosperm woody plants will be reviewed and possible ways to minimize this variation discussed. Interested readers will be referred to several comprehensive reviews for further details.

2. Somaclonal Variation

Phenotypic and genotypic variation observed among plants regenerated from *in vitro* cultured somatic cells is called *somaclonal variation* and is defined as heritable (i.e., transmitted through meiosis) genetic variability that occurs randomly, and usually irreversibly (Larkin and Scowcroft, 1981). Genetic variability in cell and callus cultures is referred to as *tissue culture instability*. In addition to these two classes of genetic variation, *epigenetic* (i.e., non-genetic) variation frequently appears in regenerated plants as a result of physiological responses, but these changes will not be expressed in the progeny (Meins, 1983). Epigenetic variation is both predictable and reversible.

Somaclonal variation may be the result of pre-existing genetic variation in somatic cells (Walbot, 1985), or may be due to alterations of nuclear or organellar genomes caused by the *in vitro* culture system. Several factors influence the level of somaclonal variation that occurs as a result of *in vitro* culture. The genotype of the explant is an important factor influencing the frequency of genetic variation (McCoy *et al.*, 1982), as is the presence of any pre-existing cytological variation (Lörz and Scowcroft, 1983). Genetic stability is also affected by stresses experienced through the *in vitro* culture regime, including subculture interval, levels of growth regulators, and state of the growth medium (i.e., solid versus liquid). Genetic variation has been shown to increase with prolonged *in vitro* culture (Bayliss, 1980; Meins, 1983). Increased duration in culture commonly results in loss of totipotency, which may be explained by either loss or mutation of genes responsible for regeneration or by changes in ploidy level. Genetic instability is more frequently associated with cell suspension or callus growth than with organized cultures from meristems (Thomas *et al.*, 1982).

Somatic embryogenesis may reduce the frequency of genetic variation in regenerated plants due to developmental constraints during embryogenesis that select against genetically altered cells. Also, somatic embryogenic clusters most often originate from a single cell, which prevents the formation of cytological mosaics and results in more uniform regenerated plantlets with a lower frequency of somaclonal variation (Thorpe, 1988; Haccius, 1978). However, this is not always the case; most of the somatic embryos differentiated on *Camellia japonica* L. cotyledons apparently had a multicellular origin from multicellular proembryonal complexes (Barciela and Vieitez, 1993). Alternatively, the formation of adventitious buds on stem or leaf explants have the potential to form cellular mosaics during proliferation *in vitro* because of their multicellular origin. Explants that retain their developmental integrity in culture, such as apical and axillary meristems, will rarely produce plantlets exhibiting somaclonal variation.

The majority of somaclonal variations arise due to changes in the nuclear genome. The most common genetic change observed in tissue culture is increased ploidy levels. Polyploidy may pre-exist in cell culture in the form of endoreduplicated cells in the explant or may originate *in vitro* through defective mitosis. Aneuploidy and haploidy can arise in a proportion of the cells through amitosis (nuclear fragmentation). Amplification or under replication of DNA has been shown to occur during *in vitro* plant cell culture (Nagl, 1990). Structural changes in chromosomes, gene mutations, and activation of transposable genetic elements also occur in cultured cells (Peschke and Phillips, 1991). Two possible theories have been proposed to explain the mechanisms leading to structural changes in chromosomes during cell culture (Lee and Phillips, 1988). The first mechanism is based on late replication of heterochromatin. If chromosome replication is disturbed during DNA synthesis and cell division the integrity of the heterochromatic regions would be vulnerable to rearrangements and breakage because these regions replicate later than euchromatic segments. The second mechanism is related to the consequences of nucleotide pool imbalance. Imbalances in the intracellular deoxyribonucleotide pools affect the fidelity of such processes as DNA repair, replication, and recombination. Consequences of these imbalances are DNA mutations, mitotic recombination, changes in chromosome structure, aneuploidy, and other serious genetic abberations.

A number of theories have been proposed to explain the increase of genetic instability during *in vitro* culture of explant tissue. Firstly, tissue culture imposes novel conditions on cellular metabolism, creating an environmental stress to which the genome attempts to adapt. For example, the genome of flax changes under different environmental conditions and it has been proposed that the genomic alteration is an adaptation to the new conditions (Cullis, 1983). Secondly, the stress of tissue culture may also activate transposons as a means for the genome to rapidly evolve to adapt to changing environmental conditions (McClintock, 1984). Thirdly, the tissue culture medium is composed of substances that are almost certainly mutagenic. Both 2,4-dichlorophenoxyacetic acid (2,4-D) and naphthaleneacetic acid (NAA) cause chromosomal aberrations when applied to cell cultures at elevated concentrations. However, at concentrations used in tissue culture these substances may cause aberrations indirectly by stimulating disorganized growth (Bayliss, 1980). For example, specific cytogenetic effects have been observed on root tips of *Pinus nigra* seedlings caused by the auxins NAA and indole-3-butyric acid (IBA). Both mitotic activity and chromosomal abberations increased with increasing concentrations of auxins. Nuclear changes such as spindle failure, chromosome fragmentation, lagging chromosomes, and polyploidy have been induced by NAA, and by IBA at lower frequencies (Papes *et al.*, 1991). The inclusion of the auxin NAA in the shoot-regeneration medium for loblolly pine increased ploidy levels in cells present in regenerated apices and callus (Renfroe and Berlyn, 1985). In addition, the presence of the auxin IBA in the rooting medium for buds regenerated from Caribbean pine explants resulted in roots with increased DNA content, although the buds remained genetically stable (Berlyn *et al.*, 1987). The auxin 2,4-D has also been implicated as the cause of heritable variation in soybean plantlets regenerated from somatic embryogenic culture, with lower concentrations of 2,4-D producing more variation than higher concentrations. Teratogenic effects on embryo morphology and development were also greater at lower 2,4-D concentrations (Shoemaker *et al.*, 1991).

In addition to these nuclear changes, genetic alterations in the cytoplasmic organelles have been detected during culture, especially in the mitochondrial genome (DeVerno *et al.*, 1994; Rode *et al.*, 1988; Hartmann *et al.*, 1989; Shirzadegan *et al.*, 1991; Gengenbach and Connelly, 1981; Kemble and Shepard, 1984) and to a lesser extent in the chloroplast genome (Zong-Xui *et al.*, 1983; Day and Ellis, 1985; Dunford and Walden, 1991). For more comprehensive reviews the reader is referred to D'Amato (1991), Karp and Bright (1985), Karp (1991), and Lee and Phillips (1988).

Somaclonal variation can be assessed by analysis of phenotype, chromosome number and structure, proteins, or direct DNA evaluation of plants regenerated from *in vitro* culture and their progeny (since the variation is transferred through meiosis to the offspring). The effect of somaclonal variation on the phenotype can be expressed as the percentage of plants that show aberrations for one or more defined characteristics. Chromosome numbers are most accurately determined by chromosome counts and DNA content measurements. Changes in the structure and organization of nuclear and organellar DNA can be directly evaluated by restriction fragment analysis or DNA sequencing. The effect of *in vitro* culture on proteins can be evaluated by isozyme electrophoresis or analysis of seed storage protein patterns. Methods chosen to evaluate somaclonal variation vary in their sensitivity to detect heritable genetic change. Phenotypic and karyotic examinations are less sensitive than biochemical methods. Direct analysis of the DNA using restriction fragment length polymorphism (RFLPs) and randomly amplified polymorphic DNA (RAPDs), as well as DNA sequencing, are more sensitive

Table 1. Summary of analyses of somaclonal variation in somatic embryogenesis of woody plants.

Woody Plant	Marker(s)	Somaclonal Variation	Reference
Abies alba	chromosome count	absent	Schuller *et al.*, 1989
Aesculus hippocastanum	chromosome count	absent	Radojevic and Stankovic, 1988
Carica hybrids	isozymes	present	Moore and Litz, 1984; Chen *et al.*, 1991
Castanea	chromosome count	absent	Vieitez, personal communication
Chimonanthus	chromosome count	present	Radojevic *et al.*, 1988
Citrus paradisi	leaf morphology isozymes	absent absent	Dhillon *et al.*, 1989
Elaeis guineensis	morphology	present	Corley *et al.*, 1986
Eucalyptus	RAPD	absent	Haque *et al.*, 1992
Fagus	chromosome count	absent	Vieitez, personal communication
Larix species	RFLP	present	DeVerno *et al.*, 1994
Liridendron tulipera	RFLP	absent	Merkle *et al.*, 1988
Picea abies	flow cytometry	absent	Mo *et al.*, 1989
Picea abies	morphology chromosome count	present present	Lelu, 1987
Picea abies	ethylene glutathione total reductants	present (epigenetic?) present	Wann *et al.*, 1986
Picea abies	RAPD	absent	Heinze *et al.*, 1992
Picea glauca × *engelmanni*	abcisic acid profile isozymes	absent absent	Eastman *et al.*, 1991
Picea glauca × *engelmanni*	RFLP	absent	Leggo and Rutledge, 1993
Picea mariana	RAPD	absent	Isabel *et al.*, 1993
Santalum album	morphology	present	Rao *et al.*, 1984

techniques for evaluating genetic change. Measuring somaclonal variation is described in a review by DeKlerk (1990).

3. Somaclonal Variation in Somatic Embryogenesis of Woody Plants

Genetic changes arising during *in vitro* somatic embryogenic culture that remain undetected in regenerated plants could lead to large investments in clonally propagated woody plants that are genetically defective. To prevent potential losses, these somatic embryo-derived plants need to be screened for genetic integrity during early stages of regeneration. Evaluations using morphological markers, chromosome counts, breeding behavior, and protein electrophoresis indicate that many somatic embryo-derived non-woody plants

electrophoresis indicate that many somatic embryo-derived non-woody plants maintain genetic integrity and clonal fidelity (Maheswaran and Williams, 1987; Karp *et al.*, 1987; Hanna *et al.*, 1984; Moore and Litz, 1984), while studies of other non-woody plants have detected somaclonal variation (Amberger *et al.*, 1992; Dale and Jones, 1982; Sondahl, 1982). Similar results have been found from the study of woody plants using morphological, biochemical, and DNA markers. A summary of somaclonal variation in somatic embryogenesis of both angiosperm and gymnosperm woody plants can be found in Table 1.

4. Somaclonal Variation in Angiosperm Woody Plants

Somaclonal variation in angiosperm trees was first detected within clones of *Populus nigra* regenerated from calli that had been initiated from shoot tips (Lester and Berbee, 1977). Variation was detected in height, number of branches, and leaf traits, with some regenerants exhibiting extremely reduced vigour. Although leaf cells remained diploid, chromosome counts of roots were highly variable. Callus culture has also been used to produce *Populus* somaclones resistant to the leaf spot fungus *Septoria musiva* (Ostry and Skilling, 1988). Similarly, somaclonal variant plants of *Populus deltoides* regenerated from callus that had been initiated from leaf explants had an increased partial resistance to leaf rust disease caused by *Melampsora medusae*. Somaclones exhibited a longer latent period relative to the parental genotype, but complete resistance was not observed in any of the somaclones and some somaclones were more susceptible to the disease than the parental genotype (Prakash and Thielges, 1989).

Isozyme analysis has been valuable in the study of genetic variation in natural populations of many tree species and this technique can also be applied to detect somaclonal variation in isozymes of tissue culture-derived material. Isozyme variation, without any associated morphological changes, has been detected among *Populus tremuloides* plants regenerated from callus cultures that had been initiated from shoots (Noh and Minocha, 1990).

In fruit trees, spontaneous morphological variation was first noted in *Citrus* callus during prolonged culture, as evidenced by a shift from shoot-bud differentiation to embryogenesis (Chaturvedi and Mitra, 1975). Somaclonal variant plantlets of peach regenerated from callus cultures that were initiated from immature embryos were selected to be more resistant than the parental genotype to the toxic metabolite(s) of the bacteria *Xanthomonas pruni*, the causative agent of leaf-spot disease (Hammerschlag, 1988). As well, somaclones of *Malus* have been produced that are resistant to fireblight caused by *Erwinia amylovora* (Donovan, 1991), and somaclones of *Prunus avium* × *pseudocerasus* (Colt cherry) that are heritably resistant to salt have been obtained from explants and protoplasts (Ochatt and Power, 1989).

5. Somaclonal Variation in Somatic Embryogenesis of Angiosperm Woody Plants

Plants regenerated from somatic embryos induced on callus cultures established from stem segments of grapefruit (*Citrus paradisi*) were shown to have no significant variations in leaf morphology, including leaf shape index, margin serration, wing petiole, and leaf apex shape. As well, no variations were detected in the relative position or number of bands for the isoenzymes of peroxidases, esterase, and acid phosphatase. These results suggested that the regenerated plants were genetically stable and uniform (Dhillon *et al.*, 1989). In contrast, the isozyme patterns for malate dehydrogenase and peroxidase of regenerated plantlets derived from somatic embryos that had either been induced directly on immature embryos of *Carica papaya* × *Carica cauliflora* hybrids (Chen *et al.*, 1991), or indirectly from somatic embryogenic cells induced from ovular callus (Moore and Litz, 1984), contained unique bands not seen in either of the parents. However, the isozyme patterns among the plantlets derived from these somatic embryos were uniform, suggesting genetic uniformity among the regenerated plants.

Tissue culture genetic instability in proembryos of *Chimonanthus* embryogenic callus cultures induced from microspores was detected by cytogenetic analysis. After 6 months in culture the proembryos showed haploid, diploid, and tetraploid chromosome numbers while, after two years, none of the cells showed haploidy or diploidy, and polyploid-hypotetraploid chromosome numbers were observed in some cells (Radojevic *et al.*, 1988). In contrast, plantlets regenerated from somatic embryogenic callus cultures of horsechestnut (*Aesculus hippocastanum* L.) initiated from immature embryos retained their diploid chromosome number (Radojevic and Stankovic, 1988).

Embryogenic cell lines of yellow poplar (*Liridendron tulipera* L.) were analyzed for RFLPs using a homologous probe of highly repeated DNA sequences. No polymorphism was found in sublines of embryogenic suspension cultures and comparison of banding patterns among plantlets derived from the same clonal line also lacked any variation in restriction fragments (Merkle *et al.*, 1988). As well, preliminary studies of somatic embryogenesis in *Eucalyptus* using RAPD markers indicated genetic stability in tissue culture lines of this species (Haque *et al.*, 1992). However, morphological characters exhibited somaclonal variation with high frequency in sandalwood (*Santalum album*) plantlets regenerated from somatic embryos that had been differentiated *in vitro* from somatic embryogenic callus initiated from stem segments (Rao *et al.*, 1984).

Preliminary studies indicate that chromosome numbers are not altered in embryogenic systems of *Fagus* and *Castanea*, as well as in root tips from the regenerated plants. Although multinucleate cells and cells that appear to express a meiotic-like phenotype (pollen-like cells, tetrads) were observed in the embryogenic cell suspension cultures of both beech and chestnut, these aberrations were not apparent in regenerated plants (F.J. Viéitez, personal communication).

6. Somaclonal Variation in Gymnosperms

Gymnosperms are generally considered genetically conservative. Only three naturally occuring polyploid species of gymnosperms are known (Wright, 1976), and polyploids in pines have reduced growth, form quality, and longevity (Berlyn *et al.*, 1986). There are several reports of gymnosperms that develop elevated ploidy levels and chromosomal aberrations as a result of *in vitro* culture of explant tissue, especially in pines, although most other conifers do not appear to exhibit these changes (see Table 2 for summary).

7. Somaclonal Variation in Somatic Embryogenesis of Gymnosperms

A chromosome count of somatic embryos of *Abies alba* derived from segments of female megagametophytes containing immature zygotic embryos indicated that all cells were diploid only, $2n = 24$ (Schuller *et al.*, 1989). Similarly, *Picea abies* plantlets regenerated from somatic embryos were shown by flow cytometry to maintain a stable DNA content of 32pg/C1 nuclei (Table 3), and they remained at the diploid stage (Mo *et al.*, 1989).

The random amplification of polymorphic DNA (RAPD) has been shown to be a sensitive technique for detecting genetic variation among individuals within and between species (Williams *et al.*, 1990). Many regions of the genome can be sampled at random with a potentially unlimited number of markers. A preliminary study to monitor the clonal fidelity in *Picea abies* somatic embryogenesis using RAPD fingerprinting also indicated genetic stability in tissue culture lines (Heinze *et al.*, 1992). However, in another study certain somatic embryos of *Picea abies* underwent a genetic change from diploid to tetraploid (Table 4) and others exhibited phenotypic aberrations (Lelu, 1987). Biochemical changes associated with somatic embryogenesis from immature embryos of *Picea abies* have also been detected (Wann *et al.*, 1986). These variations include a significant decrease in the evolution rate of ethylene, reduced levels of glutathione, and less total reductants (Table 5). However, these physiological changes were probably due to epigenetic variation.

Isozyme analysis was used to evaluate somaclonal variation in callus derived from zygotic embryos of interior spruce (*Picea glauca* × *engelmannii* complex), as well as subcloned calli generated from somatic embryos, somatic embryos produced from parental calli, and somatic embryos from subcloned calli, using 15 different enzyme assays representing a minimum of 25 loci. Results of this analysis indicated that there was no evidence of somaclonal variation during somatic embryogenesis of interior spruce (Eastman *et al.*,

Table 2. Summary of somaclonal variation in non-somatic embryogenic cultures of gymnosperms.

Plant	Marker	Somaclonal Variation	Reference
Cryptomeria japonica	chromosomes	aneuploidy, chromosome bridges, lagging chromosomes	Mehra and Anand, 1979
Ephedra foliata (haploid cultures)	chromosomes	lagging chromosomes, abnormal telophases	Konar, 1963
Ginkgo biloba (haploid culture)	chromosomes	diploidy, tetraploidy, heteroploidy	Tulecke, 1957
Larix × eurolepis	chromosomes	none	Wyman *et al.*, 1992
Picea abies	DNA content	none	Hakman *et al.*, 1984
Picea pungens	chromosomes	none	Manandhar and Gresshoff, 1980
Pinus caribaea	DNA content	increased DNA content	Berlyn *et al.*, 1987
Pinus cembra	chromosomes	tetraploidy and higher ploidy	Salmia, 1975
Pinus coulteri	DNA content, chromosomes	increase in DNA content, chromosome bridges, micronuclei, lagging chromosomes, fragmentation	Patel and Berlyn, 1982
Pinus densiflora	chromosomes	abnormal and multiple nuclei	Tominaga and Oga, 1970
Pinus gerardiana	chromosomes	none	Konar and Nagmani, 1972
Pinus lambertiana	chromosomes	polyploidy	Partanen, 1963
Pinus roxburghii	chromosomes	polyploidy, aneuploidy, chromosome bridges, lagging chromosomes	Mehra and Anand, 1983
Pinus strobus	chromosomes	minor	Gautheret, 1956
Pinus taeda (buds regenerated from callus)	DNA content, chromosomes	none	Renfroe and Berlyn, 1984
Pinus taeda (buds regenerated directly on explants)	DNA content, chromosomes	increase in DNA content and ploidy	Renfroe and Berlyn, 1985
Pinus taeda	chromosomes	none	Franklin *et al.*, 1989

Table 3. DNA content of *Picea abies* material by flow cytometric analysis.*

Plant Material	Number of Counted Nuclei	DNA Amount (pg/nucleus)
Plantlets developed from zygotic embryos	15000	32
1st cycle embryogenic culture	3000	34
Plantlets developed from 4th cycle mature somatic embryos	5000	32

* From: Mo *et al.*, 1989.

Table 4. Ploidy level of somatic proembryos in *Picea abies* obtained from cotyledons and immature embryos.*

Source Explant	Total Number of Analysed Proembryos	Total Number of Analysed Cells	Ploidy Level (2n = 24)	Ploidy Level (4n = 48)	Number of Proembryos with Abnormal Ploidy Level
Immature embryo	32	159	159	–	0
Cotyledon	24	113	103	10	2

* From: Lelu, 1987.

Table 5. Biochemical differences between embryogenic (E) and nonembryogenic (NE) Norway spruce callus.*

Assay	NE Callus	E Callus	Ratio NE/E
Ethylene evolution rate, mL/mg fresh weight/day	1.1–1.8	0.01–0.09	19–117
GSH, nmoles/g protein	1.7–2.2	0.27–0.35	5.3–7.1
Total reductants, A_{760}/g fresh weight	438–535	22–32	17–20

* From: Wann *et al.*, 1986.

1991). In addition, a characteristic abscisic acid-dependent developmental profile has been reported for embryogenic cultures of interior spruce, which was used to compare the morphologies of parent line and subclone embryogenic cultures (Webster *et al.*, 1990). The embryogenic cultures initiated from interior spruce embryos showed genetic stability with respect to the abscisic acid profile, being consistent with that of the explant genotype. These results were further substantiated by an assessment of the genomic DNA using probes that hybridized to hypervariable regions. No repeatable differences in hybridization patterns were observed between the parental lines and the subcloned lines, which indicated that these embryogenic lines of interior spruce were genetically stable in tissue culture (Leggo and Rutledge, 1993).

The RAPD technique has been used to assess genetic stability of somatic embryos derived from somatic embryogenic cell lines of *Picea mariana* (Mill.) B.S.P. Ten RAPD markers were used to evaluate the genetic stability of somatic embryos derived from three embryogenic cell lines. No variation was detected within the clones, suggesting that the somatic embryos were genetically stable (Isabel *et al.*, 1993). In addition, karyotype analysis of somatic embryogenic cell suspensions of *Picea mariana* indicated that the cell lines were cytologically stable (Nkongolo and Klimaszewska, 1994).

Aneuploidy and polyploidization have been reported in haploid embryogenic cultures of *Larix decidua* (Von Aderkas and Anderson, 1993). Somaclonal variation and tissue culture genetic instability are most often associated with changes in the nuclear genome; however, alterations in organellar genomes also occur. Restriction fragment length polymorphism analysis of *Larix* somatic embryogenic cell lines revealed amplifications of certain regions of the mitochondrial genome, as well as variations in the stoichiometry of some minor restriction fragments. Several novel restriction fragments present in the parent were not observed in the somatic embryogenic cell culture. However, these genetic changes were not apparent in the regenerated trees, which showed a mitochondrial genome organization similar to that of the parent tree. The somatic embryogenic culture may have been composed of a mixture of cells with both normal and altered mitochondrial genomes, and cells with altered genomes lacked regenerative capacity (DeVerno *et al.*, 1994).

8. Minimizing Somaclonal Variation

Somaclonal variation is a major obstacle in the production of genetically uniform plants from somatic embryos. Many factors can influence the frequency and nature of this variation but, to date, no comprehensive studies have been done in any somatic embryogenic systems to analyze the effects of manipulating all the possible parameters to minimize somaclonal variation. However, from the studies described in this review chapter it is possible to speculate on ways to reduce or control somaclonal variation in somatic embryogenesis of woody plants.

Somaclonal variation appears to increase in frequency with duration of *in vitro* culture, especially when the cells are maintained as callus or cell suspension. Reducing the length of time required to produce regenerated plantlets from the initial induction of somatic embryos, as well as attempting to induce the production of somatic embryos directly on the explant tissue without the intervening callus stage, could minimize the occurrance of somaclonal variation. In addition, cryopreservation of somatic embryogenic cultures could be used in lieu of continuous subculturing. This would reduce the duration of the *in vitro* maintenance of cells in an unorganized state and perhaps lessen the changes that would occur over prolonged periods in culture.

The complex compositions of the various culturing media may play an important role in the control of somaclonal variation. The cells have changing nutritional requirements during different developmental stages and an imbalance in the availability of certain micronutrients could affect the frequency and nature of somaclonal variation. In addition, the type of auxin, its concentration, and duration of application can each be altered to reduce the effects of these toxic compounds on the genetic stability of the cells. Finally, it is critical to screen many explant genotypes because different genotypes display differing frequencies of somaclonal variation as a result of *in vitro* culture.

9. Conclusion

Somatic embryogenesis appears to influence the genetic stability of some plant genomes, with various plant species and genotypes responding differently. Some somaclonal variation has been detected in somatic embryogenesis of both woody angiosperms and gymnosperms. However, somaclonal variation may not always be detected during somatic embryogenic culture or during regeneration of plants. For example, abnormalities appeared after several years of growth in the field in oil palms (Corley *et al.*, 1986), although it was not clear whether this was the result of genetic or epigenetic factors. This example demonstrates the necessity of establishing that genotype integrity has not been compromised during somatic embryogenic culture and plant regeneration. However, there are few reports available in either woody angiosperms or gymnosperms to determine the extent of this type of variation in somatic embryogenesis, and none that have directly compared phenotypic, biochemical, and the several methods of direct DNA analysis for detecting somaclonal variation in these plants. There is no single technique that will unequivocally demonstrate the presence or absence of somaclonal variation in cell cultures or regenerated plants. Qualitative and quantitative changes in the nuclear or organellar genomes may go undetected prior to field trials. For this reason, plants regenerated from somatic embryogenic cultures should be stringently monitored.

Somaclonal variation, however, can also be advantageous. Selection for somaclonal variants expressing important traits such as disease resistance, coupled with early, rapid screening of regenerated plants, may be a technique that can be used for improvement of important woody plants. The ability to introduce new traits such as disease resistance into elite lines via somaclonal variation eliminates the segregation of other desirable traits that occur in hybridization. Somaclonal variants could also be produced that would express traits important for adaptation to changing environmental conditions. When the factors that affect somaclonal variation are elucidated, it may then be possible to manipulate these factors to control this phenomenon.

References

Amberger, L.A., R.C. Shoemaker and R.G Palmer, 1992. Inheritance of two independent isozyme variants in soybean plants derived from tissue culture. Theor. Appl. Genet. 84: 600–607.

Armstrong, C.L. and R.L. Phillips, 1988. Genetic and cytogenetic variation in plants regenerated from organogenic and friable, embryogenic tissue cultures of maize. Crop Sci. 28: 363–369.

Barciela, J. and A.M. Vieitez, 1993. Anatomical sequence and morphometric analysis during somatic embryogenesis on cultured cotyledon explants of *Camellia japonica* L. Ann. Bot. (Lond.) 71: 395–404.

Bayliss, M.W., 1980. Chromosomal variation in plant tissue culture. Int. Rev. Cytol. 11A: 113–143.

Berlyn, G.P., A.O. Anoruo, R.C. Beck and J. Cheng, 1987. DNA content polymorphism and tissue culture regeneration in Caribbean pine. Can. J. Bot. 65: 954–961.

Berlyn, G.P., R.C. Beck and M.H. Renfroe, 1986. Tissue culture and the propagation and genetic improvement of conifers: problems and possiblities. Tree Physiol. 1: 227–240.

Brown, P.T.H., E. Gobel and H. Lorz, 1991. RFLP analysis of *Zea mays* callus cultures and their regenerated plants. Theor. Appl. Genet. 81: 227–232.

Chaturvedi, H.C. and G.C. Mitra, 1975. A shift in morphogenetic pattern in *Citrus* callus tissue during prolonged culture. Ann. Bot. 39: 683–687.

Chen, M.H., C.C, Chen, D.N. Wang and F.C. Chen, 1991. Somatic embryogenesis and plant regeneration from immature embryos of *Carica papaya* × *Carica cauliflora* cultured *in vitro*. Can. J. Bot. 69: 1913–1918.

Corley, R.H.V., C.H. Lee, I.H. Law and C.Y. Wong, 1986. Abnormal flower development in oil palm clones. Planter, Kuala Lumpur 62: 233–240.

Cullis, C.A., 1983. Environmentally induced DNA changes in plants. CRC Crit. Rev. Plant Sci. 1: 117–131.

D'Amato, F., 1991. Nuclear changes in cultured plant cells. Caryologia 44: 217–224.

Dale, P.J. and M.G.K. Jones, 1982. Studies on callus and plant regeneration from tissues and protoplasts of the forage grass *Lolium multiflorum*. In: A. Fujiwara (Ed.), Plant Tissue Culture, pp. 579–580. Jpn. Assoc. Plant Tissue Cult., Tokyo.

Day, A. and T.H.N. Ellis, 1985. Deleted forms of plastid DNA in albino plants from cereal anther culture. Curr. Genet. 9: 671–678.

De Klerk, G.-J., 1990. How to measure somaclonal variation. Acta Bot. Neerl. 39: 129–144.

DeVerno, L.L., P.J. Charest and L. Bonen, 1994. Mitochondrial DNA variation in somatic embryogenic cultures of *Larix*. Theor. Appl. Genet. 88: 727–732.

Dhillon, B.S., H. Raman and D.S. Brar, 1989. Somatic embryogenesis in *Citrus paradisi* and characterization of regenerated plants. Acta Hort. 239: 113–116.

Donovan, A., 1991. Screening for fire blight resistance in apple (*Malus pumila*) using excised leaf assays from *in vitro* and *in vivo* grown material. Ann. Appl. Biol. 119: 59–68.

Dunford, R. and R.M. Walden, 1991. Plastid genome structure and plastid-related transcript levels in albino barley plants derived from anther culture. Curr. Genet. 20: 339–347.

Eastman, P.A.K., F.B. Webster, J.A. Pitel and D.R. Roberts, 1991. Evaluation of somaclonal variation during somatic embryogenesis of interior spruce (*Picea glauca engelmannii* complex) using culture morphology and isozyme analysis. Plant Cell Rep. 10: 425–430.

Edallo, S., C. Zucchinali, M. Perenzin and F. Salamini, 1981. Chromosomal variation and frequency of spontaneous mutation associated with *in vitro* culture and plant regeneration in maize. Maydica 26: 39–56.

Evans, D.A. and W.R. Sharp, 1983. Single gene mutations in tomato plants regenerated from tissue culture. Science 221: 949–951.

Franklin, C.I., R.L. Mott and T.M. Vuke, 1989. Stable ploidy levels in long-term callus cultures of loblolly pine. Plant Cell Rep. 8: 101–104.

Gautheret, M.R., 1956. Sur les phénomènes d'histogenèse dans les cultures de tissus de *Pinus strobus* L. C.R. Acad. Sci. (Paris) 242: 3108–3110.

Gengenbach, B.G. and J.A. Connelly, 1981. Mitochondrial DNA variation in maize plants regenerated during tissue culture selection. Theor. Appl. Genet. 59: 161–167.

Grisvard, J., M. Sevignac, M. Chateau and M. Branchard, 1990. Changes in a repetitive DNA sequence during callus culture of *Cucumis melo*. Plant Sci. 72: 81–91.

Haccius, B., 1978. Question of unicellular origin on non-zygotic embryos in callus cultures. Phytomorphol. 28: 74–81.

Hakman, I., S. von Arnold and A. Bengtsson, 1984. Cytofluorometric measurement of nuclear DNA in adventitious buds and shoots of *Picea abies* regenerated *in vitro*. Physiol. Plant 60: 321–325.

Hammerschlag, F.A., 1988. Selection of peach cells for insensitivity to culture filtrates of *Xanthomonas campestris* pv. *pruni* and regeneration of resistant plants. Theor. Appl. Genet. 76: 865–869.

Hanna, W.W., C. Lu and I.K. Vasil, 1984. Uniformity of plants regenerated from somatic embryos of *Panicum maximum* Jacq. (Guinea grass). Theor. Appl. Genet. 67: 155–159.

Haque, N.S., N.W. Fish and M. Keil, 1992. Assessment of somaclonal variation in *Eucalyptus* using random amplified polymorphic DNA markers. Proceedings: Fifth Wkshp. IUFRO Wk. Party S2.04.06. Carcans-Maubisson, June 15–18, 1992. INRA, France.

Hartmann, C., Y. Henry, J. De Buyser, C. Aubry and A. Rode, 1989. Identification of new mitochondrial genome organizations in wheat plants regenerated from somatic tissue cultures. Theor. Appl. Genet. 77: 169–175.

Heinze, B., R. Westcott and J. Schmidt, 1992. Monitoring clonal fidelity in Norway spruce somatic embryogenesis using RAPD fingerprinting – preliminary results. Conifer Biotechnology Working Group Proceedings (unpubl.).

Isabel, N., L. Tremblay, M. Michaud, F.M. Tremblay and J. Bousquet, 1993. RAPDs as an aid to evaluate the genetic integrity of somatic embryogenesis-derived populations of *Picea mariana* (Mill.) B.S.P. Theor. Appl. Genet. 86: 81–87.

Jackson, J.A. and P.J. Dale, 1989. Somaclonal variation in *Lolium multiflorum* L. and *L. temulentum* L. Plant Cell Rep. 8: 161–164.

Karp, A., S.H. Steele, S. Parmar, M.G.K. Jones, P.R. Shewry and A. Breiman, 1987. Relative stability among barley plants regenerated from cultured immature embryos. Genome 29: 405–412.

Karp, A., 1991. On the current understanding of somaclonal variation. In: B.J. Miflin (Ed.), Oxford Surveys of Plant Molecular and Cell Biology, Vol. 7, pp. 1–58. Oxford Univ. Press, Oxford, U.K.

Karp, A. and S.W.J. Bright, 1985. On the causes and origins of somaclonal variation. In: B.J. Miflin (Ed.), Oxford Surveys of Plant Molecular and Cell Biology, Vol. 2, pp. 199–234. Oxford Univ. Press, Oxford, U.K.

Kemble, R.J. and J.F. Shepard, 1984. Cytoplasmic DNA variation in a potato protoclonal population. Theor. Appl. Genet. 69: 211–216.

Konar, R.N. and R. Nagmani, 1972. Chromosome number in callus cultures of *Pinus gerardiana* Wall. Curr. Sci. 41: 714–715.

Konar, R.N., 1963. Studies on submerged callus cultures of *Pinus gerardiana* Wall. Phytomorphology 13: 165–169.

Larkin, P.J. and W.R. Scowcroft, 1981. Somaclonal variation – a novel source of variability from cell cultures for plant improvement. Theor. Appl. Genet. 60: 197–214.

Lee, M. and R.L. Phillips, 1988. The chromosomal basis of somaclonal variation. Ann. Rev. Plant Physiol. Plant Mol. Biol. 39: 413–437.

Leggo, J.C. and R.G. Rutledge, 1993. Assessment of the genetic stability of interior spruce embryogenic cultures using genomic DNA blot hybridization (unpublished results).

Lelu, M.A., 1987. Variations morphologiques et génétiques chez *Picea abies* obtenues après embryogenèse somatique – etude préliminaire In: Annales de Recherches Sylviocoles, AFOCEL, pp. 35–47.

Lester, D.T. and J.G. Berbee, 1977. Within-clone variation among black poplar trees derived from callus culture. Forest Sci. 23: 122–131.

Lewis-Smith, A.C., M. Chamberlain, S.M. Smith, 1990. Genetic and chromosomal variation in *Petunia hybrida* plants regenerated from protoplast and callus cultures. Biol. Plant. (Praha) 32: 247–255.

Lörz, H. and W.R. Scowcroft, 1983. Variability among plants and their progeny regenerated from protoplasts of Su/su heterozygotes of *Nicotiana tabacum*. Theor. Appl. Genet. 66: 67–75.

Maheswaran, G. and E.G. Williams, 1987. Uniformity of plants regenerated by direct somatic embryogenesis from zygotic embryos of *Trifolium repens*. Ann. Bot. 59: 93–97.

Manandhar, A. and P.M. Gresshoff, 1980. Blue spruce (*Picea pungens*) tissue and cell culture. Cytobios 29: 175–182.

McClintock, B., 1984. The significance of responses of the genome to challenge. Science 226: 792–801.

McCoy, T.J., R.L. Phillips and H.W. Rines, 1982. Cytogenetic analysis of plants regenerated from oat (*Avena sativa*) tissue culture: high frequency of partial chromosome loss. Can. J. Genet. Cytol. 24: 34–50.

Mehra, P.N. and M. Anand, 1983. Callus of *Pinus roxburghii* (Chir pine) and its cytology. Physiol. Plant 58: 282–286.

Mehra, P.N. and M. Anand, 1979. Cytology of callus of *Cryptomeria japonica* Don. Physiol. Plant. 45: 127–131.

Meins, F., 1983. Heritable variation in plant cell culture. Ann. Rev. Plant Physiol. 34: 327–346.

Merkle, S.A., P.L. Chou and H.E. Sommer, 1988. Stability of highly repeated sequences in the DNA of embryogenic cultures of yellow poplar. In: W.M. Cheliak and A.C. Yapa (Eds.), Molecular Genetics of Forest Trees, pp. 85–88. Petawawa National Forestry Institute Information Report PI-X-80.

Mo, L.H., S. von Arnold and U. Lagercrantz, 1989. Morphogenic and genetic stability in longterm embryogenic cultures and somatic embryos of Norway spruce (*Picea abies* {L.} Karst). Plant Cell Rep. 8: 375–378.

Moore, G.A. and R.E. Litz, 1984. Biochemical markers for *Carica papaya*, *C. cauliflora*, and plants from somatic embryos of their hybrid. J. Am. Soc. Hort. Sci. 109: 213–218.

Muller, E., P.T.H. Brown, S. Hartke and H. Lorz, 1990. DNA variation in tissue-culture-derived rice plants. Theor. Appl. Genet. 80: 673–679.

Nagl, W., 1990. Gene amplification and related events. In: Y.P.S. Bajaj (Ed.), Biotechnology in Agriculture and Forestry, pp. 153–201. Springer-Verlag, Berlin.

Nkongolo, K.K. and K. Klimaszewska, 1994. Karyotype analysis and optimization of mitotic index in *Picea mariana* (black spruce) preparations from seedling root tips and embryogenic cultures. Heredity 73: 11–17.

Noh, E.-W. and S.C. Minocha, 1990. Pigment and isozyme variation in aspen shoots regenerated from callus culture. Plant Cell Tiss. Org. Cult. 23: 39–44.

Ochatt, S.J. and J.B. Power, 1989. Selection for salt and drought tolerance in protoplast- and explant-derived tissue cultures of Colt cherry (*Prunus avium* × *pseudocerasus*). Tree Physiol. 5: 259–266.

Ostry, M.E. and D.D. Skilling, 1988. Somatic variation in resistance of *Populus* to *Septoria musiva*. Plant Dis. 72: 724–727.

Papes, D., V. Besendorfer and M. Pavlica, 1991. Nuclear changes in European black pine seedlings caused by growth regulators. Acta Bot. Croat. 50: 31–36.

Partanen, C., 1963. Plant tissue culture in relation to cytology. Int. Rev. Cytol. 15: 215–243.

Patel, K.R. and G.P. Berlyn, 1982. Genetic instability of multiple buds of *Pinus coulteri* regenerated from tissue culture. Can. J. For. Res. 12: 93–101.

Peschke, V.M. and R.L. Phillips, 1991. Activation of the maize transposable element Suppressor-mutator (*Spm*) in tissue culture. Theor. Appl. Genet. 81: 90–97.

Prakash, C.S. and B.A. Thielges, 1989. Somaclonal variation in eastern cottonwood for race-specific partial resistance to leaf rust disease. Phytopathol. 79: 805–808.

Prat, D., 1983. Genetic variability induced in *Nicotiana sylvestris* by protoplast culture. Theor. Appl. Genet. 64: 223–230.

Radojevic, L.N., N. Djordjevic and M. Guc-Scekic, 1988. *In vitro* embryogenic callus formation in *Chimonanthus*. In: M.R. Ahuja (Ed.), Somatic Cell Genetics of Woody Plants, pp. 51–52. Kluwer Academic Publishers, Dordrecht.

Radojevic, L. and S. Stankovic, 1988. Plant regeneration of horse chestnut by *in vitro* culture. In: M.R. Ahuja (Ed.,) Somatic Cell Genetics of Woody Plants, p. 53. Kluwer Academic Publishers, Dordrecht.

Rao, P.S., V.A. Bapat and M. Mhatre, 1984. Regulatory factors for *in vitro* multiplication of sandalwood tree (*Santalum album*) 2. Plant regeneration in nodal and internodal stem explants and occurrence of somaclonal variations in tissue culture raised plants. Proc. Indian Nat. Acad. Sci. Biol. 50: 196–202.

Renfroe, M.H. and G.P. Berlyn, 1985. Variation in nuclear DNA content in *Pinus taeda* L. tissue cultures of diploid origin. J. Plant Physiol. 121: 131–139.

Renfroe, M.H. and G.P. Berlyn, 1984. Stability of nuclear DNA content during adventitious shoot formation in *Pinus taeda* L. tissue culture. Am. J. Bot. 71: 268–272.

Rode, A., C. Hartmann, J. De Buyser and Y. Henry, 1988. Evidence for a direct relationship between mitochondrial genome organization and regeneration ability in hexaploid wheat somatic tissue cultures. Curr. Genet. 14: 387–394.

Salmia, M.A., 1975. Cytological studies on tissue culture of *Pinus cembra*. Physiol. Plant. 33: 58–61.

Schaeffer, G.W., 1982. Recovery of heritable variability in anther-derived doubled-haploid rice. Crop Sci. 22: 1160–1164.

Schuller, A., G. Reuther and T. Geier, 1989. Somatic embryogenesis from seed explants of *Abies alba*. Plant Cell Tiss. Org. Cult. 17: 53–58.

Shirzadegan, A., J.D. Palmer, M. Christey and E.D. Earle, 1991. Patterns of mitochondrial DNA instability in *Brassica campestris* cultured cells. Plant Mol. Biol. 16: 21–37.

Shoemaker, R.C., L.A. Amberger, R.G. Palmer, L. Oglesby and J.P. Ranch, 1991. Effect of 2,4-dichlorophenoxyacetic acid concentration on somatic embryogenesis and heritable variation in soybean *Glycine max* L. Merr. In Vitro Cell Dev. Biol. Plant 27P(2): 84–88.

Sondahl, M.R., 1982. Tissue culture of morphological mutants of coffee. In: A. Fujiwara (Ed.), Plant Tissue Culture, pp. 417–418. Jpn. Assoc. Plant Tissue Cult., Tokyo.

Thomas, E., S.W.J. Bright, J. Franklin, V. Lancaster and B.J. Miflin, 1982. Variation amongst protoplast-derived potato plants (*Solanum tuberosum* cv. "Maris Bard"). Theor. Appl. Genet. 62: 65–68.

Thorpe, T.A., 1988. *In vitro* somatic embryogenesis. In: *In Vitro* Somatic Embryogenesis. ISI Atlas of Science, Animal and Plant Sciences, pp. 81–88.

Tominaga, Y. and Oga, M., 1970. Cytological studies on the calli of *Pinus densiflora in vitro*. Bull. Hiroshima Agric. Coll. 4: 8–10.

Tulecke, W. 1957. The pollen of *Ginkgo biloba*: *in vitro* culture and tissue formation. Am. J. Bot. 41: 602–608.

Von Aderkas, P. and P. Anderson, 1993. Aneuploidy and polyploidization in haploid tissue cultures of *Larix decidua*. Physiol. Plant. 88: 73–77.

Walbot, V., 1985. On the life strategies of plants and animals. Trends Genet. 1: 165–169.

Wann, S.R., M.A. Johnson, R.P. Feirer and T.L. Noland, 1986. Norway spruce as a model system for somatic embryogenesis in conifers. Proc. TAPPI Res. Devel. Conf., Forest Biotech., Raleigh, NC, Sept. 28–Oct. 1, pp. 131–135. TAPPI Press, Atlanta, GA.

Webster, F.B., D.R. Roberts, S.M. McInnis and B.C.S. Sutton, 1990. Propagation of interior spruce by somatic embryogenesis. Can. J. For. Res. 20: 1759–1765.

Williams, J.G.K., A.R. Kubelik, K.J. Livak, J.A. Rafalski and S.V. Tingey, 1990. DNA polymorphisms amplified by arbitrary primers are useful as genetic markers. Nucleic Acids Res. 18: 6531–6535.

Wright, J.W., 1976. Introduction to Forest Genetics. Academic Press, NY.

Wyman, J., N. Brassard, D. Flipo and S. Laliberté, 1992. Ploidy level stability of callus tissue,

axillary and adventitious shoots of *Larix × eurolepis* Henry regenerated *in vitro*. Plant Sci. 85: 189–196.
Zong-Xui, S., A. Cheng-Zhang, Z. Kang-Le, Q. Xui-Fang and F. Yaping, 1983. Somaclonal genetics of rice, *Oryza sativa* L. Theor. Appl. Genet. 67: 67–72.

18. Mutation Work with Somatic Embryogenesis in Woody Plants

Berthold Heinze and Josef Schmidt

Contents

1. Introduction

Mutation breeding is a common practice for many crop plants. Lists of successfully introduced new varieties grow longer year by year (Maluszynski *et al.*, 1991, 1992). For woody plants, mutation breeding is practised in fruit trees like *Citrus* spp. or apple. On the other hand, in forest trees, this approach is not practised at all. With the growing number of applications of powerful *in vitro* methods for woody plants, such as somatic embryogenesis or *Agrobacterium* DNA transfer, new possibilities have opened up for mutation breeding, too. However, up to now, mutation work has rarely been applied to woody plants in conjunction with somatic embryogenesis. In this review, we will, therefore, discuss underlying principles of mutation breeding and cite few examples where somatic embryogenesis has been involved in induced mutations (mainly in *Citrus* spp.).

Throughout this chapter, we will consider "classical" induced mutation

S. Jain, P. Gupta & R. Newton (eds.), Somatic Embryogenesis in Woody Plants, Vol. 1, 379–398.

(by chemical, physical, or biological agents) as applied *in vitro*; somatic hybridization techniques, work with haploid material (e.g., anther culture) or examples of selection for pre-existing variation among e.g., protoplasts will not be discussed.

We also present an outlook of future prospects for these techniques based on our assessment of possibilities and constraints.

2. What Is Mutation Breeding, and How Does It Work?

Mutations, together with recombinations, are the driving forces of evolution of all organisms. They occur naturally, due to a number of causes. By inducing mutations with physical, chemical, or biological means, a plant breeder tries to increase the frequency of mutations. This results in experimental populations with a larger proportion of mutants than occurring naturally, thus allowing to obtain larger numbers of mutants for screening. Mostly, spontaneous mutants occur at very low frequencies in natural populations, so that their number is too small to base breeding programs entirely upon them. It has to be stressed that there is no qualitative difference between experimentally induced and naturally occurring mutations (Wright, 1976; Yamaguchi, 1991).

2.1. *Molecular Basis*

Mutagenic agents affect the DNA molecule directly or indirectly by causing a range of responses from cell death to genome or chromosome mutations to single gene mutations. The reaction pathways differ for ionizing and UV radiations, and different groups of chemical mutagens. Biological mutagenesis, as caused by *Agrobacterium* T-DNA, will be discussed separately (see Section 2.4). Comprehensive or concise descriptions can be found in biology textbooks as well as in dedicated reviews (e.g., Yamaguchi, 1991; Brunner, 1991) and will only be summarized below.

2.1.1. *Ionizing Radiation*

Ionizing radiations applied to biological targets are characterized by the linear energy transfer (LET) along the ionizing track. Low-LET or sparsely ionizing radiations produce only a few ionizations per μm of path (for instance, X-rays or gamma rays). They favour reactions between the radicals that they induce and solute molecules (oxidations). High-LET or densely ionizing radiations (neutrons, alpha particles, protons) leave a dense ionization track in the tissue, where radical-radical reactions (peroxidations) are favoured. Low LET radiations cause the formation of radical species – either directly in the DNA or by the radiolysis of e.g., water – which then in turn react with DNA (and other biomolecules). This leads to strand break and DNA degradation. Oxygen enhances the response of tissues to low-LET radiations. As the water content affects oxygen availability, it should be under experimental control as well. For instance, gamma and X-rays induce 40% damage by direct DNA backbone breakage and 60% damage by secondary radical reactions (H. Brunner, personal communication). Conditions for dosimetry as well as nature and source of radiation have to be specified for maximum reproducibility.

Broken DNA strands may cause "bypass synthesis" without the appropriate template in surviving cells, leading to different forms of chromosome aberrations, or may be degraded at progressively higher doses of radiation. Chromosome aberrations play an important role in the case of vegetatively propagated plants.

The unit of the absorbed dose of radiation energy is 1 Gy (Gray), equivalent to 1 J kg^{-1}. Biological effects are also dependent on cell synchronisation, and therefore, on tissue irradiated, its metabolic state, and post-treatment. A standard test procedure is to apply 220 kVp (kilo Volt peak) X-ray for comparing biological effects of different treatments (Yamaguchi, 1991).

2.1.2. *Chemical Mutagenesis*

Mutagenic chemicals in practical use (also called "radiomimic substances") such as ethylene imine or sulphonates can add alkyl groups to N-bases, preferentially to guanine (G), causing loss of the base or mispairing. This further leads to incorporation of wrong bases into DNA during replication, i.e., base substitutions. The same applies for base analogues like 5-bromouracil which can pair with more than one base.

2.1.3. *Ultra-Violet Light*

UV radiation, most effective at 260 nm wavelength, induces the formation of dipyrimidine dimers (T-T). The penetrating power of UV is very limited, and so is its application limited to small biological units as pollen grains or cell cultures.

2.1.4. *DNA Repair*

DNA damage may be repaired by enzyme systems *in vivo*, working either at the replication fork or throughout the genome. In visible light, an enzymatic system called photoreactivation dissociates UV cross-linked pyrimidines. A different initial step is the excision of the dimerized or modified bases (a three-base offset leads to double strand break). Next, the broken single strand is degraded, and the gap refilled by DNA polymerases and ligases. These mechanisms require that the tissue is in an active metabolic state.

If the repair is not completed before replication, it is prone to errors with respect to the original base sequence. Repair mechanisms are adaptive traits to DNA damage. DNA is not the only biomolecule affected by mutagenic treatment. For instance, enzymes reacted to gamma and beta irradiations (Lickl *et al.*, 1987a,b).

2.2. *Methodology Basics*

For more comprehensive chapters on mutation methods, the reader is referred to the Mutation Breeding Manual (IAEA, 1977) and reviews by Brunner (1991) on general methodology, and Novak (1991) and IAEA (1986) for *in vitro* mutation.

2.2.1. *Biological Targets and the Chimera Problem*

A prerequisite for mutation breeding is the definition of the target tissue/cell type. Meristems are the most widely used targets; with *in vitro* techniques, protoplasts or callus cells can serve as additional recipients (apart from organized tissue). It is also possible to treat pollen grains or zygotes. If the targets consist of more than one cell, chimera formation will be the likely result, as mutation is a unicellular phenomenon. Only single cells within the meristematic region will undergo a certain mutation. During further differentiation, only the sectors of the plant derived from the affected cells may show alterations. If chimera formation is to be expected, the smallest possible morphogenic units for induction and different selection schemes have to be applied (IAEA, 1986, see below for selection procedures).

2.2.2. *Experimental Procedures*

Next, all the factors that are known to influence mutagenic effects should be under control to decrease experiment variability. This includes cell cycle phase, physiological state, oxygen tension, media pH and redox potential, chemical environment for most efficient uptake and action, possible benefit from carrier substances such as dimethyl sulphoxide (DMSO), post-treatment conditions and many more (Brunner, 1991).

At the beginning of the treatment, the establishment of a dose-response curve helps to define a dose with acceptable effect that still allows for a sufficient proportion of the population to survive the treatment (Fig. 1). Usually, the "half-lethal" dose, LD_{50} (+/−10%), or a dose resulting in 20% survival are good starting points for experiments. *In vitro* cultures, due to the small size of individual units (cells, protoplasts, callus pieces, somatic embryos, etc.) will often require smaller doses than whole plants or plant organs do (James, 1987).

Then, a sufficiently large population of "plant morphogenic units" is treated under the conditions established in the pilot experiments. The size of the population depends on the mutation rate and the desired number of mutants. As induced mutagenesis works randomly, treatment of larger populations increases the probability of desired effects. Clearly, maximum population size depends on the available facilities for regeneration and selection. Populations of 10^5 surviving morphogenic units are recommended (IAEA, 1986). With the help of an *in vitro* selection method, even very large initial populations can be handled.

The terminology proposed by the IAEA (IAEA, 1986) reflects the number

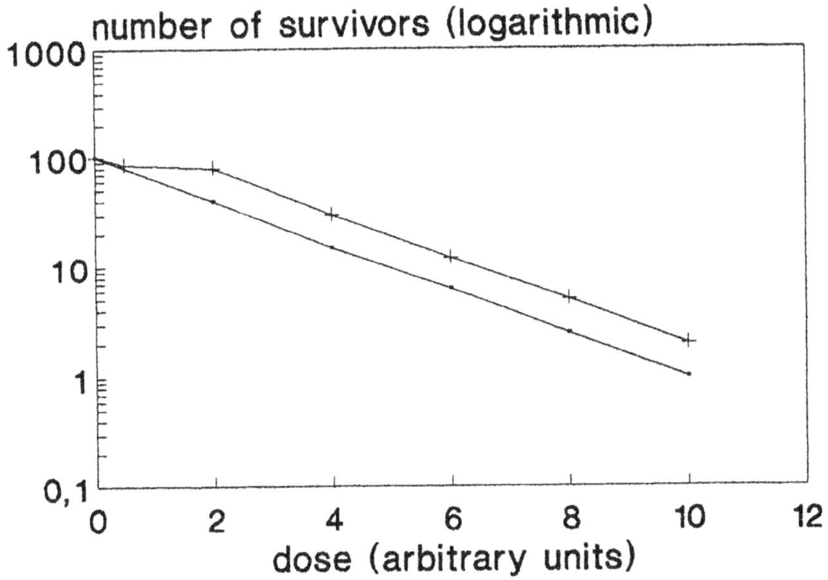

Figure 1. Typical dose-response curves for mutagenic treatments. "Shoulder" effects (upper curve) reveal a threshold level of sensitivity.

of regeneration cycles, vegetative or sexual. Counting of sexual generations starts with the mutagenized generation (M_1). Vegetative cycles of propagation after the mutagenic treatment are denoted by a V. For example, M_1V_2 specifies plant material of the second vegetative propagation cycle derived from the mutagenized generation.

2.2.3. *Regeneration and Selection*

Between treatment and selection, a recovery period depending on growth rate and physiological state of the treated material serves to manifest ("fix") mutations and "repair" other damage.

In general, induced mutations are recessive, as they result from disruptions of gene functions. Several cycles of propagation (sexual or vegetative) are necessary to obtain stable mutants. Over several generations, mutated and stable traits have to be recorded for a final evaluation of the mutants produced. For sexually propagated plants, selfing allows recessive mutations to be expressed, but they may also be lost by sexual recombination. Conventional breeding work may follow up in order to eliminate some undesired changes or to transfer single mutated genes.

In vegetatively propagated plants chimerism is a serious problem. Careful selection and separation of segments is necessary to reduce it. Systems prone to show chimerisms have to be cultivated for several vegetative generations

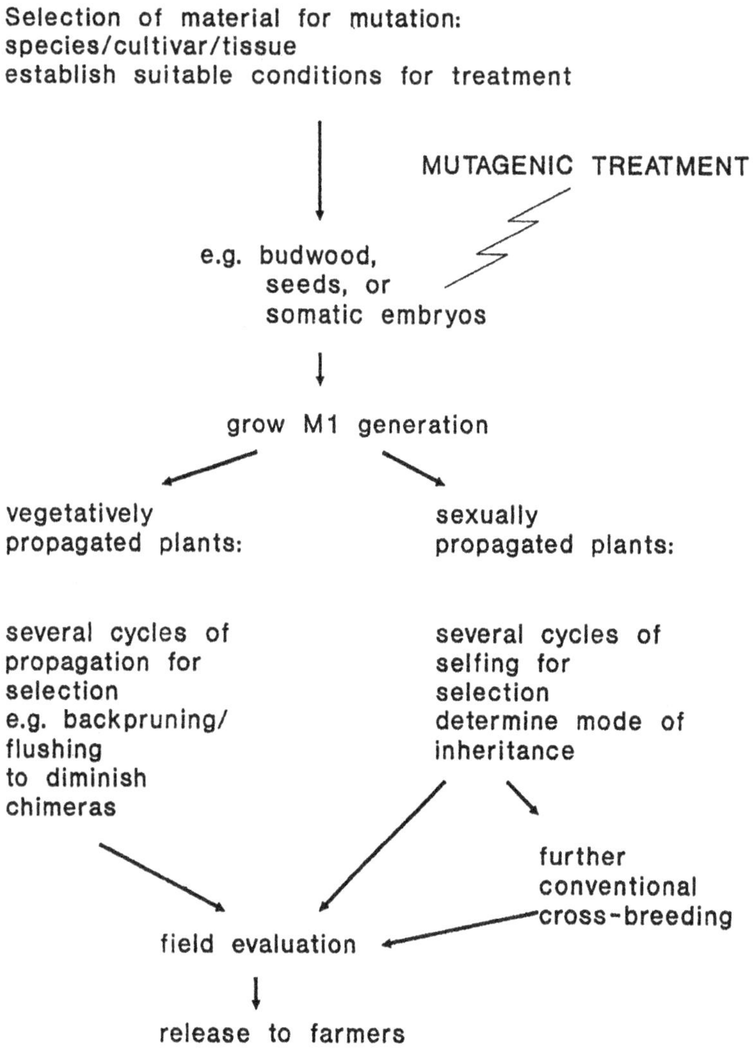

Figure 2. A general scheme of mutation breeding for woody perennials.

(e.g., M_1V_3) until fully mutated (homohistont) meristems are obtained (Fig. 2).

2.3. *Examples of Mutant Varieties*

The Mutation Breeding Newsletter published by the International Atomic Energy Agency (IAEA) in Vienna, Austria, regularly lists new varieties of

Table 1. Numbers of mutant varieties of woody plants obtained by conventional mutation breeding.[a]

Scientific Name	Common Name	Number
A) fruit trees		
Carica papaya L.	papaya	1
Citrus grandis L.	grapefruit	2
Citrus spp.	orange, mandarin	3
Cymbopogon winterianus	citronella	6
Ficus carica L.	fig	1
Malus pumila Mill.	apple	9
Morus alba L.	mulberry	3
Olea europea L.	olive	1
Prunus armenica L.	apricot	1
Prunus avium L.	sweet cherry	7
Prunus cerasus L.	sour cherry	4
Prunus domestica L.	plum	1
Prunus dulcis Webb.	almond	1
Prunus persica L.	peach	2
Punica granatum L.	pomegranate	2
Vitis vinifera L.	grape	1
B) ornamentals		
Alstroemeria sp.	alstroemeria	35
Bougainvillea sp.	bougainvillea	9
Ficus benjamina exotica	ficus	2
Forsythia × intermedia	forsythia	2
Hibiscus sp.	hibiscus	3
Hoya carnosa	hoya	4
Malus sp.	apple (flower)	1
Rhododendron sp.	azalea	15
Rosa sp.	rose	27

[a] Compiled from Maluszynski *et al.* (1991, 1992).

plants derived from induced mutagenesis released to farmers. Among them are fruit trees such as apple and cherry and several ornamental plants. Table 1 presents an overview. In fruit trees, not only new fruit characteristics are sought, but also altered growth types (e.g., dwarfism for rootstocks), self-compatibility, disease resistance, or seedlessness (in *Citrus*) are sometimes highly desired (Spiegel-Roy, 1990). For example, Brunner and Keppl (1991) developed an improved cultivar from Golden Delicious apples, termed Golden Haidegg, that had a better market value due to improved fruit characteristics. A little information is available, however, on *in vitro* mutation breeding in woody plants. Some of the rare examples of mutagenesis in temperate fruit trees *in vitro* have been summarized by James (1987). They included apple lines with altered vigor of the scion. Without respect to induced mutation, tissue culture procedures (including somatic embryogenesis) of woody plants serve to eliminate viruses from germplasms.

2.4. *Molecular Biology Approach: T-DNA from Agrobacterium*

An alternative method to the application of physical or chemical mutagens is the use of *Agrobacterium* T-DNA (Klee *et al.*, 1987). The *Agrobacterium* gene transfer system is widely used for inserting foreign DNA into plants. *Agrobacterium tumefaciens* and *A. rhizogenes* are pathogens for many plant species, causing formation of crown galls and tumors or hairy roots in infected plants, respectively. This is accomplished by the transfer of parts of a bacterial DNA plasmid (Ti, tumor inducing plasmid, or Ri, root inducing plasmid) into the host plant genome. The part of the plasmid that is integrated into the plant's own is called T-DNA. Its approximate size is 20 kb. It carries the genes responsible for tumor growth (coding for enzymes overproducing certain plant hormones), and genes for the production of unusual amino acids, which serve as nutrients exclusively to the bacteria.

T-DNA of various bacteria strains has been "disarmed" for plant transformation use, eliminating such sequences which are not essential for the integration into the plant genome. These constructs are widely used for introducing foreign DNA sequences. The integration of the T-DNA is random, so that plant genes may be disrupted and silenced in transformed cells, without concurrent tumor growth. Regeneration of whole plants from such transformed cells may lead to the identification of phenotypical mutants. Screening and selection proceeds in much the same way as described for chemical and physical mutagens.

2.4.1. *Mutant Gene Tagging*

With the help of the T-DNA "tag", the gene affected in the mutant can be studied. DNA from the mutant is subjected to "Southern blot" hybridization with a T-DNA probe. The plant DNA sequences flanking the integrated T-DNA are likely to be involved in the observed mutation (Klee *et al.*, 1987).

2.4.2. *Mutation Frequencies, Model Plants, and Woody Species*

As each transformed cell and its progeny will only contain one or a few T-DNA copies (Klee *et al.*, 1987), large numbers of transgenic plants have to be generated to identify "useful" mutants. Integration may occur in the repetitive parts of the plant genome and be ineffective. For these reasons, mutation frequencies are low, especially in plant species with a high proportion of repetitive DNA (such as many conifers [Dhillon, 1987]). Klee *et al.* (1987) described the use of a promotorless kanamycin resistance gene activated only upon integration into an active gene. Exposure to kanamycin allowed for selection of this mutation.

The T-DNA mutation system has been developed for *Arabidopsis thaliana* as a model plant (Feldmann, 1991). By transforming seed, large numbers of mutants were obtained. Yanofsky *et al.* (1990) identified a transcription factor necessary for flower development in an *Arabidopsis* mutant with the help of a T-DNA tag.

As *Agrobacterium* can also attack tree species, there is no reason why random mutation with T-DNA should not work in woody plants. Of course, there is still a lack of suitable systems for the screening of large numbers of trees for phenotype mutations. Segregation analysis, as conducted by Feldmann (1991), would take up to several decades. *In vitro* selection and molecular markers may help to save some of the time necessary for the long regeneration periods. Transgenic plants obtained after *Agrobacterium* transformation have been reported, for example, in poplar (Fillatti *et al.*, 1987), walnut (McGranahan *et al.*, 1988), hybrid aspen (Nilsson *et al.*, 1992), or larch (Huang *et al.*, 1991). In fact, host range experiments in which wild type or disarmed T-DNA was used have been successful for a large number of species (Knauf *et al.*, 1983), including conifers (Ellis *et al.*, 1989), and among them, pines (Sederoff and Stomp, 1993). Plant regeneration from transformed tissue has been achieved as well. For instance, Huang *et al.* (1991) reported on transformed adventitious shoots developing on stems of larch plants treated with wild-type *Agrobacterium rhizogenes*.

Problems specific to conifers are low infection rates (Sederoff and Stomp, 1993; Ellis *et al.*, 1989) and dependance on *Agrobacterium* strain and plant genotype. Genetic variation is high in most conifers. Genetically identical clonal material may supply sufficient numbers of susceptible plants for experiments. Furthermore, not only crown gall formation has to be demonstrated, but also regeneration of transformed plants, which is still a problem for many species and genotypes (Sederoff and Stomp, 1993). Crown galls are visible markers for transformed cells. In insertion mutagenesis with disarmed *Agrobacterium* strains, the unusual amino acids produced by introduced bacterial genes (such as nopaline or octopine) can serve for biochemical transformation assays. As common disarmed strains do not show the wide host species and genotypes range as wild type strains do (Ellis *et al.*, 1989), it may be necessary to disarm strains which proved to work. This involves a lot of effort. An alternative approach is the use of a binary system where a small extra plasmid carrying T-DNA is inserted into proven *Agrobacterium* strains (Klee *et al.*, 1987). Some plant cells attacked by these bacteria will only receive this additional DNA and possibly serve as insertional mutants. If this DNA carries genes for drug or herbicide resistance, they can be selected for *in vitro*.

T-DNA insertional mutagenesis has not yet been studied in tree species. It may be necessary to start experimental work with woody species with small genomes, for instance, with horse chestnut (*Aesculus hippocastanum* L.) or peach (*Prunus persica* L.).

3. Reasons for Applying Mutagens

Why and when is mutation breeding a method that should be taken into consideration for woody plant breeding?

In general, application of mutagens is not the first option for breeding new varieties. Classical cross-breeding will resolve many of a breeder's problems if:

- the species can be propagated sexually at all;
- large numbers of recombinants can be obtained;
- the generation time is reasonably short;
- variability in natural populations, available germplasms, or closely related species is high, and characters can be transferred from these;
- natural populations, available germplasms, or closely related species contain genes of interest;
- hybridization breeding is feasible.

However, there are cases where some of these prerequisites do not apply. Bananas and plantains (cultivated *Musa* sp.) can only be vegetatively propagated and are, therefore, not prone to sexual recombination. The same applies for other obligatory apomictic species. Many tropical trees cultivated outside their natural distribution range show markedly low genetic variation. Rubber tree plantations in Malaysia were initially started with 200 plants imported from Kew Gardens, London (Low, 1993). Many crops are threatened by pests or pathogens that they have not encountered during their evolution, and therefore no resistance genes are available. Many ornamental trees originally growing in forest surroundings have generation times too long to make cross-breeding worthwhile.

Mutation breeding, therefore, may be applied in situations where desired traits cannot be found in the existing germplasms or cannot be transferred to the species or variety in question.

3.1.1. *Fruit, Ornamental, and Forest Trees*

For woody plants, this general rule applies to many fruit trees. New fruit varieties are essential for the consumer oriented markets. Resistance to pests and pathogens can lower production costs for farmers significantly, apart from reducing environmental risk from application of chemicals. Conventional cross-breeding is a lengthy task in many fruit trees due to prolonged periods of juvenility.

A similar situation is encountered by a breeder of ornamental trees. Interesting forms or variants are only too rarely found in natural populations. Slow growth, which might be desirable for conifers in a small garden, is strongly selected against under most natural conditions. Leaf colours different from "standard" green might cause lower photosynthetic efficiency due to altered or reduced chlorophyll production, and hence less competitiveness.

Quite different is the situation for trees cultivated in forests under conditions similar to natural woodlands. Most species with a wide distribution show high levels of genetic variation due to adaption to a range of climatic and environmental conditions as well as to certain pests and diseases. In such a situation, a breeder's targets can most likely be achieved by selection and cross-breeding to combine traits in one new variety (Wright, 1976). Even if there is a low level of genetic variation in a species (take for instance silver fir, *Abies alba*, in central Europe), germplasm preservation *in* or *ex situ* and later selection is preferred.

For most tree species, the assignment to one of these two groups will be evident. There are a few exceptions, for instance, christmas trees. They are descendants from ordinary forest tree species, but their use and to some degree their cultivation is more like ornamentals. Other trees are grown for their timber value as well as for crops, like edible chestnuts, and cultivation methods vary greatly. Such multi-purpose trees sometimes have unusual characteristics of genetic variation (Chalmers *et al.*, 1992).

4. Examples of Mutation Breeding in Woody Plants Involving Somatic Embryogenesis

Examples of the application of mutagenesis to embryogenic cultures of woody plants *in vitro* are few and mainly restricted to the genus *Citrus*. This is due to several simple facts:

Citrus is the one genus of woody plants where somatic (nucellar) embryogenesis has been studied intensively over the past decades (as reviewed, *inter alia*, in Barlass and Skene, 1986), in contrast to species where somatic embryogenesis has only been achieved much more recently, e.g., apple (*Malus domestica* Borkh., Liu *et al.*, 1983), cherry (rootstock GM9, *Prunus incisa* × *serrula*, Druart, 1981), *Persea americana* (Mooney and Van Staden, 1987), horse chestnut (*Aesculus hippocastanum* L., Radojevic, 1988), oak (*Quercus acutissima* Carruth, Sasaki *et al.*, 1988), grape (*Vitis vinifera* L., Matsuta and Hirabayashi, 1989), walnut (*Juglans regia* L., Polito *et al.*, 1989), the nut tree pecan (*Carya illinoensis* Wangenh., Merkle *et al.*, 1987), several indian forest trees (Mascarenhas and Muralidharan, 1989), or conifers (Becwar, 1993). For more examples, refer to Bajaj (1986, 1989, 1991).

The long time delay in many woody plants between embryo germination and assessment of the mature plant (bearing fruit, where appropriate) might also have caused researchers to be cautious on publishing early research results (note the exceptions e.g., Omar and Novak, 1990; Adu-Ampomah *et al.*, 1988).

Still another simple explanation is that specific DNA transfer has become feasible in recent years, which has resulted in a decline of interest in random mutagenesis (see review in Donini and Ancora, 1993).

The other major point is that the nucellar (maternal) seed tissue in *Citrus* offers the opportunity to propagate a proven mature genotype with known characteristics by somatic embryogenesis. This is not usually possible in other tree species, where in general the explant to start cultures with has to be in an early juvenile condition. Exceptions are somatic embryogenesis from mango (*Mangifera indica* L.) nucellus (Litz, 1986), and from sandalwood

(*Santalum album* L.) stem callus from a 20 year-old tree (Bapat *et al.*, 1985). Only recently, protocols for propagation of a conifer, Norway spruce (*Picea abies* (L.) Karst.), by somatic embryogenesis from 7- and 20-years-old plants have been introduced (Westcott, 1992, manuscript in preparation).

4.1.1. *Citrus*

The pioneering work was carried out in Israel on "Shamouti" orange (*Citrus sinensis*) (Kochba and Spiegel-Roy, 1977). Interestingly, non-proliferating callus became embryogenic after X-ray treatment. The selection for tolerance to salt (Kochba *et al.*, 1982) and 2,4-D (Spiegel-Roy *et al.*, 1983) was conducted.

Spina *et al.* (1991) irradiated embryogenic callus of sour orange and of "Cleopatra" mandarin. The results on the recovery of mutants are not yet available.

Similarily, Wan *et al.* (1991; see also Deng *et al.*, 1989) irradiated embryogenic *Citrus* calli with gamma rays (LD_{50}, 50–70 Gy). Protoplasts were more sensitive. They also treated embryogenic cultures and protoplasts with sodium azide and ethylmethane sulphonate (EMS). The treatments resulted in salt tolerant mutants selected for *in vitro*. Plantlets could be regenerated and are under test for salt tolerance.

Goldman and Ando (1990) determined the radiation sensitivity of protoplasts of orange cv. "Pera" as a first step to mutation breeding for citrus canker (a bacterial disease) resistance. The LD_{50} was found to be 37.5 Gy gamma rays.

Matsumoto and Yamaguchi (1984) obtained salt-tolerant cell lines of trifoliate orange (*Poncirus trifoliata*) by EMS-induced mutation of embryogenic callus obtained from seedlings. Also working with trifoliate orange, Beloualy and Bouharmont (1992) produced salt-tolerant plants by selection of protoplasts without mutagenesis.

Nito *et al.* (1989) used gamma irradiation to improve regeneration frequency of somatic embryos of different *Citrus* species and cultivars and found this characteristic increased by 2.7 to 16.4 times, depending on the genotype and the radiation dose.

The genetic fidelity within nucellar tissue has been discussed by Barlass and Skene (1986). Navarro *et al.* (1985), for example, reported on a high degree of variation obtained in plants of nucellar origin in *Citrus clementina*. Effects of mutagenic treatment, somaclonal variation, and pre-existing variation should not be confounded.

4.1.2. *Other Woody Species*

Omar and Novak (1990) described the uptake of EMS by somatic embryos of the date palm (*Phoenix dactylifera* L.), but did not report on mutants yet. The work is aimed at future evaluation of possible mutant plants.

Adu-Ampomah *et al.* (1988) irradiated immature cocoa (*Theobroma cacao* L.) beans and induced somatic embryos from them. Plantlets could be regenerated when tissue culture conditions were carefully optimized. Although plant regeneration frequencies were still rather low, tissue culture served to save both time and space significantly. Selection was aimed at resistance to swollen shoot disease.

5. General (Dis)Advantages of Inducing Mutations in Embryogenic Cultures of Woody Plants

From the examples cited above, some of the main advantages and even more of the problems of induced mutation *in vitro*, especially with somatic embryos, become evident. A reliable and potent tissue culture protocol is crucial to all mutation work *in vitro*; the whole process depends on this. Only with an established regeneration system it can be successful.

Application of the actual mutagens, then, is much easier and better controllable *in vitro* than *in vivo*. Reproducibility and standardization of conditions are much more straightforward. This is due to lower organizational complexity of *in vitro* plant material and applies to both physical and chemical agents. Gamma- and X-rays can penetrate commonly used tissue culture vessels. Therefore, cultures can be treated without removing them from the sterile containers. Due to radiation effects on the nutrient media, transfer of cultures to new media is advisable after radiation treatment. Only for UV treatment, cells have to be exposed without or with a special cover because UV rays cannot penetrate regular lids, and sterile conditions have to be maintained (Novak, 1991). Due to application problems because of geometrical arrangement (i.e., size of the treated tissue), chemical mutagenesis is now mostly practised *in vitro*. This simply involves addition of the mutagen to the culture media and exposing the cultures for appropriate times.

5.1.1. *In Vitro* Selection

In general, larger numbers of cells can be treated *in vitro* because the targets for mutagens (single protoplasts, meristematic regions of somatic embryos) can be packed more densely than, e.g., budwood. *Initial* culturing, including plant regeneration, requires less space. When transferring plant material to *ex vitro* conditions, this advantage is neutralized. Therefore, a selection pressure that can be applied *in vitro* to reduce population size is a powerful instrument in conjunction with tissue culture. This has been shown for salt tolerance (Kochba *et al.*, 1982; Wan *et al.*, 1991; Deng *et al.*, 1989; Matsumoto and Yamaguchi, 1984) and tolerance to 2,4-D (Spiegel-Roy *et al.*, 1983). Other selection pressures suited for *in vitro* culture are e.g., for heavy metal tolerance, altered pH and aluminum tolerance, different light regime, toxin and herbicide tolerance, pathogen and pest resistance (if co-culturing is possible), toxin resistance, cold tolerance, and the like (IAEA, 1986). Amino acid over-production altering nutritional quality of the plant can be selected for by including toxic levels of amino acids in the culture media

(IAEA, 1986). One should not forget that the regeneration capability is an important trait selected for during regeneration.

However, *in vitro* selection might be meaningless for some traits unless it is confirmed by field studies. Cell cultures behave different from whole plants in many aspects (especially when considering structural or membrane effects). In this respect, a somatic embryogenesis system may be more suitable since germinated somatic embryos might show reactions comparable to regular seedlings.

5.1.2. *Avoiding Chimeras*

Perhaps the most important point is that structures obtained from single irradiated cells help avoiding chimeras at later selection stages. Chimerism is one of the major hurdles to be overcome in the course of the development of a particular woody plant mutant. It often requires backpruning and growing of treated scions for several vegetative cycles. To make use of this advantage offered by *in vitro* cultures, the material is mutagenized before more differentiated structures (somatic embryos, but also adventitious buds originating from single cells) develop. If this is not possible in the primary explant, secondary embryogenesis can serve to circumvent this problem. In genetic transformation of spruce (*Picea glauca*) embryos posing the same problem, Ellis *et al.* (1993) could obtain pure transformed cultures in this way.

The smaller size of *in vitro* cultures, especially somatic embryos, proembryos in embryogenic cultures, or protoplasts, allows for a precise targeting of the mutagen. While excess amounts of mutagen might become trapped in e.g., the endosperm of seeds, somatic embryos offer the opportunity to wash it off after exposure.

Tissue culture also allows for more intense "aftercare". Monitoring of the cultures is easier. Detailed histological, physiological, and genetic studies sometimes require large amounts of soft, uncontaminated material.

After the mutagenic treatment, well developed somatic embryos can be germinated into whole plants, possibly via artificial seeds. In most cases, this requires less steps than obtaining plants from adventitious or lateral buds. For mutants, it also ensures that shoot as well as root systems are of the mutated genotype. Additional adventitious rooting of *in vitro* shoots might revert some desired mutations.

Once mutant embryogenic cell lines are obtained, mass production of plants can be envisaged all year round. This steady supply can be of economical importance for newly released cultivars.

However, in some woody plants, the time lag between the *in vitro* culture and the marketable plant is substantial. Somatic embryogenesis produces juvenile material, whereas e.g., fruiting is a feature of mature plant material. Juvenile-to-mature transition is still poorly understood for woody plants. This is already a big problem at the selection stage.

When applying mutagenesis *in vitro*, somaclonal variation must be kept in mind. Although it seemingly enhances variation in the first place, in later stages of the procedure (mass propagation of the mutant), it is not desired. Molecular studies like isozyme analysis (Eastman *et al.*, 1991) or random amplified polymorphic DNA (RAPD) (Brown *et al.*, 1993; Heinze *et al.*, submitted) can help to obtain clarity.

5.1.3. *Requirements for Successful Mutation*

It follows that *in vitro* mutation using somatic embryogenesis in woody plants is a reasonable approach to breeding if some points out of the following list will apply. Some of them are compulsory prerequisites, others enhance the probability of success:

- A narrow genetic pool of the fruit crop – desired characters cannot easily be transferred from existing germplasms.
- A high demand for new varieties. This applies especially to ornamentals (compare numbers of mutant varieties in Table 1, parts A and B), but also to certain fruit crops (Spiegel-Roy, 1990).
- A reliably working tissue culture protocol is a prerequisite.
- No or just a slight time lag between mutation treatment and assessment of variations and onset of selection. Speed of transition into the market-place with a new variety requires quick propagation and yields high economic rewards, thus justifying the costs of such a program.
- The ease of selecting *in vitro* for desirable traits (or vice versa, counter-selection against undesirable traits) and a good coincidence of *in vitro* and *in vivo* selection criteria.
- The chimera problem can be effectively avoided.

A number of the above mentioned points apply to fruit trees in general, either of temperate climate or tropical origin. Consumer demand for novel tastes in fruits and longer shelf-life justify continuous breeding work.

The other major impulse for tree breeding is improvement of growth and fruiting habits like compact growth, improved wood resistance to breakage, early fruiting and many more. Of equal importance is resistance of the trees and the fruits to pests and pathogens. Both novel and existing (e.g., locally important) varieties are likely targets. Although somatic embryogenesis protocols are becoming increasingly available for fruit tree species and their different cultivars, the biggest hurdle for this approach may be the selection procedure, because for most fruits, their growth and fruiting characteristics are assessed in mature plants, not in germinating embryos and seedlings.

With its pioneer role, *Citrus* is an example that allows for assessment of some of these points. Whereas some reports on mutagenesis of embryogenic cultures date from the early seventies, descriptions of novel varieties obtained from this work are not yet found in the literature. Early work concentrated on the "Shamouti" orange only and was conducted in just one laboratory (Barlass and Skene, 1986). The situation is different now, with many diverse

Citrus species under investigation in a growing number of laboratories. Interesting results are expected for the near future.

For woody ornamental plants, tissue culture protocols are rare. Mutation experience is abundant only for few species. In species where conventional propagation is slow, somatic embryogenesis will help to increase supplies and decrease the cost. Characters assessable in seedlings (e.g., novel leaf shape and colour) will increase the acceptance of the method.

Some forest trees more recently domesticated, like christmas trees or any ornamental varieties, could benefit from quick true-to-type propagation by somatic embryogenesis in contrast to grafting (e.g., in Colorado blue spruce [*Picea pungens* Engelm.]). Whether mutation is inducing interesting changes in conifers remains to be shown. When clonally propagated trees are cultivated under forest-like conditions (with less intense care compared to gardens or parks), problems caused by pests and pathogens may arise due to their uniformity.

6. Further Roles of Somatic Embryogenesis in Mutagenesis

Even if the mutagenic treatment is not applied *in vitro*, somatic embryogenesis can support mutation breeding programs. Somatic embryos, or plants regenerated, display a high genetic uniformity (the degree of it depending on possible somaclonal variation). Large populations can be obtained even from highly heterozygous and outbreeding species. This uniform material is an ideal population for mutagenesis treatment. Especially, if certain experiments have to be repeated, the constant supply of even-aged embryos or plants is advantageous.

Germplasm conservation *in vitro* will extend the period of availability of single genotypes. With cryo-preservation techniques in liquid nitrogen, embryogenic material can be stored for decades. While new varieties are extensively tested, the reference samples are still preserved. Stability after several generations "in the field" can be compared to preserved material.

At the point where new varieties are released, somatic embryogenesis will aid in the quick and continuous supply of large numbers of plants.

7. Conclusions

Somatic embryogenesis offers a number of options for mutation breeding, the most important ones being avoidance of chimeras and handling of large populations. For woody plants, this will be rewarding if suitable selection systems, preferably *in vitro*, are available. With the uniformity of somatic embryos, and later, plants, high numbers of starting material and mutagenized propagules are supplied. As high-value mass-produced plants will offer the biggest turnover, fruit trees and woody ornamentals will be the first groups benefitting from such approaches.

With increasing experience in induction and manifestation of mutations in woody plants – mainly by evaluation at the molecular level – more efficient modification of genomes can be anticipated, be it through mutagenesis or genetic transformation. We expect interesting results from this interaction of basic science, high-tech methodology, and applied plant breeding.

Acknowledgements

We thank F.J. Novak for providing access to his extensive literature collection, before he met his tragic death, and H. Brunner for advice on mutagenesis methodology and critical reading of the manuscript.

References

Adu-Ampomah, Y., F. Novak, R. Afza, M. van Duren and M. Perea-Dallos, 1988. Initiation and growth of somatic embryos of cocoa (*Theobroma cacao* L.). Café Cacao Thé 32: 187–199.

Bajaj, Y.P.S., 1986. Biotechnology in Agriculture and Forestry, Vol. 1: Trees I. Springer-Verlag, Berlin/New York.

Bajaj, Y.P.S., 1989. Biotechnology in Agriculture and Forestry, Vol. 5: Trees II. Springer-Verlag, Berlin/New York.

Bajaj, Y.P.S., 1991. Biotechnology in Agriculture and Forestry, Vol. 16: Trees III. Springer-Verlag, Berlin/New York.

Bapat, V.A, R. Gill and P.S. Rao, 1985. Regeneration of somatic embryos and plantlets from stem callus protoplasts of sandalwood tree (*Santalum album* L.). Curr. Sci. 54: 978–982.

Barlass, M. and K.G.M. Skene, 1986. Citrus (*Citrus* species). In: Y.P.S. Bajaj (Ed.), Biotechnology in Agriculture and Forestry, Vol. 1: Trees I, pp. 207–219. Springer-Verlag, Berlin/New York.

Becwar, M.R., 1993. Conifer somatic embryogenesis and clonal forestry. In: M.R. Ahuja and W.J. Libby, Clonal Forestry, Vol. I. Genetics and Biotechnology, pp. 200–223. Springer-Verlag, Berlin/New York.

Beloualy, N. and J. Bouharmont, 1992. NaCl-tolerant plants of *Poncirus trifoliata* regenerated from tolerant cell lines. Theor. Appl. Genet. 83: 509–514.

Brown, P.T.H., F.D. Lange, E. Kranz and H. Lörz, 1993. Analysis of single protoplasts and regenerated plants by PCR and RAPD technology. Mol. Gen. Genet. 237: 311–317.

Brunner, H., 1991. Methods of induction of mutations. In: A.K. Mandal, P.K. Ganguli and S.P. Banerjee (Eds.), Advances in Plant Breeding, Vol. 1, pp. 187–220. CBS Publishers and Distributors, Delhi.

Brunner, H. and H. Keppl, 1991. Radiation induced apple mutants of improved commercial value. In: Proc. Symp. on Plant Mutation Breeding for Crop Improvement. Vol. 1, pp. 547–552. IAEA, Vienna.

Chalmers, K.J., R. Waugh, J.I. Sprent, A.J. Simmons and W. Powell, 1992. Detection of genetic variation between and within populations of *Gliricidia sepium* and *G. maculata* using RAPD markers. Heredity 69: 465–472.

Deng, Z., W. Zhang and S. Wan, 1989. *In vitro* mutation breeding for salinity tolerance in *Citrus*. Mutation Breeding Newsletter (IAEA Vienna) 33: 12–14.

Dhillon, S.S., 1987. DNA in tree species. In: J.M. Bonga and D.J. Durzan (Eds.), Cell and Tissue Culture in Forestry, Vol. 1: General Principles and Biotechnology. Martinus Nijhoff Publishers, Dordrecht.

Donini, B. and G. Ancora, 1993. Improvement of vegetatively propagated crop plants through advanced breeding technologies. In: Proc. FAO/IAEA Seminar on The Use of Induced Mutation and Related Biotechnology for Crop Improvement in the Middle East and the Mediterranean Regions, p. 26. IAEA, Vienna.

Druart, Ph., 1981. Embryogenèse somatique et obtention de plantules chez *Prunus incisa* × *serrula* (GM9) cultivé *in vitro*. Bull. Rech. Agron. Gembloux 16: 205–220.

Eastman, P.A.K., F.B. Webster, J.A. Pitel and D.R. Roberts, 1991. Evaluation of somaclonal variation during somatic embryogenesis of interior spruce (*Picea glauca engelmanni* complex) using culture morphology and isozyme analysis. Plant Cell Rep. 10: 425–430.

Ellis, D.D., D.E. McCabe, S. McInnis, R. Ramachandran, D.R. Russell, K.M. Wallace, B.J. Martinelli, D.R. Roberts, K.F. Raffa and B.H. McCown, 1993. Stable transformation of *Picea glauca* by particle acceleration. Biotechnology 11: 84–89.

Ellis, D., D. Roberts, B. Sutton, W. Lazaroff, D. Webb and B. Flinn, 1989. Transformation of white spruce and other conifer species by *Agrobacterium tumefaciens*. Plant Cell Rep. 8: 16–20.

Feldmann, K.A., 1991. T-DNA insertion mutagenesis in *Arabidopsis*: mutational spectrum. Plant J. 1: 71–82.

Fillatti, J.J., J. Sellmer, B. McCown, B. Hassig and L. Comai, 1987. *Agrobacterium* mediated transformation and regeneration of *Populus*. Mol. Gen. Genet. 206: 192–199.

Goldman, M.H.S. and A. Ando, 1990. Radiosensitivity of protoplasts of orange (*Citrus sinensis*). Mutation Breeding Newsletter (IAEA Vienna) 36: 11.

Huang, Y., A.M. Diner and D.F. Karnosky, 1991. *Agrobacterium rhizogenes* mediated genetic transformation and regeneration of a conifer, *Larix decidua*. In Vitro 27P: 201–207.

IAEA, 1977. Manual on Mutation Breeding. Technical Report Series No. 119, 2nd Ed. IAEA, Vienna.

IAEA, 1986. *In vitro* technology for mutation breeding. Tecdoc 392. IAEA, Vienna.

James, D.J., 1987. Cell and tissue culture technology for the genetic manipulation of temperate fruit trees. Biotechnol. Genet. Engineer. Rev. 5: 33–56.

Klee, H., R. Horsch and S. Rogers, 1987. *Agrobacterium*-mediated plant transformation and its further applications to plant biology. Ann. Rev. Plant. Physiol. 38: 467–486.

Knauf, V.C., M.F. Yanofsky, M.P. Gordon and E.W. Nester, 1983. Genetic analysis of host range expression by *Agrobacterium*. In: A. Puhler (Ed.), Molecular Genetics of Bacteria-Plant Interaction, pp. 240–247. Springer-Verlag, Berlin.

Kochba, J. and P. Spiegel-Roy, 1977. Embryogenesis in gamma-irradiated habituated ovular callus of the "Shamouti" orange as affected by auxin and by tissue age. Environ. Exp. Bot. 17: 151–159.

Kochba, J., G. Ben-Hayyim, P. Spiegel-Roy, S. Saad and H. Neumann, 1982. Selection of stable salt-tolerant callus cell lines and embryos in *Citrus sinensis* and *C. aurantium*. Z. Pflanzenphysiol. 106: 111–118.

Lickl, E., R.H.F. Beck and R. Ebermann, 1987a. Response of peroxidase and amylase isoenzyme activities of *Aesculus hippocastanum* and *Picea abies* to gamma and beta irradiation. Phyton (Austria) 27: 177–180.

Lickl, E., G. Alth, K. Tuma and R. Ebermann, 1987b. Veränderungen der Aktivität von Amylaseenzymen aus der Roßkastanie (*Aesculus hippocastanum* L.) nach Gamma- und Betabestrahlung. Bodenkultur (Austria) 38: 299–304.

Litz, R.E., 1986. Mango (*Mangifera indica* L.). In: Y.P.S. Bajaj (Ed.), Biotechnology in Agriculture and Forestry, Vol. 1: Trees I, pp. 267–273. Springer-Verlag, Berlin/New York.

Liu, J.R., K.C. Sink and F.G. Dennis Jr., 1983. Adventive embryogenesis from leaf explants of apple seedlings. HortScience 18: 871–873.

Low, F.-C., 1993. The examination of *Hevea brasiliensis* plants produced by *in vitro* culture and mutagenesis by DNA fingerprinting techniques. Paper presented at a Research Coordination

Meeting of the FAO/IAEA Joint Division on Nuclear Techniques in Food an Agriculture, May 10–14, 1993, Vienna.

Maluszynski, M., B. Sigurbjörnsson, E. Amano, L. Sitch and O. Kamra, 1991. Mutant varieties – data bank (FAO/IAEA database). Mutation Breeding Newsletter (IAEA, Vienna) 38: 16–49.

Maluszynski, M., B. Sigurbjörnsson, E. Amano, L. Sitch and O. Kamra, 1992. Mutant varieties – data bank (FAO/IAEA database part II). Mutation Breeding Newsletter (IAEA, Vienna) 39: 14–33.

Mascarenhas, A.F. and E.M. Muralidharan, 1989. Tissue culture of forest trees in India. Curr. Sci. 58: 606–613.

Matsumoto, K. and H. Yamaguchi, 1984. Increased variation of NaCl- tolerance in adventitious embroids of trifoliate orange using an *in vitro* technique. Rev. Brasil. Genet. (Brazil. J. Genetics) VII: 73–81.

Matsuta, N. and T. Hirabayashi, 1989. Embryogenic cell lines from somatic embryos of grape (*Vitis vinifera* L.). Plant Cell Rep. 7: 684–687.

McGranahan, G.H., C.A. Leslie, S.L. Uratsu, L.A. Martin and A.M. Dandekar, 1988. *Agrobacterium*-mediated transformation of walnut somatic embryos and regeneration of transgenic plants. Bio/Technology 6: 800–804.

Merkle, S.A., H.Y. Wetzstein and H.E. Sommer, 1987. Somatic embryogenesis in tissue cultures of pecan. HortScience 22: 128–130.

Mooney, P.A. and J. van Staden, 1987. Induction of embryogenesis in callus from immature embryos of *Persea americana*. Can. J. Bot. 65: 622–626.

Navarro, L., J.M. Ortiz and J. Juarez, 1985. Aberrant *Citrus* plants obtained by somatic embryogenesis of nucelli cultured *in vitro*. HortScience 20: 214–215.

Nilsson, O., T. Aldén, F. Sitbon, C.H.A. Little, V. Chalupa, G. Sandberg and O. Olsson, 1992. Spatial pattern of cauliflower mosaic virus 35S promoter-luciferase expression in transgenic hybrid aspen trees monitored by enzymatic assay and non-destructive imaging. Transgenic Res. 1: 209–220.

Nito, N., J.-T. Ling, M. Iwamasa and Y. Katayama, 1989. Effects of gamma-irradiation on growth and embryogenesis of *Citrus* callus. J. Japan. Soc. Hort. Sci. 58: 283–287.

Novak, F.J., 1991. *In vitro* mutation system for crop improvement. In: Proc. Symp. on Plant Mutation Breeding for Crop Improvement, Vol. 2, pp. 327–342. IAEA, Vienna.

Omar, M.S. and F.J. Novak, 1990. *In vitro* plant regeneration and ethylmethane sulphonate (EMS) uptake in somatic embryos of date palm (*Phoenix dactylifera* L.). Plant Cell Tiss Org Cult. 20: 185–190.

Polito, V.S., G. McGranahan, K. Pinney and C. Leslie, 1989. Origin of somatic embryos from repetitively embryogenic cultures of walnut (*Juglans regia* L.): implications for *Agrobacterium*-mediated transformation. Plant Cell Rep. 8: 219–221.

Radojevic, L., 1988. Plant regeneration of *Aesculus hippocastanum* L. (horse chestnut) through somatic embryogenesis. J. Plant Physiol. 132: 322–326.

Sasaki, Y., Y. Shoyama, I. Nishioka and T. Suzaki, 1988. Clonal propagation of *Quercus acutissima* Carruth by somatic embryogenesis. J. Fac. Agr. Kyushu Univ. 33: 95–101.

Sederoff, R.R. and A.M. Stomp, 1993. DNA transfer in conifers. In: M.R. Ahuja and W.J. Libby (Eds.), Clonal Forestry I: Genetics and Biotechnology, pp. 241–254. Springer-Verlag, Berlin.

Spiegel-Roy, P., 1990. Economic and agricultural impact of mutation breeding in fruit trees. Mutation Breeding Review (IAEA, Vienna) 5: 1–26.

Spiegel-Roy, P., 1991. Economic and agricultural impact of mutation breeding in fruit trees. In: Proc. Symp. on Plant Mutation Breeding for Crop Improvement, Vol. 1, pp. 215–236. IAEA, Vienna.

Spiegel-Roy, P., J. Kochba and S. Saad, 1983. Selection for tolerance to 2,4-dichlorophenoxyacetic acid in ovular callus of orange (*Citrus sinensis*). Z. Pflanzenphysiol. 109, 41–48.

Spina, P., P. Mannino, G. Reforgiato Recupero and A. Starrantino, 1991. Use of mutagenesis at the Istituto Sperimentale per l'Agrumicoltura, Acireale. In: Proc. Symp. on Plant Mutation Breeding for Crop Improvement, Vol. 1, pp. 257–261. IAEA. Vienna.

Wan, S.Y., Z.A. Deng, X.X. Deng, X.R. Ye and W.C. Zhang, 1991. Advances made in *in vitro* mutation breeding in *Citrus*. In: Proc. Symp. on Plant Mutation Breeding for Crop Improvement, Vol. 1, pp. 263–270. IAEA, Vienna.

Westcott, R.J., 1992. Embryogenesis from non-juvenile Norway spruce (*Picea abies*). In Vitro 28: 101A.

Wright, J.W., 1976. Introduction to Forest Genetics. Academic Press, New York.

Yamaguchi, H., 1991. Mutation: History, classification and theories. In: A.K. Mandal, P.K. Ganguli and S.P. Banerjee (Eds.), Advances in Plant Breeding, Vol. 1, pp. 169–186. CBS Publishers and Distributors, Delhi.

Yanofsky, M.F., H. Ma, J.L. Bowman, G.N. Drews, K.A. Feldmann and E.M. Meyerowitz, 1990. The protein encoded by the *Arabidopsis* homeotic gene *agamous* resembles transcription factors. Nature 346: 35–39.

19. Prospects and Limits of Somatic Embryogenesis of *Picea abies*

M. Pâques, J. Bercetche and M. Palada

Contents

1. Introduction

Picea abies is one of the major conifers in Northern Europe. It covers approximately 25 million hectares of European Union (EEC) and European Free Trade Association (ALE) lands.The reproduction cycle of *Picea abies* is quite long with the first flowering usually recorded 20 years after seeding. Moreover, abundant fructification frequency is low. It is observed on an average only once every 5 years, depending on the climatic conditions. Therefore, most *Picea abies* breeding programs have, until recently, mainly focused on a clonal strategy (Nanson, 1986; Kleinschmit and Svolba, 1989; Monchaux, 1989) despite the greater interest devoted to sexual reproduction and breeding (Werner *et al.*, 1986).

The clonal strategy consists of: 1) the selection of superior mature trees according to their morphology, wood characteristics, tolerance against pathogens, etc.; 2) multiclonal tests to confirm their selected properties; and finally 3) planting large-scale multiclonal mixtures for reforestation. This strategy depends on the use of vegetative propagation, usually by rooted cuttings. The main problem encountered for clonal propagation is the poor response of mature trees, the heterogeneity of rooting rate among genotypes, and often the plagiotropic development of the cuttings up to 3 years after rooting. Because of these problems with mature trees, attention has been focused on early selection of seedlings. This is possible because of the good relationship between juvenile (3 year old-seedlings) and adult traits (Nanson, 1987). This

S. Jain, P. Gupta & R. Newton (eds.), Somatic Embryogenesis in Woody Plants, Vol. 1, 399–414.

makes possible for an early selection in the progeny of open pollinated selected trees and, therefore, avoids the rooting difficulties associated with increasing chronological age of the tree. The major problem in the clonal strategy is how to maintain the selected material in a good rooting condition during the clonal test (10 to 15 years) and in a sufficient quantity to allow a large-scale propagation if the results of the clonal test are positive. It is also necessary to establish expensive clonal parks that need regular hedging to maintain and improve the rooting ability of the cuttings.

In order to reduce the cost of cuttings and shorten the time needed for rooted cutting propagation, the integration of tissue culture techniques with conventional propagation methods needs to be considered. Traditional micropropagation by axillary budding is ineffective, but somatic embryogenesis appears to be a powerful technique for large-scale propagation of *Picea abies*.

This paper summarises the state of the art of somatic embryogenesis for *Picea abies* and discusses the possibility of its use for large-scale propagation.

2. Potentialities of Somatic Embryogenesis

A general description of somatic embryogenesis will be limited in this paper because many good reviews have been, recently, published on this topic (Tautorus *et al.*, 1991; Roberts *et al.*, 1993; Gupta *et al.*, 1993). The main attention of this paper will be focused on the limitations of the technique.

Somatic embryogenesis consists of recovering embryos from sporophyte tissue without going through the gametic fusion. Somatic embryos can be successfully developed into plants (emblings) and grown in nursery (Tautorus *et al.*, 1991).

Somatic embryos can be obtained directly from the primary explant such as immature zygotic embryos, cotyledons, leaves or roots or more generally in conifers from secondary embryogenic tissues very often called *e*mbryogenic *s*uspensor *m*ass (ESM) (Gupta and Durzan, 1986).

In conifers, somatic embryogenesis is mainly initiated from early juvenile material such as immature zygotic embryos. Nevertheless the recovery of embryogenic tissues from needles has been, recently, reported in *Picea* (Ruaud *et al.*, 1992) and hybrid *Larix* (Lelu *et al.*, 1993).

Four main steps are generally recognised to induce somatic embryogenesis in conifers: 1) initiation using a culture medium containing NAA[1] and BA[2], 2) establishment and proliferation on a medium containing 2,4-D[3], 3) maturation in the presence of ABA[4], and 4) germination without growth substances

[1]NAA: Naphtalene acetic acid.
[2]BA: Benzylaminopurine.
[3]2,4-D: 2,4-Dichlorophenoxy acetic acid.
[4]ABA: Abscisic acid.

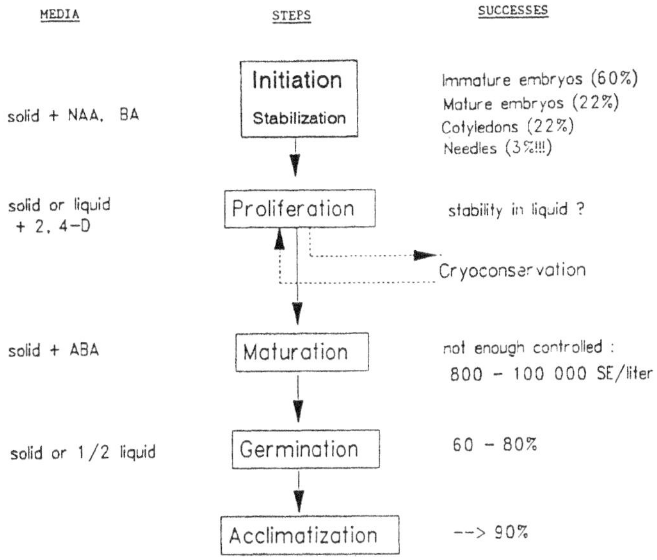

Figure 1. Main characteristics of somatic embryogenesis process: media, steps, successes.

(Fig. 1). All of these phases have been achieved on a solid medium, but proliferation can also be performed in a liquid medium using a bioreactor.

The main advantage of somatic embryogenesis is that the embryogenic tissues seem to retain their ability for embryo differentiation during several subcultures. Therefore, it is theoretically possible to produce an unlimited number of somatic embryos from one culture. The cost of those somatic plants (emblings) will be very cheap in comparison with micro cuttings, especially if bioreactor cultures are used. Indeed somatic embryo production in liquid medium can reach up to 10,000 per liter (Gupta *et al.*, 1993) without individual explant manipulation as required for micropropagation in each step of the process: micro cuttings proliferation, elongation and rooting. One other advantage is that the emblings are very similar to seedlings. They have a tap root system while micro cuttings develop only adventitious roots. Somatic embryos seem to grow like seedlings, while micro cuttings seem to grow like rooted cuttings from mature trees.

Moreover, somatic embryos can be used in synthetic seed production (Gupta *et al.*, 1993). Preliminary trials have been made by encapsulation of cotyledonary embryos in an alginate matrix enriched with fertilisers and containing fungicides and insecticides. Somatic seeds may avoid the costly embling weaning phase and could allow the seeding of "synseeds" directly in the forest as currently done with *Pinus pinaster* seeds in the Southwest of France (Chaperon, 1986).

Cryopreservation has already been successfully applied to *Picea abies*

(Bercetche *et al.*, 1990), *Picea sitchensis* (Find *et al.*, 1993), *Picea mariana* (Klimaszewska *et al.*, 1992), *Pinus taeda* (Gupta *et al.*, 1987), *Pinus çaribaea* (Laine *et al.*, 1992), *Douglas* fir (Gupta *et al.*, 1991). It will make easier the planning of *in vitro* production thus avoiding the maintenance of the selected material after intensive propagation. It also permits the maintenance in a juvenile status of future elite genotypes while they are being tested in the field (Bornman, 1993; Gupta *et al.*, 1993).

According to the method described, it appears that every step leading to embling production has been successfully developed. In actuality, few studies have combined many of these practices into a complete and an efficient propagation system. Somatic embryogenesis combined with cryopreservation would be helpful for *Picea* breeding and large-scale propagation. However, despite extensive research on this topic, no use in commercial scale exists as yet. The next point examines the main limitations of the technique.

3. Limitations of Somatic Embryogenesis

3.1. *Initiation and Culture Establishment*

The average success of embryogenic tissue initiation is 60, 22, 22 and 3% for immature embryos, mature embryos, seedling cotyledons, and needles from 14 month-old *Picea abies* plants, respectively (Ruaud *et al.*, 1992). Moreover, the embryogenic yield of emblings is 8 times greater than when compared with seedlings. The results from Association Foret-Cellulose (AFOCEL's) laboratories and the recent literature indicate that success in embryogenic tissue initiation is fully dependent on the chronological age of the sporophyte, the source of embryogenic tissue (hypocotyls, cotyledons, needles), and on the genotype (Ruaud *et al.*, 1992; Park *et al.*, 1993; Mo *et al.*, 1989).

Success in culture establishment is a critical step and is dependent on acquiring the skilled labour. It consists of harvesting and multiplication of single somatic embryos very often detectable at the surface of a non-embryogenic callus developed from the primary explant. During this step, 10 to 20 percent of the embryogenic culture (ESM) (Gupta and Durzan, 1986) can be lost (von Arnold and Woodward, 1988; Webb *et al.*, 1989).

The limited success in ESM establishment needs to be considered with attention. It will be necessary to reduce the genetic diversity of the varieties amplified by somatic embryogenesis. Moreover, initiation of somatic embryos from mature trees has not yet been demonstrated. However, we can hope for a rapid progress on the basis of the recent results obtained on mature trees of *Picea* (Westcott, 1992) and *Larix* (Bonga, 1992). Ten to twenty weeks are usually needed for ESM establishment.

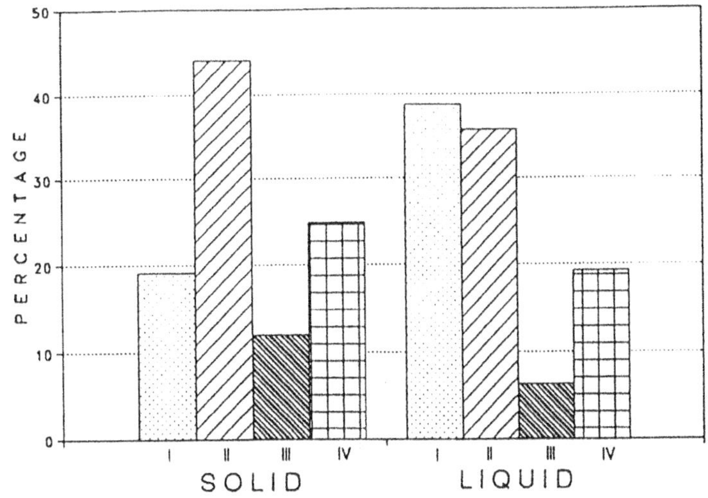

Figure 2. Relative distribution of somatic embryo classes cultured on solid or in liquid medium. Class I: initial stage; class II: precotyledonary stage; class III: globular stage; class IV: abnormal embryos.

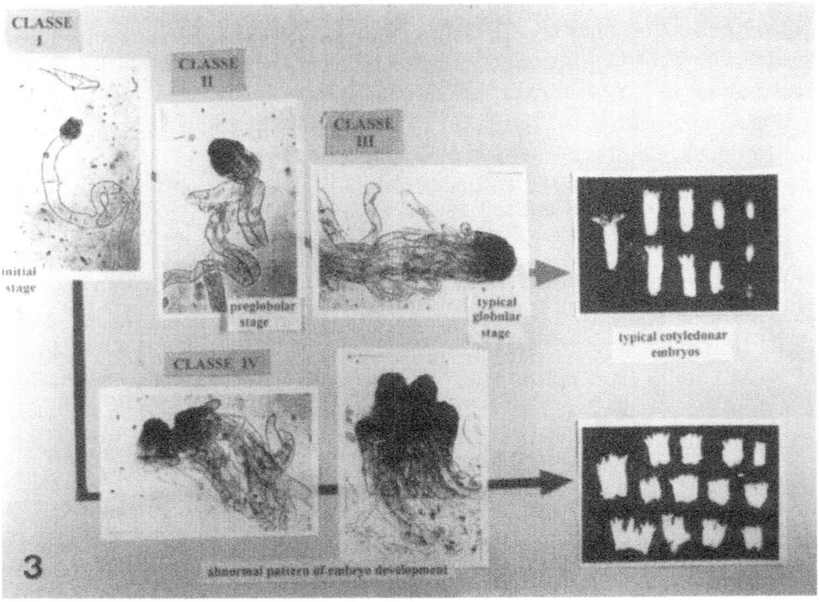

Figure 3. Evolution of the different classes observed in an ESM solid culture. Class I: initial stage; class II: preglobular stage; class III: globular stage; class IV: abnormal embryos.

3.2. *Proliferation of ESM*

The ESM are well organised and composed of elongated, vacuolated suspensor cells attached to dense cytoplasmic cells forming the "head" of the somatic embryos (Hakman *et al.*, 1985). Four major classes of somatic embryos can be recognised in ESM solid cultures (Fig. 2): class I) composed of a compact cluster of meristematic cells attached to an elongated vacuolated, unicellular cell suspensor (initial stage); class II) with a larger cluster of 16 to 64 meristematic cells attached to multicellular suspensor (preglobular stage); class III) with a typical globular head connected to a well developed suspensor (globular stage); and class IV) characterised by fasciated heads (abnormal embryos). This latter class leads to the maturation of aberrant embryos. Microscopic analysis of ESM cultures suggests that embryos from classes III and IV originate from classes I and II (Fig. 3).

After the culture establishment, the maintenance and multiplication of ESM are performed by transfer of the culture every 7 days onto a fresh solid medium. Mass propagation can also be obtained easily in a liquid medium, gently stirred in Erlenmeyer flasks, or in more sophisticated bioreactors.

ESM have been grown for two years on a solid medium without alteration of the proliferation rate. The stability in liquid cultures is lower as compared to cultures maintained on the solid medium (Tautorus *et al.*, 1990). However, the biomass production is 3 times higher in liquid as compared with solid cultures.

Despite the obvious interest in suspension cultures (homogeneity of the culture, reduction of labour, the possibility for automation . . .) for industrial production, liquid culture must be viewed with caution. Modifications in the distribution of the 4 classes of somatic embryo structures, previously described, have been observed (Fig. 2). Basic investigations should be carried out to develop better control of ESM morphogenesis.

3.3. *Maturation*

The number of cotyledonary embryos recovered from *Picea abies* ESM (one callus) varies from 8 to 30 (Becwar *et al.*, 1989; Bellarosa and von Arnold, 1992; Bercetche *et al.*, 1993) when cultured on a solid medium with ABA (2.5–100 ppm) that permits normal somatic embryo development. That means 40 to 60 cotyledonary embryos per gram of ESM. This number is low according to the number of potential embryogenic structures per gram of ESM (1,500 to 2,000).

Because maturation cannot be easily achieved in the liquid medium, many trials have been conducted to combine proliferation in the liquid medium with maturation on a solid one. Attree *et al.* (1990) obtained 3,000 to 4,000 cotyledonary embryos per litre of ESM suspension by placing ESM onto a cellulosic filter which is transferred onto a solid maturation medium. This procedure adapted by Gupta *et al.* (1993) enabled them to recover 50,000 to

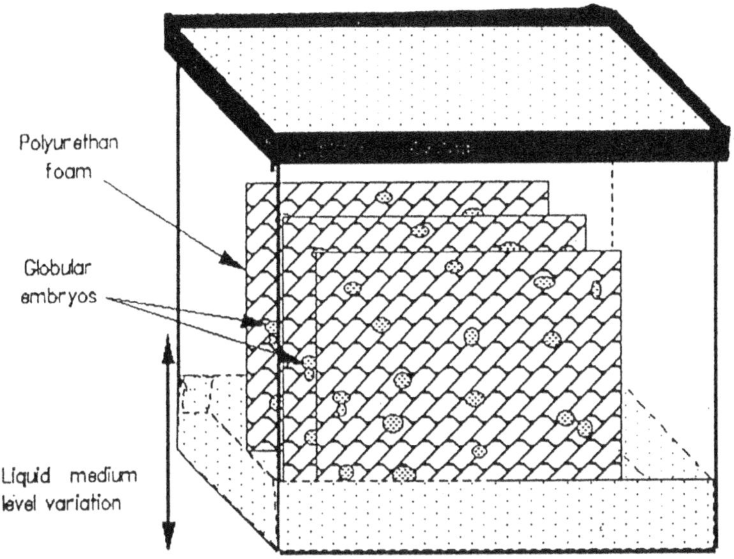

Figure 4. Bioreactor for somatic embryos maturation with a liquid medium. The level of the liquid medium and the immersion frequence can be adapted according to the embryo development.

100,000 cotyledonary embryos per litre of ESM suspension. Recently, a newly designed bioreactor has been described to recover cotyledonary embryos using liquid medium (Pâques *et al.*, 1992). The principle consists in regular immersion of ESM immobilised in polyurethane layers (Fig. 4). The level, the frequency and the period of immersion can be regulated according to the embryo development.

The yield of cotyledonary embryos can probably be increased by a better synchronisation of ESM suspension growing in controlled culture conditions. By using sophisticated bioreactors, the purpose to increase the homogeneity of the environmental conditions and to have a better understanding of the interactions between the culture medium (as an artificial endosperm) and the ESM can be achieved.

After 6 weeks of culture on a maturation medium, the cotyledonary embryos are harvested from ESM. Many studies have been performed on the mechanisation of embryo harvesting. One of the major problems is to separate somatic embryos from suspensor cells, while maintaining intact somatic embryos. Several systems have been proposed such as agitation in liquid medium with a magnetic stirrer (Gupta *et al.*, 1993) and low temperature or ultrasonic treatments (Pâques, 1993).

Presently, the best solution seems to be the harvest of embryos by hand. It provides the opportunity to combine harvesting and sorting of good embryos.

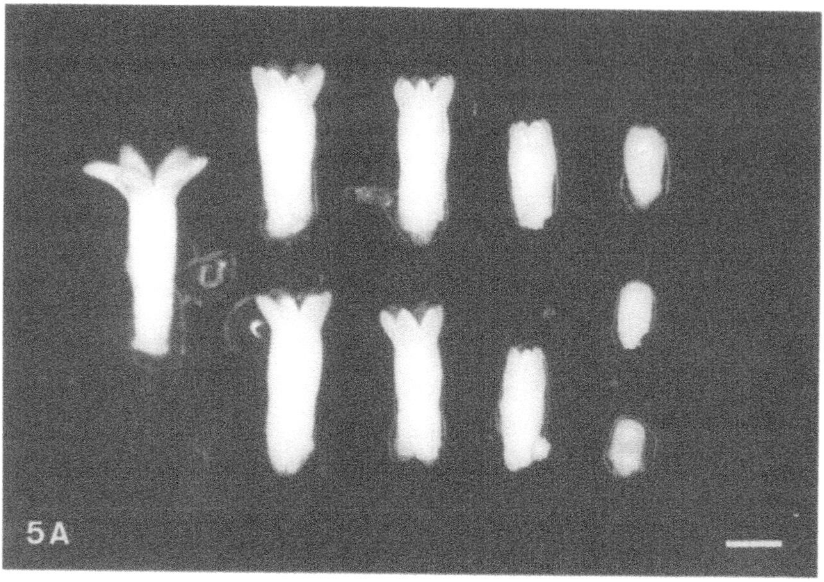

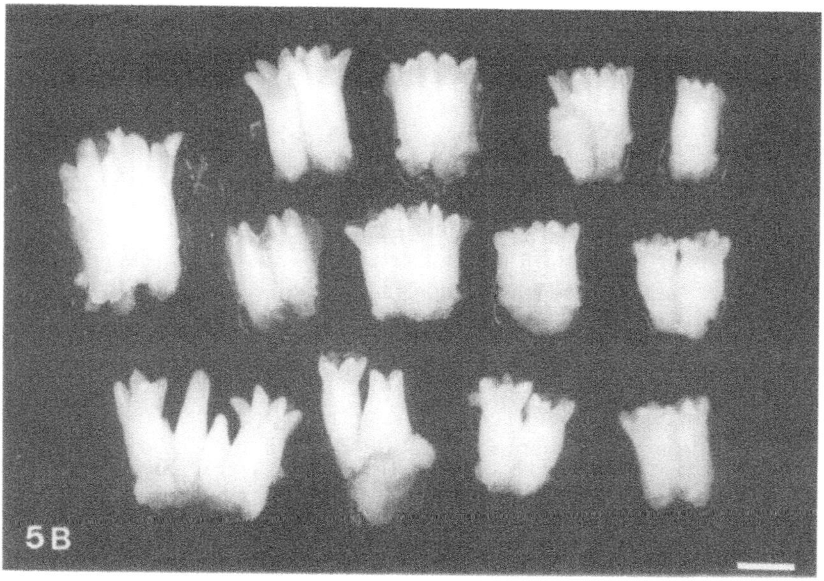

Figure 5. Non fasciated (A) and fasciated (B) cotyledonary somatic embryos observed after maturation (bar = 1 mm).

Indeed cotyledonary embryos are heterogeneous in shape, size, and sometimes attached to each other (Fig. 5).

Trials to mechanise embryo sorting are, presently, under investigation using image analysis. It has been successfully achieved for herbaceous plants (Rodriguez *et al.*, 1990), and is under development with conifers (Hämäläinen *et al.*, 1992). Some cheaper systems developed for herbaceous plants are, presently, being investigated including separation by sedimentation with a sucrose gradient (Janick *et al.*, 1993), or combining mechanical separation and sedimentation (Li, 1993). Janick *et al.* (1993) can separate celery somatic embryo stages from suspension mixtures using layered sucrose density gradients. They showed a good relation between the density and embryo size and their maturation level in relation with the lipid content. Li (1993) has developed a simple fractional method based on flow separation to separate carrot somatic embryos according to their morphology.

Maturation and embryo harvesting must be presently considered as an expensive step in the embryogenic process. Embryos of conifers are presently sorted manually one by one, according to their morphology. The incidence of sorting embryos will directly affect the germination and weaning steps.

3.4. *Germination and Weaning*

3.4.1. *Germination*

It has been determined that germination of *Picea abies* cotyledonary embryos is better in dark than in light (von Arnold and Hakman, 1988). The germination rate of selected cotyledonary embryos varies between 35 to 90 percent (von Arnold and Hakman, 1988; Becwar *et al.*, 1988; Gupta *et al.*, 1993). Germination rate of conifer embryos is strongly related to the storage protein content of the embryo (Roberts *et al.*, 1990). This can be influenced by a high relative humidity treatment (HRH) (Webster *et al.*, 1990) that avoids the rapid depletion of major storage proteins observed when somatic embryos of *Picea sitchensis* are directly transferred from maturation conditions onto a growth regulator-free medium (Roberts *et al.*, 1991).

After six days in darkness, emblings are usually placed on a sucrose medium in light, without growth regulators, under sterile conditions. After six culture weeks, emblings have developed a short epicotyle.

3.4.2. *Weaning*

Weaning is very important for the future *ex vitro* plant development. The success of weaning is dependent on the main factors : a) initial vigour of emblings, b) photoperiod length, c) relative humidity, and d) soil composition.

The success of weaning is highly dependent on the length of cotyledons and hypocotyl. The positive linear relationship between cotyledon and hypocotyl length and the success in weaning indicate that cotyledon and hypocotyl length can be considered as predictive markers of weaning success (Fig. 6)

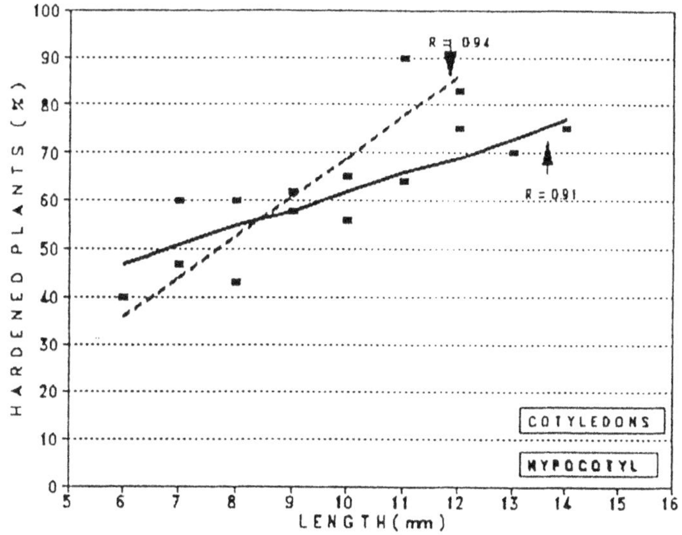

Figure 6. Relation between weaning and cotyledons-hypocotyl length.

(Bercetche *et al.*, 1993). Moreover, the presence of a well developed epicotyl is considered as a predictor of success in acclimatisation. Long photoperiod (>16 h) promotes epicotyl growth while short one (<16 h), apparently, induces apical bud "dormancy". The most appropriate substrate under our conditions is a mixture of non-composted bark and yellow peat (3/1 v/v). A high relative humidity is critical for hardening. A 90% success in weaning is possible by decreasing the relative humidity from 90% to 50% over a 2 month-period.

In order to reduce expensive transfers from *in vitro* to *in vivo* conditions, some researchers are trying to automate the process and to adapt traditional nursery trays (Gupta *et al.*, 1993). In AFOCEL (Association Foret-Cellulose), investigations are presently using the same culture matrix (a folded paper) from the cotyledonary embryos germination phase until acclimatisation. This system would reduce the number of plant transfers to one between the end of maturation and hardening.

3.5. *Nursery and Field Trials*

Until now little data are available on nursery and field trials with *Picea abies*. The first report on overwinter survival of *Picea abies* emblings was by Becwar *et al.* (1989). More recently, 3,350 emblings from seventeen genotypes have been grown in nursery by Weyerhaeuser Company (Gupta *et al.*, 1993). In 1992, 2,000 emblings from 5 genotypes were planted by AFOCEL using excised zygotic embryos and *in vitro*-grown seedlings as controls. Embling

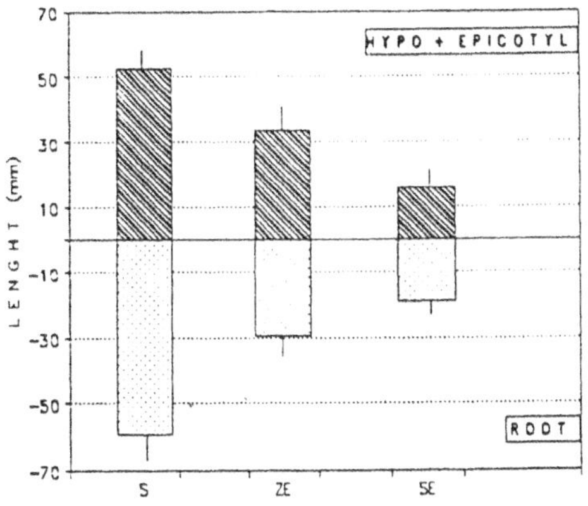

Figure 7. Comparison of seedlings (S), zygotic embryos (ZE) and emblings (SE) 2.5 months after *in vitro* germination.

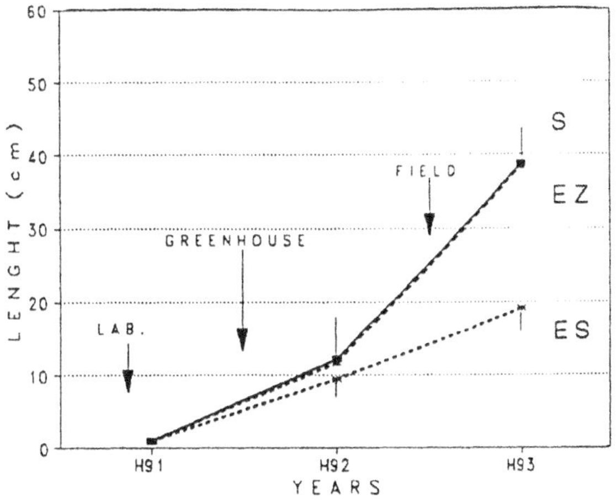

Figure 8. Ex vitro behaviour of seedlings (S), zygotic embryos (EZ) and emblings (ES).

development appears normal, which is in agreement with both Becwar's and Gupta's reports, however, after one growing season in the field, seedling development is greater than excised zygotic embryos (ZE) and emblings. Two and one half months after *in vitro* germination, the embling development is similar to the seedlings and ZEs. Indeed, the ratio between the average length of the stem and the root is similar for the 3 kinds of plants (Fig. 7). However, the length of hypocotyls and cotyledons is always greater for seedlings and equivalent to ZEs. Despite the similar growth rates of the 3 kinds of plants in the greenhouse, the stem elongation of seedlings and ZE recorded after one growing season in the field is at least two times higher as compared with emblings (Fig. 8). These results support the observations reported in *Picea glauca* (Grossnickle *et al.*, 1991). Whatever the cause, it is necessary to point out that comparisons between emblings and zygotic embryos are not really accurate. Indeed, emblings are theoretically clones while ZE and seedlings are more heterogeneous. It should be noted that some seedlings and ZE have the same growth rate as the emblings.

Until now, embling development appears normal and no somaclonal variation has been reported in conifer using chromosomal number determination (Mo *et al.*, 1989) and isozymes analysis (Eastman *et al.*, 1991). If genetic stability of somatic embryos is true, then we can consider somatic embryogenesis as a true clonal propagation method. Further investigation using restriction fragment length polymorphism (RFLP) and polymerase chain reaction (PCR) will be necessary to confirm this point. Embling behaviour in the field should be studied for at least 5 years before large-scale propagation begins.

4. Somatic Embryogenesis and Industrial Applications

Industrial application of somatic embryogenesis could be very useful for *Picea abies* tree improvement programmes. On one hand, the length of the reproductive cycle is long, and on the other hand, 400 to 500 million plants are required every year for reforestation of the Northern part of Europe.

The success reported by Gupta *et al.* (1993) ("production of 50,000 to 100,000 high-quality cotyledonary embryos per litre of settled cell suspension culture, eighty to ninety percent of germination") suggests a bright future for somatic embryogenesis.

However, somatic embryogenesis has not yet been used as a large-scale propagation system. The main problems can be summarised in 3 points: 1) very high cost of the emblings using present technology, 2) a lack of knowledge on embling field behaviour, and 3) no successful ESM initiation from mature trees.

4.1. *Cost of Production of Emblings*

Cost of embling production is strongly related to:

- the insufficient control of ESM differentiation and, as a consequence, our inability to produce mature somatic embryos in liquid medium; the very high heterogeneity of the cotyledonary embryos harvested from a solid medium, and the need to manually sort the embryos before germination,
- the high number of embling transfers required to produce a plant suitable for forest planting. This may be solved by developing artificial seed techniques. Then the major problem will be to promote somatic embryo germination in non axenic conditions. Indeed sugars seem to be required as carbohydrate source to promote germination (Gupta and Durzan, 1986).

Planning of large-scale embling production will require the ability to stop the process in order to adjust the production rates to the commercial need and eventually allow the forester to sow the material in the right season. Several possibilities exist to achieve that purpose: 1) ESM storage in liquid nitrogen regularly performed in AFOCEL allows the possibility of cotyledonary embryo recovery up to two years after freezing (Bercetche *et al.*, 1993); 2) dry and cold storage of desiccated cotyledonary embryos [desiccation tolerance to less than ten percent water content has been observed from several genotypes of *Picea abies* (Bercetche and Reymond, 1992; Gupta *et al.*, 1993)]; and 3) desiccation of artificial seeds, which has not yet been demonstrated with conifers.

4.2. *Insufficient Knowledge on Emblings Field Behaviour*

Better knowledge of emblings behaviour is needed before industrial development of the process. Experimental field plantations established by many groups will offer the opportunity to estimate the quality of emblings in terms of genotypic and phenotypic conformity. For *Picea abies*, it appears reasonable to obtain accurate field performances in the next four years. This will provide time necessary to improve the production techniques.

4.3. *ESM Induction*

ESM induction has been reported from zygotic embryos and very young seedlings. It is strongly influenced by the physiological status and the genotype of the primary explant. It is important to consider the maintaining of the genetic diversity of the genotypes that need to be bulked.

The difficulty of ESM induction from mature trees does not allow, at present, a cheap multiplication of selected, tested, superior genotypes.

Better control of ESM induction is still a major limitation to make valuable somatic embryogenesis process.

5. Conclusions and Perspectives

Despite the fact that embling recovery from *Picea abies* is well documented, and it is too early for the industrialisation of the process.

Better knowledge of the biological system is required to obtain synchronised ESM differentiation and, therefore, to be able to reduce the high production costs.

Moreover, well designed field trials are needed in order to evaluate the quality (genotype and phenotype) of emblings in comparison with seedlings and excised zygotic embryos. Foresters and the forest industry await the results of these trials. Many groups are presently working in that direction.

In order to achieve those objectives, basic investigations would be focused on the early events that occur during maturation and germination of zygotic embryos. Information on the genetic and physiological control of somatic embryogenesis is necessary to develop a strategy to improve ESM initiation from mature trees.

In the future, propagation of *Picea abies* will probably include the combination of somatic embryogenesis and cryopreservation. ESM can be induced from cotyledons of putative elite genotypes and cryopreserved for the time required for early field trials (4 years). Multiplication of the best genotypes could be implemented in 4 years provided the current embling field trials give good results.

Acknowledgements

We thank Dr. David G. Thompson who has kindly accepted to review this manuscript and Mr. P. Monchaux who has provided information on breeding programme of *Picea abies*.

References

Attree, S.M., T.E. Tautorus, D.I. Dunstan and L.C. Fowke, 1990. Somatic embryo maturation, germination and soil establishment of plants of black and white spruce (*Picea mariana* and *Picea glauca*). Can. J. Bot. 68: 2583–2589.

Becwar, M.R., S.R. Wann, M.A. Johnson, V.A. Verhagen, R.P. Feirer and R. Nagmani, 1988. Development and characterisation of *in vitro* embryogenic systems in conifers. In: M.R. Ahuja (Ed.), Somatic Cell Genetics of Woody Plants, pp. 1–18, Kluwer Academic Publishers, Dordrecht.

Becwar, M.R., L.N. Thomas and J.L. Wyckoff, 1989. Maturation, germination and conversion of Norway spruce (*Picea abies*) somatic embryos to plants. In Vitro Cell Develop. Biol. 25: 575–580.

Bellarosa, R. and S. von Arnold, 1992. The influence of auxin and cytokinin on proliferation and morphology of somatic embryos of *Picea abies* (L.) Karst. Ann. Bot. 70: 199–206.

Bercetche, J., M. Galerne and J. Dereuddre, 1990. Augmentation des capacités de régénération de cals embryogènes de *Picea abies* après congélation dans l'azote liquide. C.R. Acad. Sci. Paris 310 (III): 357–363.

Bercetche, J. and I. Reymond, 1992. Conservation of *Picea abies* somatic embryos by dehydration. Proc. IUFRO/AFOCEL Symp., Mass production technology for genetically improved fast growing forest tree species, Bordeaux, 1992, pp. 159–169.

Bercetche, J., I. Reymond and M. Pâques, 1993. Conversion des embryons de *Picea abies* en plantes: influence du support de culture. AFOCEL Annals 1992: 5–28.

Bonga, J.M., 1992. Somatic embryogenesis in explants from a 30-year-old *Larix decidua* tree. Proc. 6th International Conifer Biotechnology Working Group Conference, Raleigh, 1992 (Abstr.).

Bornman, C.H., 1993. Maturation of somatic embryos. In: K. Redenbaugh (Ed.), Synseeds – Application of Synthetic Seeds to Crop Improvement, pp. 105–113. CRC Press, Boca Raton, FL.

Chaperon, H., 1986. La culture du pin maritime en Aquitaine. 231 pp. AFOCEL (Pub.), Nangis.

Eastman, P.A.K., F.B. Webster, J.A. Pitel and D.R. Roberts, 1991. Evaluation of somaclonal variation during somatic embryogenesis of interior spruce (*Picea glauca* × *engelmannii* complex) using culture morphology and isozyme analysis. Plant Cell Rep. 10: 425–430.

Find, J.I., F. Floto, P. Krogstrup, J.D. Moller, J.V. Norgaard and M.M.H. Kristensen, 1993. Cryopreservation of an embryogenic suspension culture of *Picea sitchensis* and subsequent plant regeneration. Scand. J. For. Res. 8: 156–162.

Grossnickle, S.C., D.R. Roberts, J.E. Major, R.S. Folk, F.B. Webster and B.C.S. Sutton, 1991. Integration of somatic embryogenesis into operational forestry: comparison of interior spruce emblings and seedlings during production of + stock. Proc. Intermountain Forest Nursery Association Symposium, Park City, Utah, 1991, pp. 106–113.

Gupta, P.K. and D.J. Durzan, 1986. Somatic polyembryogenesis from callus of mature sugar pine embryos. Bio/Technology 4: 643–645.

Gupta, P.K., D.J. Durzan and B.J. Fimkle, 1987. Somatic polyembryogenesis in embryogenic cell masses of *Picea abies* (Norway spruce) and *Pinus taeda* (lobolly pine) after thawing from liquid nitrogen. Can. J. For. Res. 17: 1130–1134.

Gupta, P.K., R. Timmis, G. Pullman, M. Yancey, M. Kreitinger, W. Carlson and C. Carpenter, 1991. Development of an embryogenic system for automated propagation of forest trees. In: I.K. Vasil (Ed.), Scale-Up and Automation in Plant Propagation, Vol. 8, pp. 75–93. Academic Press, New York.

Gupta, P.K., G. Pullman, R. Timmis, M. Kreitinger, W.C. Carlson, J. Grob and E. Welty, 1993. Forestry in the 21st century. The biotechnology of somatic embryogenesis. Bio/Technology 11: 454–459.

Hakman, I., L.C. Fowke, S. von Arnold and T. Eriksson, 1985. The development of somatic embryos in tissue cultures initiated from immature embryos of *Picea abies* (Norway spruce). Plant Sci. 38: 53–59.

Hämäläinen, J.J., V. Kauppinen and J. Heilala, 1992. Automated classification of somatic embryos. Acta Hort. 319: 601–605.

Janick, J., Y.H. Kim, S. Kitto and Y. Saranga, 1993. Desiccated synthetic seed. In: K. Redenbaugh (Ed.), Synseeds – Application of Synthetic Seeds to Crop Improvement, pp. 11–46. CRC Press, Boca Raton, FL.

Kleinschmit, J. and J. Svolba, 1989. The clonal option current status and future development. Report Swedish University of Agricultural Science 11: 208–231.

Klimazewska, K., C. Ward and W.M. Cheliak, 1992. Cryopreservation and plant regeneration from embryogenic cultures of larch and black spruce. J. Expt. Bot. 43: 73–79.

Laine, E., P. Bade and A. David, 1992. Recovery of plants from cryopreserved embryogenic cell suspensions of *Pinus caribaea*. Plant Cell Rep. 11: 295–298.

Lelu, M.A., K. Klimaszewska and P.J. Charest, 1993. Somatic embryogenesis from immature and mature zygotic embryos and from cotyledons and needles of somatic plantlets of *Larix*. Can. J. Bot. (in press).

Li, X.Q., 1993. Somatic embryogenesis and synthetic seed technology using carrot as a model system. In: K. Redenbaugh (Ed.), Synseeds – Application of Synthetic Seeds to Crop Improvement, pp. 290–304. CRC Press, Boca Raton, FL.

Mo, L.H., S. von Arnold and V. Lagercrantz, 1989. Morphogenetic and genetic stability in long term embryogenic cultures and somatic embryos of Norway spruce (*Picea abies* L. Karst). Plant Cell Rep. 8: 375–378.

Monchaux, P., 1989. Les variétés polyclonales d'épicéa commun de l'AFOCEL. Informations-Forêt < numero > 3, AFOCEL fascicule < numero > 370: 155–170.

Nanson, A., 1986. The evolving seed orchard: a new type. Proc. IUFRO Conference, Williamsburg, Virginia, 1986, pp. 554–565.

Nanson, A., 1987. Juvenile-mature correlations based on *Picea abies* provenances and progeny tests. Forest Tree Improvement 20: 3–25.

Pâques, M., J. Bercetche and E. Dumas, 1992. Liquid media to improve and reduce the cost of *in vitro* conifer propagation. Acta Hort. 319: 95–100.

Pâques, M., 1993. System to harvest cotyledonary embryos from embryogenic suspensor mass culture (in preparation).
Park, Y.S., S.E. Pond and J.M. Bonga, 1993. Initiation of somatic embryogenesis in white spruce (*Picea glauca*). Genetic control treatment effects and implications for tree breeding. Theor. Appl. Genet. 86: 427–436.
Roberts, D.R., B.S. Flinn, D.T. Webb, F.B. Webster and B.C.S. Sutton, 1990. Abscisic acid and indole-3–butyric acid regulation of maturation and accumulation of storage proteins in somatic embryos of interior spruce. Physiol. Plant. 78: 355–360.
Roberts, D.R., W.R. Lazaroff and F.B. Webster, 1991. Interaction between maturation and high relative humidity treatments and their effects on germination of Sitka spruce somatic embryos. J. Plant Physiol. 138: 1–6.
Roberts, D.R., F.B. Webster, B.S. Flinn, W.R. Lazaroff and D.R. Cyr, 1993. Somatic embryogenesis of spruce. In: K. Redenbaugh (Ed.), Synseeds – Application of Synthetic Seeds to Crop Improvement, pp. 427–450. CRC Press, Boca Raton, FL.
Rodriguez, D.L., S.L. Kitto and K.M. Lomax, 1990. Mechanical purification of torpedo stage somatic embryos of *Daucus carota* L. Plant Cell Tiss. Org. Cult. 23: 9–14.
Ruaud, J.N., J. Bercetche and M. Pâques, 1992. First evidence of somatic embryogenesis from needles of 1–year-old *Picea abies* plants. Plant Cell Rep. 11: 563–566.
Tautorus, T.E., S.M. Attree, L.C. Fowke and D.I. Dunstan, 1990. Somatic embryogenesis from immature and mature zygotic embryos and embryo regeneration from protoplasts in black spruce (*Picea mariana* Mill.). Plant Sci. 67: 115–124.
Tautorus, T.E., L.C. Fowke and D.I. Dunstan, 1991. Somatic embryogenesis in conifers. Can. J. Bot. 69: 1873–1899.
von Arnold, S. and I. Hakman, 1988. Regulation of somatic embryo development in *Picea abies* by abscisic acid (ABA). J. Plant Physiol. 132: 164–169.
von Arnold, S. and S. Woodward, 1988. Organogenesis and embryogenesis in mature zygotic embryos of *Picea sitchensis*. Tree Physiol. 4: 291–300.
Webb, D.T., F. Webster, B.S. Flinn, D.R. Roberts and D.D. Ellis, 1989. Factors influencing the induction of embryogenic and caulogenic callus from embryos of *Picea glauca* and *Picea engelmanii*. Can. J. For. Res. 19: 1303–1308.
Webster, F.B., D.R. Roberts, S.M. McInnis and B.C.S. Sutton, 1990. Propagation of interior spruce by somatic embryogenesis. Can. J. For. Res. 20 : 1759–1765.
Werner, M., H. Wellendorf and H. Roulund, 1986. The development over 37 years of sixteen open pollinated families in *Picea abies*. Arsbok: 85–109.
Westcott, J.R., 1992. Somatic embryogenesis from non juvenile Norway spruce (*Picea abies*). UK patent WO 93/23990-PCT/EP93/01365.

20. Future Uses of Somatic Embryogenesis in Woody Plantation Species

Levis W. Handley

Contents

1. Introduction

Remarkable progress has been made in the commercialization of somatic embryogenesis since first described by Steward in carrot thirty-five years ago (Steward *et al.*, 1958). Today, we are on the verge of large-scale mass propagation of forest trees and other woody plants using this process. Somatic embryogenesis is no longer viewed as simply a laboratory technique to study totipotency and fundamental processes in plant morphogenesis, but is being seriously considered as a method for large-scale mass production of superior genotypes of woody plants. In the next few years somatic embryogenesis has the potential for multiplying millions of superior plants for both forestry and horticultural crops on a commercial scale. Somatic embryogenesis will be a useful tool not only for large-scale vegetative multiplication or clonal propagation, but also will prove invaluable as a tool in the genetic engineering of woody species.

Before this technology becomes a commercial reality, however, there are still hurdles that must be overcome. As in any emerging technology based on a biological system, as we move toward large-scale production, new technical and biological questions begin to arise. This chapter explores the potential roles embryogenesis may play in commercial operations and the future challenges to be faced in this rapidly developing technology. The

S. Jain, P. Gupta & R. Newton (eds.), Somatic Embryogenesis in Woody Plants, Vol. 1, 415–434.

major topics to be covered include the mass propagation of woody plantation species, automated systems for future plantations, and genetic engineering of plantation species using embryogenesis.

2. Mass Propagation of Woody Plantation Species

The single largest commercial advantage of somatic embryogenesis is its potential for producing millions of copies of superior individuals. The mass propagation of high-value genotypes may offer a significant advantage in some woody plants. Woody plantation species can be mass propagated from a genetic improvement program by either of two methods: vegetative multiplication of superior families or clonal propagation of superior individuals. The major limitations to using embryogenesis in either vegetative multiplication or clonal propagation are sometimes low culture-initiation frequencies and high labor costs, but many research programs are addressing these problems.

Current methods for mass propagating some forest species include both rooted stem cuttings (Ritchie, 1991; Denison and Quaile, 1987; Randall and Cooper, 1973) and micropropagation (Gleed *et al.*, 1991). These two technologies are currently used or are under development for mass propagation by both industrial and governmental agencies. Somatic embryogenesis is just beginning to compete with these propagation methods. In woody plants, especially some forest species, the mass propagation of juvenile material using somatic embryogenesis may become economically feasible, primarily due to the economies of scale and genetic gain potential.

In many forestry applications genetically superior individuals are usually planted in the field as open-pollinated seedlings from seed orchards or as seed from full-sib control-crosses of superior parents. These seed orchards are usually composed of superior parents derived through provenance/progeny tests or through conventional breeding programs. Additionally, superior genotypes can be deployed through two major strategies: vegetative multiplication of rare control cross seed (Ritchie, 1991) or the mass propagation of superior individual clones (Denison and Quaile, 1987; Randall and Cooper, 1973). Each of these options provides an additional increase in genetic gain over open pollinated seedlings. Each option has advantages and disadvantages, particularly depending on the species and upon the level of breeding that has been conducted.

2.1. *Vegetative Multiplication of Superior Families*

In vegetative multiplication, superior control-cross seed are bulk propagated for large-scale production. Vegetative multiplication would only be used when high-value seeds are in short supply and/or are expensive to produce. This is done operationally by rooting stem cuttings and/or micropropagation of hedged seedlings (Etheridge and Adams, 1990; Ritchie, 1991). In some forest species, these propagules are produced to shorten the time it takes to get progeny from an advanced breeding program into the field (Ritchie, 1991). Accelerated breeding programs use techniques such as gibberellin applications to shorten the breeding cycle (Greenwood *et al.*, 1988), and parents are selected for seed orchards as early as possible. Consequently the best quality seeds are usually in short supply (Schooley and Mullin, 1988). Vegetative multiplication provides a means of multiplying these seeds for rapid deployment of the best genetic material. The gain from vegetative multiplication is the same as for control-cross seed. Rare superior crosses can be quickly bulk propagated when the number of adult trees or seed production is limiting. Vegetative multiplication also has the advantage of not requiring an additional field test as does clonal propagation.

An embryogenic system could also be used to vegetatively multiply scarce seed and would offer both advantages and disadvantages. Since many woody plant embryogenesis systems start with either mature or immature zygotic embryos (Bates *et al.*, 1992; Becwar *et al.*, 1988; Fitch, 1993; Merkle *et al.*, 1990; Neuman *et al.*, 1993; Von Aderkas *et al.*, 1990) these would interface well with a vegetative multiplication system. By multiplying limited control-cross seed of parents with superior general combining ability via somatic embryogenesis, an additional gain would be realized if the number of parents limited seed production.

A major limitation in using embryogenesis for vegetative multiplication is the high cost of propagating control-cross seed using this method. This technology would be cost-prohibitive by current methods so that any increase in gain would be lost by the additional production costs. Using current technology, rooted cuttings of hedged seedings are less expensive. When automated embryogenic systems are developed where costs are equivalent to those of rooted cuttings, then such a scenario would be feasible.

2.2. *Clonal Propagation of Superior Individuals*

The production of true clones may be the most attractive use for current somatic embryogenesis systems since the additional cost of production may be partially offset by a substantial increase in genetic gain. Clonal propagation offers the best opportunity of increasing genetic gain by capturing both the additive and non-additive genetic variance in the population. This yields a genetic gain greater than that of seedling propagation in which only a portion of the additive genetic variance is exploited (McCrae *et al.*, 1993; Timmis *et al.*, 1987). In clonal deployment an additional field test is required in which the best clones are identified (Etheridge and Adams, 1990). Currently most clonal testing and deployment of forest species is conducted using either rooted cuttings or micropropagules (Etheridge and Adams, 1990; Gleed *et al.*, 1991). In short-rotation species, a clonal test typically lasts for 4 to 6 years. During these tests it is necessary to maintain all ortets in the juvenile phase. Ramets of each selection are then propagated after superior clones are identified. Typically clones are kept juvenile by hedging for rooted cuttings (Etheridge and Adams, 1990), or by maintenance in cold storage for micropropagules (Aitken-Christie and Singh, 1987). Both methods are successfully used to retard the maturation process, but maintenance of the hedges or tissue cultures during the testing period can add large labor and overhead costs to the clonal testing process. In some species, maturation of field-grown hedges also occurs, so by the time a superior clone has been identified, it is difficult to propagate large numbers of juvenile cuttings from these hedges (Arnold, 1990; Etheridge and Adams, 1990). If the clone is to be deployed for several years, maturation of hedges could continue to reduce rooting success.

2.3. *Establishment of Germplasm Banks*

Somatic embryogenesis could offer a significant advantage for clonal propagation. In an embryogenic system, maintenance of the juvenile state during the clonal test could be done by storing cultures in cryopreservation (Fig. 1) instead of using juvenile hedges. In cryopreservation, embryogenic tissues are stored in liquid nitrogen at −196°C. Cultures can be maintained indefinitely in this system without the loss of embryogenic potential. These cultures can then be thawed and placed back in tissue culture for growth and embryo development.

Embryogenic tissue cultures representing each of the clones planted in the field would be stored during the test in cryopreservation. Following the test, cultures corresponding to the selected superior clones would be removed from cold storage and mass propagated for deployment (Fig. 1). Such cryogenic storage systems are currently under development in several forest species and most studies show this to be a feasible method of storing embryogenic tissues (Kartha *et al.*, 1988; Dumet *et al.*, 1993; Laine *et al.*, 1992; Klimaszewska *et al.*, 1992). In our own research we have found cryopreservation to be very useful for the storage of embryogenic cultures of *Pinus taeda*. As embryogenesis becomes more large-scale, cryopreservation systems will become routine. However, pilot-scale production following cryostorage of a wide range of genotypes must be conducted to confirm that there are no deleterious effects due to the storage process.

2.4. *Impact of Culture Initiation Frequencies*

For somatic embryogenesis to be useful for vegetative multiplication or clonal propagation, it will be necessary to establish embryogenic cultures from many genotypes. Currently it is very difficult to initiate highly embryogenic cultures from a wide range of genotypes for many woody species. The ability of immature or mature seeds to produce embryogenic tissues when placed into tissue culture is strongly influenced by their genetic background (Becwar *et*

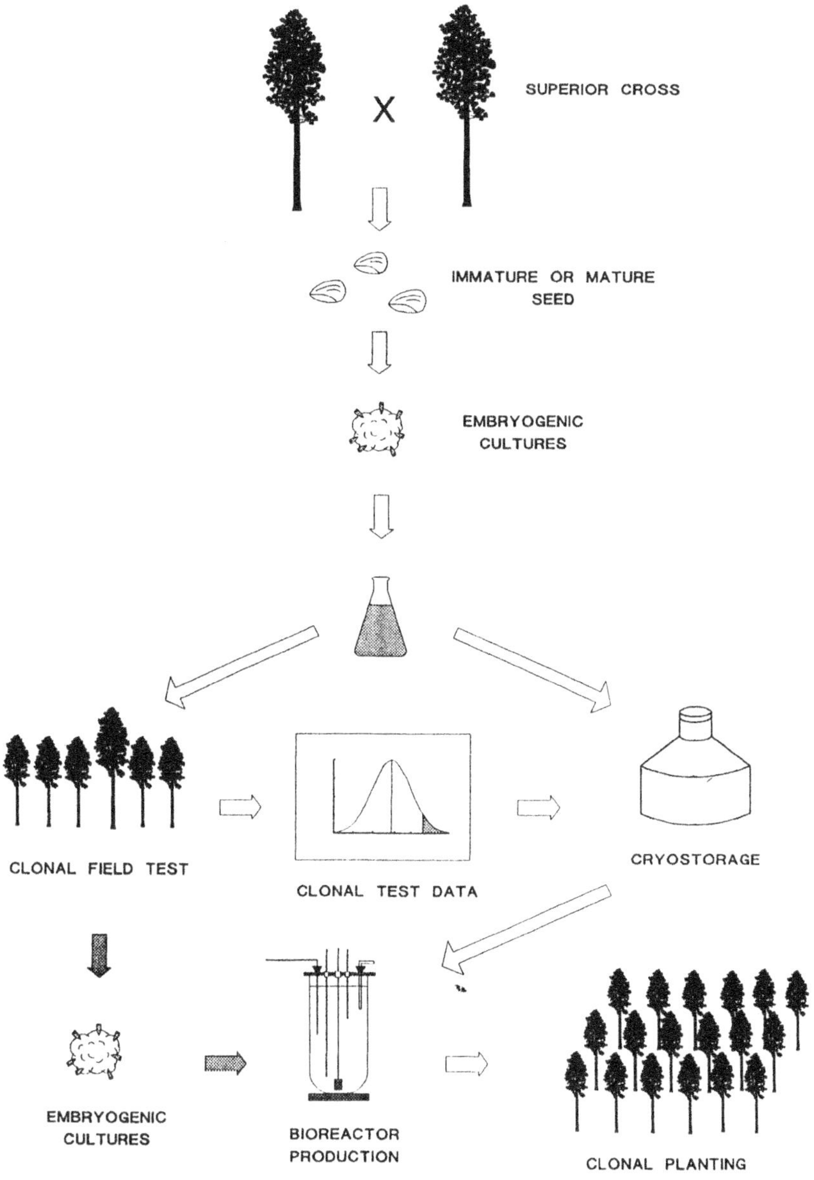

Figure 1. Possible scenario for clonal deployment of somatic embryo derived plants. Embryogenic tissues are derived from mature or immature seeds from a cross of genetically superior trees. These cultures can then be expanded in suspension culture and placed in cryostorage. At the same time somatic embryo plants are developed from these cultures and placed in a clonal field test. In short rotation species, after 4–6 years, clonal test data are used to determine which clone(s) are superior and these are retrieved from cryostorage. These cultures can be proliferated in bioreactors, and embryos developed from these are then used for clonal field plantings. If it becomes possible to establish embryogenic cultures directly from the clonal field test through a rejuvenation process, then cryostorage will not be necessary and any field-tested tree could be deployed directly through the embryogenesis process (shaded arrows).

al., 1990; Tremblay, 1990; Park *et al.*, 1993; Cheliak and Klimaszewska, 1991; Ekberg *et al.*, 1993). The response of open-pollinated families generally ranges from 0 to 60% induction, depending upon the species and tissue type used. This may or may not be an acceptable level of response.

Vegetative multiplication normally requires the production of many individuals from a given control cross. However, the limited capacity of some species to produce embryogenic cultures may not preclude using this technology if the successfully-initiated individuals can be multiplied at an extremely rapid rate. For a specific cross, embryogenesis could be successful with fewer genotypes than rooted cuttings as long as the multiplication rate is much more rapid. For example, if one could germinate 5,000 control-cross seedlings and then produce 10 rooted cuttings per seedling in a year, then 50,000 plants would be available for field planting. Using somatic embryogenesis, if one could produce embryogenic cultures from only 50 out of 5,000 control-cross seed (i.e., 10% induction), but could produce 1,000 somatic embryo plants in one year from each of those 50 cultures, then the same number of plants would be available. This assumes that selection of inferior genotypes would not occur by using only those individuals that produce somatic embryo plants and that embryogenic capacity would not be negatively correlated with growth in the field. The advantage of somatic embryogenesis in vegetative multiplication would be that numbers could be increased very rapidly, especially if a large proportion of genotypes could be initiated.

Clonal forestry captures superior individuals by selecting and developing clones from the upper ends of normal distributions for such traits as height, volume, disease resistance and other desirable characteristics. Somatic embryogenesis will be useful if a large range of genotypes are responsive to the process so that clonal tests can be established. Genotypes that produce somatic embryos must also occupy the upper end of these distributions. We currently do not know the distribution of the embryogenic response across both poor and superior families in many of our woody plantation species. Therefore, we do not know if selection occurs via the embryogenesis process, or the impact this might have on our ability to obtain superior clones from genotypes that produce high-quality plantable somatic embryos. A recent study by Ekberg *et al.* (1993) indicates that selection for embryogenic capacity is not associated with phenological traits.

We also do not know the extent to which embryogenic potential correlates with superior growth form in field-grown plants, nor do we know if the embryogenic process might produce somaclonal or epigenetic variation in the regenerated plants. Therefore, it would be prudent to continue efforts toward increasing the response of genotypes in the initiation process, and to establish field tests to evaluate the growth and form of somatic embryo plants produced by current protocols.

In the future, it will be very desirable to produce embryogenic cultures from mature tissues. Eventually, it may be possible to expand the induction of embryogenic tissues from immature or mature embryos to older seedlings or juvenile shoots from hedges. If initiation could be further extended to mature trees in the field, it would allow somatic embryogenesis to take full advantage of traditional clonal field-testing schemes. This would allow us to capture any tested clone and bypass the cryopreservation of cultures during the test (Fig. 1 – shaded arrows).

3. Automated Systems for Future Plantations

One major advantage of somatic embryogenesis is its potential for being automated. Many forest tree species are currently produced in the tens of millions per annum. In 1992, over 1.5 billion tree seedlings were produced in nurseries in the United States alone (Moulton *et al.*, 1993). The current methods for producing millions of trees by either rooted cuttings or micropropagules are still labor-intensive processes. If these systems or an embryogenic system could be automated, a significant savings in labor should be possible. In a system where millions of propagules are produced per year, economies of scale should be realized.

To be economically feasible and to compete with rooted cuttings or micropropagation, an embryogenic system will have to be highly automated. Current laboratory protocols are extremely time consuming and labor intensive. A typical harvest of somatic embryos of *Pinus taeda* from a development medium takes about 6 h/1,000 embryos in our laboratory. This includes only one step and does not include the time it takes to prepare media and maintain cultures. As each step is done by hand it adds time to the process. To make somatic embryogenesis cost-competitive with other methods of propagation a highly automated system will be necessary in the future where millions of plants will be required.

Ideally an automated system would include most if not all steps in the embryogenic process. These steps are 1) culture maintenance and scale up, 2) embryo development, and 3) conversion to autotrophic plants. Under the conversion process there are two major methods for deploying somatic embryo plants to the field. These are direct-seeding systems and artificial seed systems. Both of these are impacted by the quality of embryos produced in the embryogenic process and methods that monitor embryo quality will be a part of these systems in the future.

3.1. *Culture Maintenance*

The goal of the maintenance process is to maintain cultures in an embryogenic state on either semi-solid or liquid medium. Specifically cultures from the maintenance phase should be capable of producing fully formed (mature) embryos after placement on a development medium. The goal of this process includes the scale up of cultures to a point where large numbers of somatic embryos can be produced when placed in the development phase.

Currently there are three primary methods for maintaining cultures: 1) on solidified medium as embryogenic callus or tissues, 2) in liquid medium in shake flasks, and 3) in liquid medium in bioreactors. Most embryogenic systems for woody plants usually perform best when maintained on gelled medium; however, this is a very labor-intensive process and does not readily lend itself to automation. In a move towards easier maintenance of cultures and toward automation, embryogenic cultures of various woody plant species have been successfully established and maintained in liquid culture media. Examples include *Phoenix dactylifera* (Bhaskaran and Smith, 1992), *Elaeis guineensis* (De Touchet *et al.*, 1991), *Populus ciliata* (Cheema, 1989) *Coffea arabica* (Neuenschwander and Baumann, 1992; Zamarripa *et al.*, 1991), *Betula pendula* (Nuutila and Kauppinen, 1992), *Fagus sylvatica* (Vieitez *et al.*, 1992), *Liriodendron tulipifera* (Merkle *et al.*, 1990), *Picea glauca* (Hakman and Von Arnold, 1988), *Picea glauca* and *Picea mariana* (Lulsdorf *et al.*, 1992), *Picea abies* (Gupta *et al.*, 1991), *Pseudotsuga menziesii* (Gupta *et al.*, 1993), *Pinus radiata* (Smith *et al.*, 1991) and *Pinus taeda* (Gupta *et al.*, 1993; Handley, unpublished data). Some species continue to remain embryogenic when placed in liquid medium and produce large numbers of somatic embryos but many species appear to lose their embryogenic potential when placed in a liquid medium. Also there is a general loss in embryogenic potential over time for many embryogenic cultures, whether maintained on gelled medium or liquid medium. Cultures will often survive for only one or two years, and many show a reduced capacity to produce embryos after only a year in culture (Roberts *et al.*, 1993). Maintenance of embryogenic capacity in liquid culture and culture longevity will continue to receive research focus in the future as we move toward less labor-intensive systems.

Many researchers involved in vegetative propagation using somatic embryogenesis believe that bioreactors will become the ultimate means for automating these processes. A bioreactor system where all parameters are precisely controlled and cell or tissue culture input and output can be precisely regulated is an attractive goal. Bioreactors have been tested for the growth and multiplication of embryogenic tissues of *Pinus radiata* (Smith *et al.*, 1991), *Picea mariana*, and *Picea glauca-engelmannii* (Tautorus *et al.*, 1992) with some success. Cultures maintained in these systems usually remain capable of producing embryos when placed onto gelled medium, but often at reduced frequencies. This work is in its early stages, and to date most research has concentrated on defining the parameters for keeping cultures in an embryogenic state. Most bioreactors were not designed for the maintenance of plant cell cultures, so research now and in the future will focus on modifications of reactor design for plant embryogenic cultures.

3.2. *Embryo Development*

The goal of the development process is to produce mature embryos ready for the conversion process. As in the maintenance and scale-up stages, this step essentially remains a laboratory procedure and is very labor and time consuming. The eventual goal is to automate this step and to have as little hand labor involved as possible. Embryogenic cultures that have been maintained either on gelled medium or in liquid are usually developed on a gelled medium. This involves a large amount of labor for subculturing and harvesting embryos. Ideally the best system would have embryogenic cell cultures maintained in liquid culture and, through media changes, would induce embryos to form in liquid in the same vessel. This could be done in a bioreactor where all parameters could be controlled to optimize development.

The only report of embryos developing in liquid medium in a bioreactor from a woody plant is in *Santalum album* (Bapat *et al.*, 1990; Bapat, 1993). Two different bioreactors were used; one for the maintenance of embryogenic cells and another for the development of embryogenic cells to mature embryos. Alfalfa (*Medicago sativa*) and sweet potato (*Ipomoea batatas*) embryos have also been developed directly in bioreactors. In *Medicago* it was possible to obtain yields of embryos comparable to those on gelled medium. However, only 2 to 5% of the embryos developed in the bioreactor were capable of conversion to plants (Stuart *et al.*, 1987). Similar results have been obtained in *Ipomoea batatas* (Bieniek *et al.*, 1991, personal communication).

Mature cotyledonary embryos have not been produced in liquid in a bioreactor on a commercial scale for any plant, woody or non-woody. It has been difficult to take the optimal culture conditions developed for a gelled medium and apply them to liquid medium in a shake flask or in a bioreactor. These three culture systems have inherently different optimal conditions and thus each has to be optimized separately. Even if embryos can be developed on a small scale on solid medium it is often difficult or impossible to transfer media formulations and cultural conditions to liquid cultures for efficient production of somatic embryos.

Most bioreactor systems for plants are still incomplete. It would be most desirable for embryogenic cells to be placed in the bioreactor, proliferated and maintained in the reactor and then induced to form somatic embryos, all in the same vessel. Usually only one step is done in the reactor and the others are conducted on solid media. Future research will probably focus on accomplishing all steps in the bioreactor and optimizing the conditions for producing high-quality embryos directly in the liquid medium.

There may be other methods for automating the development process rather than directly in liquid medium. For example Attree and Fowke (1993) recently described a type of mechanical bioreactor where embryogenic suspension cultures of *Picea glauca* were plated onto a filter paper support matrix in a sterile box-like container. Development medium was continuously pumped through the system and wicked onto the filter paper matrix. This system yielded several thousand mature somatic embryos in one production run. Such a system could easily be connected to an air-lift or impeller-type

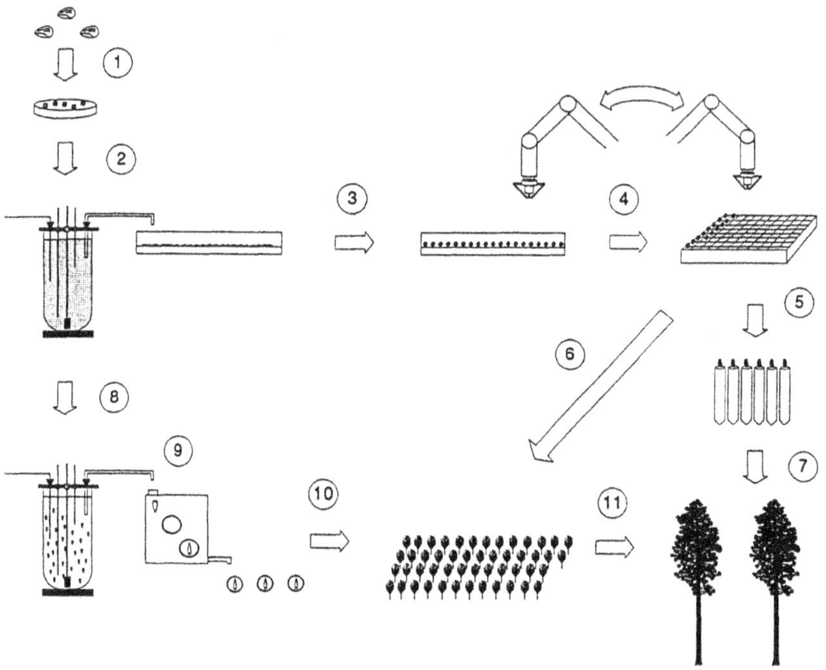

Figure 2. Potential methods for the automated deployment of somatic embryo derived plants into the field. Embryogenic cultures are initiated from seeds (1) and maintained in a bioreactor (2). In the top scheme, embryogenic cultures are delivered onto a solid medium for development of embryos (3). This development might be done in a solid-culture bioreactor. Embryos that develop are then harvested by a robot or other mechanical device and placed in mini-plug trays for further growth (4). Miniplugs can then be transplanted into larger tubes (5) for direct planting in the field (7) or can be transplanted into a nursery for further growth (6). These are then harvested and planted using traditional seedling nursery practices (11). In the bottom scheme embryogenic cultures are induced to develop embryos in liquid through medium changes in the bioreactor (8). Embryos are then encapsulated (9) and planted in a seedling nursery as artificial seeds (10). These can then be planted in the field using traditional methods (11).

bioreactor where embryogenic suspension culture cells could be automatically plated into the mechanical reactor for development. After mature embryos have developed, these could then be mechanically harvested (Fig. 2).

3.3. *Conversion to Autotrophic Plants*

The goal of the conversion process is to take mature embryos from the development step of the embryogenesis process and convert them into autotrophic plants. This step would easily lend itself to automation, and laboratory-scale systems have already been developed for this process. Somatic embryos might be delivered to the field by direct "seeding" in a soil-less container system or through artificial seeds.

3.3.1. *Direct-Seeding Systems*

Pilot-type systems have already been developed to take somatic embryo plants from the *in vitro* environment into a sterile plug system for germination (Timmis *et al.*, 1992; McElroy and Brown, 1992). Refinements in these systems will eventually allow the transfer of somatic embryo plants from the laboratory to automated plug extracting and planting systems already in use by the horticultural industry. For some horticultural crops there are now automated systems that transplant very small seedlings from germination trays into specially designed flats for further growth (Lowe, 1993). These systems are used for different bedding plant species and include automated transplant systems developed primarily in the Netherlands, France and Germany. There are different configurations of these systems with numerous tray types that are primarily designed to interface with seedling miniplug production systems. Using new imaging technology, a robot has recently been developed that automatically identifies empty plugs and fills them with young seedlings from other trays (Lowe, 1993). It is easy to imagine a system where newly germinated somatic embryos are "plucked" from a development medium and planted in a mini-plug tray (Fig. 2). Robotic miniplug transplant systems are under development for horticultural species (Ting *et al.*, 1992; Whynman, 1992), and research on image processing for grading and sorting plants grown in plugs is well underway. Modifications of this technology could be developed so that somatic embryo plants in miniplugs could be transplanted into larger tubes or containers that are currently in use for containerized field planting (Fig. 2). It also should be possible to integrate an automated plug-handling system into nursery transplant machinery whereby newly germinated somatic embryo plants could be planted into the typical forest tree nursery for further growth and hardening before field planting the next year (Gupta *et al.*, 1993). These somatic embryo plants could then be lifted and planted like seedlings.

Another method for deploying somatic embryo plants might involve the direct seeding of embryos with fluid drilling. This was demonstrated on a laboratory scale recently for sweet potato (*Ipomoea batatas*) (Schultheis and Cantliffe, 1992) and carrot (*Daucus carota*) (Kitto *et al.*, 1991) where somatic embryos were placed in hydroxyethyl cellulose gel amended with different nutrients and carbohydrates. This would be an attractive method for planting forestry species in a seedling nursery operation, provided the germination percentages were high.

3.3.2. *Artificial Seed Systems*

Direct-seeding and fluid drilling of somatic embryos do not have the advantages that true seeds possess such as long-term storage, ease of transportation, and planting with existing equipment. Therefore, a true artificial seed may be much more attractive for large-scale planting of woody plantation species where large numbers are usually sown in a short time.

Despite significant research in artificial seed technology over the past 15 years, there have been no somatic embryo derived plants deployed in commercial quantities using this technology for any plant species. The requirements of this system are complicated and are still undergoing intensive research and development. Most systems use alginate gels or other types of artificial seed coating systems (Sakamoto *et al.*, 1991; Redenbaugh 1986a,b; Fujii *et al.*, 1992), which work fairly well but only on an experimental scale.

In some woody species such as conifers, the normal zygotic seed contains the embryo along with nutritive tissues of the megagametophyte. Because these species do not have storage reserves in the cotyledons, they cannot germinate autotropically and require the support of the megagametophyte during the germination process. Thus any artificial seed for these species will require that an artificial megagametophyte be present too. There has been recent progress in the development of a nutrient source within an encapsulation matrix (Sakamoto *et al.*, 1991), but additional research is needed to make this commercially feasible, especially in woody species. The embryo must be oriented correctly within an artificial megagametophyte and the embryo/megagametophyte complex must be packaged in a matrix material that will allow water uptake without drying out. This entire package must eventually break open and be shed similar to the natural seed coat (Sakamoto *et al.*, 1992). Additional improvements may be needed in the encapsulation matrix material since it may not be ideal for some woody species. Redenbaugh (1993) points out that no new synthetic seed coatings have been developed since the early 1980s, and that this must receive more attention in the future.

3.3.3. *Influence of Embryo Quality on Conversion Systems*

The difficulty in implementing both the direct seeding or the artificial seed system is usually the very low rate of germination or lack of uniform germination. The major limitation is the inability of many embryogenic systems to produce highly uniform, high-quality embryos (Fujii *et al.*, 1993; Redenbaugh *et al.*, 1993). This may be especially true in woody species. The major problem appears to be that many somatic embryos are not physiologically ready to germinate, and/or there are physiological or morphological abnormalities of the embryos that are exaggerated in the germination process.

Recent work has concentrated on improving embryo quality and conversion frequency (Gray, 1987; Obendorf and Slawinska, 1988). This is an area where much research in somatic embryogenesis is now focusing, especially in species where embryogenesis has become routine. Research has been conducted investigating the effects of ABA (abscisic acid) (Attree *et al.*, 1992; Fujii *et al.*, 1990; Gray, 1989; Roberts *et al.*, 1990a; Von Arnold and Hakman, 1988), nitrogen form (Lai *et al.*, 1992; Bapat, 1993; McKersie and

Bowley, 1993; Tremblay and Tremblay, 1991b), osmotic agents (Misra *et al.*, 1993) and carbohydrates (Schuller and Reuther, 1993; Tremblay and Tremblay, 1991a; Avjioglu and Knox, 1989) on embryo development. By using ABA in the development medium, desiccation tolerance, accumulation of storage lipids (Attree *et al.*, 1992) and protein content (Roberts *et al.*, 1990a) can be modulated. Techniques have been developed that increase storage proteins and lipids to levels comparable to those of zygotic embryos, and these changes have significantly enhanced the germination and vigor of somatic embryos (Attree *et al.*, 1992; Roberts *et al.*, 1990b; McKersie and Bowley, 1993; Senaratna, 1992). As our understanding of the basic processes of embryo development and maturation improves, efficiencies in the cell culture process will follow.

3.3.4. *Automated Assessment of Embryo Quality*

The above factors point out that both morphological and physiological uniformity of somatic embryos derived from the embryogenic process is critical. The ability to characterize the physiological and morphological state of the embryo before conversion, will be essential to the deployment of uniform plants derived from an embryogenic process. Techniques that can rapidly measure the physiological attributes of somatic embryos may be necessary to ensure uniformity of germination and seedling establishment. On-line monitoring systems, similar to those used in more traditional bacterial bioreactor systems, may be useful. One can envision automated or semi-automated systems that can monitor the state of development of somatic embryos during the embryogenic process by measuring biochemical markers such as growth regulator(s), carbohydrate levels, storage proteins, lipids or other biological compounds from embryos or medium periodically sampled during the process. In this way the competence of embryos for the germination process could be controlled and predicted. Automated characterization of physiological parameters is an area of research that will develop as large-scale embryogenic systems become more routine.

It also will be necessary to monitor the morphological state of development of somatic embryos as they are produced to ensure the uniformity of somatic embryo plants that are deployed. Visual assessment for individual embryo quality and hand harvesting are currently practiced, but are too time consuming and subjective. Systems that can assess quality by machine coupled with some type of harvesting system will be necessary in the future. For example Harrell *et al.* (1992) are developing an image analysis system that interfaces with a bioreactor that can sample liquid embryogenic suspension cultures of sweet potato (*Ipomoea batatas*). Computer vision is used to classify somatic embryos in different states of development, and embryos that are recognized by their morphology as being in the correct state for harvest are removed. The underdeveloped embryos are returned to the bioreactor. Other computer vision classification systems have been developed for somatic embryos of birch (*Betula pendula*) (Hämäläinen *et al.*, 1993) and Norway spruce (*Picea abies*) (Hämäläinen and Jokinen, 1992). Such image analysis systems will be necessary for large-scale production of uniform plantlets, either from semi-solid culture systems or from liquid systems. This will be especially important where millions of somatic embryo plants would be produced in a single year.

The development of these automated systems will require cooperation between biologists familiar with the embryogenic system and agricultural engineers skilled in the design of mechanized handling systems. Such cooperative relationships will be necessary to reduce costs and keep survival of somatic embryo plants at acceptable levels. Biologists, engineers and computer scientists will need to interact closely in the future to develop this technology. These collaborations are occurring in a few locations in the world, but more of this type of cooperation will be necessary in the future as technology moves out of the laboratory and into commercial operation. The mechanics of these systems will become a major component of research and development.

4. Genetic Engineering of Plantation Species Using Embryogenesis

The other major application of somatic embryogenesis is as a component of genetic engineering. Somatic embryogenesis is the most often used tissue culture system in woody plants for genetic transformation. When perfected it is a highly efficient system that is an integral part of the transformation process. Many genetic transformation systems that use somatic embryogenesis are becoming routine. A question that must now be asked is which genes will be important and useful in transgenic woody species in the future.

There are many examples in both non-woody and woody plant species in which a somatic embryogenesis system has been successfully used to regenerate transgenic plants (Li *et al.*, 1993; Vain *et al.*, 1993; Vasil *et al.*, 1992; Ellis *et al.*, 1993; Fitch *et al.*, 1990; Wilde *et al.*, 1992). In woody plants *Agrobacterium* mediated transformation has been used to regenerate transformants of *Carica papaya* (Fitch *et al.*, 1993) and *Juglans* (McGranahan *et al.*, 1990). Direct DNA-mediated transfer methods such as biolistics have been used on embryogenic cultures of *Picea glauca* (Ellis *et al.*, 1993), *Liriodendron tulipifera* (Wilde *et al.*, 1992) and *Carica papaya* (Fitch *et al.*, 1990). Each of these transformation methods has used a somatic embryogenesis tissue culture system. The main prerequisites for a routine transformation system are 1) a highly regenerable tissue culture system, 2) an efficient method of DNA insertion, and 3) stable integration into the host genome. As somatic embryogenesis systems become more efficient for woody plants, routine transformation systems will follow closely behind.

The question now is not *if* genetic transformation is possible, but which genes will be used to transform woody plants when these systems become common practice. A question that researchers in both academic and commercial operations must ask themselves is which genes will be commercially important in woody plants. There is considerable interest in insect resistance, and work is in progress in several laboratories to insert *Bacillus thuringiensis* toxin genes into forest tree species (Strauss *et al.*, 1991). Transfer of chitinase genes may enhance disease resistance and genes for drought or cold tolerance may be valuable. There are also projects examining the genes that affect wood formation and cellulose chemistry (Whetten and Sederoff, 1991; Dean and Eriksson, 1992). A useful strategy may be the engineering of sterility into trees or other woody plants to prevent the release of engineered genes into natural populations (Strauss *et al.*, 1991). This may be very important in perennial species that might remain in the field for 20 or more years.

Woody plants are storehouses of many important secondary products such as lignin, fats, oils, organic acids, and resins. Perhaps adding genes to these species to produce novel chemical compounds never found in plants would have commercial potential. Similar strategies are being considered in herbaceous plants for various pharmaceutical compounds (Hiatt *et al.*, 1989). Forest trees are already a significant source of secondary chemical products, and methods for isolation and purification of these products are well established in the papermaking industry.

The major limitation to using genetic engineering to improve woody species is our lack of information on the basic physiology and biochemistry of some of our most important commercial species. Many of the above strategies hold promise but require more information before their likelihood of success can be estimated. Increased research emphasis in woody plants on the basic biochemical processes involved in wood formation, secondary product biosyntheses, and growth form would provide much needed information and provide greater opportunities (i.e., commercially important genes) for genetic transformation in woody plants.

Once genetically transformed clones are produced in the laboratory, having a deployment system that could deliver somatic embryo plants to the field in commercial quantities would be a major advantage. The deployment of these high-value, genetically transformed plants would make the somatic embryogenesis system more economically attractive. A high-value, genetically engineered clone could justify the expense of the clonal production system, similar to the way micropropagation systems are now used in the horticultural industry for high-value ornamentals.

5. Conclusions

Somatic embryogenesis holds promise in the future for the deployment of millions of genetically improved plants of commercially important woody plant species. This technology is only 35 years old but already much progress has been made. While most of these systems can be used only in the laboratory on a small scale, significant progress is being made to expand this technology to large scale production at commercial levels. However, since a biological system is at the heart of this technology there are still many unknowns, and consequently most of these systems are still not totally predictable. A greater understanding of the basic biology of the embryogenesis process will enhance our ability to use this technology on a wide range of species and genotypes and allow us to more easily control these processes. Emerging molecular biology techniques will complement this work and provide valuable information, particularly in understanding the genetic control of somatic embryogenesis and in embryo development and maturation. Robotic and automation technology will also begin to interface with these systems. As in other high technology fields, an interdisciplinary team effort will bring this technology into commercial reality in the 21st century.

References

Aitken-Christie, J. and A.P. Singh, 1987. Cold storage of tissue cultures. In: J.M. Bonga and D.J. Durzan (Eds.), Cell and Tissue Culture in Forestry, Vol. 2, Specific Principles and Methods: Growth and Developments, pp. 285–304. Martinus Nijhoff Publishers, Dordrecht.

Arnold, R.J., 1990. Control pollinated radiata pine seed – A comparison of seedling and cutting options for large-scale deployment. N.Z. For. 35: 12–17.

Attree, S.M. and L.C. Fowke, 1993. Scaling-up production of desiccation tolerant conifer somatic embryos (Abstr. SP-1001). In Vitro Cell. Dev. Biol. 29A (Suppl.): 6.

Attree, S.M., M.K. Pomeroy and L.C. Fowke, 1992. Manipulation of conditions for the culture of somatic embryos of white spruce for improved triacylglycerol biosynthesis and desiccation tolerance. Planta 187: 395–404.

Avjioglu, A. and R.B. Knox, 1989. Storage lipid accumulation by zygotic and somatic embryos in culture. Ann. Bot. 63: 409–420.

Bapat, V.A., 1993. Studies on synthetic seeds of sandalwood (*Santalum album* L.) and mulberry (*Morus indica* L.). In: K. Redenbaugh (Ed.), Synseeds – Applications of Synthetic Seeds to Crop Improvement, pp. 381–407. CRC Press, Ann Arbor.

Bapat, V.A., D.P. Fulzele, M.R. Heble and P.S. Rao, 1990. Production of sandalwood somatic embryos in bioreactors. Curr. Sci. 59: 746–748.

Bates, S., J.E. Preece, N.E. Navarrete, J.W. Van Sambeek and G.R. Gaffney, 1992. Thidiazuron stimulates shoot organogenesis and somatic embryogenesis in white ash (*Fraxinus americana* L). Plant Cell Tissue Organ Cult. 31: 21–29.

Becwar, M.R., R. Nagmani and S.R. Wann, 1990. Initiation of embryogenic cultures and somatic embryo development in loblolly pine (*Pinus taeda*). Can. J. For. Res. 20: 810–817.

Becwar, M.R., S.R. Wann, M.A. Johnson, S.A. Verhagen, R.P. Feirer and R. Nagmani, 1988. Development and characterization of invitro embryogenic systems in conifers. In: M.R. Ahuja (Ed.), Somatic Cell Genetics of Woody Plants, pp. 1–18. Kluwer Academic Publishers, Dordrecht.

Bhaskaran, S. and R.H. Smith, 1992. Somatic embryogenesis from shoot tip and immature inflorescence of *Phoenix dactylifera* cv. Barhee. Plant Cell Rep. 12: 22–25.

Bieniek, M.E., R.C. Harrell and D.J. Cantliffe, 1991. The production of mature somatic embryos in suspension cultures of *Ipomea batatas*. In Vitro Cell. Dev. Biol. 27 (Part II): 116A.

Cheema, G.S., 1989. Somatic embryogenesis and plant regeneration from cell suspension and tissue cultures of mature himalayan poplar (*Populus ciliata*). Plant Cell Rep. 8: 124–127.

Cheliak, W.M. and K. Klimaszewska, 1991. Genetic variation in somatic embryogenic response in open-pollinated families of black spruce. Theor. Appl. Genet. 82: 185–190.

De Touchet, B., Y. Duval and C. Pannetier, 1991. Plant regeneration from embryogenic suspension cultures of oil palm (*Elaeis guineensis* Jacq). Plant Cell Rep. 10: 529–532.

Dean, J.F.D. and K.E.L. Eriksson, 1992. Biotechnological modification of lignin structure and composition in forest trees. Holzforschung 46: 135–147.

Denison, N.P. and D.R. Quaile, 1987. The applied clonal eucalypt programme in Mondi forests. South African For. J. 142: 60–67.

Dumet, D., F. Engelmann, N. Chabrillange and Y. Duval, 1993. Cryopreservation of oil palm (*Elaeis guineesis* Jacq.) somatic embryos involving a desiccation step. Plant Cell Rep. 12: 352–355.

Ekberg, I., L. Norell and S. von Arnold, 1993. Are there any associations between embryogenic capacity and phenological traits in two populations of *Picea abies*? Can. J. For. Res. 23: 731–737.

Ellis, D.D., D.E. McCabe, S. McInnis, R. Ramachandran, D.R. Russell, K.M. Wallace, B.J. Martinell, D.R. Roberts, K.F. Raffa and B.H. McCown, 1993. Stable transformation of *Picea glauca* by particle acceleration. Biotechnology 11: 84–89.

Etheridge, P.G. and G.W. Adams, 1990. Vegetative propagation: Its role in applied tree improvement. In: Proc. of the Joint Meeting of Western Forest Genetics Association and IUFRO Working Parties S2.02–05,06,12 and 14, Olympia, WA, August 20, 1990, pp. 5.1–5.17.

Fitch, M.M.M., 1993. High frequency somatic embryogenesis and plant regeneration from papaya hypocotyl callus. Plant Cell Tiss Org Cult. 32: 205–212.

Fitch, M.M.M., R.M. Manshardt, D. Gonsalves, J.L. Slightom and J.C. Sanford, 1990. Stable transformation of papaya via microprojectile bombardment. Plant Cell Rep. 9: 189–194.

Fitch, M.M.M., R.M. Manshardt, D. Gonsalves and J.L. Slightom, 1993. Transgenic papaya plants from *Agrobacterium*-mediated transformation of somatic embryos. Plant Cell Rep. 12: 245–249.

Fujii, J.A.A., D. Slade, R. Olsen, S.E. Ruzin and K. Redenbaugh, 1990. Alfalfa somatic embryo maturation and conversion to plants. Plant Sci. 72: 93–100.

Fujii, J.A.A., D. Slade, J. Aguirre-Rascon and K. Redenbaugh, 1992. Field planting of alfalfa artificial seeds. In Vitro Cell. Dev. Biol. 28P: 73–80.

Fujii, J.A.A., D. Slade and K. Redenbaugh, 1993. Planting artificial seeds and somatic embryos. In: K. Redenbaugh (Ed.), Synseeds – Applications of Synthetic Seeds to Crop Improvement, pp. 183–202. CRC Press, Ann Arbor.

Gleed, J.A., R.J. Arnold and J. Siegfried, 1991. Radiata pulpwood production – an optimum approach via clonal forest plantations. Appita J. 44: 323–341.

Gray, D.J., 1987. Quiescence in monocotyledonous and dicotyledonous somatic embryos induced by dehydration. HortScience 22: 810–814.

Gray, D.J., 1989. Effects of dehydration and exogenous growth regulators on dormancy, quiescence, and germination of grape somatic embryos. In Vitro Cell. Dev. Biol. 25: 1173–1178.

Greenwood, M.S., G.W. Adams and M. Gillespie, 1988. Shortening the breeding cycle of some northeastern conifers. In: E.K. Morgenstern and T.J.B. Boyle (Eds.), Proc. of the 21st meeting of the Canadian Tree Improvement Association, Part 2, Aug. 17–21, 1987, Truro, N.S., pp. 43–52, Canadian Forestry Service.

Gupta, P.K., G. Pullman, R. Timmis, M. Kreitinger, W.C. Carlson, J. Grob and E. Welty, 1993. Forestry in the 21st Century. Biotechnology 11: 454–459.

Gupta, P.K., R. Timmis, G. Pullman, M. Yancey, M. Kreitinger, W. Carlson and C. Carpenter, 1991. Development of an embryogenic system for automated propagation of forest trees. In: I.K. Vasil (Ed.), Cell Culture and Somatic Cell Genetics of Plants, Vol. 8, pp. 75–93. Academic Press, Orlando.

Hakman, I.C. and S. von Arnold, 1988. Somatic embryogenesis and plant regeneration from suspension cultures of *Picea glauca* (White spruce). Physiol. Plant. 72: 579–587.

Hämäläinen, J.J. and K.J. Jokinen, 1992. Selection of Norway spruce somatic embryos by computer vision. SPIE Optics in Agric. For. 1836: 195–205.

Hämäläinen, J.J., U. Kurten and V. Kauppinen, 1993. Classification of plant somatic embryos by computer vision. Biotech. Bioeng. 41: 35–42.
Harrell, R.C., M. Bieniek and D.J. Cantliffe, 1992. Noninvasive evaluation of somatic embryogenesis. Biotech. Bioeng. 39: 378–383.
Hiatt, A., R. Cafferkey and K. Bowdish, 1989. Production of antibodies in transgenic plants. Nature 342: 76–78.
Kartha, K.K., L.C. Fowke, N.L. Leung, K.L. Caswell and I. Hakman, 1988. Induction of somatic embryos and plantlets from cryopreserved cell cultures of white spruce (*Picea glauca*). J. Plant Physiol. 132: 529–539.
Kitto, S.L., W.G. Pill and D.M. Molloy, 1991. Fluid drilling as a delivery system for somatic embryo-derived plantlets of carrot (*Daucus-carota* L). Sci. Hort. 47: 209–220.
Klimaszewska, K., C. Ward and W.M. Cheliak, 1992. Cryopreservation and plant regeneration from embryogenic cultures of larch (*Larix* × *eurolepis*) and black spruce (*Picea mariana*). J. Exp. Bot. 43: 73–79.
Lai, F.M., T. Senaratna and B.D. Mckersie, 1992. Glutamine enhances storage protein synthesis in *Medicago sativa* L. somatic embryos. Plant Sci. 87: 69–77.
Laine, E., P. Bade and A. David, 1992. Recovery of plants from cryopreserved embryogenic cell suspensions of *Pinus caribaea*. Plant Cell Rep. 11: 295–298.
Li, L., R. Qu, A. deKochko, C. Fauquet and R.N. Beachy. 1993. An improved rice transformation system using the biolistic method. Plant Cell Rep. 12: 250–255.
Lowe, P., 1993. Automation takes over propagation and transplanting. GrowerTalks 57: 39–45.
Lulsdorf, M.M., T.E. Tautorus, S.I. Kikcio and D.I. Dunstan, 1992. Growth parameters of embryogenic suspension cultures of interior spruce (*Picea glauca-engelmannii* complex) and black spruce (*Picea mariana* Mill). Plant Sci. 82: 227–234.
McCrae, J.B., H.E. Stelzer, G.S. Foster and T. Caldwell, 1993. Genetic test results from a tree improvement program to develop clones of loblolly pine for reforestation. Proc. of the Twenty Second Southern Forest Tree Improvement Conf., Atlanta GA, June 14–17, 1993 (in press).
McElroy, A.R. and D.C.W. Brown, 1992. A transplant plug technique for production of alfalfa (*Medicago sativa* L.) plants from somatic embryos. Can. J. Plant Sci. 72: 483–485.
McGranahan, G.H., C.A. Leslie, S.L. Uratsu and A.M. Dandekar, 1990. Improved efficiency of the walnut somatic embryo gene transfer system. Plant Cell Rep. 8: 512–516.
McKersie, B.D. and S.R. Bowley, 1993. Synthetic seeds of alfalfa. In: K. Redenbaugh (Ed.), Synseeds – Applications of Synthetic Seeds to Crop Improvement, pp. 231–255. CRC Press, Ann Arbor.
Merkle, S.A., A.T. Wiecko, R.J. Sotak and H.E. Sommer, 1990. Maturation and conversion of *Liriodendron tulipifera* somatic embryos. In Vitro Cell. Dev. Biol. 26: 1086–1093.
Misra, S., S.M. Attree, I. Leal and L.C. Fowke, 1993. Effect of abscisic acid, osmoticum, and desiccation on synthesis of storage proteins during the development of white spruce somatic embryos. Ann. Bot. 71: 11–22.
Moulton, R.J., R.M. Mangold and J.D. Snellgrove, 1993. Tree Planting in the United States 1992. USDA Forest Service Publication, May 1993.
Neuenschwander, B. and T.W. Baumann, 1992. A novel type of somatic embryogenesis in *Coffea arabica*. Plant Cell Rep. 10: 608–612.
Neuman, M.C., J.E. Preece, J.W. VanSambeek and G.R. Gaffney, 1993. Somatic embryogenesis and callus production from cotyledon explants of Eastern black walnut. Plant Cell Tiss Org Cult. 32: 9–18.
Nuutila, A.M. and V. Kauppinen, 1992. Nutrient uptake and growth of an embryogenic and a non-embryogenic cell line of birch (*Betula pendula* Roth) in suspension culture. Plant Cell Tiss Org Cult. 30: 7–13.
Obendorf, R. and J. Slawinska, 1988. Maturation of soybean somatic embryos to a desiccation tolerant state. In Vitro Cell. Dev. Biol. 24: 71A.
Park, Y.S., S.E. Pond and J.M. Bonga, 1993. Initiation of somatic embryogenesis in white

spruce (*Picea glauca*): genetic control, culture treatment effects, and implications for tree breeding. Theor. Appl. Genet. 86: 427–436.

Randall, W.K. and D.T. Cooper, 1973. Predicted genotypic gain from cottonwood clonal tests. Silvae Gen. 22: 165–167.

Redenbaugh, K., 1993. Introduction. In: K. Redenbaugh (Ed.), Synseeds – Applications of Synthetic Seeds to Crop Improvement, pp. 3–7. CRC Press, Ann Arbor.

Redenbaugh, K., J.A.A. Fujii and D. Slade, 1993. Hydrated coatings for synthetic seeds. In: K. Redenbaugh (Ed.), Synseeds – Applications of Synthetic Seeds to Crop Improvement, pp. 35–45. CRC Press, Ann Arbor.

Redenbaugh, M.K., 1986a. Analogs of botanic seed. U.S. Patent 4,562,663.

Redenbaugh, M.K., 1986b. Delivery system for meristematic tissue. U.S. Patent 4,583,320.

Ritchie, G.A., 1991. The commercial use of conifer rooted cuttings in forestry: a world overview. New For. 5: 247–275.

Roberts, D.R., B.S. Flinn, D.T. Webb, F.B. Webster and B.C.S. Sutton, 1990a. Abscisic acid and indole-3–butyric acid regulation of maturation and accumulation of storage proteins in somatic embryos of interior spruce. Physiol. Plant 78: 355–360.

Roberts, D.R., B.C.S. Sutton and B.S. Flinn, 1990b. Synchronous and high frequency germination of interior spruce somatic embryos following partial drying at high relative humidity. Can. J. Bot. 68: 1086–1090.

Roberts, D.R., F.B. Webster, B.S. Flinn, W.R. Lazaroff and D.R. Cyr, 1993. Somatic embryogenesis of spruce. In: K. Redenbaugh (Ed.), Synseeds – Applications of Synthetic Seeds to Crop Improvement, pp. 427–450. CRC Press, Ann Arbor.

Sakamoto, Y., S. Umeda and H. Ogishima, 1991. Artificial seed comprising a sustained-release sugar granule. U.S. Patent 5,010,685.

Sakamoto, Y., T. Mashiko, A. Suzuki, H. Kawata and A. Iwasaki, 1992. Development of encapsulation technology for synthetic seeds. Acta Hort. 319: 71–76.

Schooley, H.O. and T.J. Mullin, 1988. Seed production strategies: current vs. future. In: E.K. Morgenstern and T.J.B. Boyle (Eds.), Proc. of the 21st Meeting of the Canadian Tree Improvement Association, Part 2, Aug. 17–21, 1987, Truro, N.S., pp. 155–169, Canadian Forestry Service.

Schuller, A. and G. Reuther, 1993. Response of *Abies alba* embryonal-suspensor mass to various carbohydrate treatments. Plant Cell Rep. 12: 199–202.

Schultheis, J.R. and D.J. Cantliffe, 1992. Growth of somatic embryos of sweet potato (*Ipomoea batatas* (L.) Lam) in hydroxyethyl cellulose gel amended with salts and carbohydrates. Sci. Hort. 50: 21–33.

Senaratna, T., 1992. Artificial seeds. Biotech. Adv. 10: 379–392.

Smith, D.R., L.A. Donaldson and A. Warr, 1991. Bioreactor technology for mass propagation of *Pinus radiata* by somatic embryogenesis. In Vitro Cell. Dev. Biol. 27 (Part II): 43A.

Steward, F.C., M.O. Mapes and K. Mears, 1958. Growth and organized development of cultured cells. II. Organization in cultures grown from freely suspended cells. Am. J. Bot. 45: 705–708.

Strauss, S.H., G.T. Howe and B. Goldfarb, 1991. Prospects for genetic engineering of insect resistance in forest trees. Forest Ecol. Manag. 43: 181–209.

Stuart, D.A., S.G. Strickland and K.A. Walker, 1987. Bioreactor production of alfalfa somatic embryos. HortScience 22: 800–803.

Tautorus, T.E., M.M. Lulsdorf, S.I. Kikcio and D.I. Dunstan, 1992. Bioreactor culture of *Picea mariana* Mill (black spruce) and the species complex *Picea glauca engelmannii* (interior spruce) somatic embryos. Growth parameters. Appl. Microbiol. Biotech. 38: 46–51.

Timmis, R., M.M. Abo El-Nil and R.W. Stonecypher, 1987. Potential genetic gain through tissue culture. In: J.M. Bonga and D.J. Durzan (Eds.), Cell and Tissue Culture in Forestry, Vol. 1., pp. 198–215. Martinus Nijhoff Publishers, Boston.

Timmis, R., M.E. Kreitinger and M.J. Yancey, 1992. Method and apparatus for culturing autotrophic plants from heterotrophic plant material. U.S. Patent 5,119,588.

Ting, K.C., G.A. Giacomelli and P.P. Ling, 1992. Workability and productivity of robotic plug transplanting workcell. In Vitro Cell. Dev. Biol. 28P: 5–10.

Tremblay, F.M., 1990. Somatic embryogenesis and plantlet regeneration from embryős isolated from stored seeds of *Picea-glauca*. Can. J. Bot. 68: 236–242.

Tremblay, L. and F.M. Tremblay, 1991a. Carbohydrate requirements for the development of black spruce (*Picea-mariana* (Mill) BSP) and red spruce (*P. rubens* Sarg) somatic embryos. Plant Cell Tiss Org Cult. 27: 95–103.

Tremblay, L. and F.M. Tremblay, 1991b. Effects of gelling agents, ammonium nitrate, and light on the development of *Picea-mariana* (Mill) BSP (black spruce) and *Picea-rubens* Sarg (red spruce) somatic embryos. Plant Sci. 77: 233–242.

Vain, P., M.D. McMullen and J.J. Finer, 1993. Osmotic treatment enhances particle bombardment-mediated transient and stable transformation of maize. Plant Cell Rep. 12: 84–88.

Vasil, V., A.M. Castillo, M.E. Fromm and I.K. Vasil, 1992. Herbicide resistant fertile transgenic wheat plants obtained by microprojectile bombardment of regenerable embryogenic callus. Biotechnology 10: 667–674.

Vieitez, F.J., A. Ballester and A.M. Vieitez, 1992. Somatic embryogenesis and plantlet regeneration from cell suspension cultures of *Fagus sylvatica* L. Plant Cell Rep. 11: 609–613.

Von Aderkas, P., K. Klimaszewska and J.M. Bonga, 1990. Diploid and haploid embryogenesis in *Larix-leptolepis*, *L. decidua*, and their reciprocal hybrids. Can. J. For. Res. 20: 9–14.

Von Arnold, S. and I. Hakman, 1988. Regulation of somatic embryo development in *Picea abies* by abscisic acid (ABA). J. Plant Physiol. 132: 164–169.

Whetten, R. and R. Sederoff, 1991. Genetic engineering of wood. For. Ecol. Manag. 43: 301–316.

Whynman, A., 1992. Technology troubleshooters in the greenhouse. GrowerTalks 58: 75–83.

Wilde, H.D., R.B. Meagher and S.A. Merkle, 1992. Expression of foreign genes in transgenic yellow-poplar plants. Plant Physiol. 98: 114–120.

Zamarripa, A., J.P. Ducos, H. Bollon, M. Dufour and V. Petiard, 1991. Production of somatic embryos of coffee in liquid medium – effects of inoculation density and renewal of the medium. Cafe Cacao 35: 233–244.

List of Contributory Authors

Dr. F. Bekkaoui, ID Biomedical, 8855, Northbrook Court, Burnaby, BC, V5J 5J1, Canada. FAX: 604-431-5098.

Dr. J. Bercetche, AFOCEL, Station de Biotechnologies, Domaine de l'Etancon, 77370 Nangis, France.

Dr. W.C. Carlson, Weyerhaeuser Ins., Technology Center, Tacoma, WA 98477, USA. FAX: 206-924-6736.

Dr. A.M. Dandekar, Department of Pomology, University of California, Davis, CA 95617, USA. FAX: 916-752-7784.

Dr. M.R. Davey, Plant Genetic Manipulation Group, Department of Life Sciences, University of Nottingham, University Park, Nottingham, NG7 2RD, UK. FAX: 602-51-32-40.

Dr. L. L. DeVerno, Petawawa National Forestry Institute, P.O. Box 2000, Chalk River, Ontario, Canada, KOJ 1JO. FAX: 613-589-2275.

Dr. Alex M. Diner, U.S.D.A. Forest Service, Southern Forest Experiment Station, Normal, Alabama 35802, USA.

Dr. D.I. Dunstan, NRC-CNRC, Plant Biotechnology Institute, 110 Gymnasium Place, Saskatoon, Sask S7N OW9 Canada. FAX: 306-975-1839.

Dr. D. Ellis, Department of Horticulture, University of Wisconsin, 1575 Linden Drive, Madison, WI 53706, USA. FAX: 608-262-4743.

Dr. R. Feirer, Biology Department, St. Norbert College, De Pere, WI 54115, USA. FAX: 414-337-4033.

Dr. S. Garton, Department of Plant and Soil Science, Alabama A & M University, Normal, AL 35762, USA.

Dr. P.K. Gupta, Weyerhaeuser Inc., Technology Center, Tacoma, WA 98477, USA. FAX: 206-924-6736.

Dr. J.A. Grob, Weyerhaeuser Co., Technology Center, Tacoma, Washington 98477, USA.

S. Jain, P. Gupta & R. Newton (eds.), Somatic Embryogenesis in Woody Plants, Vol. 1, 435–437.

Dr. L.W. Handley, Westvaco, Forest Research, Box 1950, Summerville, SC 29484, USA. FAX: 803-875-7185.

Dr. J.E. Hartle, Weyerhaeuser Inc., Technology Center, Tacoma, Washington 98477, USA.

Dr. B. Heinze, Biotechnology Department, Austrian Research Center, Seibersdorf, A-2444, Seibersdorf, Austria. FAX: +43-2254-80-3653.

Dr. M.M. Jana, Division of Plant Tissue Culture, National Chemical Laboratory, CSIR, Pune 411008, India.

S.V. Kendurkar, Division of Plant Tissue Culture, National Chemical Laboratory, CSIR, Pune 411008, India.

Dr. Howard B. Kriebel, Ohio State University, Division of Forestry, SNR, Wooster, Ohio 44691-4096, USA. FAX: 216-263-3658.

Dr. A.F. Mascarenhas, Plant Tissue Culture Division, National Chemical Laboratory, CSIR, Pune 411008, India. FAX: 91-212-334-761.

Dr. R. Minocha, USDA, Forest Service, Northeastern Forest Experimental Station, Cor. Concord and Mast Rds., P.O. 640, Durham, NH 03824, USA. FAX: 603-868-1538.

Prof. S.C. Minocha, Department of Plant Biology, College of Life Sciences and Agriculture, Nesmith Hall, University of New Hampshire, Durham, NH 03824-3597, USA. FAX: 603-862-4757.

Dr. Santosh Misra, Department of Biochemistry and Microbiology, University of Victoria, P.O. Box 3055, Victoria, B.C., Canada V8W 3P6. FAX: 604-721-8855.

Dr. R.S. Nadgauda, Division of Plant Tissue Culture, National Chemical Laboratory, CSIR, Pune 411008, India.

Dr. R. Nagmani, Department of Soil Science, Alabama A & M University, P.O. Box 1208, Normal, AL 35762, USA. FAX: 205-851-5429.

Dr. M. Palada, Forest Research and Management Institute (ICAS), SOS, Stefanesti n 128, 72904, Bucharest-2, Romania.

Dr. M. Paques, AFO/CEL, Station de Biotechnologies, Domaine de l'Etancon, 77370 Nangis, France. FAX: 1 60 67 54 04.

Dr. C.H. Phadke, Division of Plant Tissue Culture, National Chemical Laboratory, CSIR, Pune 411008, India.

Dr. J.B. Power, Plant Genetic Manipulation Group, Department of Life Science, University of Nottingham, NG7 2RD, UK.

Dr. J.N. Ruaud, AFO/CEL, Station de Biotechnologies, Domaine de l'Etancon, 77370 Nangis, France. FAX: 1 60 67 54 04.

Emer, Prof. A. Sakai, Asbucho 1-5-23, Kitaku, Sapporo 001, Japan. FAX: 81 11 716 7711.

Dipl. Ing. J. Schmidt, Biotechnology Department, Österreichisches Forschungszentrum Seibersdorff, A-2444, Austria. FAX: 43-2254 80 3653.

Dr. S.V. Shirke, Division of Plant Tissue Culture, National Chemical Laboratory, CSIR, Pune 411008, India.

Prof. Dr. Liisa Simola, Department of Botany, University of Helsinki, FIN-00014, Unioninkatu 44, Helsinki, Finland.

Dr. T.E. Tautorus, NRC, CNRC, Plant Biotechnology Institute, 110 Gymnasium Place, Saskatoon, Sask S7N 0W9, Canada. FAX: 306-975-4839.

Dr. A. Tobok, Genetic Manipulation Group, Department of Life Sciences, University of Nottingham, Nottingham, NG7 2RD, UK.

Dr. A.E. Zipf, Department of Plant and Soil Science, Alabama A & M University, Normal, AL 35762, USA.

INDEX OF SPECIES

INDEX OF SUBJECTS

FORESTRY SCIENCES

1. P. Baas (ed.): *New Perspectives in Wood Anatomy.* Published on the Occasion of the 50th Anniversary of the International Association of Wood Anatomists. 1982 ISBN 90-247-2526-7
2. C.F.L. Prins (ed.): *Production, Marketing and Use of Finger-Jointed Sawnwood.* Proceedings of an International Seminar Organized by the Timber Committee of the UNECE (Halmar, Norway, 1980). 1982 ISBN 90-247-2569-0
3. R.A.A. Oldeman (ed.): *Tropical Hardwood Utilization.* Practice and Prospects. 1982 ISBN 90-247-2581-X
4. P. den Ouden (in collaboration with B.K. Boom): *Manual of Cultivated Conifers.* Hardy in the Cold- and Warm-Temperate Zone. 3rd ed., 1982 ISBN Hb 90-247-2148-2; Pb 90-247-2644-1
5. J.M. Bonga and D.J. Durzan (eds.): *Tissue Culture in Forestry.* 1982 ISBN 90-247-2660-3
6. T. Satoo: *Forest Biomass.* Rev. ed. by H.A.I. Madgwick. 1982 ISBN 90-247-2710-3
7. Tran Van Nao (ed.): *Forest Fire Prevention and Control.* Proceedings of an International Seminar Organized by the Timber Committee of the UNECE (Warsaw, Poland, 1981). 1982 ISBN 90-247-3050-3
8. J.J. Douglas: *A Re-Appraisal of Forestry Development in Developing Countries.* 1983 ISBN 90-247-2830-4
9. J.C. Gordon and C.T. Wheeler (eds.): *Biological Nitrogen Fixation in Forest Ecosystems.* Foundations and Applications. 1983 ISBN 90-247-2849-5
10. M. Németh: *Virus, Mycoplasma and Rickettsia Diseases of Fruit Trees.* Rev. (English) ed., 1986 ISBN 90-247-2868-1
11. M.L. Duryea and T.D. Landis (eds.): *Forest Nursery Manual.* Production of Bareroot Seedlings. 1984; 2nd printing 1987 ISBN Hb 90-247-2913-0; Pb 90-247-2914-9
12. F.C. Hummel: *Forest Policy.* A Contribution to Resource Development. 1984 ISBN 90-247-2883-5
13. P.D. Manion (ed.): *Scleroderris Canker of Conifers.* Proceedings of an International Symposium on Scleroderris Canker of Conifers (Syracuse, USA, 1983). 1984 ISBN 90-247-2912-2
14. M.L. Duryea and G.N. Brown (eds.): *Seedling Physiology and Reforestation Success.* Proceedings of the Physiology Working Group, Technical Session, Society of American Foresters National Convention (Portland, Oregon, USA, 1983). 1984 ISBN 90-247-2949-1
15. K.A.G. Staaf and N.A. Wiksten (eds.): *Tree Harvesting Techniques.* 1984 ISBN 90-247-2994-7
16. J.D. Boyd: *Biophysical Control of Microfibril Orientation in Plant Cell Walls.* Aquatic and Terrestrial Plants Including Trees. 1985 ISBN 90-247-3101-1
17. W.P.K. Findlay (ed.): *Preservation of Timber in the Tropics.* 1985 ISBN 90-247-3112-7
18. I. Samset: *Winch and Cable Systems.* 1985 ISBN 90-247-3205-0

FORESTRY SCIENCES

19. R.A. Leary: *Interaction Theory in Forest Ecology and Management.* 1985 ISBN 90-247-3220-4
20. S.P. Gessel (ed.): *Forest Site and Productivity.* 1986 ISBN 90-247-3284-0
21. T.C. Hennessey, P.M. Dougherty, S.V. Kossuth and J.D. Johnson (eds.): *Stress Physiology and Forest Productivity.* Proceedings of the Physiology Working Group, Technical Session, Society of American Foresters National Convention (Fort Collins, Colorado, USA, 1985). 1986 ISBN 90-247-3359-6
22. K.R. Shepherd: *Plantation Silviculture.* 1986 ISBN 90-247-3379-0
23. S. Sohlberg and V.E. Sokolov (eds.): *Practical Application of Remote Sensing in Forestry.* Proceedings of a Seminar on the Practical Application of Remote Sensing in Forestry (Jönköping, Sweden, 1985). 1986 ISBN 90-247-3392-8
24. J.M. Bonga and D.J. Durzan (eds.): *Cell and Tissue Culure in Forestry.* Volume 1: General Principles and Biotechnology. 1987 ISBN 90-247-3430-4
25. J.M. Bonga and D.J. Durzan (eds.): *Cell and Tissue Culure in Forestry.* Volume 2: Specific Principles and Methods: Growth and Development. 1987 ISBN 90-247-3431-2
26. J.M. Bonga and D.J. Durzan (eds.): *Cell and Tissue Culure in Forestry.* Volume 3: Case Histories: Gymnosperms, Angiosperms and Palms. 1987 ISBN 90-247-3432-0
    Set ISBN (Volumes 24-26) 90-247-3433-9
27. E.G. Richards (ed.): *Forestry and the Forest Industries: Past and Future.* Major Developments in the Forest and Forest Industries Sector Since 1947 in Europe, the USSR and North America. In Commemoration of the 40th Anniversary of the Timber Committee of the UNECE. 1987 ISBN 90-247-3592-0
28. S.V. Kossuth and S.D. Ross (eds.): *Hormonal Control of Tree Growth.* Proceedings of the Physiology Working Group, Technical Session, Society of American Foresters National Convention (Birmingham, Alabama, USA, 1986). 1987 ISBN 90-247-3621-8
29. U. Sundberg and C.R. Silversides: *Operational Efficiency in Forestry.* Vol. 1: Analysis. 1988 ISBN 90-247-3683-8
30. M.R. Ahuja (ed.): *Somatic Cell Genetics of Woody Plants.* Proceedings of the IUFRO Working Party S2.04-07 Somatic Cell Genetics (Grosshansdorf, Germany, 1987). 1988 . ISBN 90-247-3728-1
31. P.K.R. Nair (ed.): *Agroforestry Systems in the Tropics.* 1989 ISBN 90-247-3790-7
32. C.R. Silversides and U. Sundberg: *Operational Efficiency in Forestry.* Vol. 2: Practice. 1989 ISBN 0-7923-0063-7
    Set ISBN (Volumes 29 and 32) 90-247-3684-6
33. T.L. White and G.R. Hodge (eds.): *Predicting Breeding Values with Applications in Forest Tree Improvement.* 1989 ISBN 0-7923-0460-8
34. H.J. Welch: *The Conifer Manual.* Volume 1. 1991 ISBN 0-7923-0616-3
35. P.K.R. Nair, H.L. Gholz, M.L. Duryea (eds.): *Agroforestry Education and Training.* Present and Future. 1990 ISBN 0-7923-0864-6
36. M.L. Duryea and P.M. Dougherty (eds.): *Forest Regeneration Manual.* 1991 ISBN 0-7923-0960-X

FORESTRY SCIENCES

37. J.J.A. Janssen: *Mechanical Properties of Bamboo.* 1991 ISBN 0-7923-1260-0
38. J.M. Bonga and P. Von Aderkas: *In Vitro Culture of Trees.* 1992 ISBN 0-7923-1540-5
39. L. Fins, S.T. Friedman and J.V. Brotschol (eds.): *Handbook of Quantitative Forest Genetics.* 1992 ISBN 0-7923-1568-5
40. M.J. Kelty, B.C. Larson and C.D. Oliver (eds.): *The Ecology and Silviculture of Mixed-Species Forests.* A Festschrift for David M. Smith. 1992 ISBN 0-7923-1643-6
41. M.R. Ahuja (ed.): *Micropropagation of Woody Plants.* 1992 ISBN 0-7923-1807-2
42. W.T. Adams, S.H. Strauss, D.L. Copes and A.R. Griffin (eds.): *Population Genetics of Forest Trees.* Proceedings of an International Symposium (Corvallis, Oregon, USA, 1990). 1992 ISBN 0-7923-1857-9
43. R.T. Prinsley (ed.): *The Role of Trees in Sustainable Agriculture.* 1993 ISBN 0-7923-2030-1
44. S.M. Jain, P.K. Gupta and R.J. Newton (eds.): *Somatic Embryogenesis in Woody Plants,* Vol. 3: Gymnosperms. 1995 ISBN 0-7923-2938-4
45. S.M. Jain, P.K. Gupta and R.J. Newton (eds.): *Somatic Embryogenesis in Woody Plants,* Vol. 1: History, Molecular and Biochemical Aspects, and Applications. 1995 ISBN 0-7923-3035-8
46. S.M. Jain, P.K Gupta and R.J. Newton (eds.): *Somatic Embryogenesis in Woody Plants,* Vol. 2: Angiosperms. 1995 ISBN 0-7923-3070-6
    Set ISBN (Volumes 44-46) 0-7923-2939-2

KLUWER ACADEMIC PUBLISHERS – DORDRECHT / BOSTON / LONDON

GPSR Compliance

The European Union's (EU) General Product Safety Regulation (GPSR) is a set of rules that requires consumer products to be safe and our obligations to ensure this.

If you have any concerns about our products, you can contact us on ProductSafety@springernature.com

In case Publisher is established outside the EU, the EU authorized representative is:

Springer Nature Customer Service Center GmbH
Europaplatz 3
69115 Heidelberg, Germany

Batch number: 10755090

Printed by Printforce, the Netherlands